McGraw-Hill Education

# FIREFIGHTER EXAMS

McGraw-Hill Education

# FIREFIGHTER
## EXAMS

**Third Edition**

**RONALD R. SPADAFORA**

New York | Chicago | San Francisco | Athens | London | Madrid | Mexico City
Milan | New Delhi | Singapore | Sydney | Toronto

1 2 3 4 5 6 7 8 9 LHS 23 22 21 20 19 18

ISBN 978-1-260-12173-5
MHID 1-260-12173-9

e-ISBN 978-1-260-12174-2
e-MHID 1-260-12174-7

# Contents

## PART II: PREPARING FOR THE ORAL INTERVIEW AND PSYCHOLOGICAL AND PHYSICAL ABILITY TESTS

## PART III: REVIEW FOR THE WRITTEN EXAMINATION

## PART IV: PRACTICE EXAMINATIONS

# Acknowledgments

**The author would like to thank the following:**

Rhonda Roland Shearer – whose love and devotion allowed me to win the battle against leukemia.

Robert Spadafora, Jr. – my nephew, whose bone marrow donation saved my life.

Robert Spadafora, Sr. family (Carol, Nicholas, and Robert Jr.), for their ongoing support and kindness throughout my illness.

Dr. Mark Levis, for his cancer treatment at Johns Hopkins Hospital, which was instrumental in bringing me back to life.

**Additional appreciation for:**

Daina Penikas, Sr. Editing Supervisor at McGraw-Hill Education
Garret Lemoi, Editor at McGraw-Hill Education
International Association of Fire Chiefs (IAFC)
International Association of Fire Fighters (IAFF)
Richard Velez, EMT

# About the Author

**Firefighter Ronald Spadafora circa late 1970s.**

Assistant Chief Ronald R. Spadafora is a 39-year veteran in the Fire Department of New York (FDNY) and was promoted in 2015 to the Chief of Fire Prevention (4 Star). On 9/11, he responded to the World Trade Center (WTC) and supervised both rescue and fire suppression efforts at the North Tower, WTC 7, and 140 West Street (Verizon Building). He was named the WTC Chief of Safety in October 2001 for the entire recovery operation ending in late June 2002. In August 2002, he was promoted to the rank of Deputy Assistant Chief (2 Star). On August 14–15, 2003, he headed the Logistics Section for the FDNY during the New York City blackout. In the aftermath of Hurricanes Katrina and Rita in September 2005, he was designated Deputy Incident Commander of the FDNY Incident Management Team and sent to Louisiana to assist the New Orleans Fire Department in the protection of the city. He was promoted to Assistant Chief (3 Star) in 2010. In October/November 2013, Chief Spadafora once again played a major role in coordinating logistics for the FDNY during and in the months following Hurricane Sandy.

Chief Spadafora has taught fire science at John Jay College (CUNY) at the undergraduate level for 25 years, as well as emergency management at Metropolitan College of New York (MCNY) for both graduate and undergraduate programs. He is also the senior instructor for Fire Technology, Incorporated, lecturing FDNY firefighters of all ranks seeking promotion, and is a visiting instructor/advisor on urban firefighting and incident management for the Working on Fire (WoF) Programme of South Africa.

Chief Spadafora is a graduate of the first FDNY Officers Management Institute (FOMI) program (Columbia University School of Business). He holds an MPS degree in Criminal Justice from Long Island University (C. W. Post Center), a BS degree in Fire Science from John Jay College (CUNY), and a BA degree in Health Education from Queens College (CUNY). He has written more than 75 articles in professional publications such as *WNYF* (With New York Firefighters), *Fire Engineering*, *Size-Up*, *Fire Rescue*, *Fire and Rescue International*, *Firenuggets, Pass It On,* and the *American Journal of Industrial Medicine*. Books by Chief Spadafora include

*McGraw-Hill's Firefighter Exams,* Third Edition (McGraw-Hill), *Sustainable Green Design and Firefighting: A Fire Chief's Perspective* (Delmar-Cengage Learning), and *Fire Protection Equipment and Systems* (Pearson-Brady Fire Series).

His preparation, strategy, and commitment to excellence have been unsurpassed throughout his career in the fire service. With more than 34,000 applicants for the 1977 NYC Firefighter Exam, he was hired off the top of the FDNY Eligible List as a Probationary Firefighter in the first class of 121 (Cream of the Crop) on September 2, 1978. Upon successful completion of Probationary School at the FDNY Training Academy, he was assigned to Engine Company 237 in the Bushwick section of Brooklyn.

# Introduction

Do you want to be a firefighter? If so, *Firefighter Exams* offers an abundance of information about this exciting and rewarding career. It will also guide you through the conditions that must be met to become one of America's bravest.

Advice on references, résumés, the Notice of Examination, and the requirements that may be specified in a standard medical checkup can be found in Part I of this book. The material is written in an easy-to-understand format. Successfully moving through the application and health status process is the initial action that must be taken prior to getting to the next level, which involves a myriad assortment of supplementary tests.

Part II supplies insight and strategy into preparing for the oral interview, psychological assessment, and physical ability test. Performing poorly on the oral dialogue can be detrimental to your chances of getting selected. Typical questions asked by interviewers are listed along with a general summary of dos and don'ts. This insight will instill self-confidence and poise in you, the candidate, during this worrisome component of the hiring process.

Firefighting is a profession that has earned the public trust. Administrators from fire departments expect a multitude of positive personality traits from candidates that veteran firefighters already possess: common sense, competitiveness, initiative, and integrity, to name just a few. The psychological test analyzes answers that you provide to the examiner's series of written true or false questions. Personality background review questions placed in this book will prime you in answering honestly and favorably in the eyes of the decision makers.

The work performed by a firefighter is physically demanding, requiring muscular strength, stamina, superior aerobic and anaerobic capabilities, flexibility, coordination, agility, and endurance. Many heavily populated municipalities have implemented the Candidate Physical Ability Test (CPAT), consisting of a logical sequence of eight separate firefighting skill–related events. Cities and counties that do not use CPAT conduct physical aptitude tests intended to assess similar attributes. *Firefighter Exams* offers a comprehensive description of each event, as well as training tips designed to assist in preparation and achieving a high score.

Written examination study material makes up the major portion of Part III. Essential topics include mathematics, memory and visualization, chart/table/graph and diagram data displays, fire science, mechanical principles, firefighting tools and equipment, emergency medical care, reading comprehension, deductive/inductive reasoning, and more. The questions in Part III will sharpen your overall understanding of what potential firefighters need to know.

Moreover, two new chapters have been added to the third edition that will benefit your studies:

- Hydraulics, Water Supply/Distribution, and Fire Protection Systems
- Problem Sensitivity, Information Ordering, and Written Expression

Practice examinations in Part IV showcase everything presented in the first three parts of *Firefighter Exams*. These tests, each consisting of 100 multiple choice questions, are a challenge. Your goal should be to honestly measure how well you truly comprehend the subject matter. An additional practice exam is included in this third edition.

Trust me when I say that reading and studying this new edition will give you a pronounced advantage over wannabes, semi-prepared, and unprepared candidates who will not make the sacrifices necessary to do well on a firefighter exam. Rely on my experience in both firefighting and instruction to show you the correct pathway to success. Confront the challenge! Be the best you can be in all aspects of the exam. Demonstrate the positive characteristics of a modern-day firefighter in your quest to be selected at the top of the Firefighter Eligible List in your locality.

Ronald R. Spadafora
Chief of Fire Prevention (4-Star, FDNY)

*So many of our dreams at first seem impossible, then they seem improbable, and then, when we summon the will, they soon become inevitable.*

— Christopher Reeve

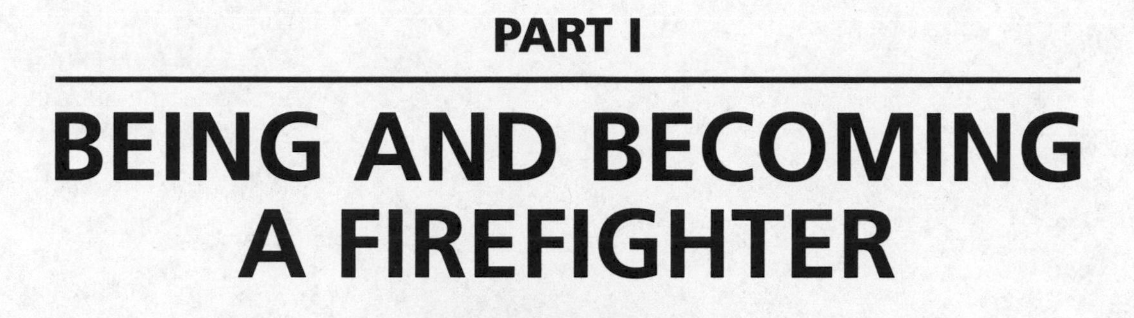

## PART I

# BEING AND BECOMING A FIREFIGHTER

CHAPTER 1

# Being a Firefighter

## Your Goals for This Chapter

- Explore the roles and duties of firefighters
- Learn about the risks of firefighting
- Understand the organization and rank system in public fire departments
- Learn about nontraditional roles such as EMT/paramedic and wildland firefighter

Every year in the United States, fires and emergencies kill and injure thousands of Americans and destroy property worth billions of dollars. A firefighter is America's first line of defense dealing with these hazardous situations. Traditionally, the firefighter's main role was to save lives; prevent loss of life and property; control, confine, and extinguish fires; and prevent unwanted fires. The role has expanded, however. Firefighters are now the first responders to major disasters and emergencies, the first to arrive on the scene to save lives, property, and the environment.

Career firefighters work in both the public sector and the private sector. According to the U.S. Department of Labor, Bureau of Labor Statistics, the number of jobs in 2014 was 327,300. For the years spanning 2014–2024, the number of job opportunities is projected to grow by 5 percent (an increase of 17,400). Keen competition for jobs is expected because this occupation attracts many qualified candidates.

Approximately 90 percent of all paid firefighters are employed by municipal or county fire departments. They work in the traditional role of firefighter or in specialized roles. This chapter deals mostly with firefighters in the public sector. Those employed in the private sector are also discussed briefly.

# FIREFIGHTERS IN THE PUBLIC SECTOR

## Trainee Programs for Firefighters

Most fire departments require new candidates who have passed the written, physical, and psychological examinations to become a firefighter to enter an apprenticeship, trainee, or probationary period. While in the trainee program for a specific period of time, the recruits are taught the firematic functions of their specific department through classroom sessions and a variety of job-related tasks during drills. They are evaluated closely to determine if they meet the required standards. Assignments include responding to alarms and assisting regular fire personnel in firefighting and medical emergency duties. Trainees typically receive less salary and fewer benefits than established firefighters in the department. After satisfactory completion of the probationary period, they are assigned to a specific unit.

## Traditional Roles and Duties of Firefighters

Firefighters commonly perform a wide array of traditional duties at operations. They usually work in coordinated teams. Assigned by a superior officer (company officer or chief officer), firefighters at a fire scene secure a positive water supply (fire hydrant), operate apparatus (engine pump, aerial ladder, tower bucket, etc.), stretch and utilize hose lines, carry and position portable ladders for ventilation, and enter and search buildings. They also use the tools and equipment of their trade to coordinate the effort to save lives, treat the injured, and extinguish the fire safely.

## Working Environment for Firefighters

During their tours of duty, firefighters live together for long periods of time in the fire station and work as a team. The fire station will usually have features similar to a residential facility (kitchen, reading room, gymnasium, bunkroom, etc.). Firefighters work long hours at irregular intervals. For example, some firefighters are on duty for 24 hours and then off for a couple of days to recuperate. Firefighters are on duty for both day and night tours. There are no guaranteed weekends and national holidays off. Work hours for firefighters are usually longer than for most. Many work more than 50 hours a week.

## Risks Involved in Firefighting

The job of firefighting involves the risk of death and serious physical injury. Each year, approximately 100 firefighters are killed and tens of thousands injured while on duty. Upon the receipt of an alarm, firefighters must stop everything they are doing and respond in an expeditious manner regardless of the time or weather conditions. This may be the reason for the leading cause of death to firefighters: heart attack. The second most common cause of death is from trauma (building collapse, falls, and motor vehicle accidents while responding to alarms). Asphyxia and burns are also high on the list of firefighter fatalities.

Burns, an injury frequently associated with firefighters, are not the major type of fire service injury today. This is true because of improvements over the years in thermal personal protection equipment. Muscle injuries (sprains and strains) caused by overexertion are the number one type of injuries firefighters sustain performing their work on the fireground. Open wounds, cuts, and bruises make up the second largest portion of fire service injuries. Thermal stress, heart attack, stroke, smoke and toxic gas inhalation, bone dislocations or fractures, and fire and chemical burns are other leading types of firefighting injuries.

## Organization and Rank (Promotional) System in Public Fire Departments

Generally, the organizational chart has a commissioner (designated by the mayor or city council) or chief of department as the leader of the department. A large municipal fire department can have both and be grouped into bureaus (operations, training, fire prevention, communications, facility maintenance, etc.) that are managed by deputy commissioners or high-ranking chief officers. The bureau of operations typically consists of divisions staffed with division or deputy chiefs. A division can consist of five or more battalions, led by battalion chiefs. Each battalion can have five or more fire companies (engines, ladders, and special units). Smaller urban areas may just have a designated chief officer as the head of the department with a minimal number of bureaus. Their operational bureau will also be greatly reduced.

Commonly, each fire company will have a captain with overall responsibility for the unit and its firefighters. Lieutenants will also be assigned to companies and assist the captain in managing the company. In a municipality, examinations for promotions are structured to include all the uniformed ranks listed above.

## Compensation (Pay and Benefits)

Wages and benefits vary considerably from city to city throughout the United States. On the average, firefighters earn approximately $23 an hour. According to U.S. Department of Labor statistics, the median annual salary for a firefighter is a little more than $48,000. Company officers with supervisory and management responsibility earn in the range of $60,000 a year base salary, with the high end reaching more than $90,000. Chief officers make considerably more money than company officers, and many reach a salary well over

$100,000 a year. Fringe benefits (overtime, holiday pay, night differential, etc.) can increase salaries greatly. Additional benefits include medical and dental coverage for firefighters and their families, sick leave, paid vacation, life insurance, opportunities for promotions, flexible work hours (24-hour shifts, tour swapping), and pension upon retirement or disability in the line of duty. Layoffs for firefighters are not common.

## Nontraditional Roles for Firefighters

Many firefighters are trained in nontraditional roles. They are generally assigned to special units such as rescue, squad, satellite, foam carrier, or marine units. Some of these specialized tasks include rescue operations (ice, water, high-angle rope, trench, structural collapse, confined space), hazardous material mitigation, and foam delivery. These firefighters respond to nontypical fires and emergencies that require their skill and expertise.

Two major roles that will be examined separately below are firefighters trained as emergency medical technicians and paramedics and firefighters who battle wildland fires.

### EMT/PARAMEDIC FIREFIGHTER

Approximately half of all fire departments nationwide provide ambulance service for victims. In most states, firefighters are also cross-trained as certified emergency medical technicians (EMTs) and paramedics. In fact, most calls to which firefighters respond involve medical emergencies requiring treatment from basic first aid to advanced life-support intervention.

Firefighters trained as EMTs or paramedics perform the initial evaluation of victims and treat and transport patients with medical problems and/or trauma. They recognize and provide initial care for a multitude of ailments and injuries, such as shock, difficulty breathing, stroke, broken bones, burns, asthma, choking, unconsciousness, epileptic convulsions, and drowning. Response time is critical in treating traumatic injuries and illnesses. As emergency responders, firefighters are usually the first on the scene. Rapid, on-scene medical intervention produces the best patient outcomes. As new lifesaving equipment and techniques are developed (oxygen delivery systems, cardiopulmonary resuscitation, portable cardiac defibrillators, etc.), fire departments will continue to train and use firefighters in their ever-expanding role of medical provider.

### WILDLAND FIREFIGHTER

When a fire breaks out in our national forests and parks, crews of firefighters are used to suppress the blaze. Wildland firefighting is unique and very different from urban firefighting. The work is performed in steep terrain and thick vegetation. Wildland firefighters are typically exposed to severe smoke and dust conditions throughout their working tours. New recruits are taught fuel management (reduction of naturally growing foliage) and fuel control techniques. They also learn forestry practices dealing with accepted fire suppression procedures for the various kinds of terrain encountered.

Wildland firefighting includes long periods of walking, climbing, shoveling dirt, chopping brush, lifting heavy objects, stretching and operating small-diameter hose lines, utilizing foam and fire retardant, driving heavy equipment, and safely using hand tools to clear vegetation and cut down trees to create fire breaks in the path of the fire to deprive it of fuel. Prescribed burning techniques (the intentional starting of fires to control the spread and direction of a forest fire) are also employed.

Elite firefighting forces called **smokejumpers** parachute from airplanes to conduct firefighting operations in strategic and inaccessible areas. **Hotshots**, specially trained wildland firefighters who ride in heavy, all-terrain vehicles to access threatened areas of the forest, are also used for vital and dangerous assignments. **Helitacks** fly inside helicopters and rappel from rope to secure and establish water-drop operations. They transport and distribute firefighting equipment and logistical supplies from prepared landing sites.

## FIREFIGHTERS IN THE PRIVATE SECTOR

Fewer than 10 percent of firefighters are employed in the private sector, working in chemical or oil processing plants, large industrial facilities, or airports.

### Industrial Fire Brigades

In the private sector, some chemical plants, airports, and large industrial facilities train personnel and establish **industrial fire brigades**. These firefighters may be responsible for incipient firefighting activities only or for interior structural firefighting, depending on their training. Sophisticated industrial fire brigades may perform interior structural firefighting as well as exterior firefighting, using apparatus-generated water streams or stationary large-volume manifold nozzles. These firefighters are trained to wear self-contained breathing apparatus (SCBA) and operate hose lines for both interior and exterior operations. The emergency work they perform includes hazardous material mitigation, spill control, decontamination of exposed victims, and medical treatment.

### Hellfighters

Hellfighters are firefighters hired by oil well control companies, based primarily in Texas and Oklahoma, to extinguish oil well fires. These firefighters are specially trained to close supply-line valves and use large-volume water streams from the perimeter of a flaming oil well to put out fires. They also detonate explosives to displace oxygen in the atmosphere surrounding the fire, thereby smothering it. Hellfighters work not only in the United States but also in other countries.

## Wildland Firefighting Companies

Wildland firefighting companies contract out paid firefighters to public agencies to supplement local fire department manpower and equipment during major wildland fires. These companies are located mainly in the northwestern United States. Company crews perform work similar to that of wildland firefighters.

CHAPTER 2

# Becoming a Firefighter

**Your Goals for This Chapter**

- Understand the requirements to become a firefighter
- Find out how to obtain and complete the application
- Learn how to prepare your résumé and cover letter
- Get valuable study and test-taking tips
- Find out about the required medical examination
- Learn about video-based firefighter examinations

This chapter provides information about the steps necessary to become a firefighter. Although specifics may differ from one fire department to another, the basic procedures are, for the most part, similar.

## REQUIREMENTS TO BECOME A FIREFIGHTER

There are several minimal qualifications or requirements to become a firefighter in most municipalities. These requirements are summarized below.

**Age**—Usually an applicant must be at least 18 years of age at the time the exam is taken and at least 21 to be hired.

**Background**—The applicant's past records are reviewed, including driving history, residency, educational transcripts, arrests or convictions, and so on.

**Character**—Applicants will need to supply references from honorable and distinguished members of the community.

**Citizenship**—Applicants must be U.S. citizens.

**Criminal Record**—An applicant's record is reviewed; any felony or misdemeanor arrests and convictions may be grounds for ineligibility.

**Discharge from the Armed Forces**—A dishonorable discharge from the armed forces may make an applicant ineligible.

**Driver's License**—Applicants should have a driver's license valid in the state in which they are taking the exam.

**Drug Screening**—Tests are conducted to discover candidate use of marijuana, amphetamine, anabolic steroids, cocaine, heroin, methadone, morphine, quaalude, or other substances.

**Education**—Usually applicants are required to have a high school diploma or general equivalency diploma (GED). More and more municipalities are requiring college credits and prehospital emergency care (EMT/paramedic) certifications.

**Language**—Applicants must speak and understand English.

**Legal Status**—Applicants should not have any legal impediments (felony conviction, for example) to their ability to perform the job functions of firefighter.

**Medical and Psychological**—Candidates undergo medical (including vision and hearing tests) and psychological examinations to determine if they can perform the functions of a firefighter. Reasonable accommodation is made to enable candidates with disabilities to take these exams.

**Polygraph**—A polygraph, or lie detector, may be used to review a candidate's qualifications and suitability.

**Proof of Identity**—Applicants must provide proof of identity (birth certificate).

**Residency**—A candidate may be required to reside within the area in which he or she is seeking employment; preferential residency examination credit may be granted for candidates living in the area of employment.

## LEARNING ABOUT THE JOB

It is important to learn about the qualifications, duties, and responsibilities of firefighters in the jurisdiction in which you wish to apply. There are several ways to learn about the job.

Consider joining one or more ancillary organizations. **Volunteer fire departments** provide the knowledge and training to perform many, if not all, firefighting duties. Membership in a volunteer fire department is an ideal way to learn the job, and, incidentally, to demonstrate to interviewers that you are serious about becoming a career firefighter.

The U.S. Department of Homeland Security's **Community Emergency Response Team (CERT)** program is another way to learn about what a career firefighter does. This program trains men and women of all ages to respond to emergencies and give critical support to first responders. CERT training includes disaster preparedness, fire suppression, basic medical operations, and light search-and-rescue operations. The Boy Scouts of America administers the **Fire Explorers**, a program for children and young adults between the ages of 14 and 21 that provides valuable insight into the firefighting profession. High schools and colleges may have a **Firefighter Cadet Program** that provides curriculum and training to school-age children and young adults in firematics. Completion of a cadet program in some jurisdictions leads directly to appointment as a probationary firefighter.

The **American Red Cross** and local **Volunteer Ambulance Corps** and affiliates also provide courses in first responder emergency medical care ranging from basic first aid to advanced life support procedures.

A visit to the **local firehouse** is also helpful. Introduce yourself to the firefighters and ask about the possibility of your being able to spend some time with them to observe training sessions and drills.

The **public library** and the **World Wide Web** are other sources of information concerning the career of a firefighter.

The **examination announcement** is the official description of the career of a firefighter. It typically lists common firefighter tasks, including response and performance at fires and utility and medical emergencies, maintenance of the firehouse, apparatus, tools, and equipment, fire prevention inspections, participation in training activities, hydrant inspection and maintenance, and public fire safety education.

# HOW TO APPLY TO BECOME A FIREFIGHTER

There are several steps in the initial application process, including obtaining and completing an application, submitting a résumé if requested, obtaining necessary information about a scheduled examination, and determining if special circumstances apply to you.

## Obtain and Complete an Application

Applications for the job of firefighter are obtained in many different ways, depending on the jurisdiction you reside in. Some municipalities make the application available on their government website, where it can be downloaded, completed, and mailed in to the agency coordinating the candidate process. Modern methods may allow candidates to fill out their application online. Follow the directions carefully on how the application should be submitted. If you mail in the application, consider having it certified with a return receipt requested in order to document its transmittal and receipt.

In other areas, applications may be acquired by visiting fire department headquarters, the local fire station, or a designated municipal office building. Always read the directions for completion of the application carefully before you attempt to fill in the required information. Should you use pen or pencil to fill out the form? Does the application need to be notarized? Is additional documentation (birth certificate, driver's license, résumé, etc.) required to be submitted with the application? And, finally, make sure you place the correct amount of postage on the envelope containing all the forms.

## Provide a Résumé, Cover Letter, and References

If a résumé is required as part of the application process, it should be one or two pages summarizing your contact information, main objective, education, military history, work background, job-related life experiences, and pertinent activities, hobbies, and interests. It should always be sent along with a cover letter. The cover letter should be geared to the particular needs and requirements of the fire service position you are seeking. A cover letter allows you to focus the reader's attention on your specific strengths and accomplishments, which are summarized in the résumé.

### COVER LETTER

Always include a cover letter when sending your résumé. It can be just as important as your résumé. The cover letter is organized into several main parts—heading, body, complimentary close, signature, and indication of any enclosures—as follows:

#### Heading

**Contact Information**—Standard contact information should include your name (in bold letters), legal address, and home and work telephone numbers. This information should come first at the top.

**Date of Letter**—Spell out the month of the year, as in January 10, 2015.

**Recipient Information**—Give the name of the person to whom you are addressing the letter and the name and address of the organization.

**Salutation**—Use the title and last name of the addressee, if available. If you do not have a specific person or office to address, use "To Whom It May Concern" as a salutation. Examples: Dear Employment Director, Dear Mr. Fulton, Dear Chief of Personnel Cummings.

#### Body

**Initial Paragraph**—Tell the reader why you are submitting your résumé (why are you interested in becoming a firefighter?), state the name of the position you are applying for, and include why you think you would be a good candidate for the job.

**Second Paragraph**—Describe your major (not all) strengths and how they are applicable to the position you are seeking. Demonstrate how your skills can be a positive addition to the organization you wish to join. Provide the reader with one or two examples of your qualifications and state that this, as well as additional pertinent information, can be found in your enclosed résumé.

**Final Paragraph**—Reiterate your interest in becoming a member of the fire service and state your eagerness to hear from or meet with the reader in the near future to discuss a possible relationship with the organization. State also that you appreciate the opportunity to submit your résumé for consideration.

## Complimentary Close

**Sincerely, Sincerely yours, Best regards**—One of these closings should appear two lines below the final paragraph.
**Signature**—Use blue or black ink.
**Identification Line**—Beneath your signature, type your name.
**Enclosure**—Type "Enclosure:" or "Encl:" and the word "Résumé." Example: Enclosure: Résumé.

## RÉSUMÉ

The résumé needs to be formatted in a uniform and concise manner. It should be self-promoting, by focusing on your positive assets and demonstrating that you are an attractive candidate to be a firefighter. One word of warning: don't put anything that is untrue or fabricated on your résumé; false information can be easily identified by fire department fact-checkers and interviewers and will result in your disqualification.

## Content and Organization

A résumé should be organized as follows:

**Contact Information**—Provide your standard contact information—name (avoid nicknames), legal address, telephone numbers (home, cellular, and work), and e-mail address, if applicable—at the top of your résumé.
**Main Objective**—Provide a clear, positive statement about your commitment to become a firefighter.
**Education**—List your educational information, starting with the most recent. Include college or postgraduate degrees earned, the year the degree was obtained, the name of the learning institution, and the area of concentrated study (major/minor). Include your grade point average (GPA) if it reflects high academic achievement (B+ or higher) and note academic honors. If a degree was not earned, include the number of college credits earned.
**Military History**—Provide the dates of enlistment and honorable discharge (if applicable), branch of service, assigned location, rank designation, duties and responsibilities, training certificates, campaign service (e.g., Gulf War) and awards, citations, and achievements.
**Work Background**—Provide employment history in reverse chronological order (last job first). If you have a wide array of work experience, confine your listing to activities that relate in some way to the work performed by firefighters (communications dispatcher, fire guard, fire safety director, peace officer/security, lifeguard, park ranger, truck/tractor-trailer driver, automotive mechanic, construction trades). For each position listed, include the name of the employer, location, job title and responsibilities, and dates of employment.
**Job-Related Life Experience**—Highlight special skills, competencies, and achievements (e.g., certified scuba diver, crane operator license) that may make you a valuable asset to the organization you wish to join. Also include any volunteer work that relates to the duties of a first responder and community service personnel [e.g., volunteer firefighter, volunteer nurse, member of

the Peace Corps, Red Cross, Salvation Army, or Community Emergency Response Team (CERT)].

**Activities, Hobbies, Interests**—Include membership in civic and fraternal organizations, as well as pertinent hobbies, interests, and sports activities that may be considered as a positive reflection of your abilities to perform as a firefighter (e.g., ham radio operation, mountain/rock climbing).

**References**—At the end of your résumé you may want to include the words, "References available on request."

**Note:** Do not include your reference information on your résumé unless specifically asked to provide it on the notice of examination or job application. Before providing the names of references to prospective employers, be sure that the persons you name have a knowledgeable understanding of you and that they are recognized as upstanding citizens of the community in which they live.

## Tips on Preparing a Résumé

To make your résumé easier to read and understand, try to adhere to the tips listed below:

Limit your résumé to one or two pages.
Avoid folding or stapling your résumé.
Use white or off-white paper only.
Use quality bond 8½ × 11-inch paper.
Use only one side of the paper.
Use a plain typeface.
Use only one typeface.
Don't use italics.
Use easy-to-read font sizes (between 10 and 14 point).
Use no more than two font sizes.
Keep margins within the 0.75- to 1.5-inch range.
Spell and grammar check your résumé.
If mailing, use a large (9 × 12-inch) envelope to protect your résumé.
If mailing, consider Priority/Certified Mail options.
Follow up mailing with a phone call to confirm delivery.

## Sample Résumé

**George West**
18 Harper Avenue
Bronx, NY 10473
718-342-8890
westg@harper.com

**Objective:** A position of firefighter in the New York City Fire Department.

**Education**
Bachelor of Arts Degree in Health Education, 2011, Queens College of the City University of New York (CUNY). Summa cum laude.
Summer research work at Jacobi Medical Center, Bronx, NY, 10461.

**Military History**
United States Naval Reserves Airman. Honorable Discharge, 2007.

**Work Background**
Jacobi Medical Center Ambulance Corps, 2011-present.
Nutritionist, Wellness Gym. Yonkers, NY, 2007-2011.

**Job-Related Life Experience**
Volunteer Firefighter, Aviation Volunteer Fire Department, Bronx, NY, 2009-present.

**Activities**
College Health Fair Coordinator, 2010-2011.
Intramural swimming, 2007-2009.

**References:** Supplied upon request

## REFERENCES

Generally, references are not included on your résumé. It is common, however, for fire service applicants to be required to furnish a list of **personal** and **professional references**. These people will have to either submit a letter of recommendation on your behalf or be available to provide meaningful information (work history, work ethics, personality, interpersonal relationships, technical abilities, life experience, ambition, education, moral character, leadership qualities, communication skills, dependability, punctuality, managerial capabilities, etc.) concerning you, either in person or over the phone. Be prepared to supply between three and five references in each category. Do not include blood relations on your list of references. A good reference is a nonrelative who has known you for at least three years and who will substantiate what you have submitted in your résumé. Good references include past and present employers, former teachers or college professors, athletic coaches, clergy and religious associates, work and volunteer-related associates, former schoolmates, or neighbors held in high esteem in the community.

Once you have chosen your references, be sure to contact them individually to verify that they are willing to provide the reference. Give each individual a copy of your résumé personally and review with them your accomplishments and strengths, relating them to the duties of a firefighter. Talking to your references before they write or speak to the fire department will allow you to confirm what positive feedback will be generated on your behalf. Let them know that you are really interested in the position of firefighter and why you want to join the fire service. Thank them for agreeing to be a reference and for the time and effort required to complete the task.

### Personal References

Provide each reference's name, home address, and phone number. If the reference has a relationship with first responder services, provide the person's job title or affiliation with the pertinent organization.

### Professional References

Include your reference's name, job title, company address and phone number, and e-mail address, if acceptable to your reference. Try to select professional references from positions in emergency service (fire and police departments, forest service, homeland security, nursing, ambulance service, etc.).

## Obtain a Notice of Examination

The notice of examination (NOE) is an informational summary of the minimum qualifications and requirements concerning the firefighter examination you will be taking in your jurisdiction. It is normally formatted and published by the civil service commission or fire department personnel office of the city conducting the exam. An NOE can commonly be accessed and reviewed in person at either of these two places or on their websites. Firehouses, post offices, libraries, and civic group agencies are other possible places to find this important initial document regarding the firefighter exam.

General information provided by the NOE is as follows:

**Name of Exam:** Firefighter.
**Type of Exam:** Phases of the exam—written test, physical ability exam, oral interview, etc.
**Exam Number**
**When to Apply/File for the Examination**
**How to Apply/File for the Examination**
**Cancellation Information**
**Website Information**
**Written Exam Study Guide:** Format and sample questions.
**Physical Ability Test Preparation Guide and Training Site(s) Information:** Overview of testing process and training locations.
**Civil Rights Protections:** Information concerning detrimental practices based on race, color, sex, national origin, age, and religion.

**Americans with Disabilities Act (ADA) Protections:** Information about discrimination for qualified candidates with disabilities.
**Protest Protocols**
**Special Test Accommodations:** Alternate exam dates for disability or religious beliefs.
**Military Veteran Preference Points**
**Application Fee:** Amount and fee payment (cash, check, credit card, etc.).
**Test Date**
**DUI/DWAI/DWI Stipulations**
**Illegal Drug Activity/Use Stipulations**
**List Promulgation**
**Eligibility List**
**Exceptions:** Age, education, disqualification, etc.
**List Termination**
**Job Functions:** Controlling and extinguishing fires, hazardous material incident mitigation, prehospital emergency medical care provider, maintenance of firehouse, apparatus, tools, and equipment, enforcement of local fire prevention laws and ordinances, fire safety educational activities, water supply and hydrant inspections, driving and operating fire apparatus, participation in training exercises and drills, etc.
**Work Schedule:** Normally long workweek hours and varied shifts.
**Salary and Benefits:** Base salaries generally start in the range between $25,000 and $50,000 for probationary firefighters. Benefits may include pension, life insurance, medical and dental coverage, prescription drug plan, paid vacation and holidays, deferred compensation plan incentives, etc.
**Probationary Period:** Generally 12 months.

## Apply for Special Considerations, if Applicable

Some applicants for a position as a firefighter may have unusual circumstances that qualify them for special consideration in taking the written or other exams or in other aspects of the application and approval process. These special considerations may apply to the religious observances of applicants, to applicants with disabilities, to veterans, and to those who may have had a change of address problem. All fire departments have procedures for handling these circumstances and usually publish a special consideration form.

## EXAMINATIONS REQUIRED

In most fire departments several different types of tests and examinations are given. Generally, the tests include a written test, an oral interview, a psychological test, a physical ability test, and a medical examination. Additional examinations can include video testing, which deals with interpersonal relationships and job scenarios.

## Written Examination

The written examination—which is the main focus of this book—is typically a multiple-choice test. It generally contains 100 questions to be answered over a period of three to four hours. Questions cover many areas, including inductive and deductive reasoning ability, mathematics, mechanical aptitude, reading comprehension, memorization and spatial orientation, visualization, interpersonal relationships, and problem sensitivity.

The written examination requires extensive preparation. This book will give the reader a comprehensive overview and understanding of the type of questions historically asked on firefighter examinations. Some basic suggestions for preparing for this examination are given below in "Successful Study Habits" and "Test-Taking Tips." A review of the content matter is provided in Part III of this book.

### SUCCESSFUL STUDY HABITS

Successful study skills can be learned. The best students use their time wisely. They first consider the amount of time they can or are willing to spend studying, and then they stick to that amount. To do this they prioritize the importance of the tasks they perform each day. Tasks that are given a low priority are marginalized or postponed until the day after the important test for which they are studying. Good students also make the most of all their time every day. They take study material with them to read while traveling, lying on the beach, eating lunch, or sitting on a park bench. They budget their time so that they meet their study time allotment each day.

For the most part, it will require a few hours a day over a three- to four-month period to comprehensively read and answer all of the questions in this firefighter preparation book. Making this long-term commitment will allow the reader to gradually get a good general understanding of the nature of firefighting and emergency service work and how to prepare for the open competitive exam leading to a career as a firefighter. You will not be able to cram all the information in this book the week before the exam. Use this book in conjunction with other firefighter informational resources found on the Internet and in print.

Most of your study time should take place in a room or area that is readily available to you and provides an environment free of distractions and interruptions. Turn off your cell phone or move it out of the room. The temperature should not be too warm or too cold. Your study space should have adequate lighting so that you can see clearly to read without straining your eyes. Keep the noise level in the room to a minimum. Your desk should be large enough to place your computer, books, manuals, dictionary, notebooks, and pens and pencils in a manner conducive to active learning, reading, and researching. A comfortable chair is another essential. If your home does not provide such a space, consider going to your local public library, school library, or community center reading room.

Before each day of studying, set goals you wish to attain that day. Try varying your topics (math, reading comprehension, memory and visualization) to provide a variety of stimulating material.

Try to schedule the bulk of your study time during daylight hours when you are the most alert and energetic. Use these daytime periods to study difficult material that requires the most concentration. Nighttime hours should be reserved for light reading and easier subject matter. Avoid staying up late to reach your study goals for the day. Lack of sleep will affect your concentration and ability to learn during this time as well as the following day. Study time should be broken up into periods no longer than one hour, followed by a ten-minute break. Use the break to get up and walk around, go outside for some fresh air, or eat a snack.

Study actively by reading aloud, taking notes, highlighting, drawing pictures or diagrams while you read, and creating study questions as you go. Ensure that all your notes are transcribed into a notebook. This will help you organize information and main ideas and allow you to review the day's learning at a later time.

Writing out questions related to what you are reading on index cards is another excellent way of studying. Write one or more questions on one side of the index card and the answers to the questions on the flip side of the card. Over the course of three to four months, you should have several hundred questions for instant review. Index cards can be carried anywhere, so you can study even when you are not in your normal study area.

Relate new information to previously learned material and to your prior knowledge to improve memory by association. Repetition and continued review are the best ways to guarantee understanding of the study material.

## TEST-TAKING TIPS

The following provides a general overview of what to do and how to handle typical multiple-choice tests. Most of the advice, suggestions, and tips, gleaned from experience in taking civil service examinations, are applicable to the firefighter examination.

### Before the Test

- Leave yourself plenty of time for traveling to the test site.
- Wear a watch so you can calculate your time during the exam.
- Carry plenty of pencils, pens, and erasers to the exam.
- Bring water and appropriate snacks for a pick-me-up when needed.
- Enter the building where the examination is taking place as soon as allowed and find your assigned room and proper seat quickly.
- Set up your test-taking materials on your desk in an easy-to-use way.
- Shut off your cell phone and leave it off during the entire test.

### During the Test

- Generally a series of bells or buzzer signals guide you through the start and conclusion of the test. Listen to the monitor in charge of your test room and read the instructions concerning what these directional sounds mean—for example, first signal: test is handed out to candidates; second signal: open test booklet and read test-taking instructions; third signal: start the test; fourth signal: end of test.
- Read the written instructions very carefully. They provide important information about the timing of the examination and its segments, whether you

can or cannot mark up the question booklet, the use or nonuse of supplemental items (calculators, scrap paper, rulers), penalties for wrong answers, and the number of questions you are required to answer.

- Examine your test booklet thoroughly before you begin the test. Ensure that you are not missing pages. Inform the monitor immediately if there are any irregularities with your test booklet.
- If you are allowed to mark up your question booklet, underline key words or phrases in each question and make notes, if necessary, in the margins.
- Review the answer sheet and ensure all required information concerning you and the examination you are taking is filled out accurately. Check to see which way your answers should be transcribed on the sheet (vertically or horizontally).
- Answer each question in the order that it appears in the question booklet.
- Apportion your time. Divide the number of questions into the number of minutes you are allocated to complete the exam to determine approximately how long you should spend on any one question. Allow at least 20 minutes for review of your answers.
- Skip over and checkmark difficult questions and come back to them later on a second or third review of the exam.
- Read each question carefully to ensure you fully understand what is being asked. Pay particular attention to words such as *all*, *most*, *fewest*, *least*, *best*, *worst*, *same*, *opposite*, and *except*.
- Answer questions on the basis of all of the information presented. Do not select an answer based on unwarranted assumptions.
- Don't get personally involved when reading the question; remain objective.
- Read all the answer options. Eliminate those that are obviously wrong and those that only partially answer the question. Select the answer you feel is best and check it against the facts presented in the question.
- If there are no penalties for wrong answers, select an answer for every question. Even if you guess, you still have a decent chance of selecting the correct answer.
- Don't change your answers once they are transcribed onto the answer sheet unless information gathered while answering questions further into the exam provides you with new information warranting such a change. Usually your first instinct in the selection of an answer is best.

## Oral Interview

Some fire departments require an oral interview; some do not. If an oral interview will be given, the examination announcement will outline what general areas it is designed to evaluate. For information on how to prepare for an oral interview, see Chapter 3.

## Psychological Test

Psychological tests are administered as part of the overall recruitment process to help the fire service determine whether candidates are mentally prepared to cope with the stressful nature of the job. A candidate cannot study for psychological testing, but knowledge of what to expect is valuable. See Chapter 4.

## Physical Ability Tests

Firefighting is a physically demanding profession, and most fire departments require a candidate to pass a physical ability test. One of the most common tests used is the Candidate Physical Ability Test (CPAT). Information about that test and tips on how to prepare and train for a physical ability test are provided in Chapter 5.

## Medical Examination

Throughout the country, medical guidelines have been established for the position of firefighter. Most of these requirements can be found in the National Fire Protection Association (NFPA) Standard 1582 (Standard on Comprehensive Occupational Medical Program for Fire Departments). In general, candidates will be examined by a physician in the employ of the fire department or civil service commission conducting the hiring process.

A medical exam is essential to ensure that potential candidates are physically capable of performing firefighting tasks. The examination includes screening for a list of possible conditions that would or could prevent a fire department member or candidate from performing as a firefighter and thereby present a significant risk to the health and safety of fellow firefighters and civilians. Any impairment that may adversely affect the ability of the candidate to perform the duties of the position could constitute grounds for disqualification. A candidate who is rejected for a medical condition that over time is corrected or improves may be able to apply for a reexamination within the time frame of the eligible list.

Requirements that may be specified in a standard medical examination are as follows:

**Visual acuity** of 20/20 binocular with or without correction or uncorrected visual acuity of 20/40 binocular for wearers of contact lenses or glasses.

**Hearing** of standard average threshold, without correction, and not worse than 25-decibel loss in the 500-, 1,000-, 2,000-, and 3,000-hertz ranges. Testing with a hearing aid is not permitted.

**Heart** of normal size, rhythm, and rate.

**Blood pressure** within normal range (systolic = 90 to 140, diastolic = 60 to 90) and normal blood conditions.

**Lungs** at normal function. A history of asthma or chronic bronchitis may constitute grounds for disqualification.

**Skin** in normal condition. Contagious skin conditions and allergies related to the job of firefighter may constitute grounds for disqualification.

**Gastrointestinal system** function normal. Chronic digestive diseases (ulcers and hemorrhoids) may constitute grounds for disqualification.

**Weight** within normal limits. Excessive body fat and obesity may constitute grounds for disqualification.

**Extremities** in normal condition. Loss of fingers and toes and joint diseases and/or disorders may constitute grounds for disqualification.

**Muscular** and **skeletal** systems that are normal. Abnormalities in these two areas of the body may constitute grounds for disqualification. For example, a candidate with a hernia, either complete or incomplete, would be disqualified until the hernia is repaired.

**Neurological** function normal. Disorders such as epilepsy may constitute grounds for disqualification.

**General good health.** Any condition in any other systems of the body that is deemed by the medical practitioner conducting the examination as having the potential to interfere with the candidate's ability to perform the duties of a firefighter may constitute grounds for disqualification.

## Video-Based Tests

As stated earlier, teamwork and interpersonal attributes and judgments are now being tested using video-based examinations specifically designed for firefighters. In fact, video-based testing has proven to be a reliable factor in the selection process, and it is currently being used in some areas of the United States as a replacement for the oral interviews.

The public expects the fire department to provide numerous services, and firefighters have been placed in the role of community leaders, looked upon as symbols of safety and protection. Potential firefighters must show competence in dealing with all types of situations and people, and video-based testing attempts to ascertain if the candidate has the attitudes and temperament to meet these requirements.

During a video-based exam, the candidate is seated with a group of candidates to watch a sequence of videos dealing with firefighters operating together to accomplish tasks on the fireground or in the firehouse or videos showing interrelationships between firefighters and civilians. The scenarios depict typical incidents (firehouse maintenance, family disputes, medical emergencies, auto accidents, summons violation issuance, fire inspections, vehicles parked in front of fire hydrants, fire drills, etc.) that probationary firefighters can expect to encounter.

While watching the video, the candidate must concentrate on the goals and objectives the firefighters are trying to attain and on any stumbling blocks presented by their behavior. Are all the firefighters in the scenario working together for the common good? Is someone in the group not cooperating and hindering the effort of the team?

At the conclusion of the video scenario, multiple-choice questions are placed on the screen for candidates to answer either electronically or manually on an answer sheet. The questions are given sequentially and must be answered in a relatively short period of time. You may be asked what the best or worst course of action is that could be taken by the firefighters in the video.

Quick judgment and keen insight are essential to answering these questions accurately and in a timely manner. No firefighting experience or training is needed to answer these questions appropriately. Base your answers on what is being presented on the screen—and on common sense and good interpersonal skills.

The questions about the videos typically assess specific performance dimensions, including the following:

Ability to take orders
Attitude and temperament
Courtesy and respect for others
Demeanor and actions during stressful situations
Flexibility
Ingenuity
Interrelationships with fellow firefighters inside and outside the firehouse
Open-mindedness
Professionalism
Public relations
Relationships with superiors
Teamwork

## OBTAINING TEST RESULTS AND LISTING INFORMATION

Written test answers can usually be found sometime after the examination on municipal government websites or in the local civil service newspaper. Candidates are allowed time to protest answers they feel are inaccurate. You should also provide a protest if you feel the answer you gave was as good as or better than the answer denoted on the official answer key.

Test results are usually mailed to participants within a short time of completion of the written examination. Along with your score, you will be informed whether you have passed or failed this portion of the exam. If you pass the written exam, you will be instructed in the mailing what steps you are required to take to complete the candidate process. If all exams are successfully completed, a list number will be given to the candidate for hiring purposes.

PART II

# PREPARING FOR THE ORAL INTERVIEW AND PSYCHOLOGICAL AND PHYSICAL ABILITY TESTS

CHAPTER 3

# Preparing for the Oral Interview

## Your Goals for This Chapter

- Explore how to prepare an interview strategy
- Find out some typical questions asked by interviewers
- Learn important interview dos and don'ts
- Monitor your social media postings and conversations

Some fire departments require an oral interview; some do not. The first place to find out if the exam you will be taking has an oral interview segment is the examination announcement. If you will be asked to attend an oral interview, the examination announcement will outline what general areas it is designed to evaluate. A candidate's prior work experience, education, skills, career interests, community activities, training, goals and ambitions, and personal characteristics are some areas that may be listed. It is your obligation, as the candidate, to gather this information and review your life's highlights (school graduations, work history, awards, and personal achievements), keep them clearly in mind, and be ready to recount them, if need be, during the interview. It is also important to have basic knowledge about the firefighter job and the skills and abilities required. It is, therefore, important to prepare for the oral interview.

In summary, the oral interview can be viewed as a strategic conversation between the candidate and interviewers (usually three or more), who have been given the task of determining whether the candidate meets the standards for entry-level selection into the fire department and will be an asset to the organization.

## FORMULATING AN INTERVIEW STRATEGY

When you take the time to prepare for the interview, you are also building up your self-confidence in the skills necessary to do well. One of the first things you should do is evaluate your personal characteristics and qualifications. Make a list of your strengths and weaknesses. Personality traits such as honesty, enthusiasm, calmness under pressure, initiative, leadership, courage, motivation, personal appearance, loyalty, self-confidence, common sense, and the ability to work with others are just some of the personal attributes required by professional firefighters. Make no mistake about it: these characteristics will be evaluated in some way during your oral interview.

When answering questions during the interview, be sure to stress some of these essential attributes. State how you acquired and developed them through prior work and life experience, sacrifice, emulation of parents or teachers, or by education. Also consider the personal characteristics that you are weak in. Consider ways to change and grow in the areas that need improvement.

Next, review your prior work experience and background to determine what job skills you possess that are relevant to the career of a firefighter. If, for example, you are proficient in the use of tools and operating machinery, think about how these skills relate to firefighting. If appropriate during the interview, cite examples of how this knowledge and proficiency will make you a good firefighter. Recall what responsibilities you were entrusted with by your previous employers and what leadership qualities you have demonstrated at work or school, and, if appropriate, articulate these in a response to a question.

Formal education also plays an important role in who you are and how you portray yourself. Be aware of current events and have knowledge of world, national, and local news that you can use, if appropriate, while answering interview questions. Demonstrate that you are a multifaceted person with many interests. If you have attended, are attending, or have graduated from college, think about how the course curriculum relates to a career in firefighting.

Finally, evaluate why you want to be a firefighter. How does it meet your short- and long-term goals? What are the key aspects of the profession that stir your interest? Do the positives (challenge, excitement, salary, pension, helping people, and saving lives) outweigh the negatives (long hours, night work, danger, health issues)?

Be honest with yourself in thinking about and preparing for the interview. Then, you will be well prepared and able to handle it, leaving the interviewers with a positive impression of a candidate willing to learn and make the sacrifices necessary to serve the community.

## TYPICAL QUESTIONS ASKED BY INTERVIEWERS

Listed below are a few of the typical questions and inquiries asked by interviewers during firefighter exam oral interviews.

Tell us a little about yourself.
Describe for us your three most important personal character traits.
How would your friends describe your character?
What is your greatest strength? Weakness?
What is the most difficult decision you have made in your life?
Why should we recruit you to be a firefighter?
What are your qualifications for the job of firefighter?
Why do you want to become a firefighter?
Where do you envision yourself five years from now? Ten years?
What do you know about our fire department?
Tell us briefly about your current job.
What are the responsibilities you have at your current job?
Are you happy in your current job? Why or why not?
Who is the main person in your life that inspires you to be successful?
Tell us about your formal education.
What is the most important lesson you have learned in school?
What is or was your major in college and why?
What is your proudest/greatest achievement?
What do you like to do in your spare time?
What was the last book you read?
What magazines or journals do you subscribe to?
What was the last movie you saw?
What is your favorite television program?
Do you have any questions you would like to ask us?

In general, the questions asked by the interviewers are designed to tell them whether you have the necessary mindset and personality traits to perform the job of a firefighter. Do you really want to be a firefighter, or are you just exploring your options? Are you motivated enough to have researched information about the organization you wish to join? Can you talk intelligently about the fire service and the work that it performs? Do you have the ability to work well with others to reach the goals of the organization? Will you fit smoothly into the way of life of the fire department?

## PRACTICING FOR THE INTERVIEW

Start your preparation early. Write down both the questions in this section and the ones you have made up on index cards. Writing down the questions is the first step toward thinking about what you want to say. At first, review the answers to the questions silently. Later on, try answering the questions aloud in a quiet place. Don't try to memorize your responses. If you do, the answers you give on the day of the interview will come out phony and canned. Just try to remember the main points you wish to convey. The rest of the information you want to communicate will come out naturally based mainly on your preparation sessions.

Sit in a chair in front of a mirror to observe your appearance and hand mannerisms during the time you are answering the questions. Nonverbal communication plays a large role in how the interviewers perceive who you are and whether or not you should be hired. Watch your body language.

Grimaces, frowns, and sad facial expressions should be avoided, as should nervous habits, such as touching your face, tapping your fingers on a table, or fidgeting with your tie or blouse. Use your hands and arms to emphasize words when appropriate, but refrain from using too many gestures, which can distract the interviewers from listening to what you are saying. Maintain eye contact with yourself while answering the questions. This practice will help you to concentrate on what you are saying and, during the actual interview, maintaining eye contact with the interviewers will show that you have self-confidence and poise.

Check your posture. Are you sitting up straight in your chair with your head up and erect? Avoid slouching, looking down at the floor, or acting too relaxed or too stiff. Try to sit slightly forward in your seat with your arms on your lap or atop the armrests of the chair. Don't cross your arms and don't cross your legs.

Start audiotaping and videotaping your responses. Play back the recordings and review how you sound and look. Are you speaking in a monotone? Try to vary your tone and pitch while speaking. Are you talking too quickly or too slowly? A candidate who speaks very fast will come across to the interviewers as being nervous and unsure. Speaking too slowly will make interviewers impatient. Is your voice loud enough to be heard clearly from a standard distance (across a table)? Avoid mumbling, slang, jargon, and using "time filler" terms such as "you know" and "OK." Evaluate your responses and make changes to improve your communication skills where needed.

Ask family members and friends to ask you the questions on your index cards. Have them give you feedback concerning what they thought was positive and negative about your responses. Try to have at least three or four different people listen to you. Multiple opinions will help you to focus in on what really needs to be maintained or improved for your oral interview. Eventually, have the participants act as role players in a mock or practice oral interview. These role-playing sessions will help you refine your oral interview skills and techniques.

Additionally, seek assistance in public speaking and oral presentation techniques by enrolling in a college or adult education course designed to help students overcome fears and inadequacies and learn how to speak clearly, intelligently, and with confidence.

## THE DAY OF THE INTERVIEW

- **Rise and shine.** Be sure you are properly groomed and wear clean, conservative clothing (no gaudy jewelry, scuffed shoes, or excessive cologne or perfume). Eat a light meal and drink plenty of fluids. Bring your index card questions with you for a final review and to keep your mind focused on your day's mission.
- **Arrive early.** Get to the interview session at least 30 minutes early. Allow extra time in case of heavy traffic or mass transit delays. This will help you

relax and focus on the job ahead. Walk around the block of the building you will enter to take the oral interview, taking deep breaths as a relaxing warm-up exercise for the main event.

- **Lose the jitters.** Remember, all the candidates taking the oral interview are in the same boat. Most, if not all, are feeling nervous like you. If, however, you have spent the appropriate amount of time practicing your verbal communication skills, there is no reason to be excessively anxious about the interview. Read some of your question index cards to help keep focused. Review just the main ideas you want to convey to the interviewers should they ask the question. Don't try to memorize replies.
- **Present a good first impression.** When called for the oral interview, enter the room with your back erect and greet the interviewers with a smile, the appropriate salutation ("good morning" or "good afternoon"), and a hearty, firm handshake. You have prepared for this moment, and now is the time to demonstrate the confidence that training can provide. Let the interviewers see what a superior firefighter candidate looks like. Maintain eye contact with your interviewers throughout the entire introductory greeting period.
- **Listen.** Follow the instructions given to you by the interviewers. Listen for information concerning where to sit, when the interview will begin, how long the interview will last, and any other particulars said prior to the start of the oral interview. Concentrate on the questions you are being asked before you reply. Don't interrupt the interviewer while he or she is asking a question, and allow the interviewers to complete each question before you begin to reply. If a question is unclear, don't be afraid to ask for a clarification. Concentrate on the question asked, then answer it completely, using examples, if appropriate, to bolster your statements. Be careful, however, not to ramble on with extraneous information. When you feel that you have answered the question sufficiently, stop talking and await the next one.
- **Sell yourself.** During the interview emphasize your strengths and, if possible, avoid mentioning your weak points. Try to provide a positive outlook on your past personal relationships, employment, education, and hobbies. Exude confidence in who you are and what you have accomplished without being cocky. Maintain eye contact on the interviewer while he or she is asking you a question and throughout your reply. If asked at the end of the oral interview to summarize your thoughts, conclude with the reasons why you would make a good firefighter. Remember, there are no right or wrong answers. Reply in the manner you feel appropriate. Don't formulate responses trying to please the interviewers.
- **Leave a good last impression.** At the conclusion of your oral interview, stand up straight with an appearance of self-satisfaction for a job well done and thank the interviewers for their time. Wish each interviewer a "good day" and provide a strong parting handshake. Listen for any final instructions concerning score results, protest protocols, or signing out. Leave the room smartly without turning around to look back at the interviewers. Do not stay inside the building in order to talk to acquaintances who may also be taking an oral interview. Leave the building forthwith; the time to discuss the interview with others is when you are a reasonable distance away from the test site.

# DOS AND DON'TS

The following list is a summary of general dos and don'ts pertaining to oral interviews.

**DO**

- relax
- smile
- be well groomed and neatly attired
- maintain a good posture
- make eye contact with the interviewers
- show confidence
- speak at an audible level and clearly
- vary the pitch and tone of your voice
- act in a professional manner
- demonstrate a positive, motivated attitude
- take time to listen to the questions
- admit what you do not know
- answer all questions honestly

**DON'T**

- be late for your oral interview
- interrupt the interviewers
- ramble on with your answers
- have anything (such as gum) in your mouth
- apologize for not knowing an answer
- take a negative attitude toward interviewers
- appear overconfident or cocky
- tell jokes or wisecracks or be sarcastic
- use slang or jargon
- use foul language
- wear dark sunglasses
- use an overabundance of gestures
- get into an argument or debate with an interviewer

# SOCIAL MEDIA

Sites like Facebook, Twitter, and LinkedIn allow employers to get a glimpse of who you are in addition to your résumé or interview. Fire department recruiters are using social media in all phases of the selection process to evaluate your character and personality in order to make hiring decisions. They want to see if you are well versed, creative, accomplished, and present yourself in an encouraging manner. Employers want to determine if you are a good fit for their department, organization, or agency. Work on building a robust social network and construct an online profile that represents your experience in the workplace. Additional positive attributes that can weigh in your favor are postings denoting your firefighter volunteer work and appropriate community service/activities.

Employers are also looking for reasons not to hire you. Provocative or inappropriate photos; evidence of drinking and/or drug use; sexual impropriety; negative remarks regarding previous employers; discriminatory comments related to race, gender, or religion; and poor communication skills are just a few of the pitfalls leading to being passed over. All photos should present you in a favorable light. Ensure that your media profile is free of typos and that the information you are posting is coherent and consistent throughout all your social media sites. Employers should want to hire you because your background reveals professional qualifications and includes outstanding references.

CHAPTER 4

# The Psychological Test

**Your Goals for This Chapter**

- Learn the purpose of firefighter psychological tests
- Find out about the occupational, personality, and polygraph components of the test

The work of first responders is difficult, to say the least. Emotions run high during fire and emergency incidents. Psychological tests are administered and interpreted by psychologists as part of the overall recruitment process to help the fire service determine whether candidates are mentally prepared to cope with the stressful nature of the job.

Psychological tests assess and evaluate information that you provide to the examiner. Generally the information is gathered through a series of written true or false questions, answers to questions read off a computer screen, and answers to interview questions. The accuracy of the test depends mainly on how pragmatically you answer the questions. Reviewing prior civil service standard professional psychological tests is difficult to do because the security of the tests must be maintained for ethical reasons. It is not so important that you know what questions you are going to be asked as it is to know what the examiner is trying to assess. This chapter provides you with a general understanding of what psychological tests are all about.

Civil service psychological tests can be broken down into three parts: occupational, personality, and polygraph. The following provides a short synopsis of these three major components of the test.

## OCCUPATIONAL

In general, an occupational test seeks to match the interests, knowledge, abilities, and other characteristics of the candidate with those of firefighters already on the job. The theory behind this part of the test is that if you demonstrate attributes similar to most firefighters, then there is a good chance that you will acclimate readily to your new profession and fit in easily with fellow workers.

It takes a particular kind of person to want to perform rescue-related tasks. First responders and persons who gravitate to rescue-related work have been deemed by some psychologists to have a "**rescue personality**." This concept is based on a hypothesis that individuals who choose to become firefighters and first responders have similar characteristics and a predisposition to be rescuers before entering the job. Psychologists have identified the following characteristics and traits as being typical of individuals with a rescue personality:

- action oriented
- easily bored
- enjoys being needed
- highly dedicated
- inner directed
- likes control (of situations and themselves)
- obsessed with high standards of performance
- socially conservative
- traditional

## PERSONALITY

A personality test attempts to measure the candidate's persona or appearance that he or she presents to the world. The most commonly used personality test given during the firefighter recruitment process is the revised Minnesota Multiphasic Personality Inventory (MMPI-2).

## The MMPI-2

The original **MMPI-1 test** was developed at the University of Minnesota Hospitals and first published in 1942. The **MMPI-2**, a revised version, was released in 1989. It is the standard used today.

The MMPI-2 test consists of more than 500 questions and takes approximately 60 to 90 minutes to complete. It is used to assess the mental status of candidates and possible abnormality in some of the following areas:

- Hypochondriasis, or abnormal concern over the body's well-being
- Paranoia, or persecution complex, characterized by rigid opinions and attitudes
- Schizophrenia, characterized by bizarre thought processes and social alienation
- Hysteria, or overreaction to stressful situations
- Depression, or dissatisfaction with one's own life
- Hypomania, or accelerated mood, speech, and motor activity
- Psychopathic deviance, or nonacceptance of authority and amorality

A psychologist interprets the information gathered from the test in conjunction with other historical data (previous employment, academic performance, letters of recommendation, etc.) and constructs a psychological profile of the candidate. The MMPI-2 is also used as a screening device to eliminate candidates with obvious mental health problems.

## THE POLYGRAPH

Also called a **lie detector**, the polygraph assesses the candidate's veracity when replying to pertinent questions of employment. The candidate is connected to a machine (polygraph) via wires and is asked a series of questions. During the question and answer session, the polygraph monitors biological responses, such as heart rate, respiration rate, blood pressure, skin temperature, and skin conductance. This information is used to assess the candidate's honesty. For example, when answering neutral questions (What is your name?), the candidate should be relaxed and exhibit uneventful physiological feedback to the machine. However, when answering real questions (Have you ever taken illegal drugs?), the candidate may exhibit abnormal physiological readings that can be analyzed to help determine if the candidate is answering truthfully. To demonstrate validity and reliability, these tests have built-in gauges that raise red flags when there are indications of lying.

**Note:** When answering true or false questions or verbally responding to questions on a psychological test, don't be untruthful in order to supply the psychiatrist with what you believe to be a healthy reply. In many cases there is no preferred answer or response. Don't make the mistake of trying to make yourself "look good" by not answering questions in an honest and forthright manner.

Psychological testing is never completely valid or reliable. If you are disqualified from the job of firefighter for failing the psychological component of the exam, you may have a behavior or attitude disorder that you don't recognize. It can also mean that the psychiatrist, psychotherapist, or polygraph examiner reviewing your test results made an evaluation mistake on your data or an inaccurate interpretation of your biological feedback results. Inquire about any recourse you may have regarding retesting and/or reevaluation. You may be required to undergo psychotherapy to discover the reasons for your failure prior to getting a second chance at participating in a new psychological test.

## PERSONALITY TRAITS AND CHARACTERISTICS

Firefighting is a profession that has earned the public trust. It is expected that firefighters perform their work at fire and emergency situations courageously and effectively. Firefighter candidates, therefore, require a multitude of positive traits and characteristics that veteran firefighters possess. Integrity and honesty are essential. Interpersonal skills and talents include self-motivation, flexibility, decisiveness, empathy, a pleasant sense of humor, common sense, cooperativeness, being a team player, initiative, and so much more. It therefore behooves the candidate to be aware of what the local fire department they seek to join deems as positive values. It should be noted that candidates who do not demonstrate the above qualities during various stages of the selection process run the risk of not being selected for hire. Examples include trouble controlling emotions, rudeness, rigid thinking, sensitivity to criticism, resentment toward others, a high degree of anxiety, and hostility.

The following list of favorable traits and characteristics is not inclusive, but it will provide valuable insight into what your potential employer is looking for in you. It is recommended that you answer questions dealing with background information and character traits decisively. Try not to answer in a manner that makes your position unclear. Although there is no right or wrong answer, it is important to understand that the fire department you seek to be a part of will invest a lot of time, effort, and money into you becoming a successful member. Examining positive traits and characteristics that have historically led to success will be part of their hiring process.

While there are a multitude of positive traits and characteristics, the ones listed below are deemed by the author as essential. A brief explanation follows each onc. Review this listing. It may provide some insight into what it takes to be a firefighter and whether or not you have what it takes to be one.

Adaptable—able to change your thinking as the situation dictates
Caring—for the well-being of your fellow workers and the citizens you serve
Common sense—use good judgment
Competitive—seeking challenges
Confidence—a belief that you can succeed in whatever you do
Decisive—able to make decisions based upon experience
Dedication—strive for excellence
Dependable—reliable and trustworthy
Empathy—supportive of others in times of need
Good sense of humor—not easily offended by comments from coworkers
Honesty—truthful
Honorable—righteous
Initiative—self-motivated
Integrity—strong moral principles
Loyalty—faithfulness
Maturity—acting like an adult
Mechanical aptitude—good at working with tools and equipment
Motivated—action oriented
Obedient—willing to comply with orders
Physically fit—able to perform the physical tasks of firefighting
Responsible—for your actions
Sacrifice—do without for the betterment of your department
Team player—work well with others

# PERSONALITY BACKGROUND INFORMATION REVIEW QUESTIONS

For all the questions below, choose the answer that best reflects your own personal opinion. Keep in mind, however, what has been discussed previously regarding the traits and characteristics fire department administrations historically seek in their firefighter candidates. Suggested answers located in the answer key at the end of the chapter are based upon the opinion of the author.

**1.** I am rigid in my thinking.

(A) agree strongly
(B) agree up to a point
(C) disagree
(D) disagree strongly

**2.** I am a self-motivated person.

(A) agree strongly
(B) agree up to a point
(C) disagree
(D) disagree strongly

**3.** My friends say I am shy.

(A) agree strongly
(B) agree up to a point
(C) disagree
(D) disagree strongly

**4.** I listen to others regardless if I disagree with their viewpoint.

(A) agree strongly
(B) agree up to a point
(C) disagree
(D) disagree strongly

**5.** Rules are made to be broken.

(A) agree strongly
(B) agree up to a point
(C) disagree
(D) disagree strongly

**6.** I am an emotionally stable person.

(A) agree strongly
(B) agree up to a point
(C) disagree
(D) disagree strongly

**7.** I always complete work assignments on time.

(A) agree strongly
(B) agree up to a point
(C) disagree
(D) disagree strongly

**8.** I tend to blame others when a team project goes wrong.

(A) agree strongly
(B) agree up to a point
(C) disagree
(D) disagree strongly

**9.** I would rather be an accountant than a mechanic.

(A) agree strongly
(B) agree up to a point
(C) disagree
(D) disagree strongly

**10.** No good deed goes unpunished.

(A) agree strongly
(B) agree up to a point
(C) disagree
(D) disagree strongly

**11.** Friends are more important than money.

(A) agree strongly
(B) agree up to a point
(C) disagree
(D) disagree strongly

**12.** Working with other firefighters of different cultures will be beneficial to me.

(A) agree strongly
(B) agree up to a point
(C) disagree
(D) disagree strongly

**13.** Homeless people are just plain lazy.

(A) agree strongly
(B) agree up to a point
(C) disagree
(D) disagree strongly

**14.** I would rather be recognized for an award individually than with a group.

(A) agree strongly
(B) agree up to a point
(C) disagree
(D) disagree strongly

**15.** I enjoy physical exercise and keeping myself fit.

(A) agree strongly
(B) agree up to a point
(C) disagree
(D) disagree strongly

**16.** I become easily confused under stressful conditions.

(A) agree strongly
(B) agree up to a point
(C) disagree
(D) disagree strongly

**17.** I often doubt if I am making the right decisions concerning my life.

(A) agree strongly
(B) agree up to a point
(C) disagree
(D) disagree strongly

**18.** I use little white lies in certain situations.

(A) agree strongly
(B) agree up to a point
(C) disagree
(D) disagree strongly

**19.** I don't wear a hat when eating at a restaurant.

(A) agree strongly
(B) agree up to a point
(C) disagree
(D) disagree strongly

**20.** Walking into an elevator makes me nervous.

(A) agree strongly
(B) agree up to a point
(C) disagree
(D) disagree strongly

**21.** I like to organize people and lead them to confront the task at hand.

(A) agree strongly
(B) agree up to a point
(C) disagree
(D) disagree strongly

**22.** On weekends I like to raise hell a little bit with my friends.

(A) agree strongly
(B) agree up to a point
(C) disagree
(D) disagree strongly

**23.** I hold no grudges against others in my past.

(A) agree strongly
(B) agree up to a point
(C) disagree
(D) disagree strongly

**24.** I have thoughts of sex all the time.

(A) agree strongly
(B) agree up to a point
(C) disagree
(D) disagree strongly

**25.** I enjoy competition and strive to win.

(A) agree strongly
(B) agree up to a point
(C) disagree
(D) disagree strongly

**26.** I like volunteering my time for good causes.

(A) agree strongly
(B) agree up to a point
(C) disagree
(D) disagree strongly

**27.** I dislike taking orders from my boss.

(A) agree strongly
(B) agree up to a point
(C) disagree
(D) disagree strongly

**28.** I avoid going to work on cold and snowy days.

(A) agree strongly
(B) agree up to a point
(C) disagree
(D) disagree strongly

**29.** I am always there for friends who need assistance.

(A) agree strongly
(B) agree up to a point
(C) disagree
(D) disagree strongly

**30.** I am not easily offended when I am the butt of a funny joke.

(A) agree strongly
(B) agree up to a point
(C) disagree
(D) disagree strongly

# Answer Key

1. D. Firefighters must be flexible in their thinking and use others to help formulate decisions.
2. A. Firefighting is an action-oriented profession.
3. D. Firefighters enjoy working and socializing with others.
4. A. Firefighters are confronted with situations where people have conflicting views.
5. D. The fire service is a quasi-military organization.
6. A. During times of stress firefighters must be able to control their emotions.
7. A. Firefighters must be dependable and responsible.
8. D. Firefighters must be accountable for their actions.
9. D. Firefighting is a profession where you work with tools and equipment.
10. D. Firefighters require a positive outlook on life.
11. A. Firefighting is a vocation focused upon giving.
12. A. Cultural sensitivity and tolerance are positive values sought by the Fire Service.
13. D. Firefighters must be caring and show empathy toward others in time of need.
14. D. Firefighters work together in teams for the greater good.
15. A. Firefighting is hard work and demands that firefighters be in tip-top shape.
16. D. Firefighters must be able to think clearly at fire and emergency incidents.
17. D. Firefighters should exude confidence in themselves and the decisions they make.
18. D. Firefighters must be honest and trustworthy.
19. A. Firefighters are looked upon as role models by society.
20. D. Firefighters often work in tight and confined spaces during fires and emergencies.
21. A. The Fire Service is looking for candidates who demonstrate leadership qualities.
22. D. The Fire Service is looking for mature candidates.
23. A. Forgiveness is a positive trait.
24. D. Firefighters must demonstrate a strong moral character.
25. A. Firefighters like challenges and try to be successful.
26. A. Sacrifice is a positive trait.
27. D. Obedience in following orders is essential in the Fire Service.
28. D. Firefighters are dependable and eager to serve 24/7/365.
29. A. Loyalty is a positive trait.
30. A. A good sense of humor is a positive trait.

# CHAPTER 5

# Preparing for the Physical Ability Test

> **Your Goals for This Chapter**
>
> - Learn about the Candidate Physical Ability Test (CPAT)
> - Get training tips for each CPAT event
> - Learn about other helpful exercise programs and good nutrition

Firefighting is a physically demanding profession. It requires flexibility, cardiopulmonary stamina, and muscular strength and endurance.

Most municipalities administer firefighter physical ability exams to ensure that candidates possess the physical capabilities to perform the duties of the firefighter efficiently and safely. The tasks that make up the physical ability exam are designed to measure a person's stamina, agility, strength, and coordination. Regular exercise and proper nutrition are very important in maintaining overall health and the ability to train for and pass the physical firefighter tests. Candidates should also practice the specific skills that are part of the exam.

Many municipalities have adopted the **Candidate Physical Ability Test (CPAT)**. This test, which was established as a joint venture by the International Association of Fire Fighters (IAFF) and International Association of Fire Chiefs (IAFC), consists of a sequence of eight separate events along a predetermined path. When administered correctly, the CPAT allows fire departments to obtain a pool of trainable candidates who are physically able to perform fireground activities.

Municipalities that do not use CPAT have historically conducted physical ability exams designed to evaluate similar attributes. Running, jumping, lifting weighted objects, handgrip strength, balance, agility, endurance, and overall conditioning are some of the areas measured. Be advised that it is in the best interest of a municipality looking for new recruits to formulate a physical ability test that clearly demonstrates applicability to the job of firefighting.

This chapter focuses on the CPAT and provides a brief discussion of ways to prepare for the CPAT through several exercise and fitness programs.

## CANDIDATE PHYSICAL ABILITY TEST (CPAT) EVENTS

The CPAT is a pass/fail test of eight sequential events to be completed in a maximum total time of 10 minutes and 20 seconds. CPAT events require the candidate to wear a 50-pound vest to simulate the weight of self-contained breathing apparatus (SCBA) and personal protective equipment (PPE). Throughout all events, candidates must wear long pants, hard hat, work gloves, and footwear. The sequential events include:

Stair climb
Hose drag
Equipment carry
Ladder raise and extension
Forcible entry
Search
Rescue
Ceiling breach and pull

The eight events are performed in a logical sequence that simulates the duties performed on the fireground, but the test allows for an 85-foot walk to recover between events. The **stair climb**, **hose drag**, and **equipment carry** are the preliminary steps required to begin fighting a fire. The **ladder raise** and **forcible entry** constitute the beginning of interior firefighting operations. The **search** and **rescue** events follow, simulating life-saving techniques and abilities. The final event, **ceiling breach and pull**, mirrors overhaul (looking for hidden fire) operations that are commonly performed subsequent to the fire being extinguished.

As stated previously, all eight events must be completed within the 10 minute and 20 second time frame of the test. If the candidate does not complete the events within that time frame, he or she fails the test.

### Event 1: Stair Climb

This event simulates the carrying of one length of bundled hose up flights of stairs. When operating at fires, firefighters often climb stairs and ladders wearing full PPE and carrying equipment.

During this event only, candidates are required to wear additional weight (two 12½-pound weights on the shoulders) and to walk on a StepMill, which is situated between a wall and an elevated platform, at a stepping rate of 60 steps per minute for three minutes. The handrail of the StepMill opposite

the wall is removed. Prior to beginning the timed event, each candidate performs a 20-second warm-up at a rate of 50 steps per minute. There is no break between this warm-up period and the actual timed test event. If the candidate falls or dismounts the StepMill three times during the warm-up period, he or she fails the test. If the candidate falls, grasps any of the test equipment, or steps off the StepMill during the timed event, the test is concluded and the candidate fails the test. The candidate is only permitted to momentarily touch the wall or handrail for balance. At the conclusion of this event, the shoulder weights are removed and the candidate walks 85 feet to the next event.

### TRAINING

This event challenges the candidate's aerobic capacity, lower body muscular endurance, and balance. Running, fast walking, stair stepping, use of a treadmill, swimming, and bicycling enhance aerobic capability. Follow an exercise program to strengthen the quadriceps, hamstrings, glutes, calf muscles, and lower back stabilizers.

Perform an actual stair-stepping exercise at the base of a staircase. Use the first step of the staircase to perform 24 complete stepping cycles within a one-minute period. A stepping cycle consists of stepping up with one foot, then the other, and down with one foot, then the other. Alternate your starting foot from right to left. Try to complete two stepping cycles within a five-second period. Step continuously for five minutes. As your fitness improves, complete a second and third five-minute exercise interspersed with several minutes of recovery time. Begin to add weight to your waist by using a knapsack while performing these step exercises. Gradually increase the weight around your waist to 50 pounds. Eventually, try carrying 10- to 15-pound dumbbells in each hand in addition to the 50 pounds around your waist. At this stage of your training, reduce the duration of the exercise intervals to three minutes.

## Event 2: Hose Drag

This event simulates firefighters stretching and deploying hose lines from the fire apparatus into the fire building and around many obstacles (doorways, furniture, stairwells) inside the building while maintaining a low posture.

During this event, the candidate grasps a six-pound nozzle attached to four lengths (200 feet) of attack hose (1½ inch diameter), places the hose over the shoulder and across the chest (maximum eight feet), and drags the hose 75 feet along a marked path (cones) to two prepositioned 55-gallon drums that are secured together and weighted. The candidate makes a 90-degree turn around the drums, continues an additional 25 feet, and then drops to at least one knee at the finish line. While kneeling, the candidate must pull the hose across the end line. If the candidate fails to go around the drum, goes outside the marked path, or does not keep one knee in contact with the ground within the marked-off area while pulling the hose across the finish line, the test is concluded and the candidate fails the test. After completing the event, the candidate walks 85 feet within the established walkway to the next event.

### TRAINING

This event requires enhancing both the aerobic and anaerobic energy systems of the body, as well as strengthening the quadriceps, hamstrings, glutes, calves, lower back stabilizers, biceps, deltoids, upper back, and forearm and hand (grip) muscles.

Attach 50 feet of rope to a weighted duffel bag. Place the rope over your shoulder and drag the duffle bag (resistance) 75 feet while running or walking quickly. Immediately drop to one knee and rapidly but steadily pull the rope hand-over-hand to bring the duffle bag to your body. Perform 8 to 10 repetitions of this event sequence exercise, with a two-minute recovery time between repetitions. As fitness improves, increase weight resistance to 60 to 80 pounds.

## Event 3: Equipment Carry

This event simulates removing tools and equipment from the apparatus and carrying them to and from a point of operation.

During this event the candidate removes two 32-pound saws from a tool cabinet, one at a time, and places them on the ground. The candidate then picks up both saws, one in each hand, and carries them 75 feet around a drum and then back to the starting point. The candidate then places the saws on the ground, picks up each saw one at a time, and replaces them in the designated space

inside the cabinet. If the candidate drops either saw on the ground during the carry, the test is concluded and the candidate fails the test. The candidate receives one warning for running; a second infraction constitutes a failure. The candidate then walks 85 feet within the established walkway to the next event.

### TRAINING

This event requires enhancing the aerobic energy system of the body, as well as strengthening the biceps, deltoids, trapezius, upper back, forearm and hand (grip) muscles, and the glutes, quadriceps, and hamstrings.

Place two 25- to 30-pound dumbbells on a shelf four feet above ground level. Remove the weights, one at a time, and place them on the ground. Pick up the weights and carry them a distance of 40 feet out and 40 feet back and replace them on the shelf. Continue this practice until it can be performed easily with 30 pounds.

## Event 4: Ladder Raise and Extension

This event simulates firefighters' use of portable ladders to reach and access windows, balconies, and roofs of fire structures.

This event uses two portable 24-foot aluminum extension ladders. One ladder is lying on the ground and hinged at one end to a wall, and the other ladder is secured in a vertical position. During the event, the candidate lifts the ladder on the ground by the unhinged end and walks underneath the ladder while raising it to a stationary position against a wall. The candidate then proceeds to the other ladder and stands in front of it with both feet inside a marked-off area and extends the fly section of the ladder hand-over-hand until it hits the stop. This concludes this event.

If a candidate misses any rung during the ladder raise, one warning is given; the second infraction constitutes a failure. If the ladder is allowed to fall to the ground or the safety lanyard is activated because the candidate completely releases the grip on the ladder, the test is concluded and the candidate fails the test. If during the ladder extension, the candidate's feet do not remain within the marked-off area, one warning is given; a second infraction constitutes a failure. The candidate walks 85 feet within the established walkway to the next event.

### TRAINING

This event requires enhancing both the aerobic and anaerobic energy systems of the body, as well as strengthening the biceps, deltoids, upper back, trapezius, forearm and hand (grip) muscles, and the glutes, quadriceps, and hamstrings.

Ideally, use an actual 24-foot aluminum extension ladder. Ensure that two adults are available to secure the ladder at the base during the ladder raise segment of this event. While practicing this skill, it is important to move safely and slowly to develop confidence in the required movements. To perform the ladder extension exercise, attach a rope to a weighted duffel bag or knapsack. Place the rope over a tree branch or horizontal bar support (playground swings) 8 to 10 feet above the ground. Use a hand-over-hand motion to steadily raise the weighted object to the top of the branch or bar and then slowly lower it to the ground using the same hand-over-hand technique. Perform 8 to 10 repetitions of this movement. Rest two minutes and repeat the exercise-rest sequence two more times. As your strength and skill improve, progressively add more weight resistance until you reach 40 to 50 pounds.

## Event 5: Forcible Entry

This event is designed to simulate the critical tasks of using force to open locked doors or breaching wood, masonry, and brick walls.

During this event, the candidate uses a 10-pound sledgehammer to strike a forcible entry machine calibrated to measure the cumulative force of 300 pounds of pressure based on the effort required to force open a door. The candidate's feet must remain outside a toe box assembly. The forcible entry machine is mounted 39 inches on center from the ground (typical location of a standard exterior doorknob). If the candidate does not maintain control of the sledgehammer and releases it from both hands while swinging, it constitutes a failure. A candidate who steps inside the toe box is warned; a second infraction constitutes a failure, the test is concluded, and the candidate fails the test. A buzzer and signal lamp denote the completion of this event. After the buzzer

is activated, the candidate places the sledgehammer on the ground. The candidate then walks 85 feet to the next event.

### TRAINING

This event requires enhancing both the aerobic and anaerobic energy systems of the body, as well as strengthening the following muscle groups: quadriceps, glutes, triceps, upper back, trapezius, and muscles of the forearm and hand (grip).

Wrap padding that has a circular target in the center around a large tree or vertical pole at a level of 39 inches above the ground. Stand to the side of the target area and swing a 10-pound sledgehammer. Practice hitting the target area in a level manner with increased velocity without sacrificing accuracy. Focus on using your legs and hips to initiate the swinging motion. Swing 15 times and rest for two minutes. Repeat the exercise two more times.

## Event 6: Search

This event simulates the firefighting task of searching for victims inside an unpredictable area (fire building) with limited visibility. Firefighters often crawl low while searching in heated areas, moving around furniture and other obstacles in total darkness (smoke environment).

During this event, the candidate crawls on hands and knees through a dark tunnel maze that is approximately 3 feet high, 4 feet wide, and 64 feet long, with two 90-degree turns. At various locations in the tunnel, the candidate

must maneuver around, over, and under obstacles. At two additional locations, the candidate is required to crawl through a narrowed space where the dimensions of the tunnel are reduced. The candidate's movement through the maze is monitored. This event ends upon exit from the tunnel maze. A candidate who requests assistance that requires the opening of the escape hatch or opening of the entrance or exit cover fails the test. The candidate who completes this event then walks 85 feet to the next event.

### TRAINING

This event requires enhancing both the aerobic and anaerobic energy systems of the body, as well as strengthening the chest, shoulder, triceps, quadriceps, abdominals, and lower back muscles. Practice crawling on hands and knees wearing loose-fitting pants and kneepads for at least 70 feet while making several right-angle turns during the crawl. Keep low (no higher than three feet above the ground) to simulate being inside the tunnel. Occasionally while crawling on hands and knees, drop to your stomach and crawl 10 feet along the ground. When comfortable with the crawling techniques, repeat the sequence with a weighted knapsack on your back. Gradually aim to increase the weight inside the knapsack to 50 pounds.

## Event 7: Rescue

This event simulates the removing of a victim or injured firefighter from the fire scene around obstacles to a safe area.

During this event, the candidate grasps a 165-pound mannequin (the minimum weight a firefighter must be able to drag to meet the physical demands of the job) by one or both of the harness shoulder handles (simulating the shoulder straps of a firefighter's SCBA) and drags it 35 feet to a prepositioned drum. The candidate then makes a 180-degree turn around the drum and continues to drag the mannequin an additional 35 feet totally across the finish line. The candidate is not permitted to grasp or rest on the drum or to drop and release the mannequin to adjust his or her grip. One warning is given if the

candidate grasps or rests on the drum at any time; a second infraction concludes the test, and the candidate fails the test. This concludes the event. The candidate walks 85 feet within the established walkway to the next event.

### TRAINING

This event requires the enhancing of both the aerobic and anaerobic energy systems of the body, as well as strengthening the quadriceps, hamstrings, glutes, abdominals, torso rotators, lower back stabilizers, trapezius, deltoids, latissimus dorsi, biceps, and muscles of the forearm and hand (grip) muscles.

Attach a short handle to a weighted duffel bag. Grasp the handle with one hand and drag the bag in a crossover, sidestepping manner. Another technique is to grasp the handle with both hands while facing the bag and moving directly backward, taking short, rapid stagger steps. Try both maneuvers and pick the technique that feels the most comfortable and effective for you. Drag the weighted duffel bag 35 to 50 feet in one direction and then turn around and drag it back to the starting point. Complete 8 to 10 repetitions of this task with a two-minute rest interval between trials. Increase the weight inside the duffel bag until you can successfully complete four repetitions (with rest intervals) with 165 pounds.

## Event 8: Ceiling Breach and Pull

This event simulates the task of breaching and pulling down a ceiling to check for hidden fire and fire extension.

During this event, the candidate removes a pike pole from a bracket and stands within an area inside the framework of the equipment. The candidate places the tip of the pike pole on the target area (a ceiling assembly eight feet above the ground containing a hinged door and handle) and pushes up (breaching action) on the hinged door with the pike pole three times. The candidate then hooks the pike pole onto the handle of the ceiling assembly and pulls downward five times. The candidate repeats the set (three pushes and five pulls) four times. The standard ceiling height of a residential structure is eight feet. The force required to breach the ceiling is 60 pounds, and the force required to pull the ceiling is 80 pounds. Three breaches followed by five pulls will provide a

four-foot by eight-foot examination opening within a structure. One warning is given if the candidate drops the pike pole to the ground. A second infraction constitutes a failure. The event and the total test time end when the applicant completes the final pull stroke repetition.

### TRAINING

This event requires enhancing both the aerobic and anaerobic energy systems of the body, as well as strengthening the following muscle groups: quadriceps, hamstrings, glutes, abdominals, torso rotators, lower back stabilizers, deltoids, trapezius, triceps, biceps, and muscles of the forearm and hand (grip).

To train for a ceiling breach, tie a short piece of rope to a dumbbell or weighted knapsack placed between your legs situated shoulder-length apart. Grasp the rope, arms slightly away from the body with one hand at upper-thigh level and the other hand at chest level. Lift upward and out from the body with an upward extension of the legs to perform an action that simulates thrusting a pole through an overhead ceiling. Complete three sets of eight repetitions with two minutes of rest between sets. Add weight as strength and technique improve.

To train for a ceiling pull, use an exercise similar to ladder extension training. The ceiling pull training, however, requires exerting power in single, repeated downward thrusts rather than in hand-over-hand movement. Grasp the rope attached to a weighted knapsack or duffel bag with hands spaced approximately one foot apart (bottom hand at your chin level). Use both hands simultaneously to provide a powerful downward motion. Lower your body to raise the weighted object several feet above the ground. Repeat 8 to 10 consecutive repetitions of the movement. Complete three sets with a two-minute recovery interval interspersed. Add weight as your strength and technique improve.

**Final Note:** Candidates who attend CPAT practice sessions and who participate in physical training prior to the actual physical ability test have a higher passing rate than candidates who have not prepared.

# PREPARING FOR THE CPAT

In its eight events, the CPAT evaluates a candidate's flexibility, cardiovascular fitness, stamina, and muscular strength and endurance. These attributes can be enhanced through exercise programs, some of which have been developed specifically for the CPAT, as well as through good nutrition.

## Flexibility

The **Flexibility Exercise Program** was developed by the IAFF and IAFC to assist firefighter candidates in preparing for the flexibility part of the CPAT. It includes a series of stretching exercises involving the muscles of the legs, chest, back, shoulders, and arms designed to increase flexibility.

# Cardiopulmonary Stamina

Cardiopulmonary stamina exercises enhance a candidate's cardiovascular fitness and general health. Cardiopulmonary endurance is the ability of the cardiovascular and respiratory systems to carry oxygen to the muscles of the body. Exercises are designed to cnhancc both thc acrobic (oxygcn utilization) and anaerobic (oxygen debt) body systems.

## AEROBIC TRAINING

Aerobic training involves moderate-intensity exercises that are sustained over a prescribed period of time (30 minutes or more). Aerobic exercise should be performed four times per week. While actively engaging in these exercises, a person's heart rate should be between 60 percent and 85 percent of his or her maximum target heart rate. (You can calculate your maximum heart rate by subtracting your age from 220.)

During aerobic exercise, your body uses oxygen to burn glycogen and fat for fuel. Cardiopulmonary benefits gained from aerobic training include increased lung capacity, enhanced oxygen efficiency, stronger heart muscle able to pump a greater quantity of blood per beat, better circulation of oxygen throughout the blood system, and a reduction in recuperation time.

Effective aerobic exercises include running, jogging, walking, dancing, cross-country skiing, swimming, bicycling, skating, aerobic exercise classes, circuit training, martial arts, treadmill, and stair climbing.

## ANAEROBIC TRAINING

Anaerobic training involves high-intensity, short-duration exercise. While actively engaged in these exercises, the heart rate should be between 75 percent and 100 percent of one's maximum target heart rate. Maximum intensity exercise is generally performed within a one- to three-minute time frame. Anaerobic training should be performed two to three times per week. Interval training involving a repeated series of exercises interspersed with rest periods (wind sprints, for example) is an excellent way to improve anaerobic endurance.

During anaerobic exercise, your body burns glycogen for fuel. Anaerobic exercise increases the amount of time that a person is able to perform at maximum intensity and also boosts the amount of glycogen that is stored in the muscles.

Examples of anaerobic exercise include heavy weight lifting, sprinting, jumping rope, racquetball, handball, boxing, and team sports (basketball, soccer, football).

# Muscular Strength and Endurance Circuit Programs

Muscular strength and endurance circuit programs that include weight training and calisthenics have been developed by the IAFF and IAFC to assist firefighter candidates in preparing for the CPAT. Circuit training—sequential activities performed with a rest period not to exceed 30 seconds between exercise stations—has been proven to be a very effective and efficient way to enhance muscular strength and endurance. It is recommended that the candidate taking a CPAT follow the IAFF/IAFC Flexibility Exercise Program as well as the Weight Training Circuit Workout and/or Calisthenics Training Circuit Workout.

The **Weight Training Circuit Workout** is designed to increase strength and endurance. The candidate is required to lift a weight resistance for 10 repetitions at each station. These exercises are designed to be performed three times per week. At first, complete only one circuit per workout. As strength and endurance increase, however, strive to complete three circuits per workout.

The **Calisthenics Training Circuit Workout** incorporates exercises designed to be performed anywhere without weights. These exercises should be performed daily. As with the Weight Training Circuit Workout, start off completing one circuit per workout and strive to complete three circuits per workout as general fitness increases.

## Nutrition

The food we eat affects our overall well-being and energy level and plays a very important role in the health and efficiency of our body systems. A body carrying excess fat forces the heart to work harder and poses many health and performance risks. The amount and type of proteins, carbohydrates, and fats consumed and the intake of vitamins and minerals are essential to maintaining a healthy body.

Good nutrition and exercise work hand in hand to help provide a lifestyle that promotes good health. This is reflected in the U.S. Department of Agriculture's dietary guidelines, which encourage people to be physically active and make healthy food choices. View the MyPlate guidelines at ChooseMyPlate.gov and refer to the links provided for information about good nutrition.

For a firefighter candidate, good nutrition provides an essential basis for the physical, psychological, and intellectual skills needed to succeed in the required examinations—and in being a successful firefighter.

## PART III

# REVIEW FOR THE WRITTEN EXAMINATION

# CHAPTER 6

# Mathematics

> **Your Goals for This Chapter**
> - Review the arithmetic, algebra, and geometry you need to know
> - Practice with sample mathematics questions

Many firefighter exams include mathematical problems pertaining to the job of firefighting. Knowledge of basic arithmetic, algebra, and geometry is necessary in the fire service. Measuring gas/oil mixtures that fuel portable power saws, mixing cleaning fluids with water during maintenance chores, determining the placement of a ladder for proper climbing angle, and monitoring gallons per minute (GPM) flowing through hose lines are just a few examples of how firefighters use numbers. The mathematics selected for this review cover the most common areas included in previous firefighter examinations, basic terminology, symbols, and order of operations.

## TERMINOLOGY (THE LANGUAGE OF MATHEMATICS)

There are many terms used in mathematics that you should understand. A list of some of the more common terms and their definitions follows:

**Real numbers**—both rational and irrational numbers
**Whole numbers**—the counting numbers that include zero

**Example:** 0, 1, 2, 3, . . .

**Natural (positive) numbers**—all whole (counting) numbers greater than zero

**Example:** 1, 2, 3, 4, ...

**Negative numbers**—all numbers less than zero

**Example:** –1, –2, –3, . . .

**Integers**—all positive and negative whole numbers, including zero, but not including fractions and decimals

**Example:** . . . –2, –1, 0, 1, 2, . . .

**Fraction**—part of a whole number

**Example:** $\frac{1}{3}$

**Mixed number**—a whole number and a fraction

**Example:** $2\frac{1}{2}$

**Rational numbers**—all numbers that can be expressed as the ratio of two integers (fractions and integers)

**Example:** $x/y$ where $x$ and $y$ are integers

**Irrational numbers**—all numbers that cannot be expressed as the ratio of two integers

**Example:** $\pi$ (3.14 . . .) and many square roots $(\sqrt{2}), (\sqrt{5}), (\sqrt{6}), (\sqrt{8})$, etc.

**Term**—a single number or the product of one or more **numerical** (number) **coefficient** and/or **literal** (letter) **coefficient** factors. **Like terms** have the same literal factor. Unlike terms do not have a common literal factor.

**Example:** (like terms) $2a$ and $3a$

**Example:** (unlike terms) 6, $3x$, and $yz$

**Factor**—an individual number (numerical) or letter (literal) in a term

**Example:** 2, $a$, and $b$ are factors of the term $2ab$

**Algebraic expression**—two or more terms connected by plus or minus signs

**Example:** $4k + 7$

**Algebraic equation**—a statement representing two things that are equal to one another

**Example:** $12x - 8x = 4x$

# MATHEMATICAL SYMBOLS AND CORRESPONDING COMMON PHRASES

| | |
|---|---|
| $+$ | sum, and, plus, add |
| $-$ | difference, remainder, subtract, minus, less than, decreased by |
| $\times$ | multiplied by, times |
| $\div$, /, — | divided by |
| : | ratio |
| $=$ | equals, is equal to, is as much as |
| $\neq$ | does not equal, is not equal to, inequality |
| $\lvert\ \rvert$ | absolute value of |

| | |
|---|---|
| $\pi$ | pi |
| [ ] | brackets, preference grouping |
| ( ) | parenthesis, preference grouping |
| $^2$ | exponent, to the power of<br>Example: $^2$ in $4^2$. The exponent 2 means that the 4 is to be taken to the second power, or squared ($4 \times 4 = 16$) |
| $\sqrt{}$ | radical sign, the square root of |
| . | decimal point, point |
| % | percent |

# ORDER OF OPERATIONS (EVALUATING EXPRESSIONS)

Many numerical expressions include two or more operations—for example, exponents, division, and addition. These operations must be performed in the correct sequential order. The acronym **PEMDAS** helps in remembering what the correct sequence is.

PEMDAS—Please Excuse (My Dear) (Aunt Sally)—stands for **P**arenthesis, **E**xponents, **M**ultiplication/**D**ivision, then **A**ddition/**S**ubtraction.

When there are two or more **parentheses**, grouping symbols, perform the innermost grouping symbol first. **Exponents** should be worked on next. **Multiplication** and **division**, as well as **addition** and **subtraction**, are grouped to denote that when these operations are next to each other, just perform the math from left to right.

PEMDAS is used when evaluating formulas, solving equations, solving algebraic expressions, and working with monomials and polynomials.

**Example:** $2 + 4[3 + (4 - 2)^2]$

$2 + 4[3 + (2)^2]$

$2 + 4[3 + 4]$

$2 + 4[7]$

$2 + 28$

$30$

**Example:** $2 + 3(6 + 1)^2$

$2 + 3(7)^2$

$2 + 3(49)$

$2 + 147$

$149$

# ODD AND EVEN NUMBERS

**Odd numbers** are whole numbers that when divided by 2 leave a remainder.

**Example:** –3, –1, 1, 3, . . . (3 ÷ 2 = 1 with 1 remainder)

**Even numbers** are whole numbers that when divided by 2 leave no remainder. Zero is included.

**Example:** –4, –2, 0, 2, 4, . . . (4 ÷ 2 = 2)

**Note:** Division by zero is undefined and therefore has no meaning regardless of whether the dividend is odd or even.

# SIGNED NUMBERS

Negative and positive numbers are **signed numbers**. A thermometer has a scale having both positive and negative numbers. The numbers run along a vertical number scale with positive numbers, indicated by the (+) sign, being above zero and negative numbers, indicated by the (–) sign, found below zero.

## Absolute Value of a Number

The **absolute value** of a number is its value when the sign is not taken into consideration. The symbol for absolute value is | |.

**Example:** |7| = 7 and |–7| = 7

## Adding Negative Numbers

Adding a negative number is the equivalent of subtracting a positive number.

**Example:** 8 + (–4) = 4 is equivalent to 8 – 4 = 4

Conversely, adding a positive number to a negative number is the same as subtracting the negative number.

**Example:** –4 + (+8) = 4 is equivalent to 8 – 4 = 4

To add two negative numbers (like signs), add their absolute values and prefix the sum with their common sign.

**Example:** –6 + (–9) = –(|6| + |9|) = –15

## Subtracting Negative Numbers

Subtracting a negative number from a positive number is the equivalent of adding a positive number.

**Example:** 8 – (–4) = 12 is equivalent to 8 + 4 = 12

Conversely, subtracting a positive number from a negative number is the equivalent of adding the absolute value of two negative numbers and prefixing the answer with a minus sign.

**Example:** $-4 - (+2) = -6$ is equivalent to $-(|-4| + |+2|) = -(4 + 2) = -6$

## Multiplying and Dividing Signed Numbers

When multiplying or dividing two signed numbers, the **product** (multiplication) or **quotient** (division) will be positive if the signs are the same. If the signs are different, however, the product or quotient will be negative.

**Example:** $(-5) \times (-7) = +35$

**Example:** $(-6) \div (-3) = +2$

**Example:** $(-5) \times (7) = -35$

**Example:** $(-6) \div (3) = -2$

**Example:** $\frac{(-4)(2)}{(-2)(-2)} = \frac{-8}{4} = -2$

When multiplying more than two numbers, the product is always a negative number if there are an odd number of negative factors. Conversely, if there is an even number of negative factors, the product is always a positive number.

**Example:** $(2)(4)(-4) = -32$

**Example:** $(8)(2)(-3)(-4) = 192$

The product of two or more numbers is always zero if any of the numbers is zero.

**Example:** $(-9)(-3)(0) = 0$

When multiplying numbers in **exponential** form (a number being multiplied by itself), odd **exponents** (powers) of negative numbers result in a product that is negative, and even exponents of negative numbers have positive products.

**Example:** $(-3)^3$ (to the third power) $= -3 \times -3 \times -3 = -27$

**Example:** $(-6)^2$ (to the second power) $= -6 \times -6 = 36$

**Note:** Division by zero is undefined and therefore has no meaning regardless of whether the dividend (numerator) is a positive or negative number.

# PRIME NUMBERS

Unique numbers whose sole factors are 1 and themselves are called **prime numbers**. All prime numbers except 2 are odd.

**Example:** 2, 3, 5, 7, 11, 13, 17, 19, 23, 29, 31, 37, . . .

To calculate the **prime factorization** of a number, divide it and all its factors until every remaining integer of the group is a prime number.

**Example:** $24 = 2 \times 12 = 2 \times 2 \times 6 = 2 \times 2 \times 2 \times 3$. The numbers 2 and 3 are both prime numbers.

**Example:** $12 = 2 \times 6 = 2 \times 2 \times 3$. The numbers 2 and 3 are both prime numbers.

To find the **greatest common factor** (GCF) of the two examples above, determine the intersection of prime numbers of the two prime factorizations. Because both prime factorizations contain $(2 \times 2 \times 3)$, the GCF is 12.

**Multiples** of a number are its products with the natural numbers.

**Example:** $4 \times 1 - 4$, $4 \times 2 = 8$, $4 \times 3 = 12$, $4 \times 4 = 16$

The multiples of 4 are therefore 4, 8, 12, 16, and so on.

The **least common multiple** (LCM) is also found using prime factorization. The LCM is equal to the multiplication of each factor by the *maximum* number of times it appears in *either* number. Using the two prime factorization examples above, once again, the number 2 appears three times for the number 24 and two times for the number 12, and the number 3 appears one time for both 24 and 12. Therefore, the LCM of 24 and 12 is the product of $2 \times 2 \times 2 \times 3 = 24$.

# FRACTIONS

A **fraction** is a part of a whole number and consists of two expressions, an upper number (**numerator**) and a lower number (**denominator**), which are called the **terms** of the fraction. The fraction represents a division process whereby the numerator is divided by the denominator.

**Example:** $\frac{3}{5} = 3 \div 5$

A fraction with a smaller numerator than denominator is a **proper fraction**.

**Example:** $\frac{7}{8}$

A fraction with a larger numerator than denominator is an **improper fraction**.

**Example:** $\frac{9}{4}$

## Equivalent Fractions

To determine if two fractions are **equivalent** (equal), multiply the numerator and denominator of one fraction by the same number so that the denominators of the fractions are equal.

**Example:** Determine if $\frac{1}{2}$ is equivalent to $\frac{4}{8}$.

To have the denominators of both fractions the same—in this case, 8—you need to multiply the denominator of the first fraction by 4 and multiply the numerator by the same number, 4. Multiplying the numerator and denominator of the first fraction by 4 gives the second fraction. Therefore, the two fractions are equivalent.

**Example:** $\frac{1\times 4}{2\times 4}=\frac{4}{8}$

**Example:** Determine if $\frac{1}{2}$ and $\frac{7}{10}$ are equivalent.

To have the denominators of both fractions the same, you need to multiply the denominator of the first fraction by 5. Then, multiply the numerator by the same number, 5.

**Example:** $\frac{1\times 5}{2\times 5}=\frac{5}{10}$

The result, $\frac{5}{10}$ is not the same as $\frac{7}{10}$, so the two fractions—$\frac{1}{2}$ and $\frac{7}{10}$—are not equivalent.

**Note:** If you multiply or divide both the numerator and the denominator of a fraction by the same number (excluding zero), you will not change the value of the fraction. Remember, if the numerator and denominator of a fraction are both the same number, the fraction is equal to 1, so multiplying the numerator and denominator by the same number is the same as multiplying the fraction by 1.

## Reducing Fractions

To **reduce** a fraction to its lowest form, divide the numerator and the denominator by their greatest common factor (GFC). The GCF of two numbers is the largest factor that is common to both.

**Example:** Reduce $\frac{25}{60}$ to its lowest form.

Prime factorization of $25 = 5 \times 5$

Prime factorization of $60 = 2 \times 30 = 2 \times 5 \times 6 = 2 \times 5 \times 2 \times 3 = 2 \times 2 \times 3 \times 5$

GCF of 25 and 60 = 5

Divide the numerator and denominator by the GCF, 5, to reduce the fraction.

$\frac{25 \div 5}{60 \div 5}=\frac{5}{12}$

The fraction $\frac{5}{12}$ is at its lowest form since there is no number except 1 that can divide both 5 and 12.

## Comparing Fractions

If the denominators of two fractions are equal, then the fraction with the larger numerator is bigger. If the numerators of the two fractions are equal, then the fraction with the smaller denominator is bigger.

When comparing fractions that have different numerators and denominators, cross multiply the numerator of each fraction by the denominator of the other fraction. Write the product of each multiplication next to the numerator you used to get it. Compare the products on each side of the fractions to determine which of the two fractions is greater.

**Example:** Compare $\frac{4{,}000}{60{,}000}$ and $\frac{1}{3}$.

Cross multiply:

$(12{,}000)\frac{4{,}000}{60{,}000}$ and $\frac{1}{3}(60{,}000)$

12,000 is less than 60,000

therefore $\frac{4{,}000}{60{,}000}$ is less than $\frac{1}{3}$

## Adding and Subtracting Fractions

Fractions having the same denominators require only the adding or subtracting of the numerators.

**Example:** $\frac{3}{5}+\frac{2}{5}=\frac{5}{5}=1$

**Example:** $\frac{3}{5}-\frac{2}{5}=\frac{1}{5}$

Fractions having different denominators require first finding a **common denominator** or **least common denominator (LCD)** before you can either add or subtract them.

**Example:** $\frac{1}{5}+\frac{2}{3}$

The least common multiple of 5 and 3 is 15; therefore the LCD is 15.

Multiply the numerator *and* denominator of the first fraction by the number required to change the denominator to 15—in this case, 3.

$$\frac{1}{5}\times\frac{3}{3}=\frac{3}{15}$$

Repeat the process for the second fraction by multiplying the numerator *and* denominator by 5 (the number required to change the denominator to 15).

$$\frac{2}{3}\times\frac{5}{5}=\frac{10}{15}$$

Then add the numerators and place the sum over the common denominator:

$$\frac{3}{15}+\frac{10}{15}=\frac{13}{15}$$

**Note:** Multiplying the denominators together to get a common denominator allows you to omit having to find the LCD, and sometimes will, in fact, give you the least common denominator. However, there are two drawbacks to this approach: first, you may be working with larger numbers since you may not be using the least common denominator; and second, you will have to reduce your answer to the lowest form of the fraction.

## Multiplying and Dividing Fractions

To multiply fractions, multiply their numerators and denominators.

**Example:** $\frac{4}{9} \times \frac{5}{7} = \frac{20}{63}$

To divide fractions, simply invert the second fraction and multiply the numerators and denominators. Remember: multiplication and division are **inverse operations** (as are addition and subtraction).

**Example:** $\frac{1}{3} \div \frac{4}{7} = \frac{1}{3} \times \frac{7}{4} = \frac{7}{12}$

## MIXED NUMBERS

A mixed number consists of an integer and a fraction. Mixed numbers must be converted into fractions prior to performing addition, subtraction, multiplication, and division operations. To convert a mixed number to a fraction, multiply the integer of the mixed number by the denominator of the fraction and then add the product to the numerator of the fraction and put that number over the denominator of the original fraction.

**Example:** $4\frac{3}{5} = \frac{(4 \times 5) + 3}{5} = \frac{20 + 3}{5} = \frac{23}{5}$

## DECIMALS

Numbers that contain a **decimal point** are called **decimals.** The number to the left of the decimal point is a **whole number**, and the number to the right of the decimal point is called the **tenths** digit. The tenths digit gives you the number of tenths that are added to the whole number. If there are two numbers to the right of the decimal point, then both numbers to the right are the number of **hundredths** to be added to the whole number. Three numbers to the right of the decimal point would indicate that all three numbers to the right are the number of **thousandths** to be added to the whole number, and so forth.

## Converting Decimals into Fractions

Decimals, indicating parts of a whole number, can be converted into fractions and vice versa.

**Example:** Convert 0.4 to a fraction

Count the number of decimal places—in this case, tenths. Use 10 as the denominator of the fraction and place the decimal digits without the decimal point as the numerator. The fraction is $\frac{4}{10}$. Reduce it to $\frac{2}{5}$.

## Converting Decimals into Mixed Numbers

**Example:** Convert 2.3 into a mixed number.

Count the number of decimal places and use that number to find whether the denominator should be tenths, hundredths, or thousandths as explained above. If the decimal number is greater than 1, keep the whole number and write the decimal digits without the decimal points as a numerator. Therefore, $2.3 = 2\frac{3}{10}$.

## Converting Fractions and Mixed Numbers into Decimals

To convert a proper fraction into a decimal, divide the numerator by the denominator to obtain the **decimal equivalent**.

**Example:** $\frac{45}{50} = 0.9$

To convert an improper fraction into a decimal, first convert the improper fraction into a mixed number; keep the whole number and divide the remainder (the numerator by the denominator).

**Example:** $\frac{23}{10} = 2\frac{3}{10} = 2.3$

## Rounding Off Decimals

Reducing a decimal to fewer decimal places is known as **rounding off**. If the digit in the decimal place to be eliminated is 5 or greater, increase the digit in the next decimal place to the left by 1. If the digit to be eliminated is less than 5, leave the retained digits unchanged.

**Examples:** 0.2344 rounded to the nearest thousandth is 0.234

3.14 rounded to the nearest tenth is 3.1

## Adding and Subtracting Decimals

When adding and subtracting decimals, write the numerals one above the other with their decimal points lined up.

**Examples:**

| | | | |
|---|---|---|---|
| 1.5 | 5.83 | 22.09 | 2.9 |
| +4.2 | +1.1 | −13.02 | −0.7 |
| 5.7 | 6.93 | 9.07 | 2.2 |

## Multiplying and Dividing Decimals

When multiplying two decimals, initially ignore the decimal points and simply multiply the two numbers to get the product. Then, add together the number of digits to the right of the decimal point in each number and put the decimal in the product so that there is the same number of decimal places as in the numbers multiplied.

**Example:**

$$\begin{array}{r} 1.3 \\ \times 2.2 \\ \hline 26 \\ 260 \\ \hline 2.86 \end{array}$$

Count one digit to the right of the decimal point in 1.3 plus one digit to the right of the decimal point in 2.2 and place the decimal point so that there are two digits to the right of the decimal point in the product.

To divide one decimal into another, move the decimal in the divisor to the right until it is a whole number and then move the decimal to the right the same number of places in the dividend.

**Example:** $4.5 \div 1.5$

$= 45 \div 15$

$= 3$

## PERCENTAGE

A **percent** is defined as one part in a hundred. The whole amount can be defined as 100 percent (%) and can be divided into 100 equal portions.

## Converting Decimals to Percent and Vice Versa

To convert a decimal to a percent, move the decimal point two places to the right and insert the % sign at the end of your answer.

**Example:** 0.567 = 56.7%

To convert a percent to a decimal, move the decimal point two places to the left and remove the % sign at the end of your answer.

**Example:** 56.7% = 0.567

## Converting Fractions and Mixed Numbers to Percent and Vice Versa

To convert a fraction to a percent, multiply the fraction by 100%. You can express 100% as a fraction by placing it over 1.

**Example:** Convert $\frac{1}{2}$ to a percent.

$$\frac{1}{2} \times \frac{100\%}{1} = \frac{100\%}{2} = 50\%$$

To convert a percent to a fraction, remember that the definition of percent is one part in a hundred. Remove the % sign and multiply the percent value by 1/100 and then reduce the fraction to its lowest terms.

**Example:** Convert 40% to a fraction.

$$40 \times \frac{1}{100} = \frac{40}{100} = \frac{4}{10} = \frac{2}{5}$$

To convert a mixed number percent to a fraction, eliminate the % sign and convert the mixed number to an improper fraction. Then multiply the improper fraction by 1/100 to obtain the fraction.

**Example:** Convert $4\frac{3}{4}\%$ to a fraction.

$$4\frac{3}{4} = \frac{19}{4}$$

$$\frac{19}{4} \times \frac{1}{100} = \frac{19}{400}$$

## Percent Problems—Finding the Part, Whole, or Percent

To find the part when given the whole and the percent, use the formula:

Part = % · Whole ÷ 100

**Example:** What is 54% of 50?

Part = % · Whole ÷ 100

Part = (54)(50) ÷ 100

Part = 2,700 ÷ 100 = 27

To find the % when given the whole and the part, use the formula:

% = Part ÷ Whole · 100

**Example:** What percent of 60 is 12?

% = 12 ÷ 60 · 100

% = 0.2 · 100

% = 20

To find the whole when given the part and %, use the formula:

Whole = Part ÷ % · 100

**Example:** 6 is 15% of what number?

Whole − Part ÷ % · 100

Whole = (6) ÷ (15) · 100

Whole = 0.4 · 100 = 40

## Percent Problems—Increase and Decrease

**Example:** What is 60 increased by 10%?

Part = % · Whole ÷ 100

Part = (10)(60) ÷ 100

Part = 600 ÷ 100 = 6

Add the percentage increase to the original number.

60 + 6 = 66

**Example:** What is 12 decreased by 25%?

Part = % · Whole ÷ 100

Part = (25)(12) ÷ 100

Part = 300 ÷ 100 = 3

Subtract the percentage decrease from the original.

12 − 3 = 9

# RATIO AND PROPORTION

A **ratio** is a comparison of two numbers or objects. The symbol : is used to separate the values in the ratio. To compare ratios, convert them to fractions by placing the number to the left of the symbol used to separate the two numbers as the numerator and the number to the right of the symbol as the denominator. The order of values in the given expression is important when notating the ratio.

**Example:** In a group of 30, there are 24 men and 6 women. The ratio of men to women is 24:6 or $\frac{24}{6}$.

$$\frac{24}{6} = \frac{4}{1}$$

Therefore, the expressed ratio of men to women can be simplified to 4:1 (for every 4 men in the group, there is 1 woman).

Equivalent (equal) ratios have the same proportion when both sides of the ratio are multiplied or divided by the same number.

**Example:** 1:6 and 5:30 and 20:120 are all equivalent ratios, but 1:6 is the simplest form.

A **proportion** is an equation of two ratios that are equal to each other. A basic property of a proportion is that the product of the means is equal to the product of the extremes.

**Example:** $a/b = c/d$ or $ad = bc$ ($b$ and $c$ are the means and $a$ and $d$ the extremes)

$2:x = 4:12 \quad \frac{2}{x} = \frac{4}{12} \quad 24 = 4x \quad 6 = x$

**Note:** Two quantities are in **direct proportion** when they decrease or increase by the same factor.

# PROBABILITY, COMBINATIONS, AND THE COUNTING PRINCIPLE

## Probability

Probability attempts to quantify the notion of *probable*. The probability of an event occurring, **P(E)**, is generally represented as a real number between 0 and 1. The more likely the probability of an event occurring, the closer the probability is to 1. It should be understood, however, that a probability of 0 is not impossible, nor a probability of 1 a certainty.

**Note:** If P(E) is the probability that an event will occur, then P(E) cannot be a negative number.

**Example:** A coin is thrown in the air five times. If the coin lands with the tail up on the first four tosses, what is the probability that the coin will land with the tail up on the fifth toss?

**Answer:** The fifth toss is independent of the first four tosses, and therefore the probability remains one out of two.

## Combinations and the Counting Principle

For any word problem that involves two or more actions or objects, each having a number of choices, and asks for the number of **combinations**, use the **counting principle** formula:

Number of ways $= x \cdot y$

**Example:** A firefighter, during her meal period, wants a sandwich and a seltzer from the corner delicatessen. If the deli has six choices of sandwiches and four choices of seltzer to choose from, how many different ways can she order her lunch?

Number of ways = (6 sandwiches)(4 seltzers)

Number of ways = 24

**Example:** How many different five-letter arrangements can be formed using the letters FIRES, if each letter is used only once?

Five letters can be used to fill the first position in the new word, then four letters to fill the second position, three for the third position, and so on. Therefore, the number of ways is: $5 \times 4 \times 3 \times 2 \times 1 = 120$.

## ARITHMETIC SEQUENCES

An ordered list of terms in which the difference between consecutive terms is constant is called an **arithmetic sequence**. If you subtract any two consecutive terms of the sequence you will obtain the same difference, known as the **constant interval** between the terms.

**Example:** 2, 5, 8, 11, 14, 17, 20, . . .

$20 - 17 = 3$; $17 - 14 = 3$; $5 - 2 = 3$

## EXPONENTS

The **exponent** in an expression such as $3^2$ shows how many times the **base**—in this case, 3—is multiplied by itself to find the result, known as the **power**.

**Examples:** $3^2 = 3 \times 3 = 9$; 9 is the second power of 3

$6^3 = 6 \times 6 \times 6 = 216$; 216 is the third power of 6

Any base raised to the **power of zero** is equal to 1.

**Examples:** $6^0 = 1$; $4^0 = 1$

Any base raised to the **power of 1** is equal to itself.

**Example:** $7^1 = 7$; $-55^1 = -55$

A negative base number raised to an even exponent will be positive, while a negative base number raised to an odd exponent will be negative.

**Examples:** $(-3)^2 = 9$; $(-3)^3 = -27$

Any base raised to a negative power is equal to the reciprocal of the base number (a fraction with 1 placed over the number) raised to the opposite power.

**Example:** $2^{-3} = \left(\frac{1}{2}\right)^3 = \frac{1}{8}$

# Multiplying Numbers with Exponents

When multiplying two like bases, each with a positive integer exponent, add the exponents.

**Example:** $(4^2) \times (4^3) = 4^{(2+3)} = 4^5 = 1{,}024$

To verify this: $(4^2 = 16) \times (4^3 = 64) = 16 \times 64 = 1{,}024$

When multiplying a base with two exponents, multiply the exponents.

**Example:** $(2^3)^4 = 2^{(3 \times 4)} = 2^{12} = 4{,}096$

# Dividing Numbers with Exponents

When dividing two like bases, each with an exponent, subtract the exponents.

**Example:** $\frac{3^3}{3^2} = 3^{(3-2)} = 3^1 = 3$

To verify this: $(3^3 = 27) \div (3^2 = 9) = 27 \div 9 = 3$

# SQUARE ROOTS AND RADICALS

Roots and exponents (powers) are reciprocal. The **principal square root** of a nonnegative real number is denoted by the symbol $\sqrt{x}$ where $x$ represents the nonnegative real number ($x$ is greater than zero) whose **square** (the result of multiplying the number by itself) is $x$. The symbol $\sqrt{\ }$ is called the **radical sign** and the number under the radical sign is called the **radicand**.

**Examples:** $\sqrt{64} = 8$ $\sqrt{9} = 3$ $\sqrt{144} = 12$

The first 10 square numbers and their roots are listed below.

| Square numbers | 1 | 4 | 9 | 16 | 25 | 36 | 49 | 64 | 81 | 100 |
|---|---|---|---|---|---|---|---|---|---|---|
| Square roots | 1 | 2 | 3 | 4 | 5 | 6 | 7 | 8 | 9 | 10 |

# INEQUALITIES

A statement that one expression is greater than or less than another expression is called an **inequality**.

Several symbols are used in statements and word problems involving inequalities. A list of these symbols and their meaning follows:

$<$ $a < b$; $a$ is less than $b$ (point is always toward the lesser expression)
$>$ $b > a$; $b$ is greater than $a$
$\leq$ less than or equal to
$\geq$ greater than or equal to
$\neq$ not equal to

# ALGEBRA—AN INTRODUCTION TO VARIABLES

The origins of algebra can be traced to the ancient Egyptians and Babylonians who used an early type of algebra more than 3,000 years ago. This section reviews the fundamentals of algebra and its application in solving word problems asked in previous firefighter exams.

When a number's value is unknown, a symbol (such as $a$, $b$, $x$, or $y$), called a **variable**, is used to represent it.

**Example:** $5x + x = 5x + 1x = 6x$

**Note:** If no number appears in front of the variable, the number (coefficient) is 1.

Variables can be canceled in fractions the same way whole numbers are, but only factors, not terms, can be canceled.

**Examples:**

$$\frac{5x}{x} = \frac{5}{1} \cdot \frac{\cancel{x}}{\cancel{x}} = 5$$

$$\frac{(3+6x)(x-4)}{3+6x} = \frac{\cancel{3+6x}}{\cancel{3+6x}} \cdot \frac{x-4}{1} = x-4$$

In the example above, $3 + 6x$ is a factor of $(3 + 6x)(x - 4)$ and therefore can be canceled.

## Negative Variables

When working with an equation dealing with a **negative variable** $(-x)$, you must change the equation in terms of $(x)$ using multiplication or division.

**Example:** $-x = 5$ Multiply both sides of the equation by $-1$

$(-1)(-x) = 5(-1)$

$x = -5$

**Note:** If the term containing the variable in the algebraic equation has a negative **coefficient** (factors being multiplied), multiply both sides of the equation by the **reciprocal** (inversion) of the coefficient, which will also be negative.

**Example:** $-2x = 8$

Multiply both sides by $-\frac{1}{2}$

$-\frac{1}{2} \cdot -2x = 8 \cdot -\frac{1}{2}$

$x = -4$

## Addition and Subtraction of Fractions with Variables

When adding and subtracting fractions with variables, treat the variables as if they were prime numbers (to find the LCD).

**Example:** $\frac{3}{50}+\frac{x}{2}\cdot$ The least common denominator is 50.

$$\frac{3}{50}+\frac{x}{2}=\frac{3}{50}+\frac{25x}{2\times 25}=\frac{3}{50}+\frac{25x}{50}=\frac{3+25x}{50}$$

## Multiplication of Fractions with Variables

Multiply fractions with variables the same way you would multiply fractions without variables. Multiply the numerators first and then multiply the denominators.

**Examples:** $\frac{5}{7}\cdot\frac{2x}{3}=\frac{10x}{21}$

$$\frac{2}{3y}\cdot\frac{6x}{7}=\frac{12x}{21y}=\frac{4x}{7y}$$

## Division of Fractions with Variables

Dividing fractions with variables is similar to dividing fractions without variables. Invert the second fraction and multiply the fractions.

**Example:** $\frac{4}{5}\div\frac{x}{9}=\frac{4}{5}\cdot\frac{9}{x}=\frac{36}{5x}$

# MONOMIALS AND POLYNOMIALS

A **monomial** is a single term that has no plus or minus sign between entries. Examples of monomials are $3x$; $-22y^2$; $-3xy^2$; $380xy^2$. But $4x + 2$ and $6y^2 - 3y$ are not monomials because they have a plus or minus sign between entries.

A **polynomial** is two or more monomials attached with plus signs and/or minus signs.

**Examples:** $4x + 2$; $4y - 3$; $2x^2 + 3x$; $7x^2 - 9x + 1$

# FACTORING

Expressing an algebraic expression as a product of certain factors is called **factoring**. It generally involves finding one factor of a product and using it as a divisor to find other factors. Factoring is important in algebra and useful in simplifying expressions and solving equations.

# FOIL Method of Expanding Expressions Under the Distributive Law

**FOIL** is a method of multiplying terms and subsequently adding or subtracting like terms to get a polynomial result.

**Example:** $(x + 5)\ (3x - 7)$

| | | | |
|---|---|---|---|
| | | **F** **F** | |
| **F** | First · First | $(x + 5)\ (3x - 7)$ | $x(3x) = \mathbf{3x^2}$ |
| | | **O** **O** | |
| **O** | Outer · Outer | $(x + 5)\ (3x - 7)$ | $x(-7) = \mathbf{-7x}$ |
| | | **I** **I** | |
| **I** | Inner · Inner | $(x + 5)\ (3x - 7)$ | $(5)(3x) = \mathbf{15x}$ |
| | | **L** **L** | |
| **L** | Last · Last | $(x + 5)\ (3x - 7)$ | $(5)(-7) = \mathbf{-35}$ |

$(x + 5)\ (3x - 7) = \mathbf{3x^2 - 7x + 15x - 35} = 3x^2 + 8x - 35$

**Example:** $(4x - 8)\ (2x + 6)$

F O I L

$8x^2 + 24x - 16x - 48 = 8x^2 + 8x - 48$

**Note:** The distributive laws allow us to execute the often easier task of completing two smaller multiplications and calculating the sum, rather than performing one multiplication involving large numbers.

The following sections show applications of arithmetic and algebraic equations.

# COMPUTATION PROBLEMS (ADDITION, SUBTRACTION, MULTIPLICATION, DIVISION)

**Example:** George rents a car for a day from a company charging \$60 a day, plus a lateness fee of \$10 an hour for cars not brought back on time. If George returns the car the next day two hours late, how much would he be obligated to pay the rental company?

**Answer:** \$60 for the day, plus \$20 for two-hour late fee = \$80.

**Example:** A beachcomber collects \$43.20 off the sand on Saturday and \$27.67 on Sunday. How much more money did the searcher collect on Saturday compared to Sunday?

**Answer:** \$43.20 – \$27.67 = \$15.53.

**Example:** An EMT checks the pulse of a rescued fire victim. The victim's pulse count is 19 for the 15 seconds that the pulse was examined. What is the number of beats per minute?

**Answer:** Since 15 seconds is ¼ of a minute, multiply the pulse rate in 15 seconds by 4 to get the pulse rate per minute: $19 \cdot 4 = 76$.

**Example:** Four businesses share the cost of purchasing a building valued at \$5,000,000. If all four businesses pay equal amounts, how much would each business have to pay?

**Answer:** \$5,000,000 ÷ 4 = \$1,250,000.

# TEMPERATURE SCALE CONVERSION PROBLEMS

**Example:** Given the temperature on the Fahrenheit scale is 41 degrees, what is the corresponding temperature in degrees on the Celsius scale?

Formula: $C = \frac{5}{9}(F - 32)$

$C = \frac{5}{9}(41 - 32)$. Perform the subtraction in the parentheses first.

$C = \frac{5}{9}(9)$

$C = 5$ degrees

**Example:** Given the temperature on the Celsius scale is 20 degrees, what is the corresponding temperature in degrees on the Fahrenheit scale?

Formula: $F = \frac{9}{5}(C) + 32$

$F = \frac{9}{5}(20) + 32$

$F = \frac{180}{5} + 32$

$F = 36 + 32$

$F = 68$ degrees

## MIXTURE PROBLEM

**Example:** How much 20% acid solution should be added to 50 liters of 45% acid solution to obtain a 30% acid solution?

Let $x$ = amount of 20% (0.20) acid solution

0.45(50) = amount of acid in 50 liters of 45% acid solution

$(50 + x)$ = total amount of acid solution

$0.30(50 + x)$ = amount of acid in 30% (0.30) acid solution

Therefore $0.20x + 0.45(50) = 0.30(50 + x)$

Multiply terms in parentheses first: $0.20x + 22.50 = 15 + 0.30x$

Add/subtract like terms and bring the variable over to one side of the equation.

$7.50 = 0.10x$

75 liters = $x$ = amount of 20% acid solution to be added.

## CONSECUTIVE INTEGER PROBLEM

**Example:** The sum of two consecutive integers is 47. What are the numbers?

Let $x$ = the first number and $(x + 1)$ = the second number.

Therefore

$x + (x + 1) = 47$

$2x + 1 = 47$

$2x = 46$

$x = 23$ = the first number

$(x + 1) = 24$ = the second number

## SUM OF TWO NUMBERS PROBLEM

**Example:** The sum of two numbers is 60. One number is four more than the other number. Find the two numbers.

Let $x$ = the smaller number and $(x + 4)$ = the larger number

Therefore

$x + (x + 4) = 60$

$2x + 4 = 60$

$2x = 56$

$x = 28$ = the smaller number

$(x + 4) = 32$ = the larger number

## AVERAGE PROBLEMS

To **average** a number of values, take the sum of all the values and divide the sum by the total number of values.

**Example:** Find the average of the following numbers: 10, 20, 50, 25, and 75.

Use the formula: $\text{avg} = \dfrac{\text{total sum of values } (10+20+50+25+75)}{\text{number of values}}$

$$\text{avg} = \frac{180}{5} = 36$$

**Example:** A student has grades of 68, 70, 75, and 80. What will the next grade have to be for the student to obtain a 75 average?

Let the next grade $= x$

Therefore there will be a total of 5 grades and

$\dfrac{68+70+75+80+x}{5} = 75$

$\dfrac{293+x}{5} = 75$

Multiply both sides by 5: $293+x = 375$

Subtract 293 from each side: $x = 82$

## AGE PROBLEM

**Example:** Veteran firefighter John is twice as old as firefighter Joyce, and Joyce is five years older than probationary firefighter Jane. The sum of their ages is 95. What are the ages of all three firefighters?

Let $x$ = Joyce's age

$2x$ = John's age

$x - 5$ = Jane's age

Therefore

$x + (2x) + (x - 5) = 95$

$x + 2x + x - 5 = 95$

$4x - 5 = 95$

$4x = 100$

$x = 25$ = Joyce's age

$2x = 50$ = John's age

$x - 5 = 20$ = Jane's age

# WORK PROBLEMS

**Note:** If it takes $x$ hours to perform a task, $\frac{1}{x}$ of the job is done each hour.

**Example:** Probationary firefighter John can clean all the hose on the apparatus by himself in six hours. Veteran firefighter Raymond can complete the task in only four hours. If they work together, how long will it take?

$x$ = time it takes to perform the work together.

John does $\frac{1}{6}$ the work in one hour.

Raymond does $\frac{1}{4}$ the work in one hour.

**Note:** If several persons work together, the amount of work done is assumed to be the sum of the individual amounts.

Therefore

$$\frac{1}{x} = \frac{1}{6} + \frac{1}{4}$$

Find a common denominator ($24x$) by multiplying the denominators.

$$24 = 6x + 4x$$

$$24 = 10x$$

$$\frac{24}{10} = x$$

$$2\frac{2}{5} \text{ hours} = x$$

It will take John and Raymond $2\frac{2}{5}$ hours, or 2 hours and 24 minutes, working together.

**Example:** Randy, a new recruit, can wash the fire apparatus alone in 60 minutes. If Randy works together with firefighter Clay, they can wash the apparatus in 20 minutes. How long would Clay need to wash the fire apparatus by himself?

**Note:** Use the formula $W$ (Work done) = $r$ (rate) · $t$ (time). Basic-level work problems usually have $W$ equal to 1.

Therefore

$$W = r \cdot t \text{ and } \frac{W}{r} = t; \frac{W}{t} = r$$

Let $W = 1$ (job of washing the fire apparatus)

$\frac{1}{60}$ = rate at which Randy performs the work

$\frac{1}{20}$ = rate at which both Randy and Clay perform the work

$\frac{1}{t}$ = rate at which Clay performs the work

Therefore

$$\frac{1}{60}+\frac{1}{t}=\frac{1}{20}$$ find LCD = 60$t$

$$(60t)\frac{1}{60}+(60t)\frac{1}{t}=(60t)\frac{1}{20}$$ multiply each term by LCD (60$t$)

$$t+60=3t$$ bring the variable over to one side of the equation

$$60=2t$$

$$30=t$$ number of minutes Clay needs to wash the fire apparatus by himself

## DISTANCE, RATE, AND TIME PROBLEMS

When we look at the relationship between distance ($d$), rate ($r$), and time ($t$), we can use the formula: $d = rt$ to find the answer not only for distance but for rate of speed and time as well.

$$d = r \cdot t \quad \text{and} \quad \frac{d}{t} = r; \ \frac{d}{r} = t$$

**Note:** If in a word problem, parts of the trip are at different rates of speed, then to find the total time or total distance you must separate the parts of the trip.

**Example:** A plane can fly at a rate of 400 miles per hour (mph) from point A to point B, a distance of 1,600 miles. A high-speed train, however, makes the same trip back from point B to point A at a rate of speed of 160 mph. How many hours was the whole trip to and from points A and B?

$$(t)=\frac{(d)}{(r)}$$

($t$) going = 1,600 miles ÷ 400 mph = 4 hours

($t$) returning = 1,600 miles ÷ 160 mph = 10 hours

Total ($t$) of whole trip = 14 hours

**Example:** A man can walk 4 mph in a still wind. When a favorable wind is at his back he can walk 15 miles in 3 hours. How fast is the wind?

**Note:** Use the formula: $d = r \cdot t$

($r$) = wind speed

($r$ + 4) = speed walking with wind at his back

15 miles = ($r$ + 4) (3 hours) = 3$r$ + 12

$15 = 3r + 12$

$3 = 3r$

1 mph = $r$

**Example:** An automobile travels through an intersection going north at 30 mph. Ten minutes later, a second automobile passes through the same intersection traveling in the same direction going 60 mph. How long will it take for the second automobile to overtake the first automobile?

**Note:** Use the formula: $d = r \cdot t$

Let the first automobile = $(30 \text{ mph})\dfrac{(10 \text{ minutes})}{(60 \text{ minutes})}$ 10 minute head start at 30 mph

$(30)\frac{1}{6} = 5$ miles traveled prior to second automobile reaching same intersection

The second automobile is gaining at a rate of 60 mph – 30 mph = 30 mph

How long ($t$) will the second automobile gaining at a rate ($r$) of 30 mph take to cover the ($d$) 5 mile head start of the first automobile?

$d = r \cdot t$

$5 = 30(t)$ Divide both sides of the equation by 30

$\frac{1}{6} = t$

$\frac{1}{6}$ of an hour = 10 minutes = $t$ = time required for the second automobile to overtake the first automobile

**Example:** Two trains pass in the night. One train is traveling west at 60 mph while the second train is going east at 80 mph. When will the trains be 420 miles apart?

**Note:** Use the formula: $d = r \cdot t$

**Note:** When objects are moving in opposite directions, regardless of whether the objects are moving toward each other or away from each other, the rate at which the distance between them is changing is the sum of their individual rates.

Therefore 60 mph + 80 mph = 140 mph

Let

$t$ = number of hours the trains will travel after passing each other

420 miles = 140 mph $\cdot\ t$

$\dfrac{420 \text{ miles}}{140 \text{ mph}} = t$ Divide by 140 on both sides of the equation to isolate the variable

3 hours = $t$ = number of hours trains will travel after passing each other to be 420 miles apart

## COIN PROBLEM

**Example:** Lance has $10 in quarters and dimes. He has a total of 55 coins. How many of each coin does he have?

Let $x$ = number of dimes; $0.10x$ = dollar amount in dimes

$55 - x$ = number of quarters; $0.25(55 - x)$ = dollar amount in quarters

Therefore

$0.10x + 0.25(55 - x) = 10$
$0.10x + 13.75 - 0.25x = 10$

Add/subtract like terms: $-0.15x = -3.75$

Divide both sides by 0.15: $-x = -25$

Divide both sides by $-1$

$x = 25$ = number of dimes

$55 - x = 30$ = number of quarters

Check:

25 dimes × 0.10 = $2.50

30 quarters × 0.25 = $7.50

Total: 55 coins = $10.00

## HOURLY WAGE PROBLEM

**Example:** Senior paramedic Kathy earns two times more per hour than Eric, a junior paramedic, while Eric earns $4 more per hour than EMT Alex. Together they earn $48 per hour. How much is each medic's hourly wage?

Let $x$ = Alex; $(x + 4)$ = Eric; $2(x + 4)$ = Kathy

Therefore

$x + (x + 4) + 2(x + 4) = 48$

$x + x + 4 + 2x + 8 = 48$

$4x + 12 = 48$
$4x = 36$

$x = 9$ = Alex's hourly wage

$(x + 4) = 13$ = Eric's hourly wage

$2(x + 4) = 26$ = Kathy's hourly wage

# PERCENT PROBLEMS

**Example:** A $200 rebar cutter is reduced 20% for a catalog sale. What will be the sale price?

Let $x$ = sale price

$200 = original price

(200)(0.20) = original price reduced 20%

Therefore

$(200) - (200)(0.20) = x$

$200 - 40 = x$

$\$160 = x =$ sale price

**Example:** Dave's monthly budget is $2,400. His monthly rent is $600. What percent of his monthly budget is spent on rent?

Use formula: % = Part ÷ Whole · 100

Therefore

$\% = 600 \div 2{,}400 \cdot 100$

$\% = 0.25 \cdot 100 = 25\%$

# INVESTMENT PROBLEMS

Use this simple formula to compute interest on dollars at an interest rate percent per year for a given number of years.

$$i = prt$$

where $i$ = interest, $p$ = principal, $r$ = rate, and $t$ = time

**Example:** What is the interest on $200 invested at 6% for two months?

$i = prt$

$i = 200\left(\frac{6}{100}\right)\left(\frac{2}{12}\right)$

$i = 200\left(\frac{3}{50}\right)\left(\frac{1}{6}\right)$ reduced to lowest form

$i = \$\left(\frac{600}{300}\right)$ multiply both numerators and denominators

$i = \$2.00$

**Example:** Quinn deposits \$3,500 into two savings deposit accounts. One account yields 4.5% annual earnings, while the second account yields 5.0%. Total interest earnings for the first year for both accounts is \$180.00. How much money did Quinn deposit in each account?

Let $x$ = account yielding 4.5% (0.045)

$(3{,}500 - x)$ = account yielding 5.0% (0.050)

Therefore

$0.045x + 0.050(3{,}500 - x) = 180$

Multiply terms in parentheses first: $0.045x + 175 - 0.050x = 180$

Add/subtract like terms: $0.005x = 5$

Divide both sides by 0.005:

$x$ = \$1,000 deposited into 4.5%-yielding account

$(3{,}500 - x)$ = \$2,500 deposited into 5%-yielding account

## PROBABILITY PROBLEMS

**Example:** If a die is rolled once, determine the probability of rolling a 3.

Use the formula: $\text{Probability} = \dfrac{\text{event (set of outcomes)}}{\text{possible outcomes}}$

Therefore

$$\text{Probability} = \frac{\text{rolling a 3 (set of outcomes is 1)}}{\text{possible outcomes (set of outcomes is 6)}}$$

$$\text{Probability} = \frac{1}{6}$$

**Example:** A deck of cards contains 52 cards, 13 of which are from each of 4 suits. If a card is randomly flipped over after a thorough shuffling, what is the probability that the suit will be a diamond?

Use the formula: $\text{Probability} = \dfrac{\text{event (set of outcomes)}}{\text{possible outcomes}}$

Therefore

$$\text{Probability} = \frac{\text{flipping over a diamond (set of outcomes is 13)}}{\text{possible outcomes (set of outcomes is 52)}}$$

$$\text{Probability} = \frac{13}{52} = \frac{1}{4}$$

## COUNTING PRINCIPLE PROBLEM

**Example:** An ice cream cone comes in eight flavors with five possible combinations of toppings. How many different ice cream cones can be made with one flavor of ice cream and one topping?

**Note:** Rather than list the entire sample of all possible combinations, use the simple multiplication of factors.

Therefore $8 \times 5 = 40$ different ice cream cones can be made.

## RATIO PROBLEMS

**Example:** In a classroom of 48 students, the ratio of passing grades to failing grades is 8 to 4. How many of the students had failing grades?

Let 8 passing grades + 4 failing grades = 12 (representative set).

Therefore $\frac{4}{12}$ of the class failed.

$\frac{4}{12}(48) = 16$ students had failing grades.

**Example:** An automobile travels 180 miles using 9 gallons of gasoline. Express the terms as a ratio in its simplest form to determine the miles traveled per gallon.

180:9 Convert to a fraction: $\frac{180}{9} = 20$

Simplify the ratio: 20 miles per gallon

## PROPORTION PROBLEMS

**Example:** Find the unknown value in the proportion: $7{:}x = 3{:}6$

Convert to fraction form: $\frac{7}{x} = \frac{3}{6}$

Multiply the means and the extremes: $3x = 42$

Isolate the variable by dividing by 3 on both sides of the equation:

$x = \frac{42}{3}$

$x = 14$

**Example:** A wooden floor joist weighs 120 pounds and is 10 feet long. What is the weight of a similar floor joist that is 8 feet long?

Set up the proportion: $\frac{\text{weight}}{\text{length}} : \frac{120}{10} : \frac{x}{8}$

Multiply the means and the extremes:

$10x = 120(8)$

$10x = 960$

Isolate the variable by dividing by 10 on both sides of the equation:

$x = 96$ pounds

## GEOMETRY—BASIC TERMINOLOGY

**Geometry**, one of the oldest branches of mathematics, is the study of shapes. The properties of points, lines, surfaces, and solids were first recognized by the early Babylonians and Egyptians over 4,000 years ago, and the rules of geometry helped in the construction of the pyramids.

This section is a very much abbreviated review of geometry. It focuses on the basics of geometry that will prove useful in answering mathematical questions dealing with the measurement of **length**, **angles**, **perimeter**, **surface area**, **area**, and **volume**. These measurements and calculations are useful to firefighters when performing their duties.

**Point**—In mathematics, a point represents position. It has no size or dimension. A point is represented by a dot followed by a capital letter.

**Example:** · A

**Line**—A line contains at least two points. It is a one-dimensional, straight line figure that extends indefinitely in two opposite directions. A line has no width. It is represented by arrowheads at both ends to denote that it has no endpoints.

**Example:** <——•(A)————•(B)——>

**Line segment**—A line segment is a one-dimensional set of points containing point A and point B and all the points lying between A and B. A line segment has a finite length and does not extend forever.

**Example:** •(A)————•(B)

**Ray**—A ray is drawn as a one-dimensional, straight line that has one endpoint. A ray is represented by an arrow starting at an endpoint and a designated point that the line passes through.

**Example:** •(A)————•(B)————>

# ANGLES

The intersection of two rays with a common endpoint forms an **angle**. The symbol for angle is $\angle$.

## Types of Angles

**Acute angle**—An angle that measures less than 90°.

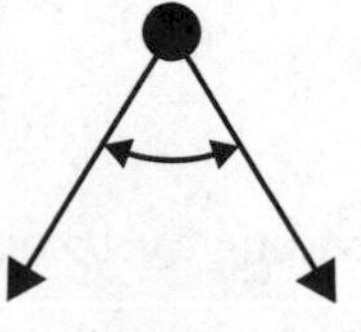

**Right angle**—An angle that measures exactly 90°. All right angles are equal.

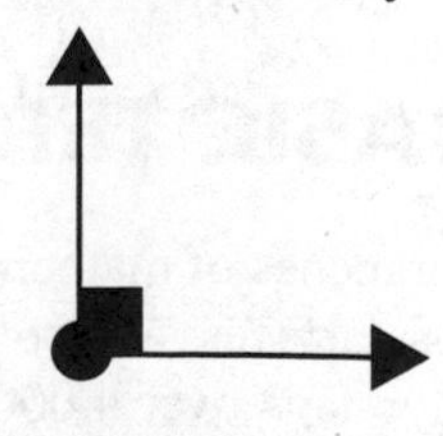

**Obtuse angle**—An angle that measures more than 90° but less than 180°.

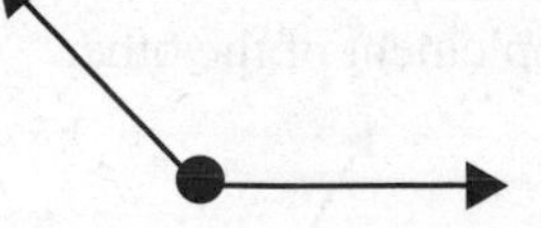

**Straight angle**—An angle whose sides are opposite rays and that measures exactly 180°. All straight angles are equal.

# Pairs of Angles

Pairs of angles may have several different relationships.

**Adjacent angles**—Two angles that share a common side and endpoint (**vertex**), but the rays are not common to both angles.

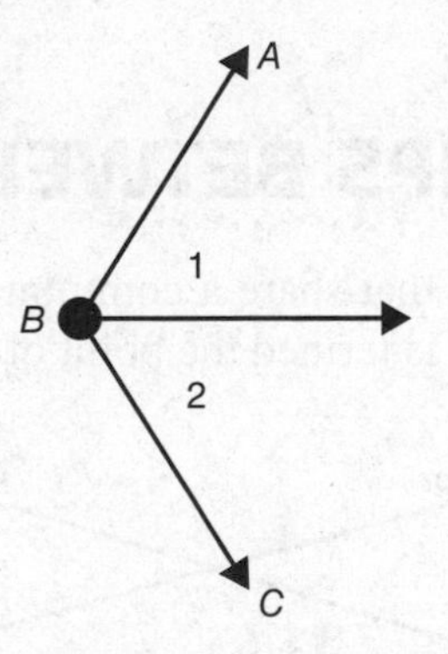


**Vertical angles**—Formed by intersecting lines whose sides form two pairs of opposite rays. Vertical angles are located across from each other in the corners of the "X" formed by the intersection of the lines. They are not adjacent. Vertical angles are always equal in measure.

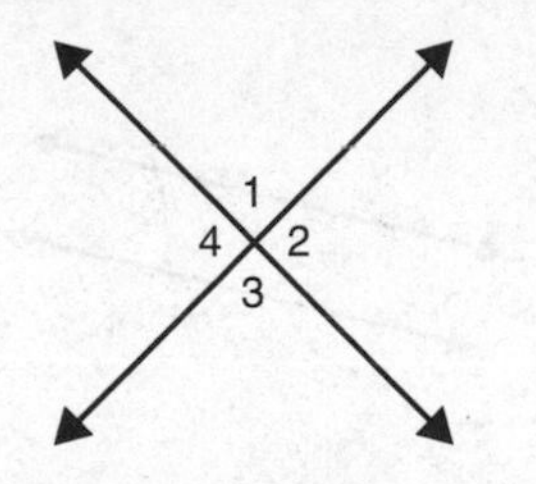


**Complementary angles**—Two angles the sum of whose measures is 90°. Complementary angles are placed so that they form perpendicular lines. Each angle is called the complement of the other.

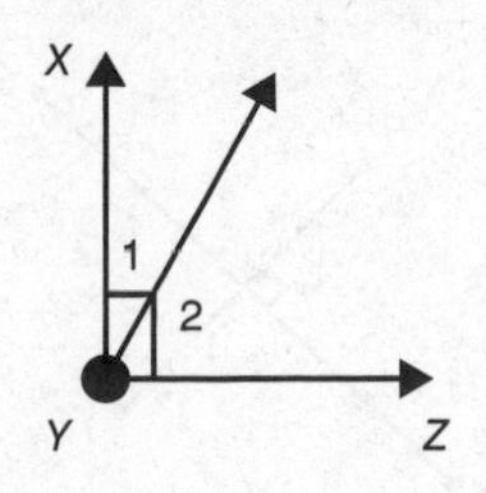

**Supplementary angles**—Two angles the sum of whose measures is 180°. Supplementary angles are placed so that they form a straight line. Each angle is called the supplement of the other.

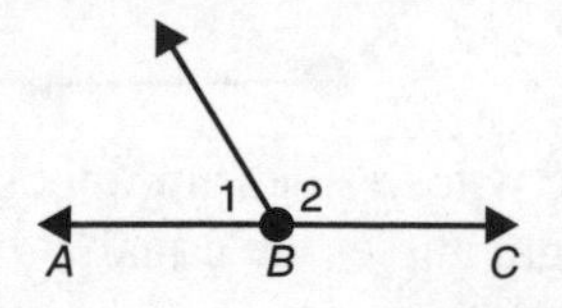


**Equal angles**—Angles that have the same measure.

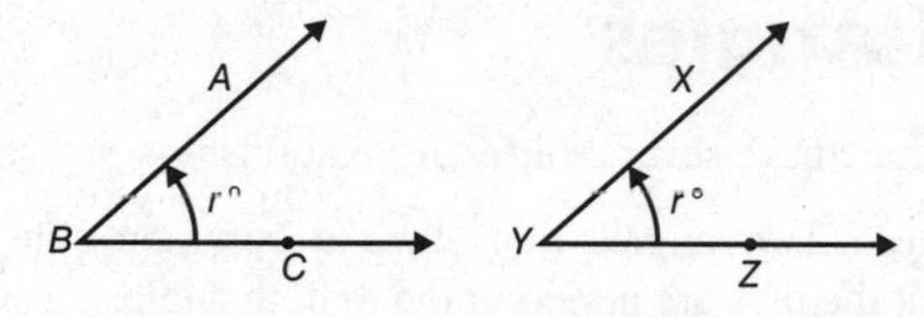


# RELATIONSHIPS BETWEEN LINES

**Intersecting lines**—Lines that share a common point are said to **intersect**, and the common point is termed the point of intersection.

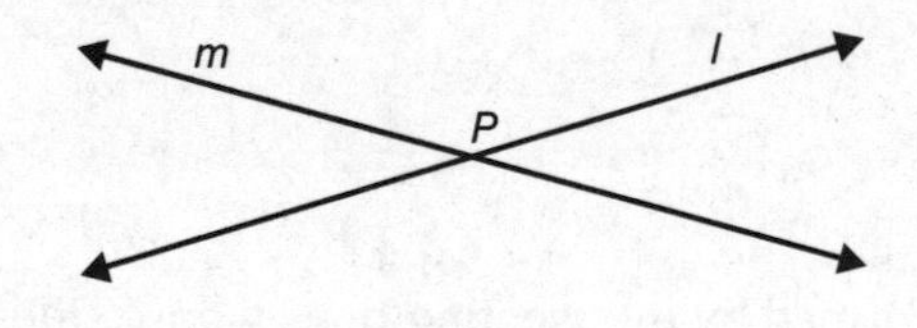


**Parallel lines**—Two lines are **parallel** when they lie in the same plane and do not share a common point. The symbol || denotes two lines as being parallel.

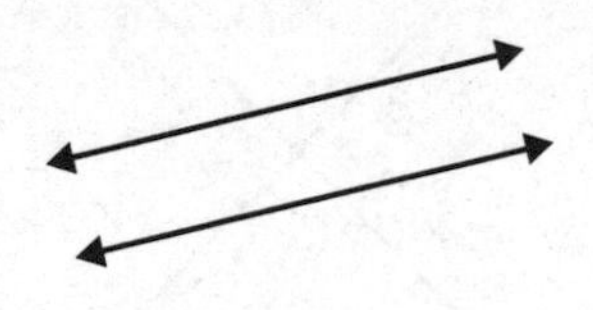

**Perpendicular lines**—When two lines intersect to form equal adjacent angles, the lines are said to be **perpendicular**. The angles formed are right angles (90°). The symbol ⊥ denotes two lines as being perpendicular.

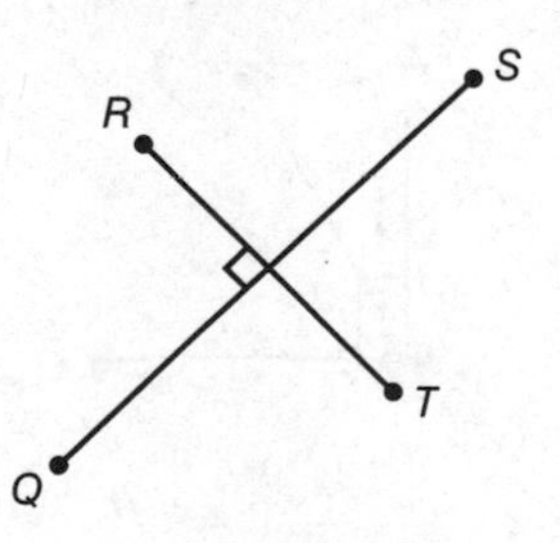

**Transversal lines**—A **transversal** is a line that intersects two or more other lines at different points. The transversal is generally denoted by the letter *t*.

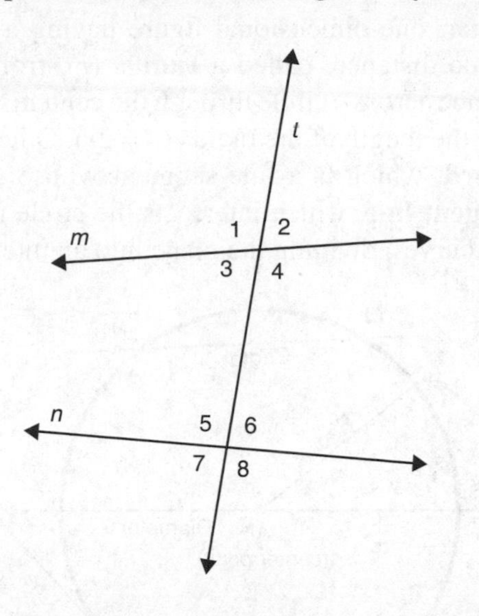


# GEOMETRIC SHAPES AND THEIR MEASUREMENT

Geometric shapes can be classified as one dimensional, two dimensional, or three dimensional. An example of a one-dimensional object is a circle. Common two-dimensional objects—a plane, a polygon, a parallelogram, a rhombus, a trapezoid, and a square—have length or vertical length (height) and width, but they lack depth. Common three-dimensional objects—cube, cylinder, sphere, and cone—have depth in addition to length or vertical length (height) and width. Each object has its own formulas for calculating perimeter, area, and volume, if applicable.

**Perimeter (P)** is the distance around a two-dimensional object. It is one dimensional and measured in linear units: millimeters (mm), centimeters (cm), meters (m), kilometers (km), inches (in), feet (ft), yards (yd), and so on. The perimeter of a circle is known as its circumference.

**Area (A)** is a physical quantity denoting the size of a part of a two-dimensional surface. It has units of distance squared.

**Surface area (SA)** is a summation of the areas of the exposed sides of a three-dimensional object. It has units of distance squared.

**Volume (V)** is how much space a three-dimensional object displaces. It is commonly measured in units of distance cubed such as cubic millimeters, centimeters, or meters or cubic inches, feet, or yards. When used to find how much a three-dimensional object can hold, it is called **capacity**. Common fluid measurements include milliliters, liters, ounces, and gallons.

The remainder of this section gives the formulas for calculating the perimeter, area, and volume of some common types of objects.

## A One-Dimensional Figure—Circle

A **circle** is a planar, one-dimensional figure having a set of points all of which are the same distance, called a **radius** ($r$), from a fixed point (the **center**). The distance across a circle through the center is called the **diameter** ($d$), and it is twice the length of the radius ($d = 2r$). Other important terms to know are the **chord**, which is a line segment whose endpoints lie on the circle and the **tangent line**, which intersects the circle in exactly one point. Circles are closed curves, dividing the plane into an interior and an exterior.

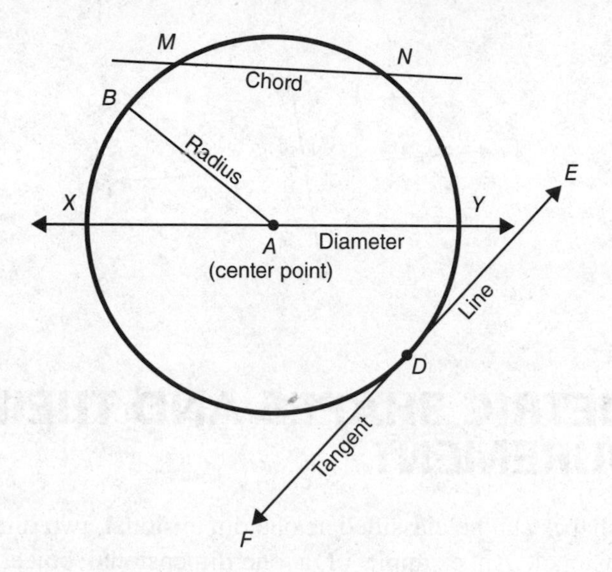


The **circumference of a circle** is the distance around or the length of a circle. It is basically the perimeter of the circle.

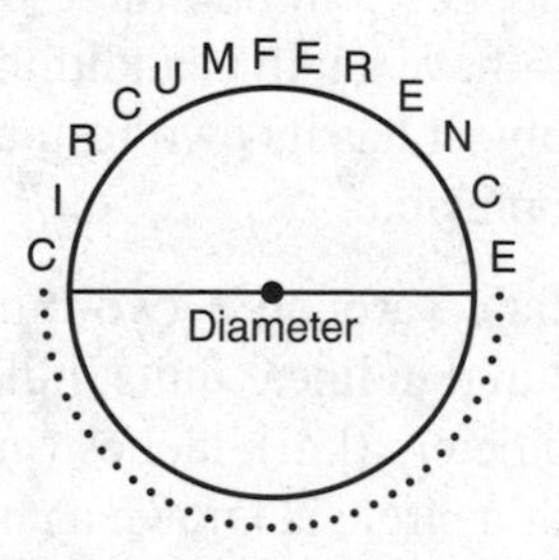


To determine the circumference (perimeter) of a circle, the mathematical constant known as pi ($\pi$) is used. It represents the ratio of the circumference to the diameter. When you divide the circumference by the diameter you get $\pi$ or the approximate value of 3.14. Therefore, $C/d = \pi$. The circumference ($C$) of a circle is calculated by using the following formula:

$$C = \pi \cdot d \text{ or } C = 2 \cdot \pi \cdot r$$

The **area of a circle** can be calculated by multiplying $\pi$ ($\pi = 3.14$) by the square of the radius. The formula is as follows:

$$A = \pi r^2$$

# Two-Dimensional Figures—Polygons

A **polygon** is a closed, two-dimensional planar shape composed of a fixed number of straight line segments. Polygons are named according to the number of sides they have. Common examples include the **triangle** (3 sides), **quadrilateral** (4 sides), **pentagon** (5), **hexagon** (6), and **octagon** (8).

Quadrilaterals include a **parallelogram**, **rhombus**, **trapezoid**, and **square**.

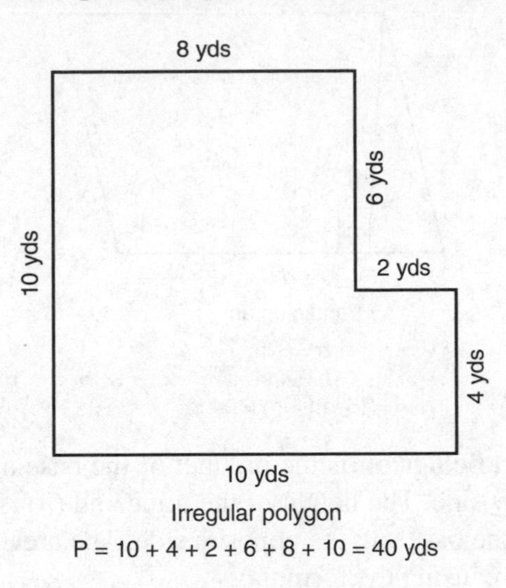


Irregular polygon
P = 10 + 4 + 2 + 6 + 8 + 10 = 40 yds

The **perimeter of irregularly shaped polygons** is calculated by just adding all the lengths of the sides.

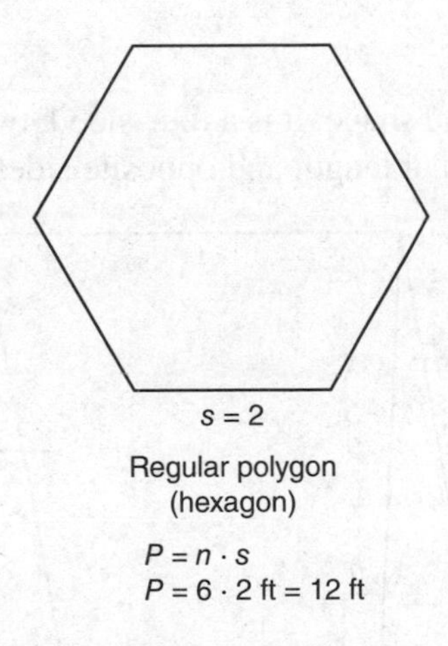


Regular polygon
(hexagon)

$P = n \cdot s$
$P = 6 \cdot 2$ ft = 12 ft

The **perimeter of regular polygons** (all angles and all sides coincide, or are **congruent**) is calculated by multiplying the number ($n$) of sides in the polygon by the length of a given side ($s$). The formula is: $P = n \cdot s$.

The **area of different polygons** is calculated using different formulas that are given below.

## PARALLELOGRAM

A parallelogram is a four-sided, two-dimensional figure that has two sets of opposite parallel sides. The **diagonals** (straight line segments that cut across polygons joining two vertices) of a parallelogram **bisect** (cut in half) each other.

The **perimeter of a parallelogram** is calculated by adding the lengths of the four sides.

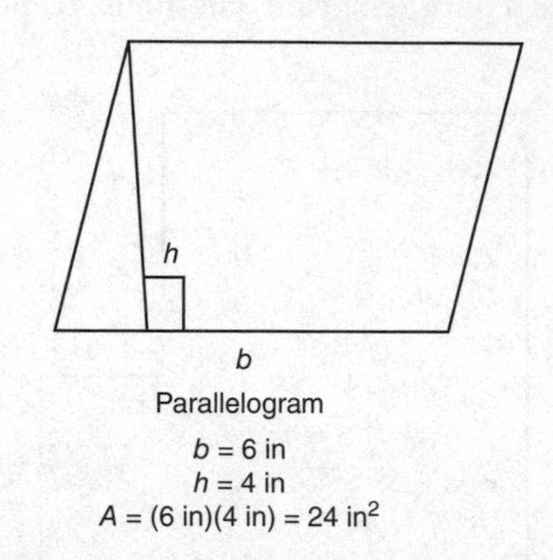


Parallelogram

$b = 6$ in
$h = 4$ in
$A = (6 \text{ in})(4 \text{ in}) = 24 \text{ in}^2$

The **area of a parallelogram** is the product of the base and the height. The base ($b$) can be any side. The height of the trapezoid ($h$) is the perpendicular distance between the base and the opposite side. The area of a parallelogram can be calculated by using the formula:

$$A = b \cdot h$$

## RHOMBUS

A rhombus has a diamond shape. It is a four-sided, two-dimensional figure in which all sides are of equal length and opposite sides are parallel.

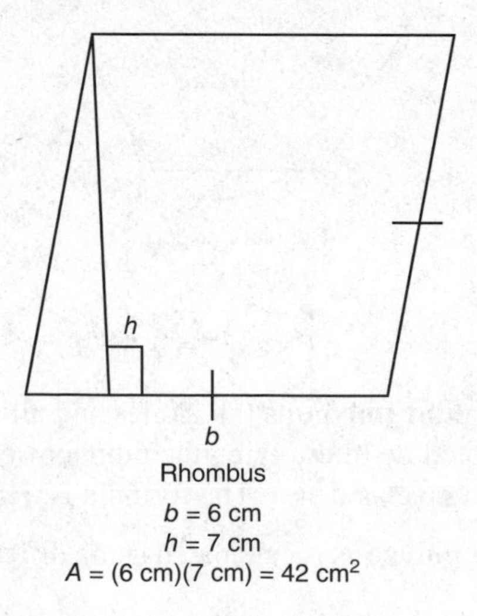


Rhombus

$b = 6$ cm
$h = 7$ cm
$A = (6 \text{ cm})(7 \text{ cm}) = 42 \text{ cm}^2$

The **perimeter of a rhombus** is calculated using the same formula as used to get the perimeter of a regular polygon ($P = n \cdot s$).

The **area ($A$) of a rhombus** can be calculated by multiplying the base (which can be any side) times the height ($A = b \cdot h$) or by finding half of the product of the lengths of its diagonals, $A = 1/2(d_1 \cdot d_2)$.

## TRAPEZOID

A trapezoid is a four-sided, two-dimensional shape with two parallel sides. The parallel sides of a trapezoid are called the bases ($b_1$ and $b_2$). The height of the trapezoid is the perpendicular distance between the bases ($h$).

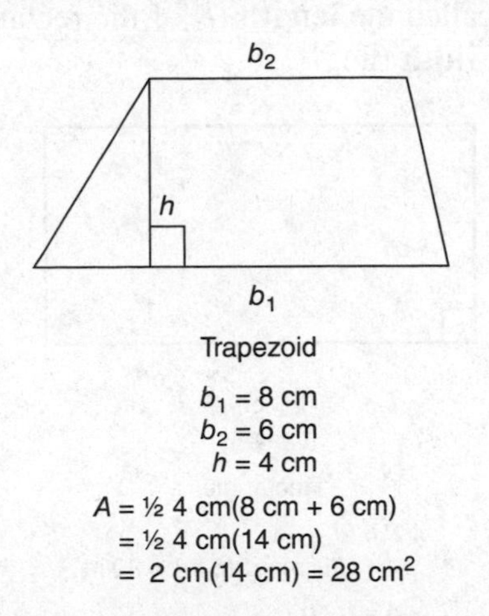


The **perimeter of a trapezoid** is calculated by adding the lengths of the four sides.

The **area (A) of a trapezoid** is calculated by adding the lengths of the two bases, multiplying that by the height, and then dividing by 2. The formula is: $A = \frac{1}{2}h(b_1 + b_2)$.

## SQUARE

The square is a four-sided, two-dimensional shape (polygon) with four equal length sides and four equal (right) interior angles. The diagonals of a square are also equal to each other and perpendicular to each other, and they are $\sqrt{2}$ (approximately 1.41) times the length of a side of a square.

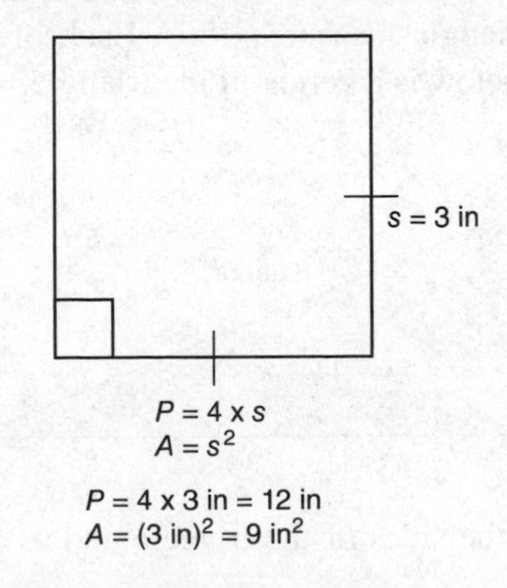


The **perimeter of a square** whose sides have length ($s$) is calculated by multiplying any given side by 4. The formula is: $P = 4 \cdot s$.

The **area of a square** is calculated by raising any given side ($s$) to the second power. The formula is: $A = s^2$.

## RECTANGLE

A rectangle is a four-sided, two-dimensional shape (polygon) where all four of its interior angles are 90° (right) angles. The lengths of its diagonals are equal. It has two pairs of opposite sides that are of equal length. The length of the longer side is called the **length** ($l$) of the rectangle, and the length of the shorter side is its **width** ($w$).

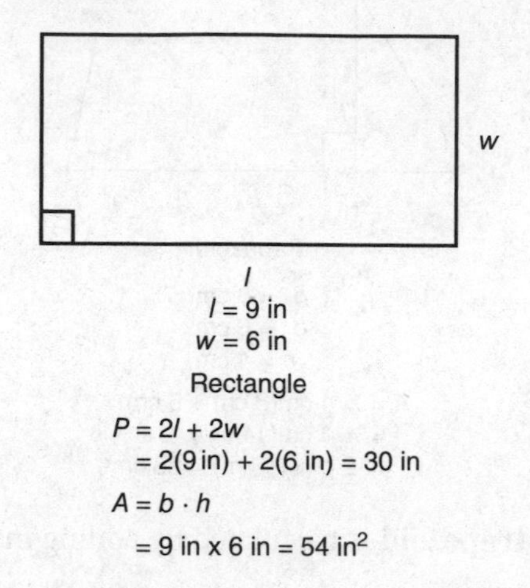


The **perimeter of a rectangle** is calculated by multiplying both the length and width by 2 and then adding the products. The formula is: $P = 2l + 2w$ or $P = 2\ (l + w)$.

The **area of a rectangle** is the product of the base (length) and the height (width). The base ($b$) can be any side. Either side perpendicular to the base is called the height ($h$). The terms *length* and *width* can also be used in the same manner to find area. The formula is: $A = b \cdot h$ or $A = l \cdot w$.

## TRIANGLE

A triangle is a multisided, two-dimensional shape (polygon) consisting of three vertices and three sides that are straight line segments. The sum of the internal angles of a triangle measures 180°. Each of the points (A, B, C) shown in the triangle below is a vertex of the triangle.

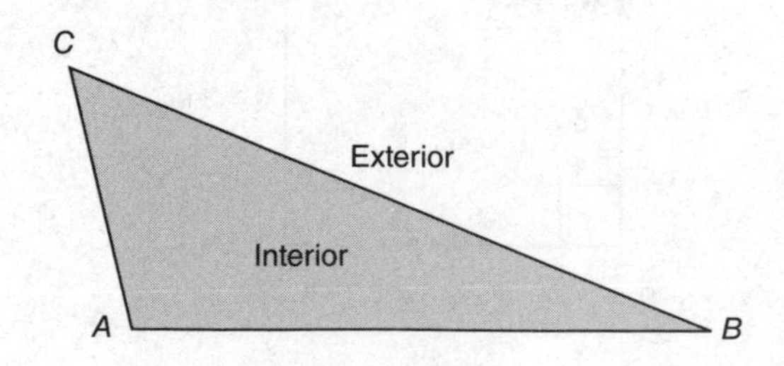


$\overline{AB}$, $\overline{AC}$, and $\overline{BC}$ are the sides of the triangle and ∠A, ∠B, and ∠C are the angles of the triangle.

Triangles can be categorized according to the length of their sides and by the size of their largest internal angle.

An **equilateral triangle** has all three sides of equal length. An equilateral triangle is also equiangular, in that all the internal angles are equal (60°).

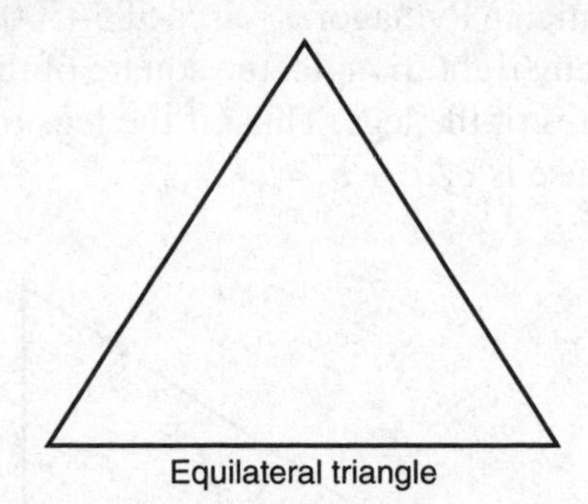

Equilateral triangle

An **isosceles triangle** is a triangle in which at least two sides are of equal length. In the isosceles triangle, C is the **vertex**, ∠C is the **vertex angle**, $\overline{AB}$ is the **base**, and ∠A and ∠B are the **base angles** of the triangle. ∠A and ∠B are equal.

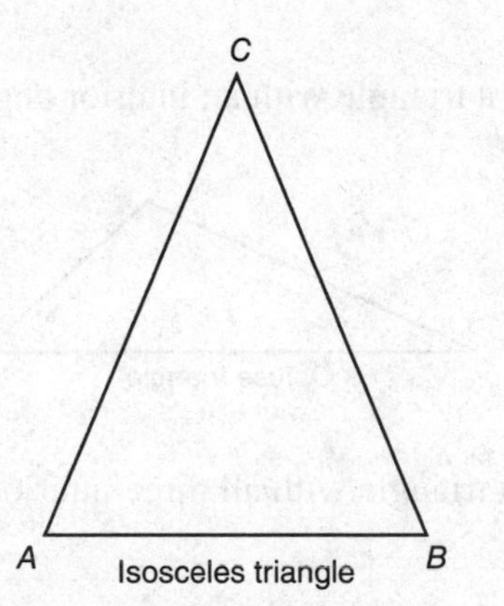


Isosceles triangle

In a **scalene triangle** all sides of the triangle are of different lengths and the internal angles of the triangle are all different.

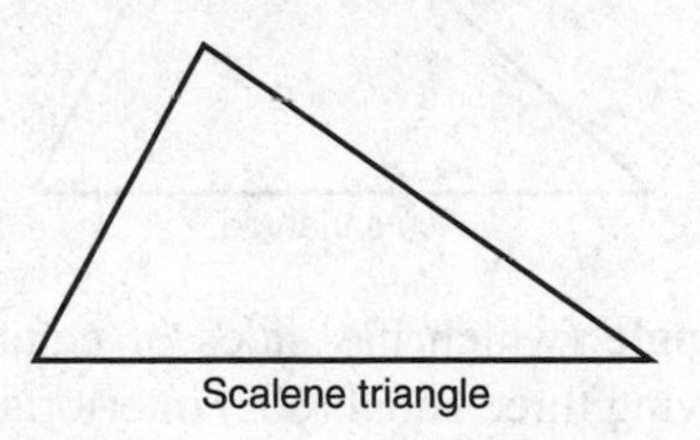

Scalene triangle

A **right triangle** has one interior right (90°) angle. The side opposite the right angle is called the **hypotenuse** and the other two sides are the **legs**.

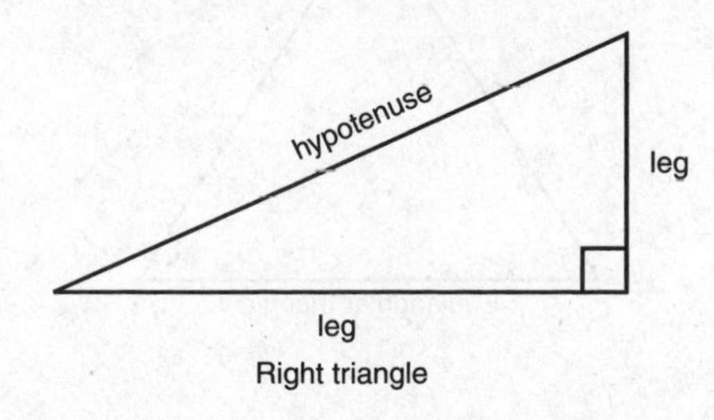


Right triangle

The relationship between the lengths of the legs and the length of the hypotenuse of a right triangle is stated in the **Pythagorean theorem**. Named for the Greek mathematician Pythagoras (circa 585–500 BC), the Pythagorean theorem states that in any right triangle, the square of the hypotenuse is equal to the sum of the squares of the legs. Thus, if the legs of a right triangle are *a* and *b* and the hypotenuse is *c*, $a^2 + b^2 = c^2$.

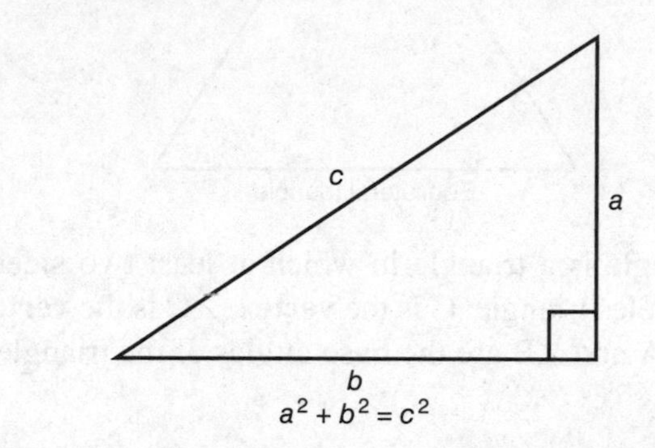


An **obtuse triangle** is a triangle with an interior angle greater than 90°.

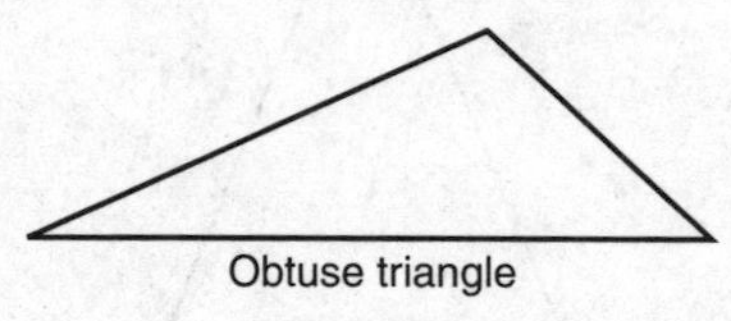
Obtuse triangle

An **acute triangle** is a triangle with all three interior angles less than 90°.

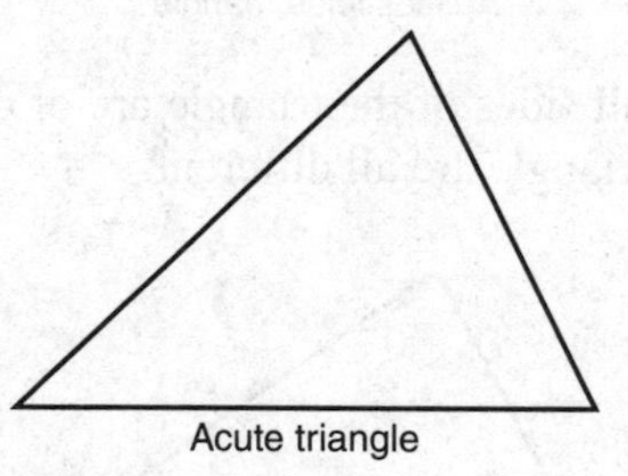
Acute triangle

An **equiangular triangle** (which has sides of equal length as mentioned above) is a triangle having three equal (60°) interior angles.

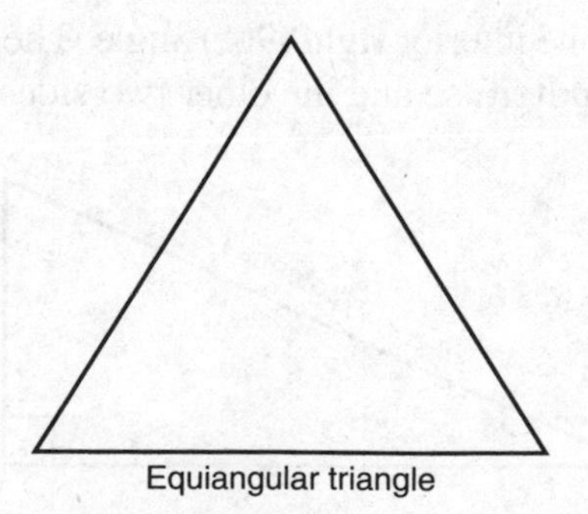
Equiangular triangle

The **perimeter of a triangle** is calculated by adding the length of its three sides and has the formula: $P = a + b + c$.

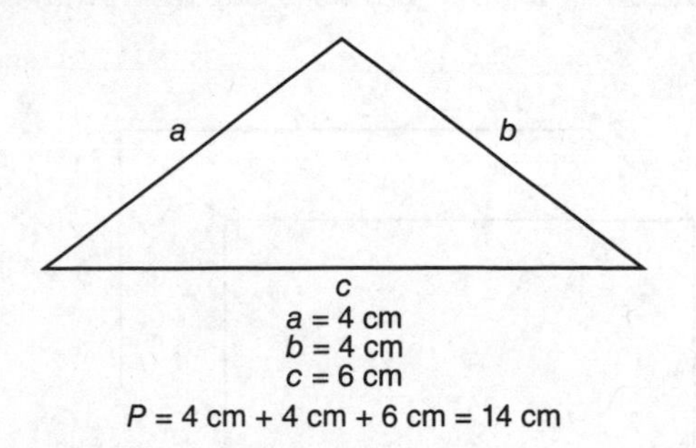


The **area of a triangle** is calculated by multiplying the **base** by the **height** and then dividing by 2. The formula is: $\frac{1}{2}(b \cdot h)$.

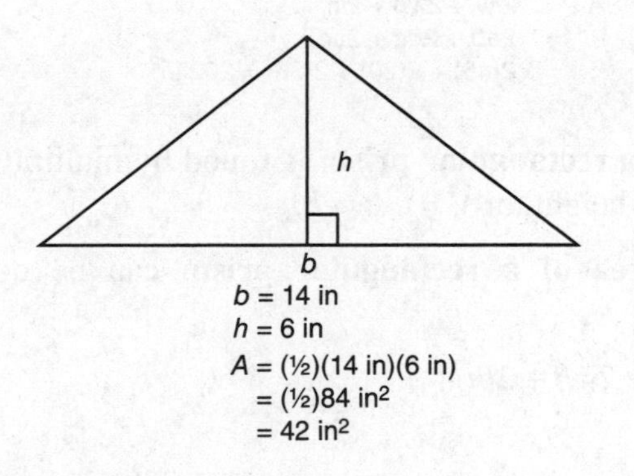


# Three-Dimensional Figures

## CUBE

A cube is a three-dimensional figure having six square-shaped sides. The **volume of a cube** is calculated by raising the length of one side to the third power. The formula is: $V = L^3$.

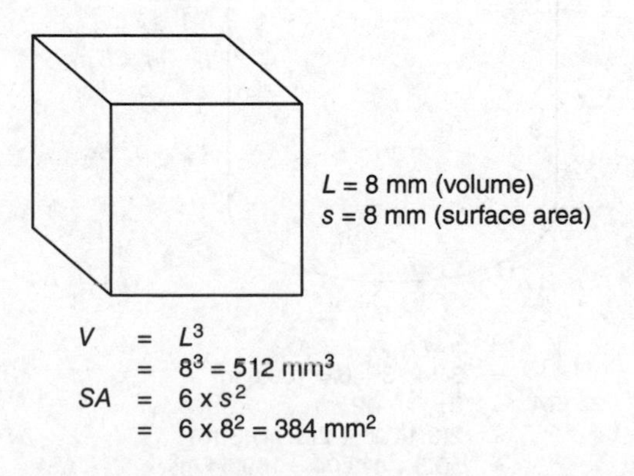


The **surface area (SA) of a cube** is calculated by multiplying 6 times the length of one side to the second power: $SA = 6 \cdot s^2$.

## RECTANGULAR PRISM

A **rectangular prism** is a three-dimensional figure having six rectangular sides.

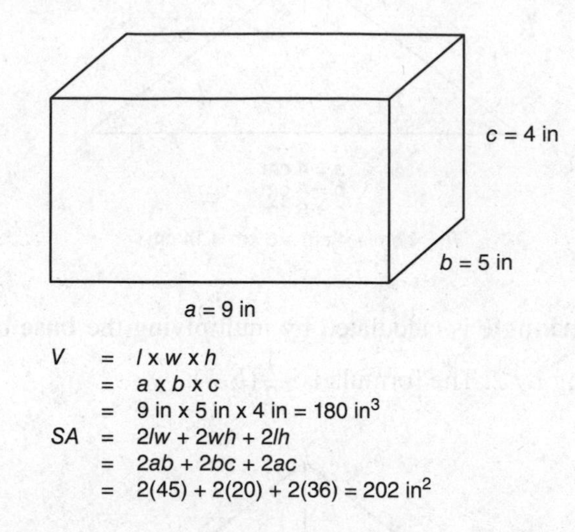


$V = l \times w \times h$
$= a \times b \times c$
$= 9 \text{ in} \times 5 \text{ in} \times 4 \text{ in} = 180 \text{ in}^3$
$SA = 2lw + 2wh + 2lh$
$= 2ab + 2bc + 2ac$
$= 2(45) + 2(20) + 2(36) = 202 \text{ in}^2$

The **volume of a rectangular prism** is found by multiplying the length by the width by the height, or $V = l \cdot w \cdot h$.

The **surface area of a rectangular prism** can be determined by the formula:

$$SA = 2lw + 2wh + 2lh$$

## CYLINDER

A cylinder is a three-dimensional figure having two congruent circular bases that are parallel.

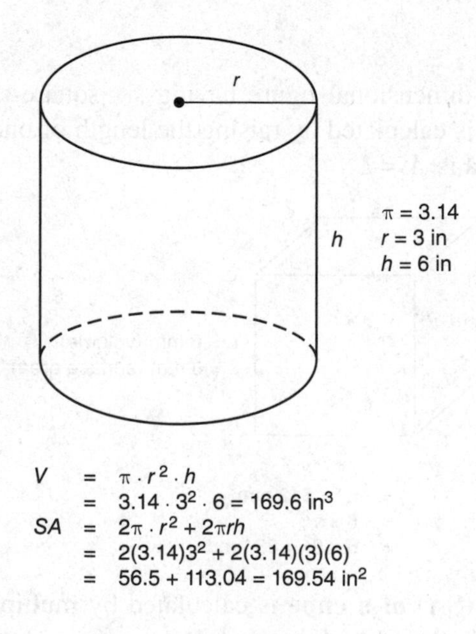


$V = \pi \cdot r^2 \cdot h$
$= 3.14 \cdot 3^2 \cdot 6 = 169.6 \text{ in}^3$
$SA = 2\pi \cdot r^2 + 2\pi rh$
$= 2(3.14)3^2 + 2(3.14)(3)(6)$
$= 56.5 + 113.04 = 169.54 \text{ in}^2$

The **volume of a cylinder** is given by the formula: $V = \pi \cdot r^2 \cdot h$.

The **surface area of a cylinder** is given by the formula:

$$SA = 2\pi r^2 + 2\pi rh$$

## SPHERE

A sphere is a three-dimensional figure having all its points the same distance from its center.

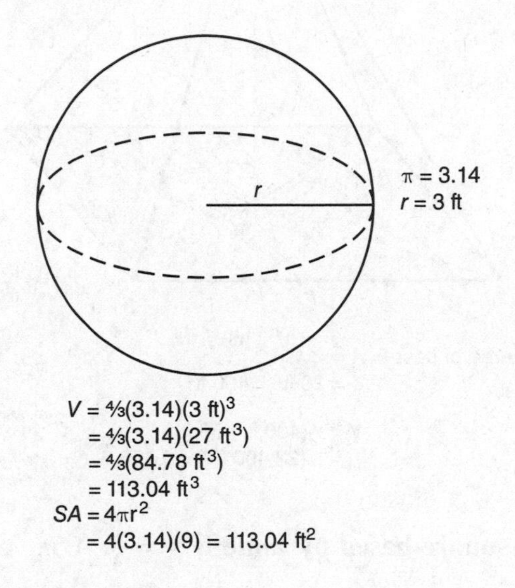


$V = \frac{4}{3}(3.14)(3\text{ ft})^3$
$= \frac{4}{3}(3.14)(27\text{ ft}^3)$
$= \frac{4}{3}(84.78\text{ ft}^3)$
$= 113.04\text{ ft}^3$
$SA = 4\pi r^2$
$= 4(3.14)(9) = 113.04\text{ ft}^2$

The **volume of a sphere** is given by the formula: $V = \frac{4}{3} \cdot \pi \cdot r^3$.

The **surface area of a sphere** is given by the formula: $SA = 4 \cdot \pi \cdot r^2$.

## CONE

A cone is a three-dimensional figure having a circular base and a single vertex.

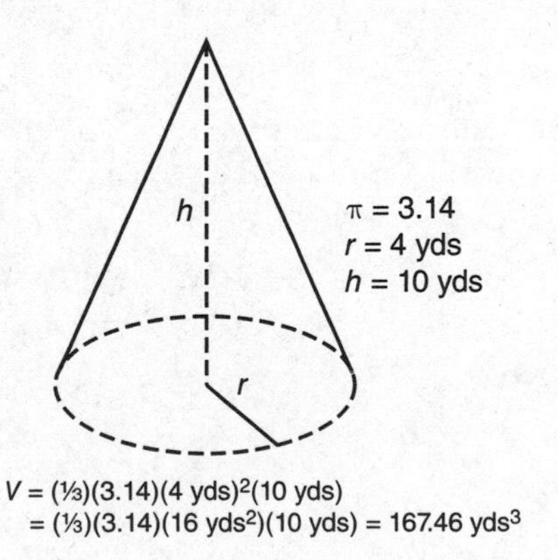


$V = (\frac{1}{3})(3.14)(4\text{ yds})^2(10\text{ yds})$
$= (\frac{1}{3})(3.14)(16\text{ yds}^2)(10\text{ yds}) = 167.46\text{ yds}^3$

The **volume of a cone** is $V = \frac{1}{3} \cdot \pi \cdot r^2 \cdot h$.

## PYRAMID

A square-based pyramid is a three-dimensional figure having a square base and four triangular-shaped sides.

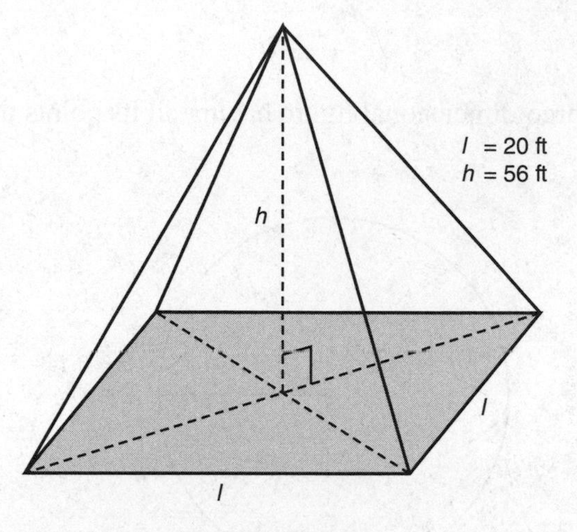


Area of base = $A = l^2$
$= 20 \text{ ft}^2 = 400 \text{ ft}^2$

$V = ⅓(400 \text{ ft}^2)(56 \text{ ft})$
$= ⅓(22{,}400 \text{ ft}^3) = 7{,}466.6 \text{ ft}^3$

The **volume of a square-based pyramid** is $V = \frac{1}{3} \cdot A \cdot h.$

# REVIEW QUESTIONS

*Circle the letter of your choice.*

**1.** The mathematical symbol | | means

(A) is equal to
(B) ratio
(C) inequality
(D) absolute value of

**2.** Find the correct total: using the PEMDAS system to perform the mathematical operations in the correct sequence, evaluate the expression $8 + 7[5 + (9 - 2)^2]$

(A) 810
(B) 1,016
(C) 386
(D) 271

**3.** Subtracting one odd number from another odd number always leaves a remainder that is

(A) even
(B) odd
(C) either even or odd
(D) a negative number

**4.** Even numbers are numbers that when divided by 2

(A) leave no remainder
(B) always leave a remainder
(C) leave a remainder that is always an odd number
(D) always leave a negative result

**5.** Division by zero is

(A) possible only if the number being divided is odd
(B) possible only if the number being divided is even
(C) undefined and therefore has no meaning
(D) always equal to 1

**6.** A thermometer uses what type of numbers along its scale?

(A) absolute values
(B) ratios
(C) irrational
(D) signed

**7.** The absolute value of $-8$, represented by $|-8|$ is

(A) $-8$
(B) 64
(C) 8
(D) 4

**8.** $-4 + (-3) =$

(A) $-7$
(B) 1
(C) $-1$
(D) 7

**9.** $(-9)(-2) =$

(A) 7
(B) 18
(C) $-18$
(D) $-7$

**10.** $(3)(-4)(-7)(-1)(9) =$

(A) 756
(B) 675
(C) $-675$
(D) $-756$

**11.** $(96) \div (-12) =$

(A) $-6$
(B) $-8$
(C) 8
(D) 6

**12.** $(-5)^3 =$

(A) $-125$
(B) 25
(C) $-25$
(D) 125

**13.** In the decimal number 42.089, the three numbers to the right of the decimal point indicate most accurately

(A) the number of tenths to be added to the whole number
(B) the number of hundredths to be added to the whole number
(C) the number of thousandths to be added to the whole number
(D) that the whole number should be rounded off to 43

**14.** $\frac{3}{18}$ equals which of the following decimals, rounded to a thousandth?

(A) 0.166
(B) 0.167
(C) 0.1666
(D) 0.1667

**15.** $4.726 + 32.88 =$

(A) 80.140
(B) 37.154
(C) 37.606
(D) 37.506

**16.** $1.42 - 0.578 =$

(A) 0.825
(B) 0.824
(C) 0.844
(D) 0.842

**17.** $6.9 \times 3.72 =$

(A) 25.876
(B) 25.668
(C) 24.998
(D) 25.658

**18.** $27.45 \div 0.3 =$

(A) 9.15
(B) 91.5
(C) 93.0
(D) 9.45

**19.** $2\frac{3}{4} =$

(A) $\frac{10}{4}$
(B) $\frac{11}{4}$
(C) $\frac{24}{4}$
(D) $\frac{9}{4}$

**20.** Which of the following is not a prime number?

(A) 2
(B) 3
(C) 4
(D) 5

**21.** Which of the following lists the first four multiples of 9?

(A) 9, 18, 27, 36
(B) 9, 81, 729, 6,561
(C) 1, 9, 18, 27
(D) 1, 9, 81, 729

**22.** The fraction 14/35 reduced to lowest terms equals

(A) 5
(B) $\frac{1}{5}$
(C) $\frac{2}{5}$
(D) 7

**23.** Which of the following fractions is greater than $\frac{1}{2}$?

(A) $\frac{9}{19}$
(B) $\frac{7}{16}$
(C) $\frac{18}{28}$
(D) $\frac{34}{70}$

**24.** The sum of $\frac{2}{3}+\frac{5}{6}=$

(A) $\frac{5}{9}$
(B) $\frac{7}{9}$
(C) $1\frac{1}{2}$
(D) None of the above

**25.** $\frac{5}{9}-\frac{2}{7}=$

(A) $\frac{17}{63}$
(B) $\frac{3}{2}$
(C) $\frac{10}{63}$
(D) $\frac{3}{63}$

**26.** The product of $\frac{7}{8} \times \frac{2}{3}$ reduced to its lowest terms equals

(A) $\frac{7}{12}$

(B) $1\frac{1}{2}$

(C) $\frac{9}{11}$

(D) $\frac{14}{24}$

**27.** $\frac{1}{3} \div \frac{2}{5} =$

(A) $\frac{2}{15}$

(B) $\frac{1}{2}$

(C) 1

(D) $\frac{11}{15}$

**28.** 0.275 equals which percent?

(A) 00275%
(B) 0.0275%
(C) 2.75%
(D) 27.5%

**29.** 96.9% equals which decimal?

(A) 9.69
(B) 0.969
(C) 96.9
(D) None of the above

**30.** What is 38% of 25?

(A) 9
(B) 10
(C) 9.5
(D) 10.3

**31.** What percent of 22 is 13?

(A) 57.90
(B) 59.09
(C) 58.37
(D) 58.73

**32.** 16 is 47% of what number?

(A) 34
(B) 32
(C) 35
(D) 30

**33.** What is 33 increased by 72%?

(A) 57.67
(B) 56.84
(C) 23.76
(D) 56.76

**34.** What is 74 decreased by 9%?

(A) 6.66
(B) 67.58
(C) 6.80
(D) 67.34

**35.** $\frac{7}{13}$ equals which percent?

(A) 56.1%
(B) 53.8%
(C) 55.3%
(D) 57.2%

**36.** 34% equals which fraction, reduced to lowest terms?

(A) $\frac{3}{4}$

(B) $\frac{13}{20}$

(C) $\frac{17}{50}$

(D) $\frac{1}{3}$

**37.** In the arithmetic sequence of terms shown below, the difference between consecutive terms is constant. What is the absolute value of the constant interval between the terms?

4, 9, 14, 19, 24, 29, 34, 39, . . .

(A) 4
(B) 9
(C) 5
(D) 3

**38.** On a farm, there are 12 hens and 3 roosters. What is the ratio of hens to roosters in its simplest form?

(A) 12:3
(B) 3:12
(C) 1:3
(D) 4:1

**39.** Which ratio listed below is in its simplest form?

(A) 10:50
(B) 5:25
(C) 1:5
(D) None of the above

**40.** If four tickets to a show cost $18, which is the cost of 18 tickets?

(A) $78
(B) $81
(C) $88
(D) $76

**41.** For a house assessed at $20,000, the owner pays $500 in property tax. What should the property tax be on a house assessed at $35,000?

(A) $800
(B) $825
(C) $900
(D) $875

**42.** If a die is rolled once, what is the probability of rolling a 4 or higher number?

(A) 1/6
(B) 1/4
(C) 1/3
(D) 1/2

**43.** A movie theater sells three sizes of popcorn (small, medium, and large) with two choices of toppings (no butter and butter). How many different popcorn choices are available?

(A) 8
(B) 10
(C) 6
(D) 12

**44.** A state issues license plates containing all the letters of the alphabet as well as the numbers 0 through 9. If all of the letters and digits may be repeated, how many possible license plates can be issued with two letters followed by two numbers?

(A) 69,000
(B) 73,400
(C) 67,600
(D) 68,220

**45.** Factor the following algebraic expression: $x^2 - 3x - 18$

(A) $(x - 6)(x + 3)$
(B) $(x + 6)(x - 3)$
(C) $(x + 6)(x + 3)$
(D) $(x - 9)(x + 2)$

**46.** $9x - (-3x) =$

(A) $6x$
(B) $12x$
(C) $6x^2$
(D) $12x^2$

**47.** $(2x)(6x)(3x) =$

(A) $36x$
(B) $36x^2$
(C) $12x^2 + 3x + 36$
(D) $36x^3$

**48.** $(x)^3(x)^2(x)^3 =$

(A) $x^{18}$
(B) $x^3$
(C) $x^8$
(D) None of the above

**49.** Solve: $(92)^3 =$

(A) 8,464
(B) 716,929.6
(C) 778,688
(D) None of the above

**50.** $7^4 =$

(A) 28
(B) 49
(C) 2,401
(D) 343

**51.** $(-4)^3 =$

(A) $-64$
(B) 64
(C) $-12$
(D) 12

**52.** $(8)^0 =$

(A) 0
(B) 1
(C) 8
(D) None of the above

**53.** $\sqrt{81} =$

(A) 6
(B) 8
(C) 9
(D) 18

**54.** Solve for $x$: $22x - 68 = 12 + 2x$

(A) 2
(B) 4
(C) 6
(D) 8

**55.** Solve for $x$: $4(5x + 10) - 9(x + 3) = 68$

(A) 3
(B) 4
(C) 5
(D) 2

**56.** A $400 firefighter bunker coat sells for how much money after it is discounted 5%?

(A) $350
(B) $370
(C) $380
(D) $395

**57.** Six pairs of work boots cost as much as three pairs of dress shoes. Which ratio compares the cost of work boots to the cost of dress shoes?

(A) 2:1
(B) 5:4
(C) 6:2
(D) 5:2

**58.** Find the pattern in the two columns below. Then solve for $x$.

| a | b |
|---|---|
| 2 | 10 |
| 3 | 16 |
| 4 | 22 |
| 5 | $x$ |

(A) 24
(B) 26
(C) 28
(D) 32

**59.** $1.2 \times 2.6 =$

(A) 3.61
(B) 2.87
(C) 3.12
(D) 3.88

**60.** The sum of three consecutive multiples of 3 is 27. What is the largest of the three numbers?

(A) 18
(B) 15
(C) 12
(D) 9

**61.** If 30,000 candidates compete for firefighter jobs in a municipality, and only 15% of those candidates are expected to be hired over a four-year period, how many candidates will be hired?

(A) 3,000
(B) 1,500
(C) 6,000
(D) 4,500

**62.** A newly hired firefighter buys three uniform shirts, four pairs of pants, and two pairs of work shoes at the quartermaster prior to the first day on the job. How many different outfits can be worn if an outfit consists of any shirt worn with any pair of pants and either pair of shoes?

(A) 8
(B) 9
(C) 12
(D) 24

**63.** If two packages of latex gloves cost $2.50, how much would five packages cost?

(A) $5.50
(B) $6.25
(C) $6.50
(D) $6.75

**64.** What is 75 decreased by 30%?

(A) 52.5
(B) 53
(C) 53.5
(D) 54.5

**65.** John drives his vehicle 60 mph for 3.5 hours to get to the street fair in the big city. How many miles did he travel?

(A) 160 miles
(B) 210 miles
(C) 300 miles
(D) 320 miles

**66.** If 3 and $x$ have the same average as 3, 7, and 20, what is the value of $x$?

(A) 14
(B) 17
(C) 20
(D) 22

**67.** A horse weighs 781 pounds, rounded to the nearest pound. Which weight listed below cannot be the actual weight of the horse?

(A) 780.6 pounds
(B) 781.1 pounds
(C) 781.4 pounds
(D) 781.7 pounds

**68.** A total of $360 is divided into equal shares. John receives four shares, David receives three shares, and Allen receives the remaining two shares. How much money did David get?

(A) $160
(B) $80
(C) $120
(D) $100

**69.** A basketball league has 100 players on eight different teams. Each team has at least 12 players. What is the largest possible number of players on any one team?

(A) 14
(B) 16
(C) 15
(D) 17

**70.** You have $80 to spend on a pair of shoes. The sales tax is 7%. What is the maximum amount the shoes can cost?

(A) $75.46
(B) $78.13
(C) $74.40
(D) $77.89

**71.** The formula for converting Celsius to Fahrenheit is $F = \frac{9}{5}C + 32$. Use the formula to convert 10° Celsius into Fahrenheit degrees.

(A) 50
(B) 48
(C) 52
(D) 56

**72.** The formula for converting Fahrenheit to Celsius is $C = \frac{5}{9}(F - 32)$. Use the formula to convert 77° Fahrenheit into Celsius degrees.

(A) 22
(B) 29
(C) 23
(D) 25

**73.** Right angles measure

(A) less than 90°
(B) more than 90°
(C) 90°
(D) 180°

**74.** Two lines are parallel when

(A) they intersect
(B) they form equal adjacent angles
(C) they do not share a common point
(D) the angles formed are right angles

**75.** The angles formed by two perpendicular lines are

(A) right angles
(B) acute angles
(C) obtuse angles
(D) more than 180°

**76.** The distance around a two-dimensional object that is measured in linear units is called the

(A) perimeter
(B) area
(C) surface area
(D) volume

**77.** The amount of space a three-dimensional object displaces is called the

(A) surface area
(B) perimeter
(C) area
(D) volume

**78.** The sum of the areas of the exposed sides of a three-dimensional object is called the

(A) perimeter
(B) volume
(C) surface area
(D) area

**79.** The radius of a circle is

(A) twice as long as the diameter
(B) half as long as the diameter
(C) the same length as the diameter
(D) None of the above

**80.** The distance around a circle is called

(A) $\pi$
(B) the area
(C) the circumference
(D) a chord

**81.** Find the approximate circumference of the circle. Use the formula: $C = \pi \cdot d$ or $C = 2 \cdot \pi \cdot r$.

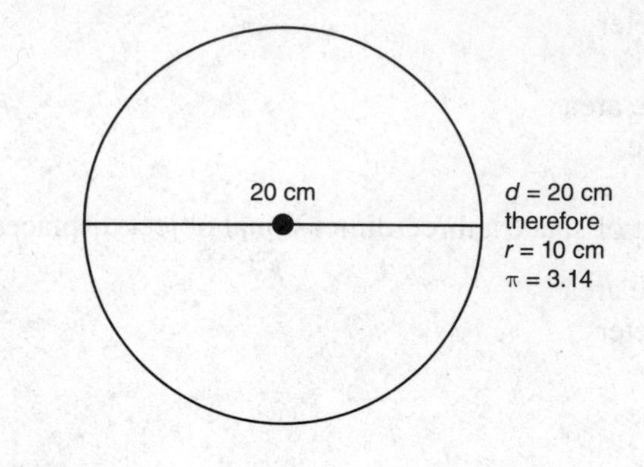


(A) 67.6 cm
(B) 65.3 cm
(C) 63.4 cm
(D) 62.8 cm

**82.** Find the area of the circle in question 81. Use the formula: $A = \pi \cdot r^2$.

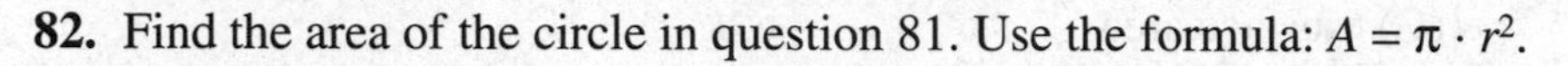

(A) 3.14 $cm^2$
(B) 31.4 $cm^2$
(C) 314 $cm^2$
(D) 3140 $cm^2$

**83.** Find the perimeter ($P$) of the irregular polygon below.

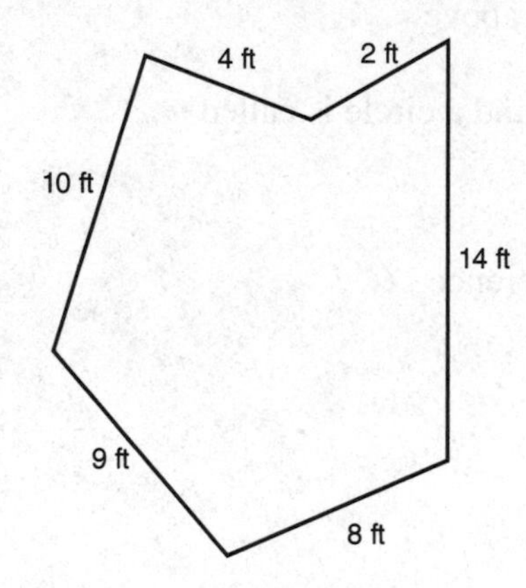


(A) 40 ft
(B) 47 ft
(C) 51 ft
(D) 53 ft

**84.** Find the perimeter of a regular six-sided polygon (hexagon) whose sides are 7 cm. Use the formula: $P = ns$.

(A) 28 cm
(B) 36 cm
(C) 42 cm
(D) 62 cm

**85.** Find the height ($h$) of the parallelogram shown below whose area is 200 in$^2$ and base is 20 in. Use the formula: $A = b \cdot h$.

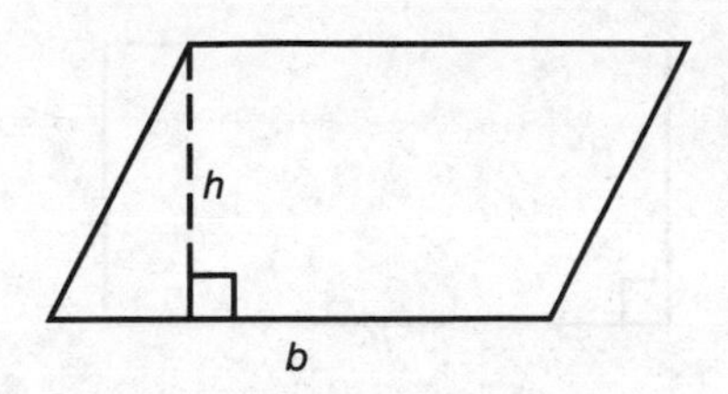


(A) 10 in
(B) 9 in
(C) 12 in
(D) 8 in

**86.** Find the perimeter of the square shown below with each side measuring 5 in. Use the formula: $P = 4s$.

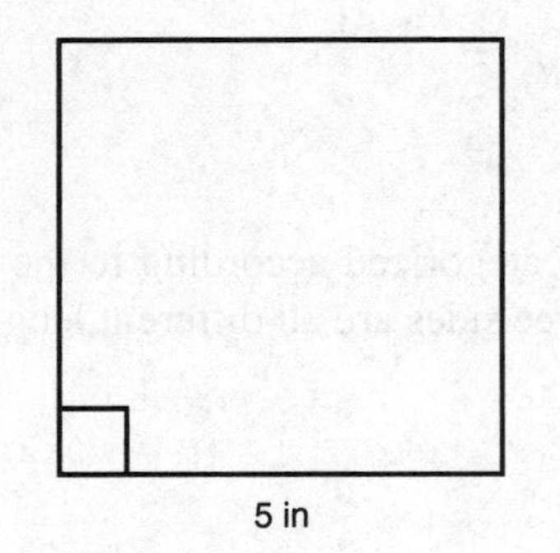


(A) 16 in
(B) 20 in
(C) 24 in
(D) 25 in

**87.** Find the area of the square in question 86. Use the formula: $A = s^2$.

(A) 16 in$^2$
(B) 20 in$^2$
(C) 24 in$^2$
(D) 25 in$^2$

**88.** Find the perimeter of the rectangle shown below having a length ($l$) of 24 cm and a width ($w$) of 14 cm. Use the formula: $P = 2(l) + 2(w)$.

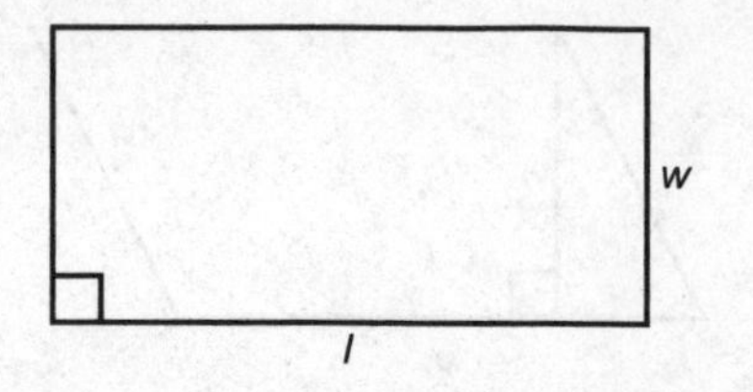


(A) 38 cm
(B) 62 cm
(C) 76 cm
(D) 91 cm

**89.** The sum of the internal angles of a triangle is

(A) 90°
(B) 120°
(C) 160°
(D) 180°

**90.** Triangles can be categorized according to the length of their sides. A triangle whose three sides are all different lengths is called

(A) equilateral
(B) scalene
(C) isosceles
(D) None of the above

**91.** Triangles can also be categorized according to the size of their largest interior angle. A triangle with three interior angles measuring less than 90° is called

(A) acute
(B) right
(C) obtuse
(D) None of the above

**92.** In an equilateral triangle, all three interior angles measure how many degrees?

(A) 30
(B) 45
(C) 60
(D) 65

**93.** Find the perimeter of the triangle shown below. Use the formula: $P = a + b + c$.

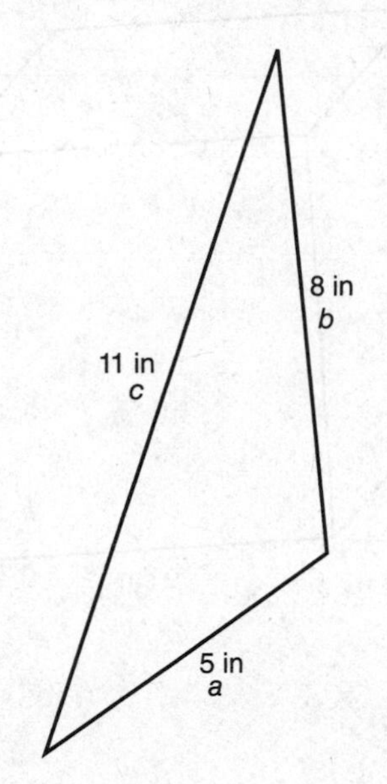


(A) 20 in
(B) 24 in
(C) 30 in
(D) 34 in

**94.** Find the area of the triangle shown below. Use the formula: $A = \frac{1}{2} b \cdot h$.

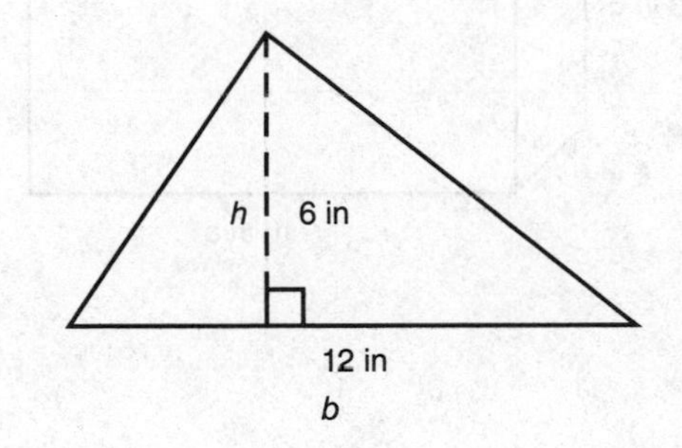


(A) 36 in$^2$
(B) 42 in$^2$
(C) 48 in$^2$
(D) 54 in$^2$

**95.** Find the volume of the cube shown below. Use the formula: $V = L^3$.

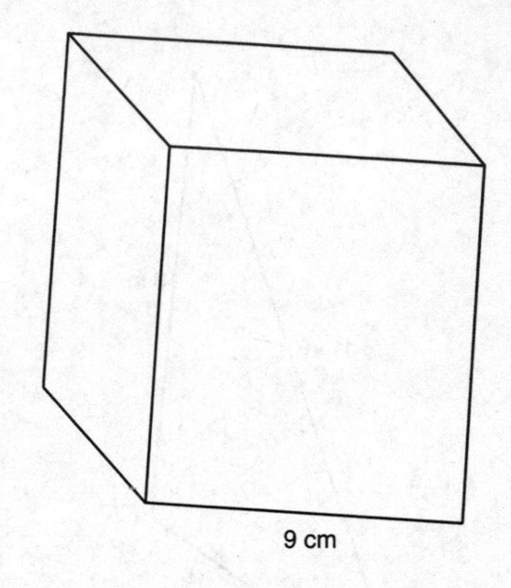


(A) 81 $cm^3$
(B) 729 $cm^3$
(C) 917 $cm^3$
(D) 936 $cm^3$

**96.** Find the surface area of the rectangular prism shown below. Use the formula: $2ab + 2bc + 2ac$ or $2lw + 2wh + 2lh$.

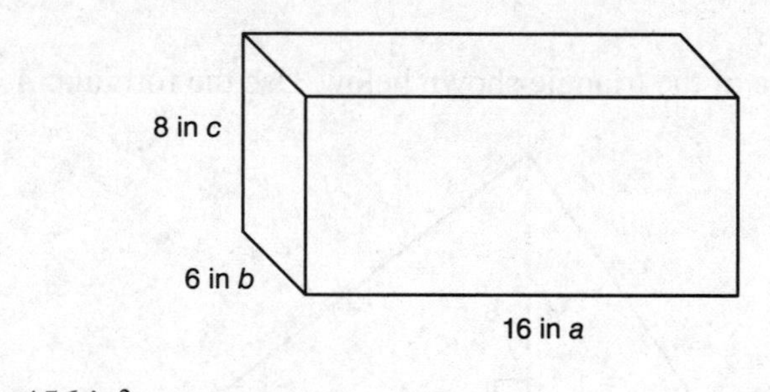


(A) 456 $in^2$
(B) 523 $in^2$
(C) 544 $in^2$
(D) 572 $in^2$

**97.** Find the approximate volume of the cylinder shown below. Use the formula: $V = \pi \cdot r^2 \cdot h$.

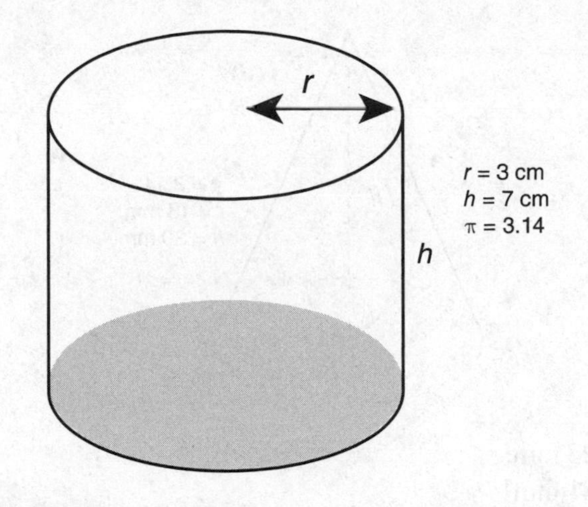


(A) 123.37 $cm^3$
(B) 571.95 $cm^3$
(C) 319.18 $cm^3$
(D) 197.82 $cm^3$

**98.** Find the approximate volume of the sphere shown below. Use the formula: $V = \frac{4}{3}\pi \cdot r^3$.

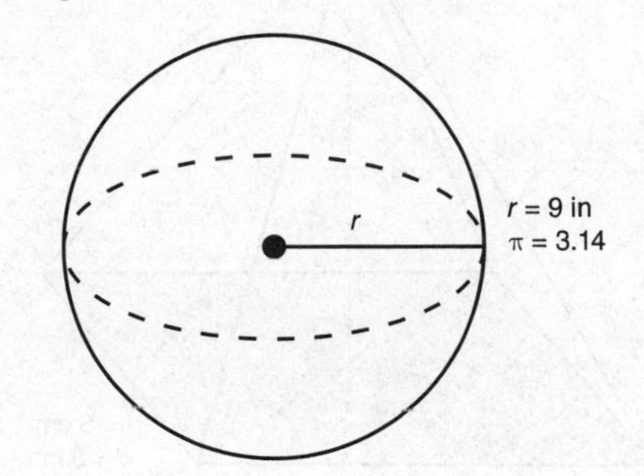


(A) 3,053 $in^3$
(B) 3,180 $in^3$
(C) 3,510 $in^3$
(D) 3,795 $in^3$

**99.** Find the approximate volume of the cone shown below. Use the formula: $V = 1/3\ \pi \cdot r^2 \cdot h$.

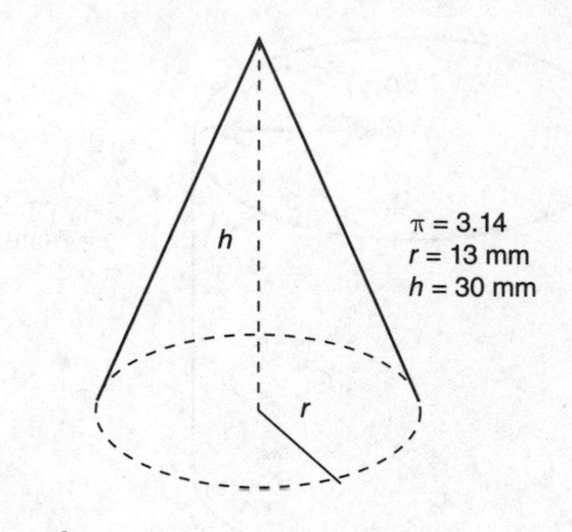


(A) 4,723 $mm^3$
(B) 4,981 $mm^3$
(C) 5,169 $mm^3$
(D) 5,256 $mm^3$

**100.** Find the approximate volume of the square-based pyramid shown below. Use the formula: $V = 1/3 \cdot A \cdot h$.

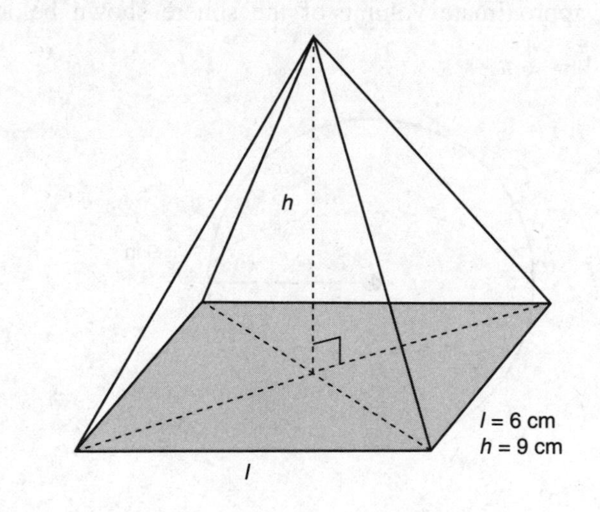


(A) 96 $cm^3$
(B) 102 $cm^3$
(C) 107 $cm^3$
(D) 117 $cm^3$

# Answer Key

**1.** D
**2.** C
**3.** A
**4.** A
**5.** C
**6.** D
**7.** C
**8.** A
**9.** B
**10.** D
**11.** B
**12.** A
**13.** C
**14.** B
**15.** C
**16.** D
**17.** B
**18.** B
**19.** B
**20.** C
**21.** A
**22.** C
**23.** C
**24.** C
**25.** A
**26.** A
**27.** D
**28.** D
**29.** B
**30.** C
**31.** B
**32.** A
**33.** C
**34.** D
**35.** B
**36.** C
**37.** C
**38.** D
**39.** C
**40.** B
**41.** D
**42.** D
**43.** C
**44.** C
**45.** A
**46.** B
**47.** D
**48.** C
**49.** C
**50.** C
**51.** A
**52.** B
**53.** C
**54.** B
**55.** C
**56.** C
**57.** A
**58.** C
**59.** C
**60.** C
**61.** D
**62.** D
**63.** B
**64.** A
**65.** B
**66.** B
**67.** D
**68.** C
**69.** B
**70.** C
**71.** A
**72.** D
**73.** C
**74.** C
**75.** A
**76.** A
**77.** D
**78.** C
**79.** B
**80.** C
**81.** D
**82.** C
**83.** B
**84.** C
**85.** A
**86.** B
**87.** D
**88.** C
**89.** D
**90.** B
**91.** A
**92.** C
**93.** B
**94.** A
**95.** B
**96.** C
**97.** D
**98.** A
**99.** D
**100.** C

## ANSWER EXPLANATIONS

Following you will find explanations to specific questions that you might find helpful.

1. **D** The absolute value of a number is its value when the sign is not taken into consideration. The symbol for absolute value is | |.

   **Example:** $|7| = 7$ and $|-7| = 7$

2. **C** PEMDAS—stands for **P**arenthesis, **E**xponents, **M**ultiplication/**D**ivision, then **A**ddition/**S**ubtraction. When there are two or more parentheses (grouping symbols), perform the innermost grouping symbol first. Exponents should be worked on next. Multiplication and division, as well as addition and subtraction, are grouped to denote that when these operations are next to each other, just perform the math from left to right.

   **Example:** $8 + 7[5 + (9 - 2)^2]$
   $8 + 7[5 + (7)^2]$
   $8 + 7[5 + 49]$
   $8 + 7\ [54]$
   $8 + 378 = 386$

3. **A**

   **Examples:** $7 - 3 = 4$; $13 - 7 = 6$

4. **A**

   **Examples:** $8 \div 2 = 4$; $16 \div 2 = 8$

5. **C** Division by zero is undefined and therefore has no meaning, regardless of whether the numerator is a positive or negative number.

6. **D** A thermometer has a scale with both positive and negative numbers. The numbers run along a vertical number scale, with positive numbers, indicated by the (+) sign, being above zero and negative numbers, indicated by the (–) sign, found below zero.

7. **C** The absolute value of a number is its value when the sign is not taken into consideration. The symbol for absolute value is | |.

   **Example:** $|-8| = 8$

8. **A** To add two negative numbers (like signs), add their absolute values and prefix the sum with their common sign.

   **Example:** $-4 + (-3) = -(|4| + |3|) = -7$

9. **B** When multiplying two signed numbers, the product will be positive if the signs are the same.

   **Example:** $(-9)\ (-2) = 18$

10. **D** When multiplying more than two numbers, the product is always a negative number if there are an odd number of negative factors. In the question there are an odd number (3) of negative factors.
    **Example:** $(3)(-4)(-7)(-1)\ (9) = -756$

**11.** B When dividing two signed numbers, the quotient will be negative if the signs are different.

**Example:** $(96) \div (-12) = -8$

**12.** A When multiplying numbers in exponential form (a number being multiplied by itself), odd **exponents** (powers) of negative numbers result in a product that is negative.

**Example:** $(-5) \times (-5) \times (-5) = -125$

**13.** C 42.089 Numbers that contain a decimal point are called decimals. The number to the left of the decimal point is a whole number, and the number to the right of the decimal point is called the tenths digit. The tenths digit gives you the number of tenths that are added to the whole number. If there are two numbers to the right of the decimal point, then both numbers to the right are the number of hundredths to be added to the whole number. Three numbers to the right of the decimal point would indicate that all three numbers to the right are the number of thousandths to be added to the whole number, and so forth.

**14.** B To convert a proper fraction into a decimal, divide the numerator by the denominator to obtain the decimal equivalent.

**Example:** $3/18 = 0.167$ rounded to a thousandth.

**15.** C When adding decimals, write the numerals one above the other with their decimal points lined up.

**Example:**
$$\begin{array}{r} 4.726 \\ +\ 32.88 \\ \hline 37.606 \end{array}$$

**16.** D When subtracting decimals, write the numerals one above the other with their decimal points lined up.

**Example:**
$$\begin{array}{r} 1.42 \\ -\ 0.578 \\ \hline 0.842 \end{array}$$

**17.** B When multiplying two decimals, initially ignore the decimal points and simply multiply the two numbers to get the product. Then add together the number of digits to the right of the decimal point in each number, and put the decimal in the product so that there is the same number of decimal places as in the numbers multiplied.

**Example:** $6.9 \times 3.72 =$
$$\begin{array}{r} 6.9 \\ \times\ 3.72 \\ \hline 25.668 \end{array}$$

**18.** B To divide one decimal into another, move the decimal in the divisor to the right until it is a whole number, and then move the decimal to the right the same number of places in the dividend.

**Example:** $27.45 \div 0.3 = 2745 \div 300 = 91.5$

**19.** B To convert a mixed number to a fraction, multiply the integer of the mixed number by the denominator of the fraction, and then add the product to the numerator of the fraction and put that number over the denominator of the original fraction.

**Example:** $2\frac{3}{4} = \frac{(2 \times 4) + 3}{4} = \frac{8 + 3}{4} = \frac{11}{4}$

**20.** C Prime numbers are unique numbers whose sole factors are 1. All prime numbers except 2 are odd.

**Examples:** 2, 3, 5, 7, 11, 13, 17, 19, 23, 29, 31, 37. . .

**21.** A Multiples of a number are its products with the natural numbers.

**Example:** $4 \times 1 = 4$, $4 \times 2 = 8$, $4 \times 3 = 12$, $4 \times 4 = 16$.

The multiples of 9 are therefore 9, 18, 27, 36, and so on.

**22.** C To reduce a fraction to its lowest form, divide the numerator and the denominator by their greatest common factor (GFC). The GCF of two numbers is the largest factor that is common to both.

**Example:** Reduce $\frac{14}{35}$ to its lowest form.

Prime factorization of $14 = 7 \times 2$

Prime factorization of $35 = 5 \times 7$

GCF of 14 and 35 = 7

Divide the numerator and denominator by the GCF, 7, to reduce the fraction.

$14 \div 7 = 2$

$35 \div 7 = 5$

Fraction in lowest form $= \frac{2}{5}$

**23.** C $\frac{9}{19} = 0.47$; $\frac{7}{16} = 0.43$; $\frac{18}{28} = 0.64$; $\frac{34}{70} = 0.48$

**24.** C Fractions with different denominators require first finding a common denominator or least common denominator (LCD) before you can add them.

**Example:** $\frac{2}{3} + \frac{5}{6}$

The least common multiple of 3 and 6 is 6; therefore, the LCD is 6. Multiply the numerator *and* denominator of the first fraction by the number required to change the denominator to 6—in this case, 2.

$\frac{2}{3} \times \frac{2}{2} = \frac{4}{6}$

Repeat the process for the second fraction by multiplying the numerator *and* denominator by 1; the number required to change the denominator to 6.

$\frac{5}{6} \times \frac{1}{1} = \frac{5}{6}$

Then add the numerators and place the sum over the common denominator:

$\frac{4}{6} + \frac{5}{6} = \frac{9}{6} = 1.5 = 1\frac{1}{2}$

**25** A Fractions with different denominators require first finding a common denominator or least common denominator (LCD) before you can subtract them.

**Example:** $\frac{5}{9}-\frac{2}{7}$

The least common multiple of 9 and 7 is 63; therefore, the LCD is 63. Multiply the numerator *and* denominator of the first fraction by the number required to change the denominator to 63—in this case, 7.

$\frac{5}{9}\times\frac{7}{7}=\frac{35}{63}$

Repeat the process for the second fraction by multiplying the numerator *and* denominator by 9, the number required to change the denominator to 63.

$\frac{2}{7}\times\frac{9}{9}=\frac{18}{63}$

Then subtract the numerators and place the sum over the common denominator:

$\frac{35}{63}-\frac{18}{63}=\frac{17}{63}$

**26.** A To multiply fractions, multiply their numerators and denominators.

**Example:** $\frac{7}{8}\times\frac{2}{3}=\frac{14}{24}$

Reduced to lowest terms, divide the numerator and denominator by 2: $\frac{7}{12}$

**27. D** Fractions with different denominators require first finding a common denominator or least common denominator (LCD) before you can add them.

**Example:** $\frac{1}{3}+\frac{2}{5}$

The least common multiple of 3 and 5 is 15; therefore, the LCD is 15. Multiply the numerator *and* denominator of the first fraction by the number required to change the denominator to 15—in this case, 5.

$\frac{1}{3}\times\frac{5}{5}=\frac{5}{15}$

Repeat the process for the second fraction by multiplying the numerator *and* denominator by 3, the number required to change the denominator to 15.

$\frac{2}{5}\times\frac{3}{5}=\frac{6}{15}$

Then add the numerators and place the sum over the common denominator:

$\frac{5}{15}+\frac{6}{15}=\frac{11}{15}$

**28.** D To convert a decimal to a percent, multiply by 100. The short way to convert from decimal to percent is to move the decimal point two places to the right and add a percent sign.

**Example:** 0.275 = 27.5%

**29.** B To convert from percent to decimal, divide by 100 and remove the % sign. The short way to convert a percentage to a decimal is to move the decimal point two places to the left and drop the % sign.

**Example:** $\frac{96.9}{100} = 0.969$

**Example:** 96.9% = 0.969 (short way)

**30.** C To find the part when given the whole and the percent, use the formula: Part = % · Whole ÷ 100

**Example:** Part = 38 · 25 ÷ 100

**Example:** Part = 950 ÷ 100 = 9.5

**31.** B To find the percent when given the whole and the part, use the formula: % = Part ÷ Whole · 100

**Example:** % = 13 ÷ 22 · 100

**Example:** % = 0.59 · 100 = 59.09

**32.** A To find the whole when given the part and percent, use the formula: Whole = Part ÷ % · 100

**Example:** Whole = 16 ÷ 47 · 100

**Example:** Whole = 0.34 · 100 = 34

**33.** C Use the formula: Part = % · Whole ÷ 100

Part = % · Whole ÷ 100

**Example:** Part = 72 · 33 ÷ 100 = 2376 ÷ 100 = 23.76

**34.** D Use the formula: Part = % · Whole ÷ 100

(9)(74) ÷ 100
666 ÷ 100 = 6.66

Subtract the percentage decrease from the original.

74 – 6.66 = 67.34

**35.** B To convert a fraction to a percent, multiply the fraction by 100%. You can express 100% as a fraction by placing it over 1.

**Example:** $\frac{7}{13} \cdot \frac{100\%}{1} = \frac{700\%}{13} = 53.\ 8\%$

**36.** C To convert a percent to a fraction: $34 \cdot \frac{1}{100} = \frac{34}{100}$. Reduced to lowest terms: $\frac{17}{50}$.

**37.** C An ordered list of terms in which the difference between consecutive terms is constant is called an arithmetic sequence. If you subtract any two consecutive terms of the sequence, you will obtain the same difference, known as the constant interval, between the terms.

**Example:** 4, 9, 14, 19, 24, 29, 34, 39…
34 – 29 = 5; 19 – 14 = 5; 9 – 4 = 5

**38.** D A ratio is a comparison of two numbers or objects. The symbol : is used to separate the values in the ratio. To compare ratios, convert them to fractions by placing the number to the left of the symbol used to separate the two numbers as the numerator and the number to the right of the symbol as the denominator. The order of values in the given expression is important when notating the ratio.

**Example:** $12{:}3 = \frac{12}{3}$ . Divide by $\frac{3}{3} = 4{:}1$

**39.** C Equivalent (equal) ratios have the same proportion when both sides of the ratio are multiplied or divided by the same number.

**Example:** 10:50 and 5:25 and 1:5 are all equivalent ratios, but 1:5 is the simplest form.

**40.** B $18 \div 4 = \$4.5$ per ticket. $18 \cdot 4.5 = \$81$

**41.** D Property tax for a house assessed at $20,000 is $500.
Property tax for a house assessed at $15,000 is $375
$500 + $375 = $875

**42.** D Die = 6 possible numbers; rolling 4, 5, 6 $= \frac{3}{6} = \frac{1}{2}$

**43.** C For any word problem that involves two or more actions or objects, each having a number of choices, and asks for the number of combinations, use the counting principle formula: Number of ways $= x \cdot y$
**Example:** popcorn (3) · toppings (2) = 6

**44.** C There are 26 choices of letters: 26 choices for the first letter and 26 choices for the second letter = 676 pairs of letters. Ten possibilities for the first digit and 10 possibilities for the second digit = 100. 676 · 100 = 67,600.

**45.** A $x^2 - 3x - 18$
$(x - 6)(x + 3)$

**46.** B Add like signs: $9x + 3x = 12x$

**47.** D $(2r)(6r)(3r) = (12r)^2(3r) = 36r^3$

**48.** C $(x)^3(x)^2(x)^3$ = Add the exponents $= x^8$

**49.** C $(92)^3 = 778{,}688$
$(92) \cdot (92) = 8464 \cdot 92 = 778{,}688$

**50.** C $7{\cdot}7{\cdot}7{\cdot}7 = 2{,}401$

**51.** A $(-4)^3 = -64$

**52.** B $(8)^0 = 1$ If the exponent is zero, then you get 1.

**53.** C $\sqrt{81} = 9$

**54.** B Simplifying: $22x + -68 = 12 + 2x$
Reorder the terms: $-68 + 22x = 12 + 2x$
Solving: $-68 + 22x = 12 + 2x$

Solving for variable $x$: Move all terms containing $x$ to the left and all other terms to the right.

Add $-2x$ to each side of the equation: $-68 + 22x + -2x = 12 + 2x + -2x$

Combine like terms:

$22x + -2x = 20x$

$-68 + 20x = 12 + 2x + -2x$

Combine like terms:

$2x + -2x = 0$

$-68 + 20x = 12 + 0$

$-68 + 20x = 12$

Add 68 to each side of the equation. $-68 + 68 + 20x = 12 + 68$

Combine like terms:

$-68 + 68 = 0$

$0 + 20x = 12 + 68$

$20x = 12 + 68$

Combine like terms:

$12 + 68 = 80$

$20x = 80$

Divide each side by 20: $x = 4$

**55.** C Simplifying: $4(5x + 10) + -9(x + 3) = 68$

Reorder the terms:

$4\ (10 + 5x) + -9\ (x + 3) = 68$

$(10 * 4 + 5x * 4) + -9\ (x + 3) = 68$

$(40 + 20x) + -9\ (x + 3) = 68$

Reorder the terms:

$40 + 20x + -9\ (3 + x) = 68$

$40 + 20x + (3 - 9 + x - 9) = 68$

$40 + 20x + (-27 + -9x) = 68$

Reorder the terms:

$40 + -27 + 20x + -9x = 68$

Combine like terms:

$40 + -7 = 13$

$13 + 20x + -9x = 68$

Combine like terms:

$20x + -9x = 11x$

$13 + 11x = 68$

Solving: $13 + 11x = 68$

Solving for variable x: Move all terms containing x to the left and all other terms to the right. Add $-13$ to each side of the equation.

$13 + -13 + 11x = 68 + -13$

Combine like terms:

$13 + -13 = 0$

$0 + 11x = 68 + -13$

$11x = 68 + -13$

Combine like terms:
$68 + -13 = 55$
$11x = 55$
Divide each side by 11: $x = 5$

**56.** C $\frac{\$400}{1} \cdot \frac{5}{100} = 4 \cdot 5 = 20$
$\$400 - 20 = \$380$

**57.** A 6:3; 2:1

**58.** C Column b increases by 6 as you move from Column a, which increases by 1.

**59.** C
$$\begin{array}{r} 1.2 \\ \times\ 2.6 \\ \hline 3.12 \end{array}$$

**60.** C 12, 9, 6

**61.** D $\frac{30000}{1} \cdot \frac{15}{100} = 300 \cdot 15 = 4{,}500$

**62.** D 3 shirts, 4 pairs of pants, and 2 pairs of shoes $= 3 \cdot 4 \cdot 2 = 24$

**63.** B $\$1.25 \cdot 5 = \$6.25$

**64.** A Part = % · Whole ÷ 100
Part = (30) (75) ÷ 100 = 22.5
75 – 22.5 = 52.5

**65.** B $d = r \cdot t$
$d = 60$ mph $\cdot\ 3.5 = 210$ miles

**66.** B 3, 7, 20 = 30
$\frac{30}{3} = 10$
$x = 17;\ 17 + 3 = 20$
$\frac{20}{2} = 10$

**67.** D 780.6 rounded = 781; 781.1 rounded = 781; 781.4 rounded = 781; 781.7 rounded = 782

**68.** C $\frac{\$360}{9}$ shares = \$40 per share; $\$40 \cdot 3 = \$120$

**69.** B $\frac{100}{8} = 12.5;\ 8 \cdot .5 = 4;\ 12 + 4 = 16$

**70.** C Part = % · Whole ÷ 100
Part = (7)(80) ÷ 100 = 560 ÷ 100 = 5.6; 80 – 5.6 = 74.40

**71.** A $F = \frac{9}{5}(C) + 32 = \frac{9}{5}(10) + 32 = 18 + 32 = 50$

**72.** D $C = \frac{5}{9}(F - 32) = \frac{5}{9}(77 - 32) = \frac{5}{9}(45) = 25$

**73.** C A right angle measures exactly 90 degrees

**74.** C Two lines are parallel when they lie in the same plane and do not share a common point

**75.** A When two lines intersect to form equal adjacent angles, the lines are perpendicular. The angles formed are right angles (90 degrees)

**76.** A The perimeter is the distance around a two-dimensional shape. It is measured in linear units (millimeters, centimeters, meters, kilometers, inches, feet, yards and so on

**77.** D Definition of volume (amount of space a three-dimensional object displaces)

**78.** C Definition of surface area (sum of the areas of the exposed sides of a three-dimensional object). It has units of distance squared.

**79.** B $d = 2r$

**80.** C Definition of circumference (distance around a circle)

**81.** D $C = 2 \cdot \pi \cdot r$
$C = 2 \cdot 3.14 \cdot 10 \text{ cm} = 62.8 \text{ cm}$

**82.** C $A = \pi \cdot r^2$
$A = 3.14 \cdot (10 \text{ cm})(10 \text{ cm})$
$A = 3.14 \cdot 100 \text{ cm}^2 = 314 \text{ cm}^2$

**83.** B 4 ft +2 ft + 14 ft + 8 ft + 9 ft + 10 ft = 47 ft

**84.** C $P = ns = P = (6)(7 \text{ cm}) = 42 \text{ cm}$

**85.** A $A = b \cdot h$
$200 \text{ in}^2 = 20 \text{ in} \cdot h$; divide by 20 in on both sides of the equation
$10 \text{ in} = h$

**86.** B $P = 4s$; $P = (4)(5 \text{ in}) = 20 \text{ in}$

**87.** D $A = s^2$
$A = 5 \text{ in}^2 \ = 25 \text{ in}$

**88.** C $P = 2(l) + 2(w)$
$P = 2(24 \text{ cm}) + 2(14 \text{ cm}) = 48 \text{ cm} + 28 \text{ cm} = 76 \text{ cm}$

**89.** D The sum of the interior angles of a triangle add up to a straight angle or 180 degrees

**90.** B Scalene

**91.** A An acute triangle has all three interior angles less than 90 degrees

**92.** C An equilateral triangle is a triangle having three equal 60 degree angles. It also has sides of equal length.

**93.** B $P = a + b + c$
$P = 5 \text{ in} + 8 \text{ in} + 11 \text{ in} = 24 \text{ in}$

**94.** A $A = \frac{1}{2}\, b \cdot h$
$A = \frac{1}{2}(12 \text{ in}) \cdot 6 \text{ in} = 6 \text{ in} \cdot 6 \text{ in} = 36 \text{ in}^2$

**95.** B $V = L^3$
$V = (9 \text{ cm})^3 = 729 \text{ cm}^3$

**96.** C $2ab + 2bc + 2ac = 2(16 \text{ in})(6 \text{ in}) + 2(6 \text{ in})(8 \text{ in}) + 2(16 \text{ in})(8 \text{ in}) =$
$192 \text{ in}^2 \quad + 96 \text{ in}^2 \quad + 256 \text{ in}^2 \quad = 544 \text{ in}^2$

OR

$2(lw) + 2(wh) + 2(lh) = 2(16 \text{ in})(6 \text{ in}) + 2(6 \text{ in})(8 \text{ in}) + 2(16 \text{ in})(8 \text{ in})$
$= 192 \text{ in}^2 + 96 \text{ in}^2 \quad + 256 \text{ in}^2 = 544 \text{ in}^2$

**97.** D $V = \pi \cdot r^2 \cdot h = 3.14 \cdot (3 \text{ cm})^2 \cdot 7 \text{ cm} = 3.14 \cdot 9 \text{ cm}^2 \cdot 7 \text{ cm} = 197.82 \text{ cm}^3$

**98.** A $V = \frac{4}{3}\pi \cdot r^3 = \frac{4}{3} \cdot 3.14 \cdot (9 \text{ in})^3 = 1.33 \cdot 3.14 \cdot 729 \text{ in}^3 = 3053 \text{ in}^3$

**99.** D $V = \frac{1}{3}\pi \cdot r^2 \cdot h = .33 \cdot 3.14 \cdot 13 \text{ mm}^2 \cdot 30 \text{ mm} = 5256 \text{ mm}^3$

**100.** C $V = \frac{1}{3} \cdot A \cdot h \qquad [A = 6 \text{ cm}^2 = 36 \text{ cm}^2]$
$V = .33 \cdot 36 \text{ cm}^2 \cdot 9 \text{ cm} = 106.92$, rounded off to $107 \text{ cm}^3$

CHAPTER 7

# Charts, Tables, Graphs, and Diagrams

**Your Goals for This Chapter**

- Understand pie chart design concepts
- Examine table and graph display information
- Review Venn diagram concepts
- Practice with sample pie chart, table, graph, and Venn diagram questions

This chapter will examine the relationship among pie charts, tables, graphs, and Venn diagrams. It will review and help explain the similarities and differences. Each visual format is useful for showcasing data depending upon what kind of information is being evaluated. The various visual representation techniques will also help you select the most appropriate data display when giving a presentation. Review questions test your ability to interpret the information correctly. The answers to the questions are given at the end of the chapter.

## PIE CHARTS

A **pie chart** is in the shape of a circle. It is a visual design that represents a whole with each segment portrayed as a sector or slice. It is a graphical way to divide items into segments and to compare parts to a whole. Additionally, in a pie chart, the arc length of each sector is proportional to the quantity it represents. A full circle, for example, has 360 degrees. Each sector, therefore, would represent a certain proportion of the circle. The following pie chart questions will provide insight into pie chart design concepts.

## Review Questions

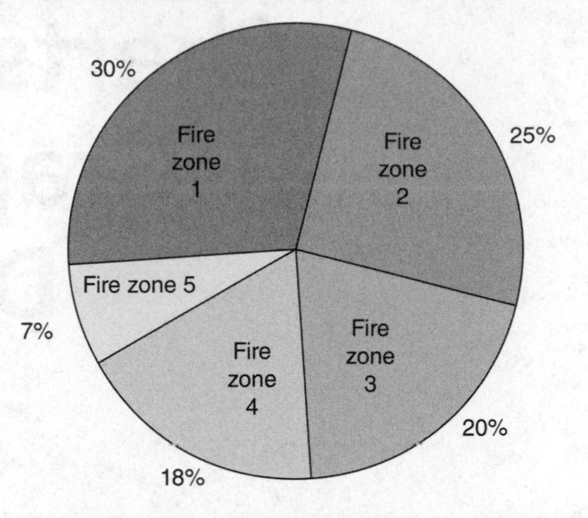


**DISTRIBUTION OF FIRES: PAXTON TOWNSHIP**

**1.** According to the pie chart above, what three fire zones combined had half the total number of fires in Paxton Township?

(A) Zone 1, Zone 4, and Zone 5
(B) Zone 2, Zone 4, and Zone 5
(C) Zone 3, Zone 4, and Zone 5
(D) None of the above

**2.** If the total number of fires in Paxton Township was 620, how many fires were in Zone 3?

(A) 108
(B) 116
(C) 124
(D) 130

**3.** What two fire zones combined represent a quarter of all fires inside Paxton Township?

(A) Zone 1 and Zone 3
(B) Zone 3 and Zone 4
(C) Zone 2 and Zone 5
(D) Zone 4 and Zone 5

**4.** Using the information in question 2, how many fires were in Zone 1?

(A) 186
(B) 180
(C) 177
(D) 168

**5.** If there were 330 fires in Zone 2, how many total fires would there be in Paxton Township?

(A) 1,280
(B) 1,300
(C) 1,320
(D) 1,375

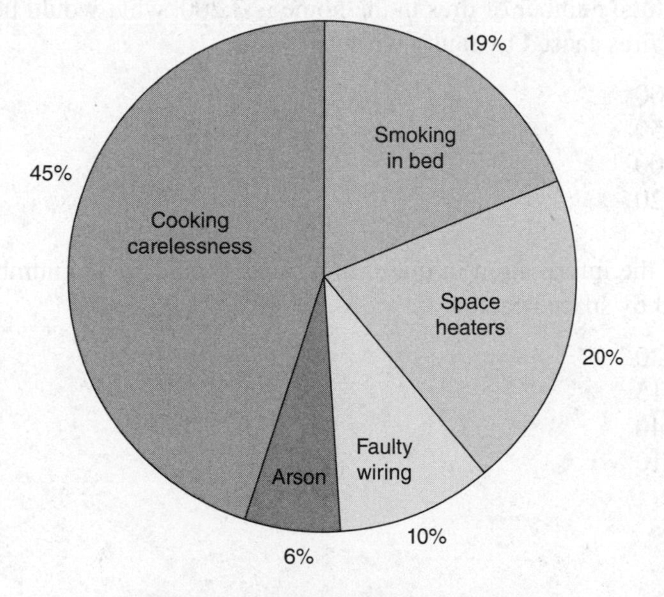


**FIRES IN THE HOME (COMMON CAUSES)**

**1.** Based on the pie chart above, what percentages of fires in the home are deemed incendiary?

(A) 25
(B) 20
(C) 10
(D) 6

**2.** The majority of the fires listed are started due to

(A) fire code violations
(B) cooking carelessness
(C) malicious actions
(D) cold weather

**3.** Based upon the percentages for fire, a fire extinguisher is needed most in what room of the house?

(A) Kitchen
(B) Bedroom
(C) Den
(D) Living room

**4.** Which of the following pairs of common causes of fire in the home accounts for the greatest percentage of fires?

(A) Arson and faulty wiring
(B) Faulty wiring and space heaters
(C) Arson and smoking in bed
(D) Faulty wiring and smoking in bed

**5.** If the total number of fires in the home is 3,200, what would be the number of fires caused by faulty wiring?

(A) 100
(B) 250
(C) 300
(D) 320

**6.** Using the information in question 5, what would be the number of fires caused by space heaters?

(A) 730
(B) 715
(C) 640
(D) 610

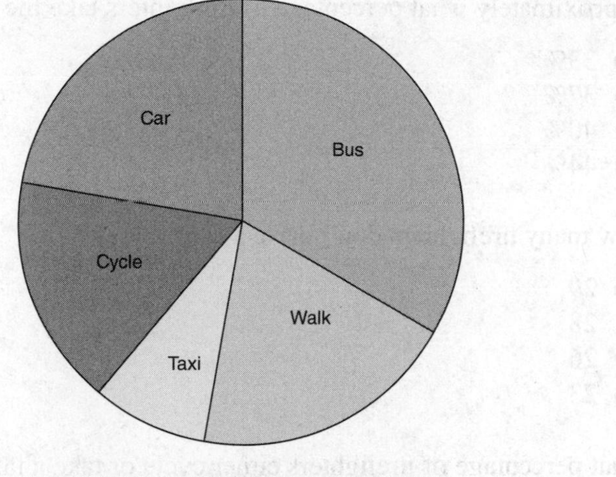

**MEANS OF TRAVEL TO THE FIREHOUSE**

Firefighters at a firehouse use many means of travel to get to work. View the pie chart above and legend below to answer questions 1 through 6.

**Legend**

Car = 8

Bus = 12

Walk = 7

Taxi = 3

Cycle = 6

**1.** What is the total number of firefighters at the firehouse?

(A) 35
(B) 36
(C) 38
(D) 40

**2.** What is the least common mode of travel?

(A) Cycle
(B) Walk
(C) Taxi
(D) Bus

**3.** What fraction of firefighters cycle to work?

(A) 1/3
(B) 1/4
(C) 1/5
(D) 1/6

**4.** Approximately what percentage of firefighters take the bus to work?

(A) 33%
(B) 37%
(C) 40%
(D) 44%

**5.** How many firefighters don't drive a car to work?

(A) 29
(B) 28
(C) 26
(D) 23

**6.** What percentage of firefighters either cycle or take a taxi to work?

(A) 30%
(B) 28%
(C) 25%
(D) 22%

# TABLES

A **table** is a means of arranging numbers or data in rows and columns. Tables are pervasive throughout the fire service to help firefighters better understand statistical information within reports. They are ordered visual displays of data that enhance analysis. The following questions using tables provide examples on how this is accomplished.

# Review Questions

| **Winchester County: Annual Arson Statistics** | | |
|---|---|---|
| **Town** | **Arson Fires** | **Status*** |
| Quincy | 19 | Level H, 3 FM |
| Chase | 2 | Level L, 1 FM |
| Hampton | 11 | Level M, 2 FM |
| Douglas | 8 | Level M, 2 FM |
| Melville | 22 | Level H, 3 FM |
| Tanner | 8 | Level M, 2 FM |
| *L = low; M = moderate; H = high; FM = fire marshal. | | |

**1.** According to the table above, how many total arson fires were there in Winchester County?

(A) 60
(B) 70
(C) 80
(D) 85

**2.** What three towns combined had half the arson fires in Winchester County?

(A) Douglas, Melville, and Tanner
(B) Quincy, Chase, and Hampton
(C) Chase, Hampton, and Melville
(D) Hampton, Melville, and Tanner

**3.** How many fire marshals (FM) were assigned to Level M towns?

(A) 27
(B) 13
(C) 7
(D) 6

**4.** What is the total number of fire marshals assigned?

(A) 13
(B) 11
(C) 9
(D) 70

**5.** How many arson fires were there in Level H towns?

(A) 41
(B) 45
(C) 48
(D) 51

**6.** What percentage of fires in Winchester County occurred in Level H towns?

(A) 48%
(B) 58%
(C) 64%
(D) 66%

| **Victory City: Comparison of Fire Companies** | | | |
|---|---|---|---|
| **Fire Companies** | **1990** | **2005** | **% Change** |
| Engines | 20 | 25 | +25 |
| Ladders | 12 | 18 | +50 |
| Marine | 4 | 2 | −50 |
| Squad/rescue | 10 | 8 | −20 |
| **Totals** | 46 | 53 | +14.5 |

**1.** On the basis of the figures given in the table above, it can be accurately stated that between 1990 and 2005:

(A) The number of squad/rescue companies increased.
(B) Engine companies increased by a greater percentage than ladder companies.
(C) Marine companies suffered the highest percentage of cuts.
(D) More marine companies were eliminated than squad/rescue companies.

**2.** The total increase in fire companies from 1990 to 2005 is a result of

(A) the increase in engine companies
(B) the increase in ladder companies
(C) the decrease in marine companies
(D) increases and decreases in fire companies

3. In 2005, what type of fire company represented approximately 33 percent of the total number of companies?

   (A) Engines
   (B) Ladders
   (C) Marine
   (D) Squad/rescue

4. In 1990, engines were five times as numerous as

   (A) marine companies
   (B) ladder companies
   (C) squad/rescue companies
   (D) None of the above

5. What was the total number of engines and ladders added to the Victory City department from 1990 to 2005?

   (A) 25
   (B) 18
   (C) 11
   (D) 8

# GRAPHS

A **line graph** visually summarizes how two items of information are related. The numbers along the side of the line graph are called the scale.

A **bar graph** consists of an axis and vertical or horizontal bars that represent items of information. The bars denote various values for each item. The numbers along the side of a bar graph are also called the scale.

## Review Questions

Review the following line and bar graph questions to become more familiar with these two visual design concepts.

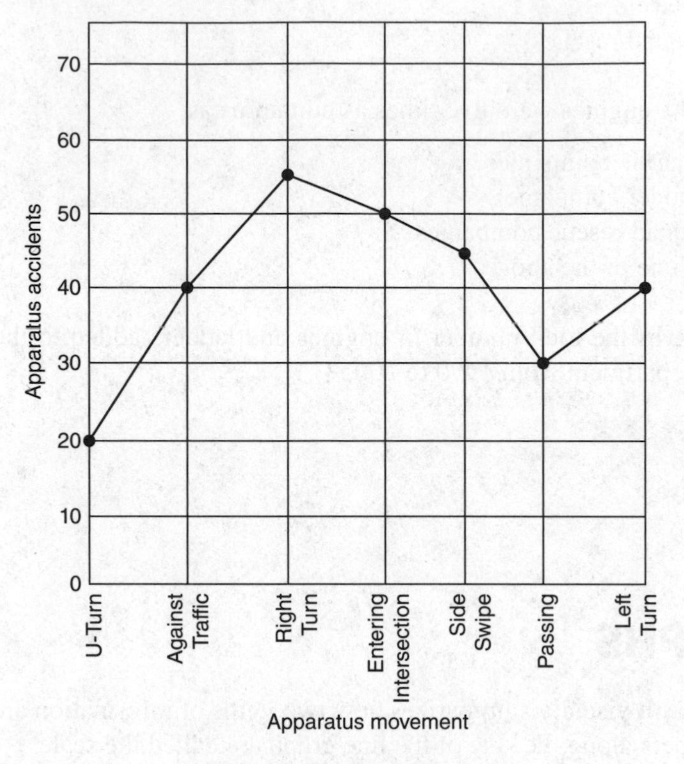


**SAFETY BATTALION STATISTICS**

1. According to the line graph above, what kind of apparatus maneuver is most prone to being involved in an accident?

   (A) U-turn
   (B) Right turn
   (C) Entering an intersection
   (D) Left turn

2. The number of accidents occurring while entering an intersection is equal to the number of accidents occurring during

   (A) both right turn and U-turn
   (B) both left turn and side swipes
   (C) both U-turn and passing
   (D) both left turn and against traffic

3. What two apparatus movements have the same number of accidents?

   (A) Against traffic and left turn
   (B) U-turn and passing
   (C) Right turn and side swipe
   (D) Entering an intersection and against traffic

**4.** The total number of apparatus accidents is

(A) 300
(B) 290
(C) 280
(D) 270

**5.** How many more right turn accidents were there than U-turn accidents?

(A) 25
(B) 35
(C) 45
(D) 50

**6.** What percentage of total accidents was against traffic?

(A) 20%
(B) 18%
(C) 14%
(D) 12%

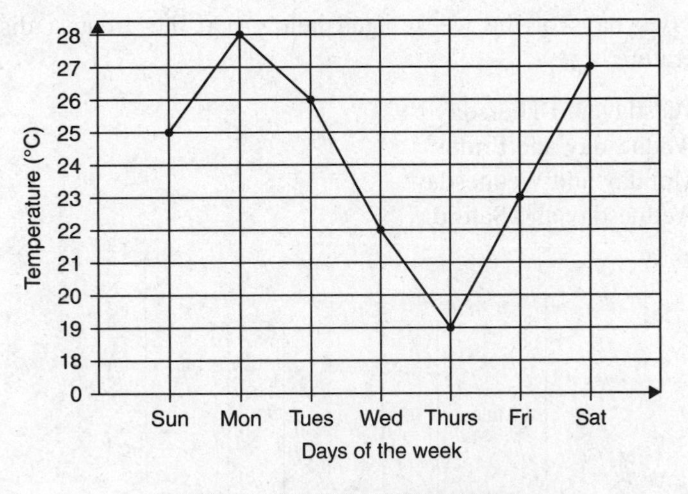


**DAYS OF THE WEEK: TEMPERATURE**

**1.** According to the line graph above, what day of the week had the highest temperature reading?

(A) Sunday
(B) Monday
(C) Tuesday
(D) Wednesday

**2.** What temperature reading was the lowest recorded for the week?

(A) 28
(B) 26
(C) 22
(D) 19

3. What part of the week had the highest recorded temperatures?

(A) Beginning of the week
(B) End of the week
(C) Middle of the week
(D) All of the above

4. What was the average temperature for the last three days of the week?

(A) 19
(B) 21
(C) 23
(D) 27

5. What is the approximate average temperature for the week?

(A) 26
(B) 24
(C) 21
(D) 20

6. What two days of the week listed below had the greatest disparity in temperature?

(A) Tuesday and Thursday
(B) Wednesday and Friday
(C) Monday and Wednesday
(D) Wednesday and Saturday

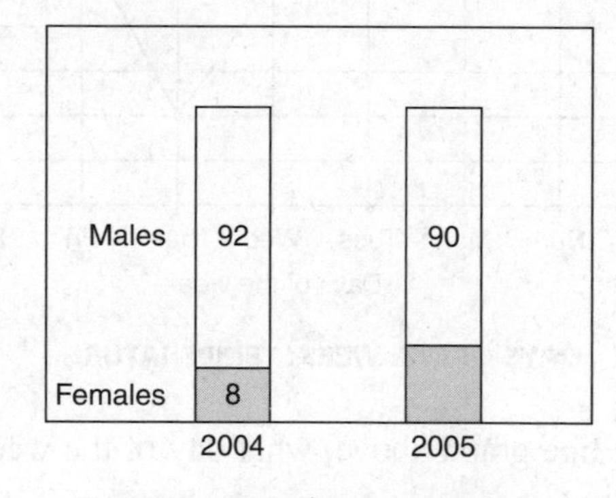


**FIRE ACADEMY RECRUITS**

1. According to the bar graph above, the total number of new recruits entering the fire academy was the same in 2004 and 2005. What was the percent increase for female enrollment from 2004 to 2005?

(A) 2%
(B) 8%
(C) 25%
(D) 20%

2. What percent decrease in male recruits did the fire academy have from 2004 to 2005?

(A) 2.1%
(B) 3.3%
(C) 3.7%
(D) 4.2%

3. What was the total number of female recruits during 2004 and 2005?

(A) 8
(B) 10
(C) 18
(D) 20

4. In 2006, the same number of recruits is brought into the fire academy. The number of female recruits, however, increases to 14. How many male recruits would there be?

(A) 88
(B) 86
(C) 84
(D) 82

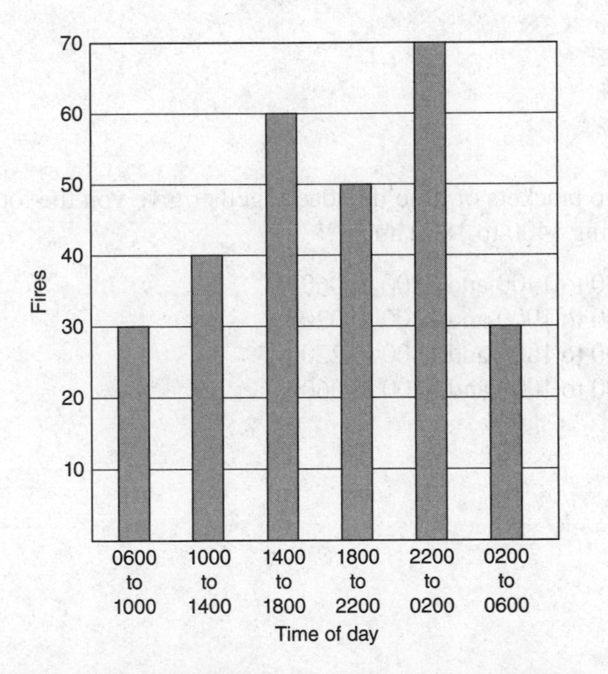


**BUREAU OF OPERATIONS: MONTHLY ACTIVITY REPORT**

1. According to the bar graph above, the most fires occur during what time of day?

(A) Morning (0600–1000)
(B) Noon time (1000–1400)
(C) Late afternoon (1400–1800)
(D) Late night (2200–0200)

**2.** The total number of fires for a 24-hour period is most nearly

(A) 300
(B) 280
(C) 270
(D) 250

**3.** During which time period have the fewest fires occurred?

(A) 6 p.m. to 6 a.m. (1800–0600)
(B) 2 p.m. to 2 a.m. (1400–0200)
(C) 6 a.m. to 6 p.m. (0600–1800)
(D) 10 a.m. to 10 p.m. (1000–2200)

**4.** How many fires occurred from 6 p.m. to 10 a.m.?

(A) 180
(B) 190
(C) 200
(D) 240

**5.** What percentage of fires occurred between the hours of 2200 and 0200 hours?

(A) 22%
(B) 25%
(C) 28%
(D) 29%

**6.** What two brackets of time if added together give you the total number of fires during 1400 to 1800 hours?

(A) 1000 to 1400 and 0200 to 0600
(B) 0600 to 1000 and 2200 to 0200
(C) 1400 to 1800 and 1800 to 2200
(D) 0600 to 1000 and 0200 to 0600

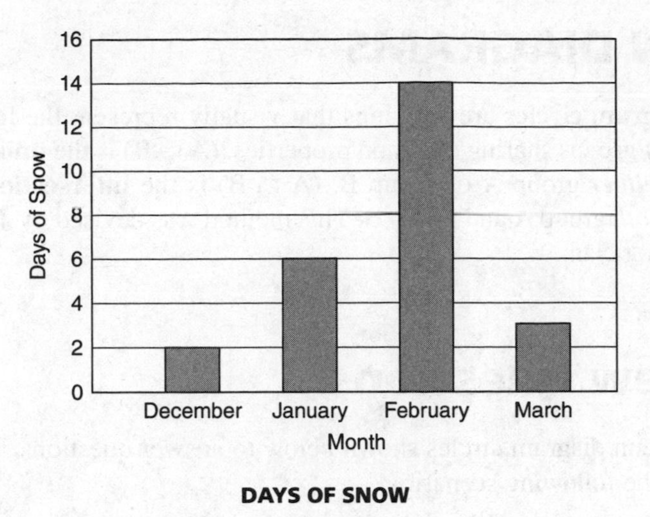


**DAYS OF SNOW**

**1.** According to the bar graph above, what month had the most snow days?

(A) February
(B) March
(C) December
(D) January

**2.** How many total snow days were there during the four-month period?

(A) 22
(B) 25
(C) 28
(D) 30

**3.** Which two months had the greatest discrepancy in snow days?

(A) December and March
(B) January and February
(C) February and March
(D) December and February

**4.** What percentage of snow days for the four-month period were in January?

(A) 18%
(B) 20%
(C) 24%
(D) 30%

**5.** If the total number of snow days for the year was 30, what percentage of snow days is represented by the four months on the bar graph?

(A) 84.6%
(B) 83.3%
(C) 81.2%
(D) 79.7%

# VENN DIAGRAMS

**Venn diagram circles** are drawings that visually represent the logical relationship of groups sharing common properties. ($A \cup B$) is the **union** of items found in *either* group A or group B. ($A \cap B$) is the **intersection** of items found in *both* group A and group B. This method was devised by **John Venn**, a British logician.

# Review Questions

Use the Venn diagram circles shown below to answer questions 1 through 4 based on the following scenario:

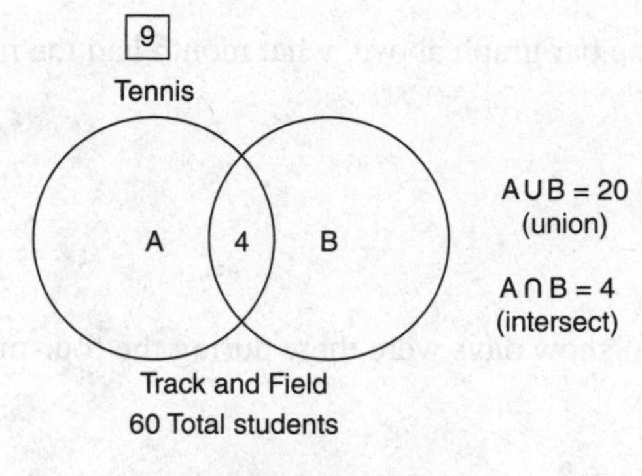


**BOARDING SCHOOL**

At a small boarding school with an enrollment of 60 students, 9 students are currently on the tennis team and 4 students are taking both tennis and track and field ($A \cap B$). The union of sets A and B ($A \cup B$) is 20 students.

**1.** How many students are only playing tennis?

(A) 10
(B) 9
(C) 5
(D) 4

**2.** How many students are only participating in track and field?

(A) 10
(B) 11
(C) 12
(D) 20

**3.** What is the total number of students participating in track and field?

(A) 15
(B) 16
(C) 20
(D) 21

**4.** How many students are not participating in either tennis or track and field?

(A) 15
(B) 25
(C) 30
(D) 40

Use the Venn diagram circles shown below to answer questions 1 through 4 based on the following scenario.

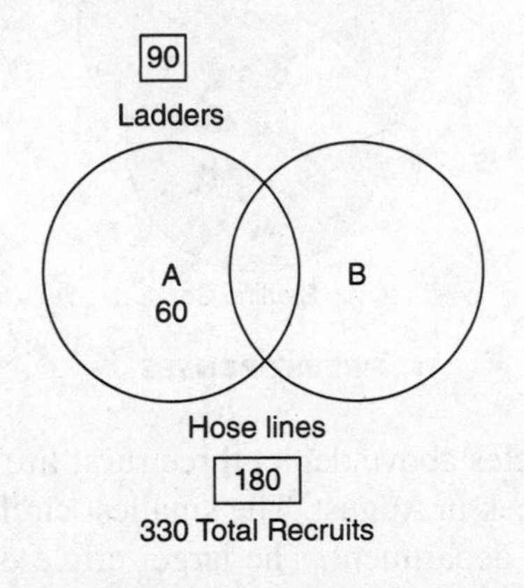


**FIRE DEPARTMENT TRAINING ACADEMY**

There are 330 recruits at the fire department training academy. Training includes ladder placement and hose line stretching. Ninety recruits are using ladders while 180 are carrying hose lines. Sixty recruits are only using ladders.

**1.** How many recruits are working with both ladders and hose lines ($A \cap B$)?

(A) 30
(B) 60
(C) 90
(D) 100

**2.** How many recruits are only working with hose lines?

(A) 180
(B) 150
(C) 130
(D) 100

**3.** How many recruits are working in at least one of the ladder or hose line training components ($A \cup B$)?

(A) 90
(B) 180
(C) 210
(D) 240

**4.** How many recruits are not involved in working with ladders or hose lines?

(A) 60
(B) 90
(C) 100
(D) 105

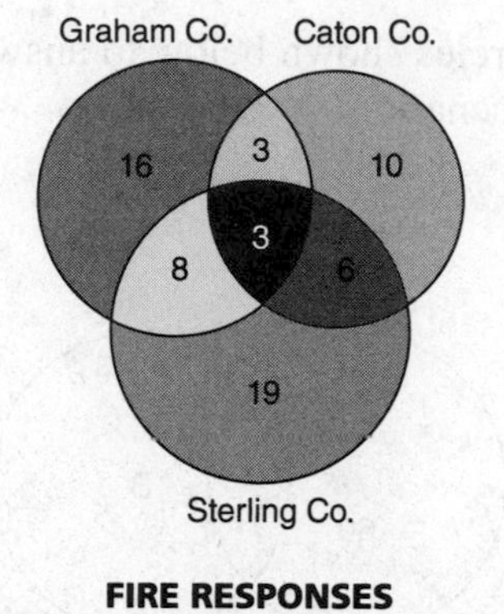


**FIRE RESPONSES**

The Venn diagram circles above depict three rural fire departments and their fire responses for a week in August. The smallest circle on the upper right is the Caton County fire department. The larger circle on the upper left is the Graham County fire department. The largest circle at the bottom represents the Sterling County fire department. Overlapping sections denote fire responses where two or all three county fire departments responded in a joint mutual aid effort. Answer questions 1 through 8 based upon the Venn diagram.

**1.** What county fire department had the most fire responses for the week?

(A) Graham
(B) Sterling
(C) Caton
(D) Tie between Caton and Sterling

**2.** How many fire responses did the Graham Co. fire department have for the week?

(A) 34
(B) 32
(C) 30
(D) 28

**3.** What was the total number of fire responses where just the Graham and Caton County fire departments responded together?

(A) 3
(B) 6
(C) 9
(D) 12

**4.** How many fire responses involved all three counties?

(A) 3
(B) 6
(C) 8
(D) 9

**5.** What was the total number of single responses for all three county fire departments combined?

(A) 35
(B) 45
(C) 55
(D) 58

**6.** How many fire responses involved only two county fire departments?

(A) 6
(B) 8
(C) 10
(D) 34

**7.** How many total fire response incidents were there involving any of the three county fire departments for the week?

(A) 55
(B) 60
(C) 65
(D) 88

**8.** What county fire department had the least mutual aid fire responses for the week?

(A) Sterling
(B) Caton
(C) Graham
(D) All of the above

# Answer Key

## CHARTS, TABLES, GRAPHS, AND DIAGRAMS ANSWER KEY

### Charts

**Distribution of Fires: Paxton Township**

1. B
2. C
3. D
4. A
5. C

**Fires in the Home (Common Causes)**

1. D
2. B
3. A
4. B
5. D
6. C

**Means of Travel to the Firehouse**

1. B
2. C
3. D
4. A
5. B
6. C

### Tables

**Winchester County Annual Arson Statistics**

1. B
2. C
3. D
4. A
5. A
6. B

**Victory City: Comparison of Fire Companies**

1. C
2. D
3. B
4. A
5. C

# Graphs

### Safety Battalion Statistics

1. B
2. C
3. A
4. C
5. B
6. C

### Days of the Week: Temperature

1. B
2. D
3. A
4. C
5. B
6. A

### Fire Academy Recruits

1. C
2. A
3. C
4. B

### Bureau of Operations: Monthly Activity Report

1. D
2. B
3. C
4. A
5. B
6. D

### Days of Snow

1. A
2. B
3. D
4. C
5. B

# Venn Diagrams

### Boarding School

1. C
2. B
3. A
4. D

### Fire Department Training Academy

1. A
2. B
3. D
4. B

### Fire Responses

1. B
2. C
3. A
4. A
5. B
6. D
7. C
8. B

# ANSWER EXPLANATIONS

Following you will find explanations to specific questions that you might find helpful.

## Charts

### Distribution of Fires: Paxton Township

**1.** B Zone 2 = 25%, Zone 4 = 18%, Zone 5 = 7% = 50%

**2.** C What is 20% of 620?
Part = % · Whole ÷ 100
Part = (20)(620) ÷ 100
Part = 12,400 ÷ 100 = 124

**3. D** Zone 4 = 18% and Zone 5 = 7% = 25%

**4. A** What is 30% of 620?
Part = % · Whole ÷ 100
Part = (30)(620) ÷ 100 = 18,600 ÷ 100 = 186

**5. C** Zone 2 = 330; Zone 2 = 25%
330 · 4 = 1320

### Fires in the Home (Common Causes)

**2.** B Cooking carelessness = 45%

**4.** B Faulty wiring and space heaters = 30%
Arson and faulty wiring = 16%
Arson and smoking in bed = 25%
Faulty wiring and smoking in bed = 29%

**5.** D Total number of fires = 3,200; faulty wiring = 10%
What is 10% of 3,200?
Part = % · Whole ÷ 100
Part = 10 · 3,200 ÷ 100 = 32,000 ÷ 100 = 320

**6.** C Space heaters = 20%
What is 20% of 3,200?
Part = % · Whole ÷ 100
Part = 20 · 3,200 ÷ 100 = 64,000 ÷ 100 = 640

### Means of Travel to the Firehouse

**1.** B Car = 8, Bus = 12, Walk = 7, Taxi = 3, Cycle = 6
8 + 12 + 7 + 3 + 6 = 36

**2.** C Taxi = 3

**3.** D 6 firefighters cycle/36 firefighters = 6/36 = 1/6

**4.** A What % of 36 is 12?
Percent = Part ÷ Whole · 100
Percent = 12 ÷ 36 · 100 = .33 · 100 = 33

5. B Bus = 12, Walk = 7, Taxi = 3, Cycle = 6 = 28

6. C Cycle = 6, Taxi = 3 = 9
What percent of 36 is 9?
Percent = Part ÷ Whole · 100
Percent = 9 ÷ 36 · 100 = .25 · 100 = 25

# Tables

**Winchester County Annual Arson Statistics**

1. B Quincy = 19, Chase = 2, Hampton = 11, Douglas = 8, Melville = 22, Tanner = 8
19 + 2 + 11 + 8 + 22 + 8 = 70

2. C Chase = 2, Hampton = 11, Melville = 22 = 35; 35/70 = 1/2

3. D Hampton = 2, Douglas = 2, Tanner = 2 = 6

4. A Quincy = 3, Chase = 1, Hampton = 2, Douglas = 2, Melville = 3, Tanner = 2 = 13

5. A Quincy = 19, Melville = 22 = 41

6. B Total Arson Fires, 70; Quincy = 19, Melville = 22 = 41
What percent of 70 is 41?
Percent = Part ÷ Whole · 100
Percent = 41 ÷ 70 · 100 = .58 · 100 = 58

**Victory City: Comparison of Fire Companies**

1. C A decreased; B lesser percentage; D less

3. B Engines = 25, Ladders = 18, Marine = 2, Squad/rescue = 8 = 53
What percent of 53 is 18 (Ladders)?
Percent = Part ÷ Whole · 100 = 18 ÷ 53 · 100 = .33 · 100 = 33

4. A 1990 Engines = 20; Marine = 4
20/4 = 5

5. C 1990 to 2005, Engines 20 to 25 = +5; Ladders 12 to 18 = + 6 = 11

# Graphs

**Safety Battalion Statistics**

1. B Right turn = 55 apparatus accidents

2. C Entering intersection = 50
U-turn = 20 + Passing = 30 = 50

3. A Against traffic = 40; Left turn = 40

4. C U-turn = 20, Against traffic = 40, Right turn = 55, Entering intersection = 50, Side swipe = 45, Passing = 30, Left turn = 40 = 280 total accidents

**5.** B Right turn = 55, U-turn = 20. 55 – 20 = 35

**6.** C Total number of accidents = 280; Against traffic = 40
What percent of 280 is 40?
Percent = Part ÷ Whole · 100 = 40 ÷ 280 · 100 = .14 · 100 = 14

### Days of the Week: Temperature

**1.** B Monday = 28°C

**2.** D 19°C

**4.** C Thursday = 19, Friday = 23, Saturday = 27 = 69 ÷ 3 = 23°C

**5.** B Sunday = 25, Monday = 28, Tuesday = 26, Wednesday = 22, Thursday = 19, Friday = 23, Saturday = 27 = 170 ÷ 7 = 24.2°C (approximately 24°C)

**6.** A Tuesday = 26, Thursday 19 = 7°C difference
Wednesday = 22, Friday = 23 = 1°C difference
Monday = 28, Wednesday = 22 = 6°C difference
Wednesday = 22, Saturday = 27 = 5°C difference

### Fire Academy Recruits

**1.** C Total number or recruits entering fire academy was the same for 2004 and 2005.
2004 = 92 males, 8 females; 2005 = 10 females; 2 female increase
2 ÷ 8 = 0.25 · 100 = 25% increase

**2.** A 2004 = 92 males; 2005 = 90 males; 2 male decrease
2 ÷ 92 = 0.021 · 100 = 2.1% decrease

**3.** C 2004 = 8 females, 2005 = 10 females = 18

**4.** B Total number of recruits in 2006 = 100
Females increase from 10 to 14; males, therefore, decrease from 90 to 86.

### Bureau of Operations: Monthly Activity Report

**1.** D Late night 70
A Morning 30; B Noon time 40; C Late afternoon 60

**2.** B Morning 30, Noon time 40, Late afternoon 60, 1800–2200 50, Late night 70, 0200–0600 30
30 + 40 + 60 + 50 + 70 + 30 = 280

**3.** C 0600–1800 = 130
A 1800–0600 = 150; B 1400–0200 = 180; D 1000–2200 = 150

**4.** A 1800–1000 = 180
1800–2200 = 50, 2200–0200 = 70, 0200–0600 = 30, 0600–1000 = 30 =180

**5.** B 2200–0200 = 70
What percent of 280 is 70?
Percent = Part ÷ Whole · 100
Percent = 70 ÷ 280 · 100 = 0.25 · 100 = 25

**6.** D 1400–1800 = 60
0600–1000 = 30 and 0200–0600 = 30 = 60

### Days of Snow

**1.** A February 14
December 2, January 6, March 3

**2.** B 2 + 6 + 14 + 3 = 25

**3.** D December 2 and February 14

**4.** C January = 6
What percent of 25 is 6?
Percent = Part ÷ Whole · 100
Percent = 6 ÷ 25 · 100 = 0.24 · 100 = 24

**5.** B Total number of snow days for year = 30
What percent of 30 is 25?
Percent = Part ÷ Whole · 100
Percent = 25 ÷ 30 · 100 = .833 · 100 = 83.3

# Venn Diagrams

### Boarding School

**1.** C Nine students are on the tennis team, and four students are taking both tennis and track and field.
Students only playing tennis = 9 – 4 = 5

**2.** B Students only participating in track and field
($A \cup B$) is 20. 20 – 9 = 11.

**3.** A ($A \cup B$) is 20; A = 5 + B = X = 20
B = 15

**4.** D Total number of students = 60
($A \cup B$) is 20; 60 – 20 = 40

### Fire Department Training Academy

**1.** A Ninety recruits are using ladders, of which 60 are only using ladders; this means that 30 are using both ladders and hose lines ($A \cap B$).

**2.** B ($A \cap B$) = 30; 180 – 30 = 150

**3.** D ($A \cup B$) = 60 + 180 = 240

**4.** B 330 – 240 = 90

### Fire Responses

**1.** B Sterling 19

**2.** C Graham 16 + 8 + 3 + 3 = 30

**3.** A Graham and Caton = 3

**4.** A  Graham, Sterling, and Caton = 3

**5.** B  Sterling 19, Graham 16, Caton 10 = 45

**6.** D  Graham 8 + 3 = 11; Sterling 8 + 6 = 14; Caton 3 + 6 = 9 = 34

**7.** C  Sterling 19, Graham 16, Caton 10 = 45; Combinations 8 + 3 + 3 + 6 = 20
45 + 20 = 65

**8.** B  Caton 12
Sterling 17, Graham 14

CHAPTER 8

# Memory and Visualization

### Your Goals for This Chapter

- Find out how to answer memorization questions
- Learn how to answer visualization questions
- Test your memorization and visualization skills with review questions

Firefighter exams test the candidate's ability to memorize and recall information and to visualize and analyze information. In typical memorization questions, the candidate is presented with a fireground pictorial scene to be studied for a short period of time, generally three to five minutes. Then, the pictorial is taken away and the test taker is asked questions about the scene. In visualization questions, the candidate is asked to view maps, sketches, and drawings and then to analyze the information shown to answer directional or orientation perspective questions.

## MEMORIZATION QUESTIONS

In some memorization questions a building or row of buildings of different heights and dimensions fronting on one or more streets or avenues is shown. A **fire scenario** is usually depicted through the roof or inside a window, and people are shown in distress in various parts of the building. A **legend** is usually provided to explain what some of the important symbols represent (occupants, victims, staircase, doorways, windows, fire escape, detectors, fire hydrant, fire department connection, alarm box). Firefighters, fire apparatus, fire suppression systems (sprinkler and standpipe systems), infrastructure (telephone poles, electrical lines, lampposts, and traffic lights), pedestrians, compass directions, and wind direction and speed may also be included in the pictorial.

Other memory problems involve studying **building layouts** and **apartment floor plans**. Key features to focus on when looking at building layouts is the numbering and lettering system used to denote floors, apartments, stairways, and elevator banks. Also note surrounding properties, entrances and exits, fire escapes, roof obstructions, security gates or window bars, and fire hydrants. When looking at a fire situation superimposed on an apartment floor layout, look for people or victims inside the apartment and try to determine the best way for firefighters to rescue the trapped occupants—through windows, an entrance door, or stairs. Also note in what room the fire is located, the sequence of rooms, where smoke and carbon monoxide detectors are located, if and where there are fire escapes, and any secondary access and egress routes. It is helpful to become familiar with typical building layouts and floor plans such as access roadways, parking lots, courtyards, hallways, elevator banks, staircases, doorways, passageways, foyers, closets, kitchen fixtures, and so on. Architecture and building construction books can be helpful in doing this.

## VISUALIZATION QUESTIONS

Some visualization questions involve the use of maps and roadway grids in an urban setting. The candidate is allowed to use the schematic when answering the questions. Various structures and occupancies, including the local firehouse, are usually marked on the schematic. Generally, questions involve situations requiring a fire apparatus to leave the firehouse and get from point A to point B quickly and without breaking any traffic laws, but some questions will involve a response to a structure or occupancy while the firefighters are out of the firehouse.

The first thing the candidate should look for is the directional indicator on the map or road grid. Most times, north will be drawn at the top of the page, but not always, and you may have to rotate the map or grid to get a better understanding of the directions. Remember, once you locate north, you then know all other directions. Street direction will normally be denoted by arrows.

Visualization questions also examine how well you can orient yourself to seeing objects and structures from different observational positions (rear, side, interior, aerial). Note the spatial relationship between windows and doors when studying a structure. Are the windows above, below, or even with the top of the door? Look at the shape of all exterior components. How many panes do the windows have? Are the panes square or rectangular? Is there an even number of windows on both sides of the door? Is the door situated in the middle of the structure or off to the right or left side? Is the doorknob located on the right or left side? Examine roof design and projections and where they are located. Review the surrounding properties and grounds (carport, garage, fences). What is their spatial relationship to the main structure?

# EXAMPLES OF MEMORIZATION QUESTIONS

**Directions:** Memorize the fireground situation below for three minutes. At the end of the allotted time, cover the pictorial and answer the following questions (1–16). Circle the letter of your choice.

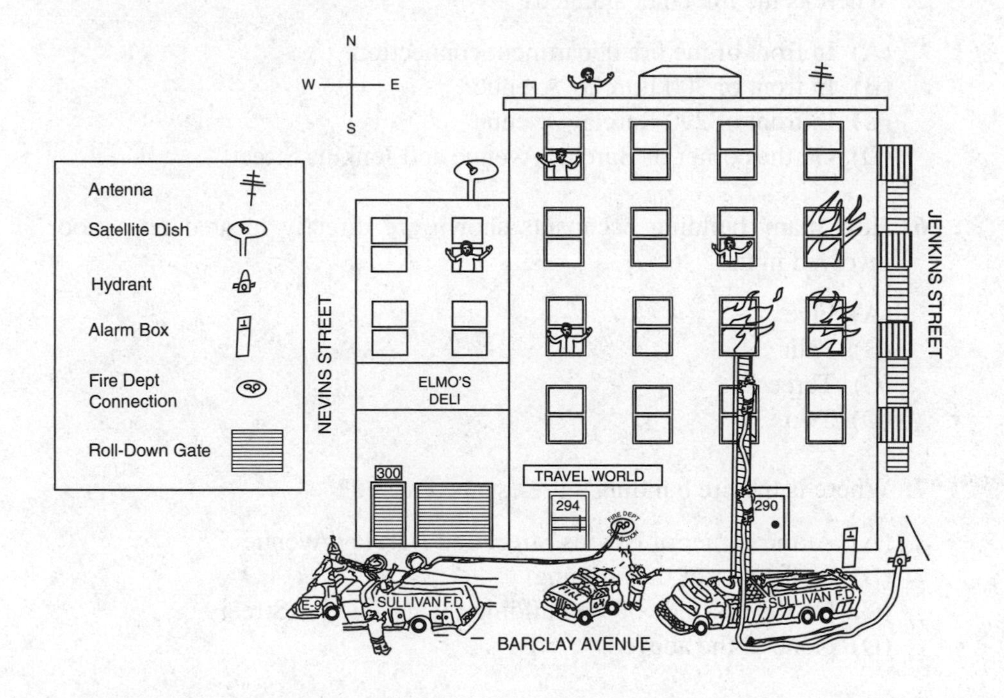


**1.** How many people inside the fire building are in need of rescue?

(A) Two
(B) Three
(C) Five
(D) Six

**2.** What is the name of the fire department at the scene?

(A) Singleton
(B) Sultan
(C) Melvan
(D) Sullivan

**3.** How many firefighters are on the aerial ladder?

(A) One
(B) Two
(C) Three
(D) Four

4. What are the firefighters doing on the aerial ladder?

(A) Assisting a building occupant down the ladder
(B) Stretching a hose line
(C) Ventilating a window
(D) Performing a rope rescue

5. Where is the fire chief situated?

(A) In front of the fire department connection
(B) In front of 300 Barclay Avenue
(C) In front of 290 Barclay Avenue
(D) On the corner of Barclay Avenue and Jenkins Street

6. How many building occupants shown are directly located on a floor involved in fire?

(A) Five
(B) Four
(C) Three
(D) Two

7. Where is the fire building's fire escape located?

(A) At the corner of Nevins Street and Barclay Avenue
(B) At the rear of the building
(C) On the east side of the building facing Jenkins Street
(D) None of the above

8. In your opinion, who is the most seriously endangered occupant?

(A) The person on the roof of the fire building
(B) The person on the top floor of 300 Barclay Avenue
(C) The person on the top floor of the fire building
(D) The person on the fourth floor of the fire building

9. What is the number of the apparatus in front of 300 Barclay Avenue?

(A) 9
(B) 6
(C) 8
(D) 11

10. What action is the apparatus in front of 300 Barclay Avenue performing?

(A) Stretching an attack hose line into the fire building to extinguish the fire
(B) Climbing the aerial ladder
(C) Discussing strategy with the fire chief
(D) Supplying water to the fire department connection

11. The apparatus in front of 300 Barclay Avenue is connected to a fire hydrant located where?

(A) At the corner of Jenkins Street and Barclay Avenue
(B) At the corner of Nevins Street and Barclay Avenue
(C) At the front of 290 Barclay Avenue
(D) In front of Travel World

12. How many occupants are looking out windows at 300 Barclay Avenue?

(A) One
(B) Two
(C) Three
(D) Four

13. What action is the fire chief performing at this fire operation?

(A) Stretching supply hose into the fire department connection
(B) Giving orders on a handheld radio
(C) Using a flashlight to illuminate the sidewalk
(D) Driving the chief's vehicle

14. Select the item that is not seen on the roof of any building.

(A) Building occupant
(B) Antennae
(C) Satellite dish
(D) Clothesline

15. Where is a roll-down gate found?

(A) Travel World
(B) The door to 290 Barclay Avenue
(C) Elmo's deli
(D) None of the above

16. Where is an alarm box located?

(A) Adjacent to the fire hydrant situated on the corner of Jenkins and Barclay
(B) Adjacent to the Travel World entrance door
(C) Adjacent to the fire hydrant situated on the corner of Nevins and Barclay
(D) In front of Elmo's deli

**Directions:** Study the fireground situation below for three minutes. At the end of the allotted time, cover the picture and answer questions 17–26.

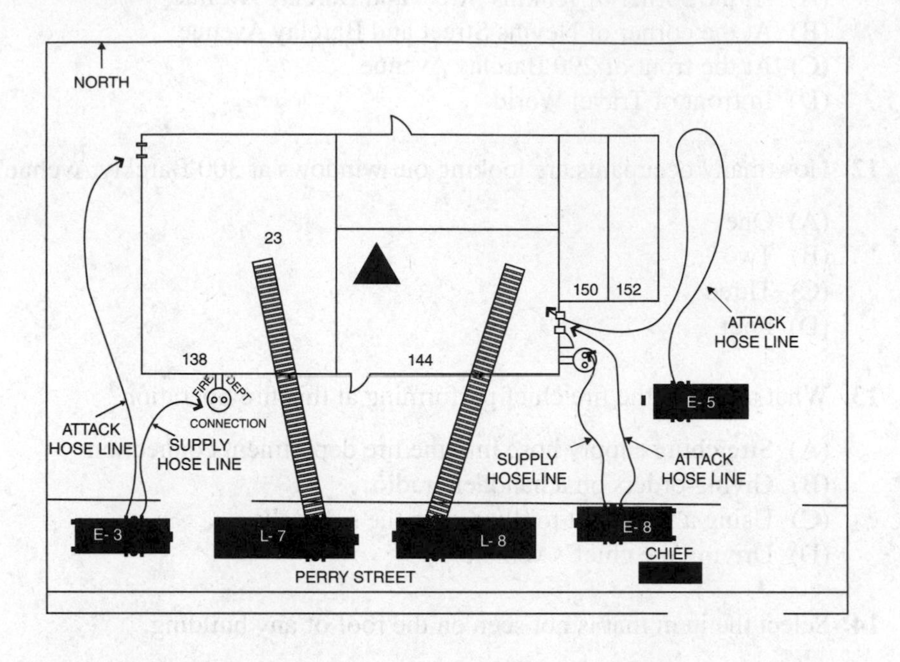


**17.** Not counting the chief's car, how many fire apparatus are at the fire scene?

(A) Three
(B) Four
(C) Five
(D) Six

**18.** How many fire engines (pumpers) are at the fire scene?

(A) One
(B) Two
(C) Three
(D) Four

**19.** What is the correct address of the building on fire?

(A) 138 Perry Street
(B) 144 Perry Street
(C) 23 Huron Street
(D) 23 Huron Avenue

**20.** What apparatus has placed a ladder to the fire building?

(A) Ladder 8
(B) Engine 5
(C) Ladder 7
(D) Engine 8

**21.** Which apparatus is connected to the fire building's fire department connection?

(A) Engine 5
(B) Engine 8
(C) Ladder 8
(D) Engine 3

**22.** The chief's car is situated across from what building at the fire scene?

(A) 138 Perry Street
(B) 144 Perry Street
(C) 150 Perry Street
(D) 152 Perry Street

**23.** Which engines have attack hose lines stretched to the building on fire?

(A) Engine 3 and Engine 8
(B) Engine 3 and Engine 5
(C) Engine 5 and Engine 8
(D) Only Engine 8

**24.** Engine 3 is performing what actions on the fireground?

(A) Stretching an attack hose line to the fire building
(B) Stretching an attack hose line to the fire building and supplying an FD connection
(C) Stretching an attack hose line to 138 Perry Street and supplying an FD connection
(D) Stretching an attack hose line up Ladder 7's aerial ladder

**25.** Where is the fire department connection for the fire building located?

(A) At the southeast corner of the building
(B) At the northwest corner of the building
(C) At the southwest corner of the building
(D) At the northeast corner of the building

**26.** Which building on the fireground does not front on Huron Avenue?

(A) 23 Huron Avenue
(B) 150 Perry Street
(C) 152 Perry Street
(D) 144 Perry Street

**Directions:** Study the floor plan diagram below for three minutes. At the end of the allotted time, cover the picture and answer questions 27–33.

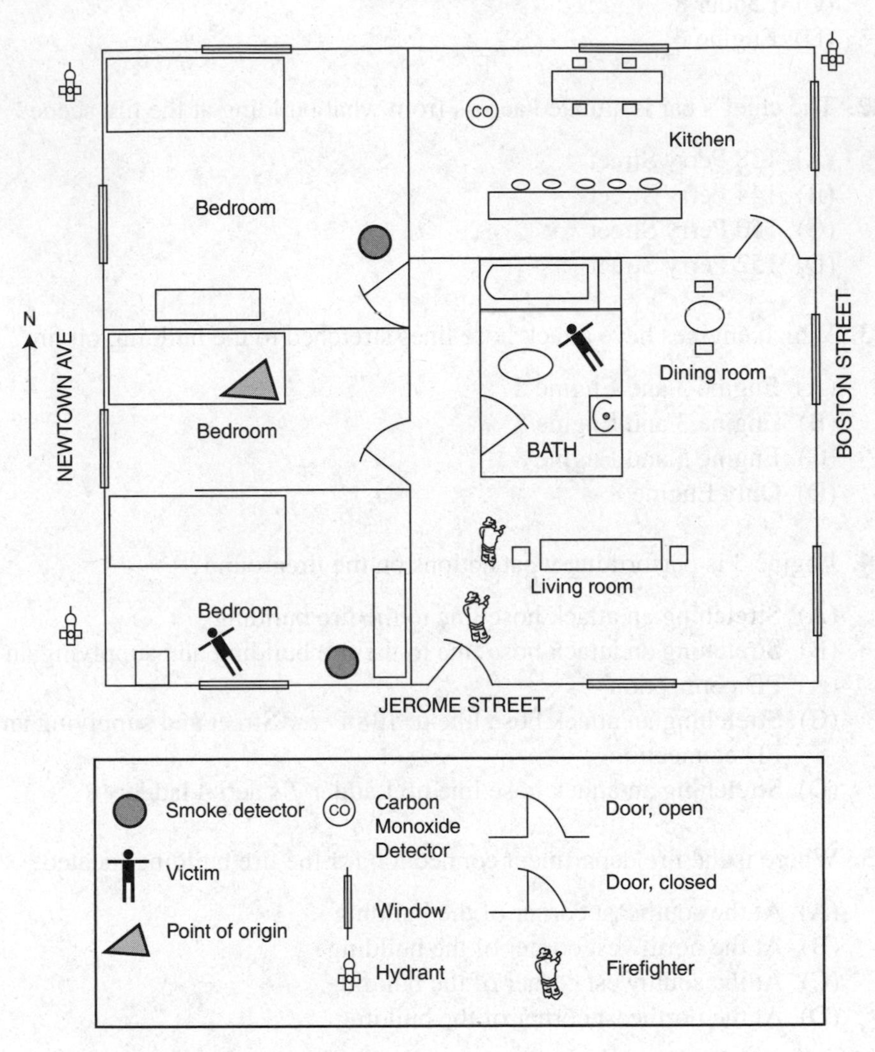


**27.** Where has the fire originated?

(A) The kitchen
(B) The middle bedroom
(C) The bedroom adjacent to Monitor Street
(D) The living room

**28.** Firefighters have just entered the apartment via the

(A) window fronting on Newtown Avenue
(B) window fronting on Boston Street
(C) doorway fronting on Jerome Street
(D) kitchen window

**29.** Which fire hydrant would the firefighters most likely use to get the initial attack hose line into operating position?

(A) The hydrant near the corner of Monitor Street and Newtown Avenue
(B) The hydrant near the corner of Jerome Street and Newtown Avenue
(C) The hydrant near the corner of Monitor Street and Boston Street
(D) None of the above

**30.** Where are the victims located inside the apartment?

(A) The bedroom adjacent to Jerome Street and the bathroom
(B) The middle bedroom and the dining room
(C) The bathroom and the bedroom adjacent to Monitor Street
(D) The living room and kitchen

**31.** Which room does not have a smoke detector or a carbon monoxide detector installed?

(A) The bedroom adjacent to Monitor Street
(B) The dining room
(C) The bedroom adjacent to Jerome Street
(D) The kitchen

**32.** Which room has a carbon monoxide detector installed?

(A) The bathroom
(B) The living room
(C) The dining room
(D) The kitchen

**33.** What is the most likely cause of the fire that occurred inside the apartment?

(A) Food on the stove
(B) Smoking in bed
(C) A defective electric curling iron
(D) A faulty chimney

# EXAMPLES OF VISUALIZATION QUESTIONS

**Directions:** Study the road map, then answer the following questions (34–37). All responses by the fire apparatus must be performed in a legal manner as designated by road map directional signs.

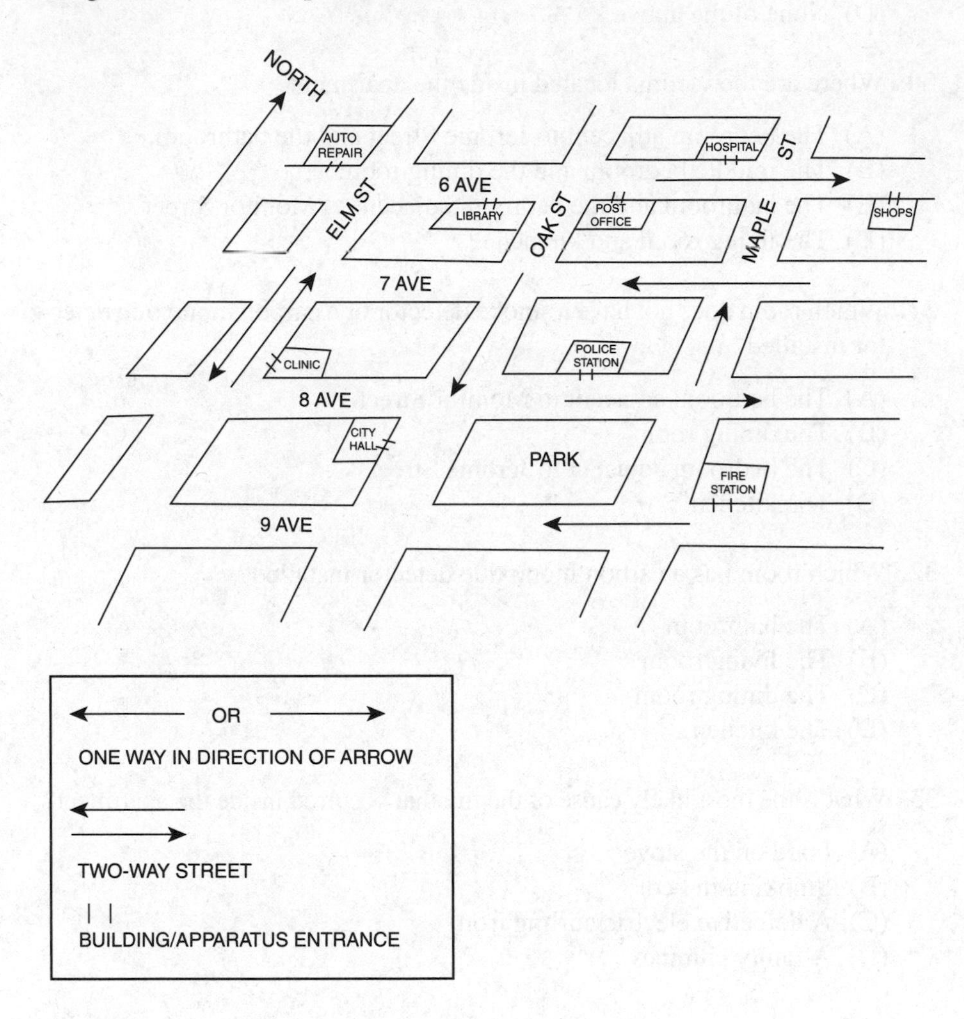


**34.** Leaving quarters to respond to an alarm received for a fire in the auto repair shop, firefighters would turn

(A) left out of quarters, right onto Ninth Avenue, left on Elm Street
(B) right out of quarters, left onto Seventh Avenue, right on Elm Street
(C) right out of quarters, right on Elm Street
(D) left out of quarters, right onto Ninth Avenue, right on Elm Street

**35.** A fire marshal visiting the firehouse needs to walk over to the police precinct to complete a criminal report concerning an arson fire. Firefighters should instruct the fire marshal to

(A) take Ninth Avenue west to Elm Street, go north to Eighth Avenue, and then go east
(B) take Maple Street north to Eighth Avenue and then go west
(C) take Maple Street north to Eighth Avenue and then go east
(D) go west on Ninth Avenue, north on Oak Street, and then east on Eighth Avenue

**36.** An ambulance parked on the corner of Sixth Avenue and Oak Street is dispatched to respond to a medical emergency at the clinic. The ambulance should travel in what direction to arrive at the clinic entrance in the quickest possible fashion?

(A) East on Sixth Avenue, south on Maple Street, west on Eighth Avenue
(B) South on Oak Street, west on Eighth Avenue, north on Elm Street
(C) South on Oak Street, west on Seventh Avenue, south on Elm Street
(D) West on Sixth Avenue, south on Elm Street

**37.** Which building entrance listed below is farthest for the fire apparatus to reach when responding from the firehouse? Remember, obey all traffic regulations.

(A) City hall
(B) Library
(C) Post office
(D) Hospital

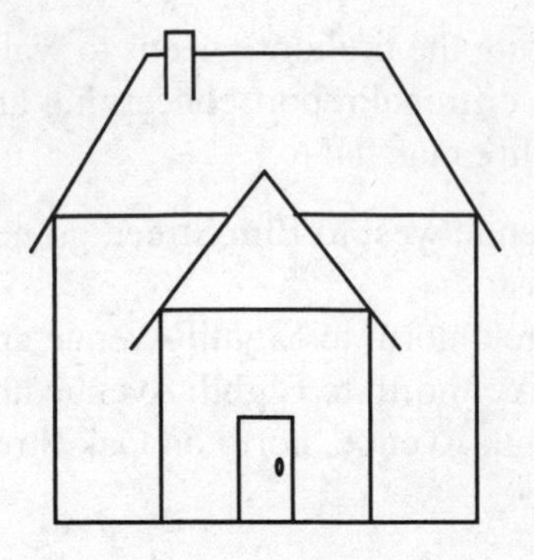

**38.** The house shown above, if viewed from an aerial perspective, would look most like which of the following?

(A)

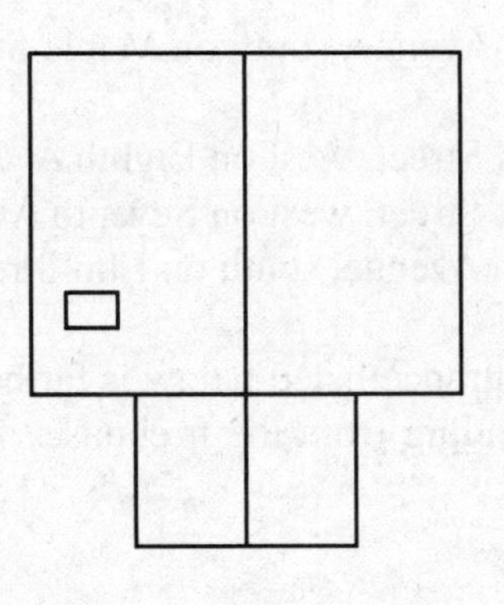

(B)

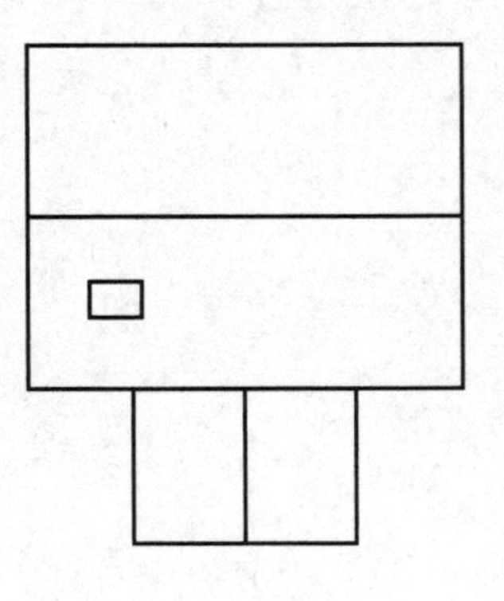

(C)

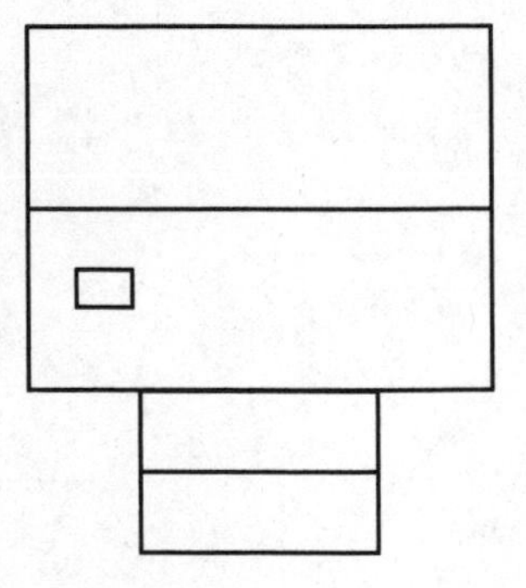

(D)

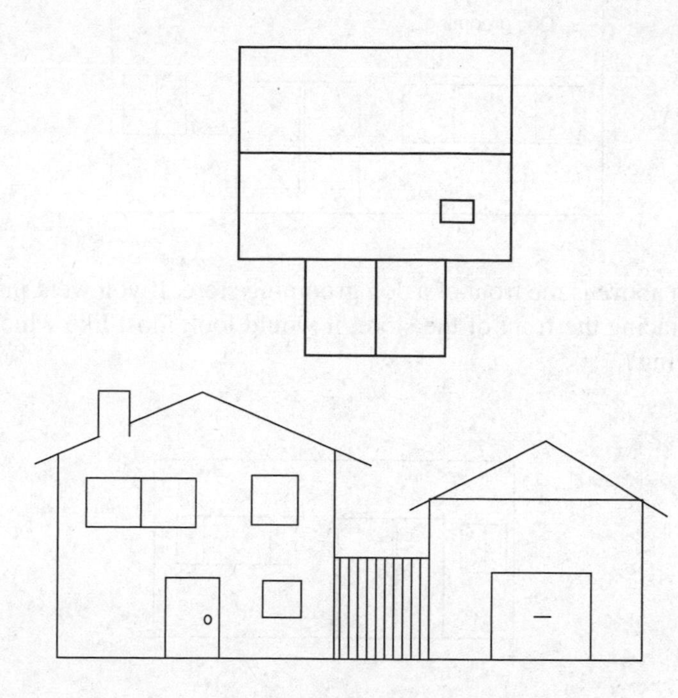

**39.** Shown above is the front of a house. If you went around to the rear of the house, it would look like which of the following?

(A)

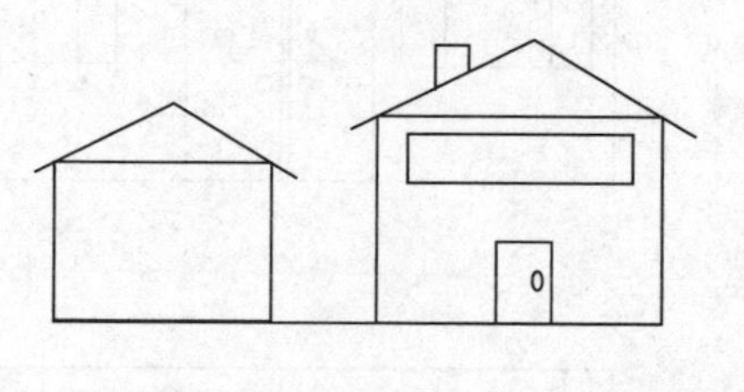

(B)

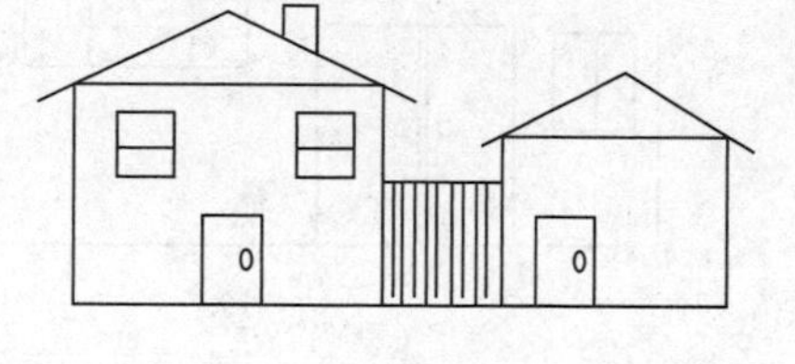

(C)

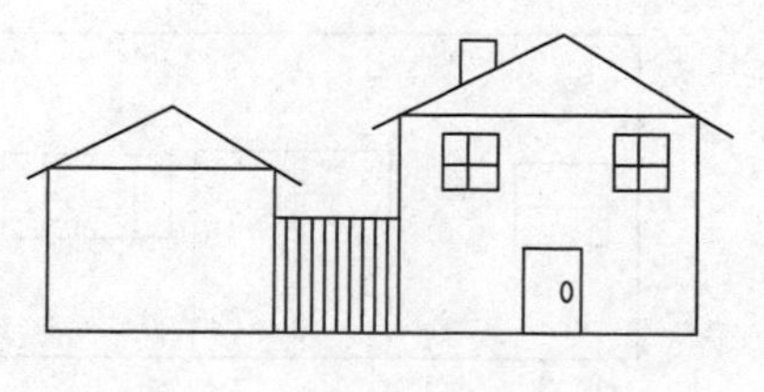

(D)

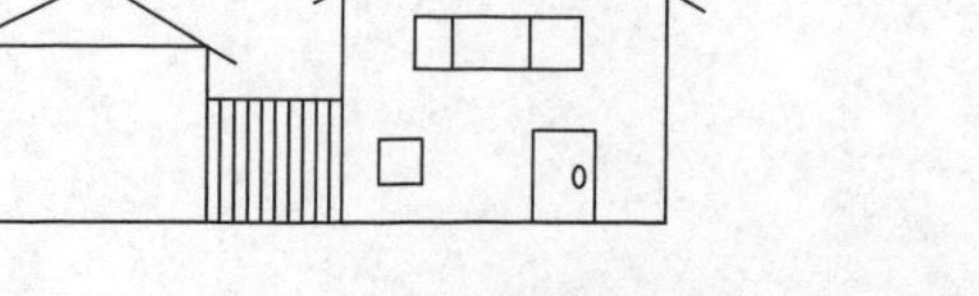

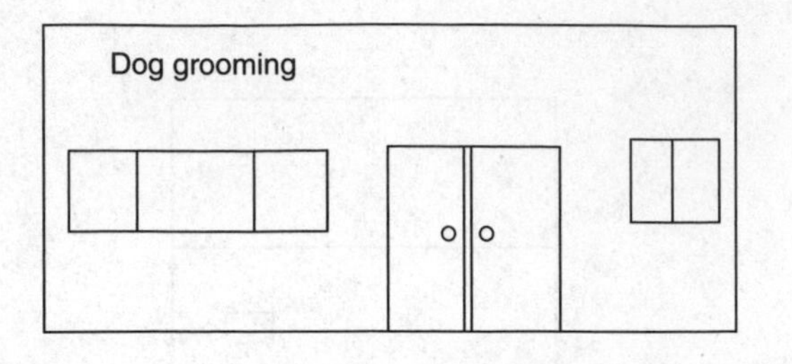


**40.** Shown above is the front of a dog grooming store. If you went inside and stood facing the front of the store, it would look most like which of the following?

(A)

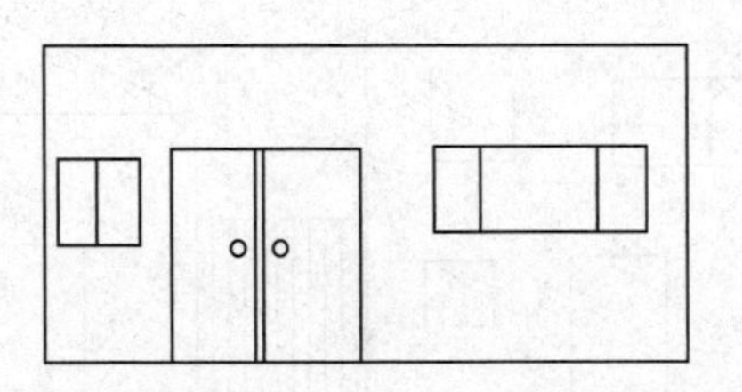

(B)

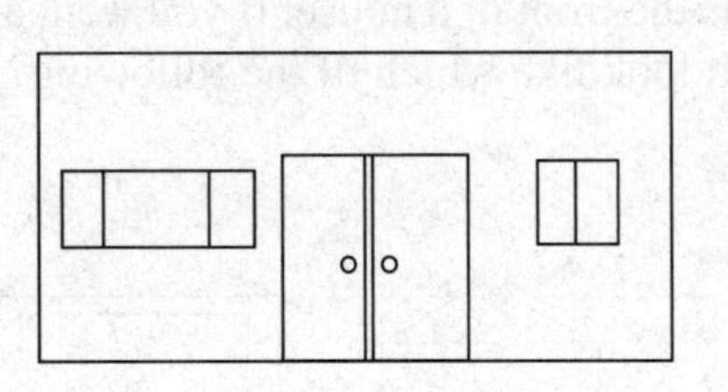

(C)

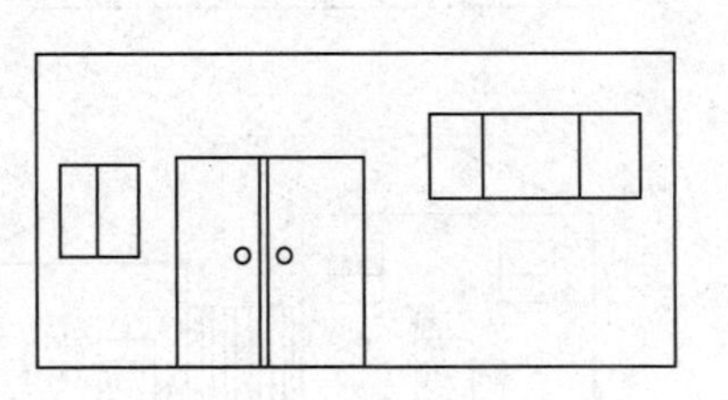

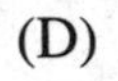
(D)

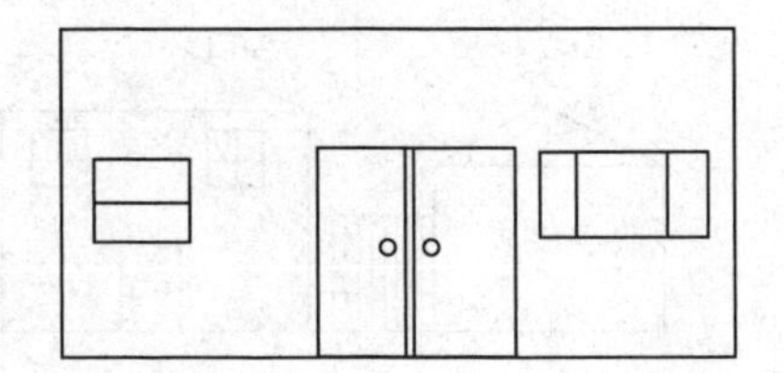

# Answer Key

## Memorization Questions

**1.** B
**2.** D
**3.** B
**4.** B
**5.** A
**6.** D
**7.** C
**8.** D
**9.** A
**10.** D
**11.** B
**12.** A
**13.** B
**14.** D
**15.** C
**16.** A
**17.** C
**18.** C
**19.** B
**20.** A
**21.** B
**22.** D
**23.** C
**24.** C
**25.** A
**26.** D
**27.** B
**28.** C
**29.** B
**30.** A
**31.** B
**32.** D
**33.** B

## Visualization Questions

**34.** C
**35.** B
**36.** C
**37.** D
**38.** B
**39.** D
**40.** A

## ANSWER EXPLANATIONS

Following you will find explanations to specific questions that you might find helpful.

**38** B Peaked roof over entranceway runs front to rear, while ridge pole of the main roof runs parallel to the front of the house.

**39** D Flip the drawing over from right to left. House will now be to your right with chimney located at the far right. Garage is located on your left.

**40** A Flip the drawing over from left to right. Three-pane window will be to your right and two-pane window with vertical divider will be to your left. Front doors shift over to the left of center.

CHAPTER 9

# Fire Science Basics

### Your Goals for This Chapter

- Understand the characteristics and behavior of fire
- Learn how to confine, control, and extinguish fires
- Test your fire science knowledge with review questions

A basic understanding of the characteristics and behavior of fire is important to anyone aspiring to be a firefighter. This section reviews key concepts relating to fire. Understanding fire behavior requires a good working knowledge of both chemistry and physics. This information is designed to give the potential firefighter candidate a good idea of what the firefighter profession is all about—the confinement, control, and extinguishment of fire.

The following definitions, grouped into major categories, provide basic knowledge all firefighters need to know and understand.

## FIRE

Fire is a rapid, self-sustaining oxidation process generating heat and light. The components of a fire and the types of fire are discussed below.

**Oxidation**—a chemical reaction between an oxidizer and fuel.

**Oxidizer**—generally, a substance containing oxygen that will chemically react with fuel to start and/or feed a fire. Examples include oxygen in the air, fluorine gas, chlorine gas, bromine, and iodine.

**Fuel**—materials that burn. Most common fuels contain carbon, hydrogen, and oxygen. Examples include wood, paper, propane gas, methane gas, and plastics.

**Combustion**—a rapid oxidation reaction that can produce fire. The oxygen in the air (21 percent) is generally the oxidizer, chemically reacting with the fuel. All combustion reactions give off heat and are therefore exothermic reactions. Combustion is commonly called fire. If combustion is confined and a rapid pressure rise occurs, it is called an explosion.

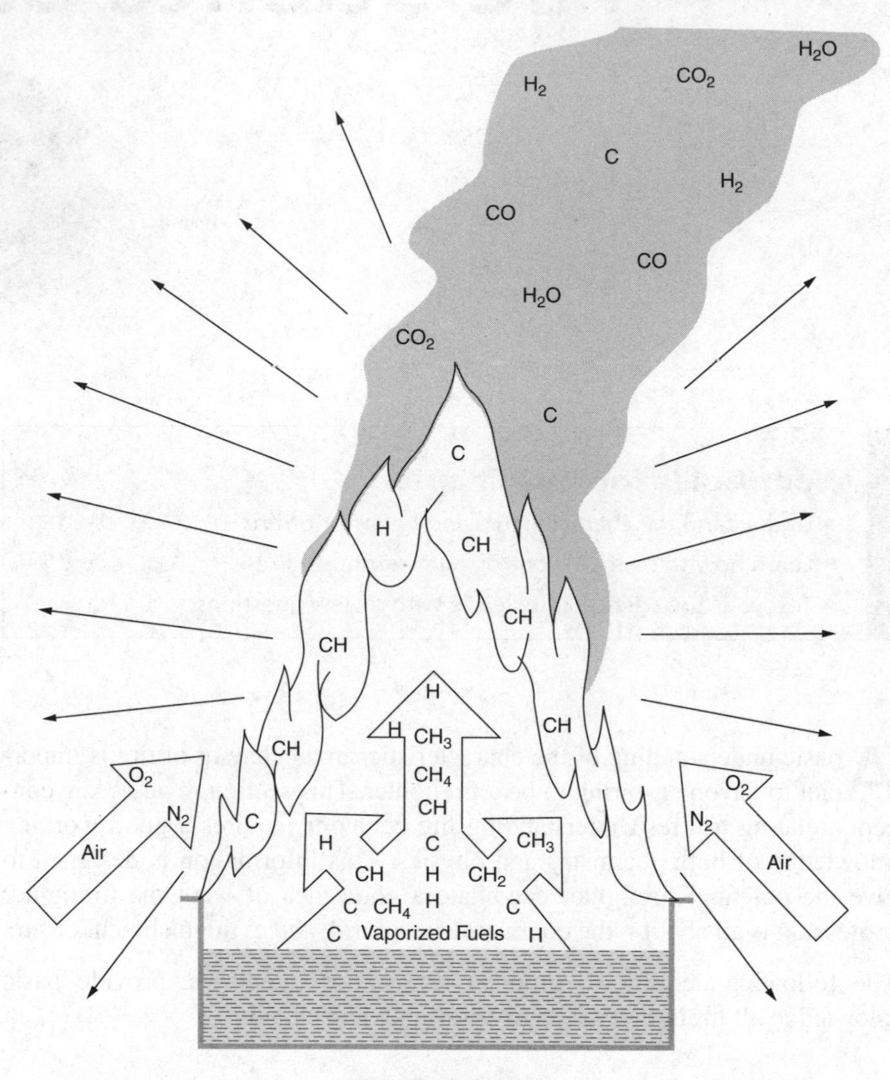


**FIRE COMPONENTS**

**Explosion**—a rapid expansion of gases (fuel and oxygen) that have mixed prior to ignition. Common explosions encountered by firefighters are chemical and mechanical explosions.

**Chemical explosion**—a rapid combustion reaction classified as a detonation or deflagration, depending on the rate of propagation.

**Detonation**—a reaction that propagates at the speed of sound (1,088 feet per second in air) producing a shock wave. Examples include high explosives (dynamite, blasting agents).

**Deflagration**—a reaction that propagates at less than the speed of sound. Examples include low explosives (gunpowder) and combustible gases and dusts.

**Backdraft**—an explosion caused by the sudden influx of air into an oxygen-starved area filled with a mixture of combustible gases (primarily carbon monoxide) that are heated above their ignition temperature.

**Mechanical explosion**—a physical explosion. Examples include a boiler explosion and a boiling liquid expanding vapor explosion (BLEVE).

**BLEVE**—a container failure in the form of an explosion caused by the weakening of the container shell from the heat from a fire, corrosion, or mechanical damage. If the contents inside the container are flammable, a dramatic fireball results.

**Pyrolysis**—a decomposition reaction in a solid material, not fast enough to be self-sustaining, usually brought on by the introduction of heat. It is the precursor to combustion. Characteristics of pyrolysis include the discoloration or browning of the surface of the material and the emission of smoke vapors.

**Exothermic reaction**—a chemical reaction that generates heat. New substances formed have less heat energy than was in the reacting materials. An example is combustion.

**Endothermic reaction**—a chemical reaction causing the absorption of heat. New substances formed by the chemical reactions contain more heat energy than prior to the reaction. An example is spontaneous combustion.

**Spontaneous combustion**—an endothermic chemical reaction causing self-ignition. Examples include a pile of rags dipped in linseed oil, alkyd enamel resins, or drying oils not properly stored or discarded, and wet hay inside a barn loft.

**Fire triangle**—a model used to help in the understanding of the three major elements necessary for ignition: **heat (thermal energy)**, **fuel**, and **oxidizer (oxygen)**. It visually depicts the ignition sequence.

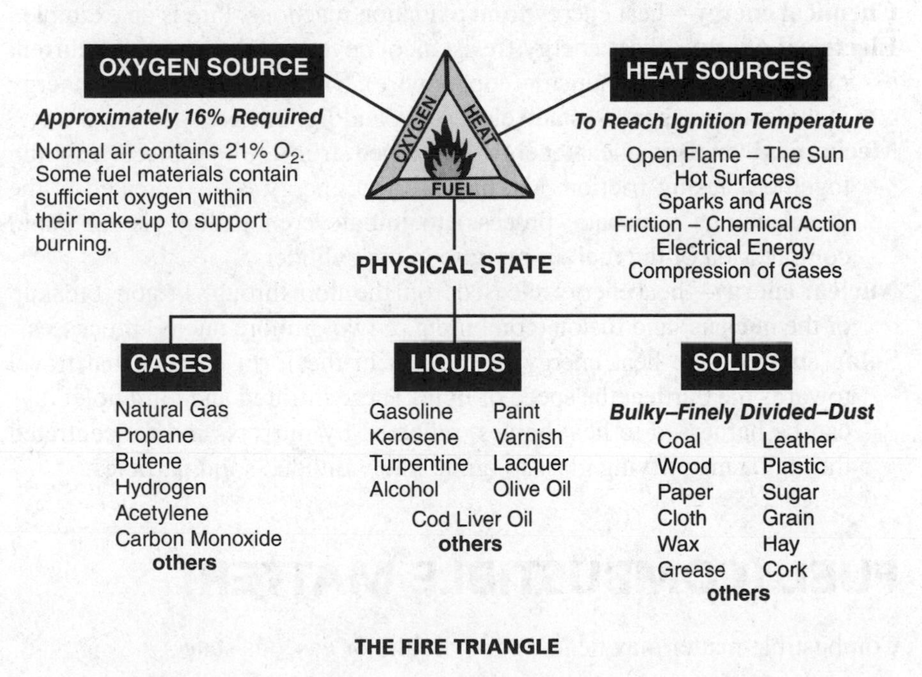


**THE FIRE TRIANGLE**

**Fire tetrahedron**—a model that expands on the one-dimensional fire triangle. The fire tetrahedron visually shows the interrelationship among the three components of the fire triangle and further clarifies the definition of combustion by adding a fourth component, **chemical chain reaction**, depicting the concept of the rapid, self-sustaining oxidation reaction. The fire tetrahedron depicts the growth of ignition into a fire.

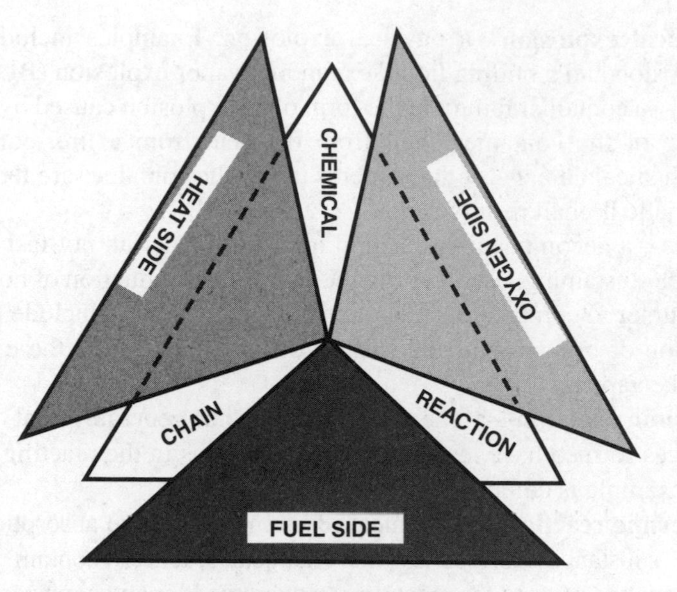


THE FIRE TETRAHEDRON

# HEAT (THERMAL ENERGY)

Heat is defined as thermal energy. There are several types of heat, or thermal, energy.

**Chemical energy**—heat energy from oxidation reactions. Fire is an example.

**Electrical energy**—heat energy (resistance) developed by electrical current moving through a conductor (copper wire). Examples of electrical energy include arcing, sparks, static electricity, and lightning.

**Mechanical energy**—heat energy developed from solid objects rubbing together causing friction. Mechanical heat energy is also created in the diesel engine (adiabatic process) to initiate combustion via the rapid compression of the fuel-air mixture in the cylinders.

**Nuclear energy**—heat energy released from the atom through fission (breakup of the nucleus) and fusion (combining of two or more nuclei) processes.

**Solar energy**—the heat energy of the sun in the form of rays that travel towards the Earth at the speed of light. These infrared and ultraviolet rays can be harnessed to heat homes, reflected by mirrors, and concentrated through a magnifying glass to ignite finely divided solid particles.

# FUEL (COMBUSTIBLE MATTER)

Combustible matter may be in a solid, liquid, or gaseous state.

## Solids

Solids are materials with defined volume, size, and shape at a given temperature. Examples are wood and wood products (paper, cardboard), carbon-containing materials (coal, charcoal), plastics (polyvinyl chloride, epoxies), textiles (cotton, wool, rayon), and combustible metals (magnesium, aluminum).

## WOOD AND WOOD PRODUCTS

Wood and wood products are the most common solids encountered by firefighters. They are considered Class A materials and require water or water solutions to cool them below their ignition temperature and extinguish them. The average ignition temperature of wood is approximately 400°F. Major components are carbon, hydrogen, and oxygen.

### Factors Affecting the Ignition and Combustibility of Wood and Wood Products

Many factors influence the ignition and combustibility of wood and wood products, the most important of which are cited below.

- **Physical form (size, form, shape, mass)**—The greater the mass in relation to surface area, the more heat energy will be required to ignite it and the slower the rate of burning will be once ignited.
- **Thermal inertia**—Resistance to heating, generally based on the specific gravity and density of the material, is known as its thermal inertia. Materials with a low thermal inertia (low specific gravity and density) will heat up and ignite more readily than materials with a high thermal inertia, specific gravity, and density.
- **Moisture content**—Wet wood is more difficult to ignite than dry wood. Wood is very difficult to ignite when the moisture content rises above 15 percent.
- **Species**—Low-density softwoods (pine) will ignite at lower temperatures than high-density hardwoods (oak).
- **Ignition temperature**—The minimum temperature to which a material must be heated for it to ignite and be self-sustaining without an external input of heat is known as the ignition temperature.
- **Piloted-ignition temperature**—Ignition temperature caused with the assistance of an external heat source (flame, spark) is known as the piloted-ignition temperature. It is usually considerably lower than the ignition temperature.
- **Arrangement**—The term *arrangement* refers to the spacing of the fuel material. Tightly stacked lumber is much more difficult to ignite and will burn at a slower rate than lumber loosely arranged.
- **Time**—Wood and wood products must be exposed to heat for a certain period of time before combustible vapors are produced and ignite.
- **Heat source**—A heat source provides heat. For wood products, heat sources include steam pipes, matches, and a blowtorch.
- **Rate of heating**—The rate or speed at which a substance becomes heated may be constant or sporadic.
- **Oxygen**—Oxygen-enriched atmospheres (greater than 21 percent in air) enhance burning, whereas oxygen-deficient atmospheres (less than 15 percent in air) will generally not support combustion.

## CARBON AND CARBON-CONTAINING MATERIALS

Carbon and carbon-containing materials have ignition temperatures in the range of 600° to 1,400°F, depending on the amount of carbon in them. Coal and charcoal burn hotter than wood and wood products, and they generate large quantities of toxic and flammable carbon monoxide gas. They are classified as **Class A** materials.

## PLASTICS

Plastics are other common combustible solids, although they may be produced as a liquid or foam. Most plastics are petroleum based (hydrocarbons). They can be soft or hard and be electrically conductive or nonconductive (insulators). Manufactured plastics usually contain additives (colorants, stabilizers, lubricants) that change the chemical nature and combustibility of the original plastic. Pyrolysis doesn't occur as readily in plastics as in wood, and, therefore, plastics tend to have a higher ignition temperature than wood and wood products.

Plastics can be divided into two categories: thermoplastics and thermosets. **Thermoplastics** (polypropylene, polyvinyl chloride) are formed by heat and pressure and can be reshaped repeatedly by heat and pressure. In a fire, they will melt and flow like a liquid. **Thermosets** (alkyds, epoxies), on the other hand, may only be formed by heat and pressure once. When subjected to the heat from a fire, they will decompose and burn. Both are classified as **Class A** materials.

## TEXTILES

Textiles include clothing, bedding, upholstery, and carpeting. In general, all textile fibers are combustible. Textiles can be divided into two categories: natural fiber and synthetic fiber.

**Natural fiber textiles** can be divided into fibers derived from plants (cotton, linen, hemp) and those derived from animals (wool, mohair, camel hair). Plant fiber is composed mostly of cellulose, which consists of carbon, hydrogen, and oxygen. During a fire situation, plant fiber will decompose and burn but will not melt. The ignition temperature of cotton is approximately 750°F. Animal fiber, however, is chemically different from plant textile material. Protein is the major component. Animal fiber, with an ignition temperature of approximately 1,100°F, will not ignite as readily as plant fiber. Natural fibers are Class A materials.

**Synthetic fiber** (rayon, nylon, polyester) is material woven from artificial fiber (plastic, hydrocarbon, metal, glass). Burning characteristics of synthetic fiber include decomposition, burning, and melting. Synthetic fiber can, however, be made flame retardant and various kinds of synthetic fiber (Nomex, Kevlar) are used in the production of "fireproof" clothing for firefighters. Synthetic fibers are classified as **Class A** materials.

### Factors Affecting Ignition and Combustibility of Textile Products

Various factors affect the ignition and combustibility of textile products. Some of these are listed below.

- Chemical composition
- Weight of the fabric
- Type of weave
- Finishing treatments

### COMBUSTIBLE METALS

The elements that will combine with oxygen, reach their ignition temperature, and burn are known as combustible metals. Metals do not, however, undergo pyrolysis to produce combustible vapors when heated. They burn on their surface with no flaming combustion. Metals that burn produce an abundance of heat energy. When water is applied and the water molecules separate, steam and hydrogen explosions can occur. For this reason, water, unless in large amounts, is not recommended as an extinguishing agent on combustible metals. Specific extinguishing agents (graphite, salts) have been developed to cover the surface of the burning metal and exclude oxygen. Combustible metals are classified as **Class D** materials.

## Liquids

Liquids make up the stage of matter between solids and gases. A liquid has definite volume but takes the shape of the container it is being stored in. Liquids that produce vapors that burn can be divided into two categories: combustible liquids (kerosene, diesel, heavy fuel oils) and flammable liquids (gasoline, methyl alcohol, acetone). Liquids can present other hazards to firefighters besides fire (corrosiveness and toxicity). In general, liquids that burn are classified as Class B materials; however, vegetable oils used in cooking and the preparation of foods are classified as Class K materials. Some key characteristics to understand concerning liquids that burn are the flash point, boiling point, specific gravity, solubility, and viscosity.

- **Flash point**—The flash point is the minimum temperature of a liquid at which it emits vapors to form an ignitable mixture with air. For firefighters, the flash point is the most important property of liquids that burn. The degree of hazard will be determined by the flash point of the liquid because it is the vapors of the liquid that burn, not the liquid itself. Liquids are classified as **combustible** (flash point of 100° F or more) and **flammable** (flash point of less than 100°F).
- **Boiling point**—The boiling point is the temperature of the liquid at which it will liberate the most vapors. It is the temperature at which the vapor pressure of the liquid equals atmospheric pressure. The **normal** boiling point of a liquid is the temperature at which it boils at sea level, usually recorded as 14.7 pounds per square inch absolute (psia). It is impossible to raise the temperature of a liquid above its boiling point, except if it is under pressure.
- **Specific gravity**—The specific gravity of a liquid is the ratio of the weight of the liquid to the weight of an equal volume of water. The specific gravity

of water is 1. A liquid with a specific gravity less than water (gasoline, 0.8) will float on water, whereas a liquid with a specific gravity more than 1 (sulfuric acid, 1.8) will sink.

- **Solubility**—The solubility of a liquid is the percentage by weight of the liquid that will dissolve in water. The solubility of a liquid ranges from **negligible** (less than one-tenth of 1 percent) to complete (100 percent).
- **Viscosity**—Viscosity is a measure of a liquid's flow (through an opening or into a container) in relation to time. Thick liquids (molasses, asphalt, wax) are on the borderline between liquids and solids and are considered viscous.

# Gases

Gases are the third stage of matter. The volume of a given amount of gas is dependent on its temperature and the surrounding pressure. An important concept for firefighters to understand regarding gases and vapors being emitted from a liquid is vapor density. **Vapor density** is the relative density of the gas or vapor as compared to air. The vapor density of air is 1. A gas or vapor with a vapor density more than 1 (butane, 2.1) will be heavier than air and travel along the ground surface where it may encounter an ignition source. A gas or vapor with a vapor density less than 1 (methane, 0.55) will rise and disperse readily into the air. Gases are classified as **Class B** materials.

## CHEMICAL PROPERTIES OF GASES

Gases can be classified according to their chemical properties as **flammable** (will burn in air), **inert** (will not burn in air or in any concentration of oxygen and will not support combustion), **oxidizer** (will not burn in air or in any concentration of oxygen but will support combustion), **toxic** (poisonous or irritating when inhaled), and **reactive** (can rearrange chemically when exposed to heat or shock and explode or can react with other materials and ignite).

- **Flammable**—A gas that will burn in normal concentrations of oxygen in air is a flammable gas. When discussing flammable gases (or flammable vapors boiling off a liquid) mixing with air, the concept of **flammable range** must be understood. The flammable range is defined as the ratio of gas or vapor in air that is between the upper and lower flammable limits. The **upper flammable limit** is the maximum ratio of flammable gases or vapors to air above which ignition will not occur; it is too rich a mixture. The **lower flammable limit** is the minimum ratio of flammable gases or vapors in air below which ignition will not occur; it is too lean a mixture. Examples of flammable gases include acetylene, hydrogen, and propane.
- **Inert**—An inert gas is a nonflammable gas that will not support combustion. Examples include helium, nitrogen, and argon.
- **Oxidizer**—A nonflammable gas that will support combustion is known as an oxidizer. Examples include oxygen and chlorine.
- **Toxic**—Gases that cause harm to living tissue via chemical activity are called toxic gases. They can endanger the lives and health of those who inhale or come into skin contact with them. Examples include hydrogen cyanide, carbon monoxide, and ammonia.

• **Reactive**—Gases that react internally and with other materials are reactive gases. They can be heat sensitive and shock sensitive and also react with organic and inorganic substances to cause combustion. Examples include fluorine and vinyl chloride.

# OXIDIZER

Oxygen gas is the most common oxidizer that firefighters deal with during fire operations. The atmosphere consists of 21 percent oxygen, 78 percent nitrogen, and 1 percent other elements. An oxygen-enriched atmosphere (greater than 21 percent) will enhance the rate and intensity of burning. Conversely, an oxygen-deprived atmosphere (less than 15 percent) will not be able to sustain combustion.

# CHEMICAL CHAIN REACTION

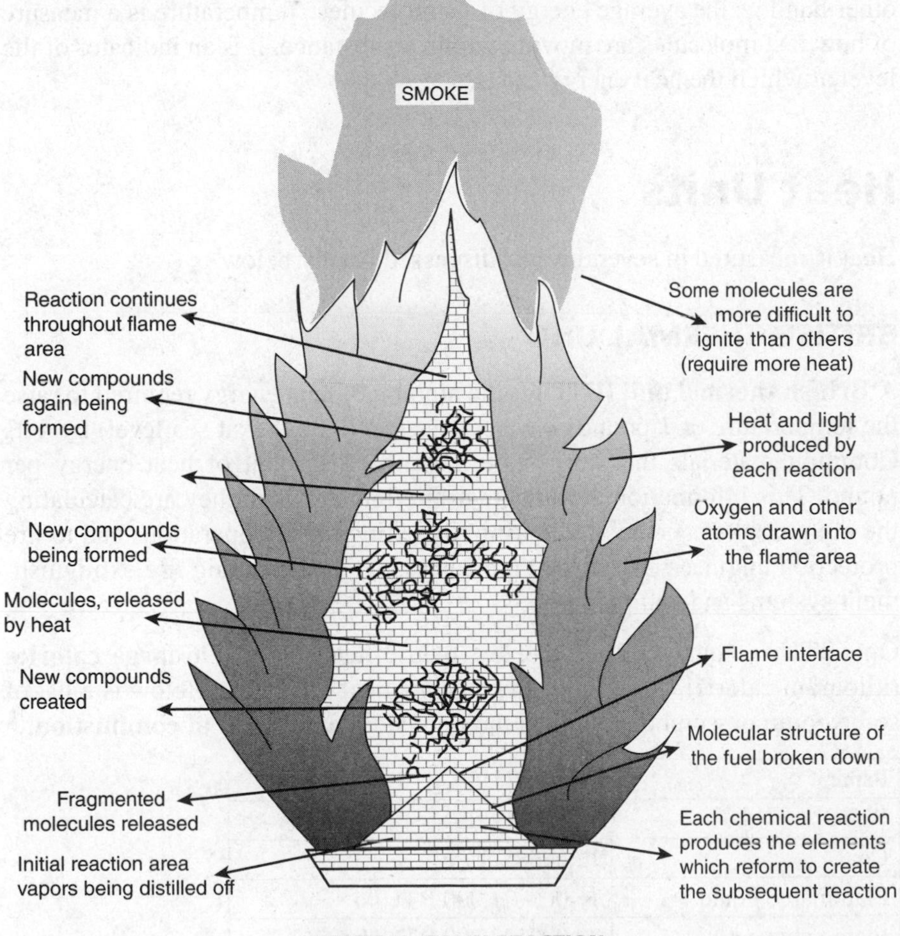


**CHEMICAL CHAIN REACTION**

The chemical chain reaction process that occurs during flaming combustion is the fourth component that was added to the fire triangle model to form the fire tetrahedron model. It depicts self-sustaining combustion with an ample amount of fuel and oxygen chemically interacting. As fuel burns it generates radiant heat traveling in all directions. Heat directed back onto the burning substance (**radiated feedback**) helps to raise more fuel to its ignition temperature and generate more vapors to mix with air and form a combustible mixture. Additional oxygen is then drawn (**entrainment**) into the zone of chemical reaction. This addition of oxygen also increases the heat of burning. The chemical reaction between the components of the fuel vapors and oxygen is called an **oxidation reaction**. The chemical reaction we know as fire continues until all the fuel is consumed, the oxygen in a confined area is diminished, or heat dissipates beyond the zone of the chemical reaction, causing the temperature of the fuel to drop below its ignition temperature.

# HEAT AND TEMPERATURE

Heat and temperature are two distinct but closely related concepts. Heat is a measure of the quantity of energy contained in a substance. It is the total amount of molecular vibration (energy) in a material. Temperature, on the other hand, is the average energy of its molecules. Temperature is a measure of how fast molecules are moving within a substance. It is an indicator of the level at which the heat energy exists.

## Heat Units

Heat is measured in several ways, discussed briefly below.

### BRITISH THERMAL UNIT

A **British thermal unit (BTU)** is the amount of heat energy required to raise the temperature of 1 pound of water (measured at 60°F at sea level) by 1°F. Common materials that burn store a standard amount of heat energy per pound. This information is valuable to firefighters when they are calculating the amount of water required during fire extinguishing operations and to fire protection engineers when they are designing and installing fire extinguishment systems and equipment.

One BTU is equal to 252 **calories** (metric heat unit), 3.96 **large calories** (kilogram calorie), or 1,055 **joules** (mechanical heat unit). Below is a list of some common combustibles and their equated **latent heat of combustion**:

| | |
|---|---|
| Paper | 6,000 BTU/lb |
| Wood | 7,000 BTU/lb |
| Coal | 12,500 BTU/lb |
| Flammable liquid | 16,000–21,000 BTU/lb |
| Flammable gas | 20,000–23,000 BTU/lb |

### CALORIE

A calorie is the amount of heat energy required to raise the temperature of 1 gram of water (measured at 15 degrees Celsius [°C] at sea level) by 1°C. One calorie is equivalent to 4.184 joules.

### JOULE

The joule is the heat energy unit in the **International System of Units (SI)**. It is the amount of heat energy provided by 1 watt flowing for 1 second.

## Temperature Units

Temperature units can be used to compare the difference in heat energy levels between two materials. Temperature is measured by monitoring how much an object expands from its size at a given starting point (the freezing point of water, for example) and defining a unit of measurement (1 degree). All temperatures are then multiples of that defined unit of measurement.

### FAHRENHEIT DEGREE

The **Fahrenheit (F)** degree is named for the German scientist Daniel Gabriel Fahrenheit, who invented the thermometer at the beginning of the eighteenth century. There are 180 increment degrees between the temperature of melting ice (32 degrees) and the boiling of water (212 degrees) on the Fahrenheit temperature scale. 1°F is equal to 5/9 degrees Celsius.

To convert (approximately) a temperature on the Fahrenheit scale to the Celsius or centigrade scale, you first subtract 32 degrees from the Fahrenheit temperature and then multiply by 5/9.

$$C = \frac{5}{9}(F - 32)$$

For example, if a person's body temperature is 98.6°F, his or her temperature in Celsius is

$$C = \frac{5}{9}(98.6 - 32)$$

$$C = \frac{5}{9}(66.6)$$

$$C \approx 37^\circ$$

### CELSIUS DEGREE

The **Celsius (C)** degree is a metric unit of temperature measurement. It is named for the Swedish professor Anders Celsius, who invented the centigrade temperature scale in the 1720s using the freezing point of water as 0 degrees and the boiling point of water as 100 degrees. This unit is approved by the SI.

To convert (approximately) normal body temperature on the Celsius scale to the Fahrenheit scale, first multiply the Celsius temperature by 1.8, or 9/5, and then add 32.

$$F = \frac{9}{5}(C) + 32$$

$$F = \frac{9}{5}(37) + 32$$

$$F \approx 66.6 + 32$$

$$F \approx 98.6^\circ$$

### RANKINE DEGREE

The **Rankine (R)** degree is a traditional unit of absolute temperature. The temperature units for Rankine and Fahrenheit are equal (1 degree Rankine represents the same temperature difference as 1 degree Fahrenheit), but the zero points differ. The zero point on the Rankine scale is set at absolute zero, which is –457.6°F, the hypothetical point at which all molecular movement ceases. The unit is named for British physicist and engineer William Rankine (1820–1872).

To convert degree units from the Rankine scale to the Fahrenheit scale and the Fahrenheit scale to the Rankine scale use the following formulas:

$$F = R - 457$$
$$R = F + 457$$

### KELVIN DEGREE

The **Kelvin degree (K)** is equal to the Celsius degree, but the Kelvin scale has its zero point set at absolute zero, which is –273.1°C. This unit is approved by the SI. The Kelvin degree is named for British inventor and scientist William Thompson, who was knighted by Queen Victoria in 1866 and named Baron Kelvin of Largs in 1892.

To convert degree units from the Kelvin scale to the Celsius scale and the Celsius scale to the Kelvin scale, use the following formulas:

$$C = K - 273$$
$$K = C + 273$$

## HEAT TRANSFER

Heat can be transferred to other materials through conduction, convection, radiation, and direct flame contact.

## Conduction

Conduction is the transfer of heat energy through a medium (usually a solid). Heat causes molecules within the material to move at a faster rate and transmit their energy to neighboring molecules. The heat of conduction can also be transferred from one material to another via direct contact in the same fashion as internal molecular movement. The amount of heat transferred and rate of travel is dependent on the **thermal conductivity** of the material. Dense materials (metals) are good conductors of heat energy. Fibrous materials (wood, paper, cloth) and air are poor conductors. In a fire situation, heat can be conducted via steel columns and girders to abutting wood floor joists, causing them to smolder and eventually ignite.

## Convection

Convection is the transfer of heat energy through a circulating medium (liquids and gases). During firefighting operations, hot air expands and rises, as do the products of incomplete combustion. Fire spread by convection is mostly in an upward and outward direction through corridors, stairwells, and shafts from floor to floor via hot air currents.

## Radiation

Radiation is the transfer of heat via infrared or ultraviolet waves or rays. These heat waves travel in a straight line through space at the speed of light in all directions and are not affected by the wind. Objects exposed to radiated heat will absorb and reflect a certain amount of heat energy, depending on certain factors. The darker and duller the object, the more heat it will absorb and the greater chance it will reach its ignition temperature and burst into flames. Light-colored, shiny objects tend to reflect radiated heat, absorb less energy, and are less likely to reach their ignition temperature. Radiated heat waves will travel through space until they are absorbed by an opaque object. These waves will pass through air, glass, transparent plastics, and water. Large amounts of radiated heat can travel large distances (50 to 100 feet) to ignite nearby buildings and structures.

## Direct Flame Contact

Direct flame contact is the transfer of heat energy via direct flame impingement or auto-exposure, such as occurs with a flame traveling upward and outward from a roof, window, or doorway to a neighboring building or exposure.

# PHASES OF FIRE

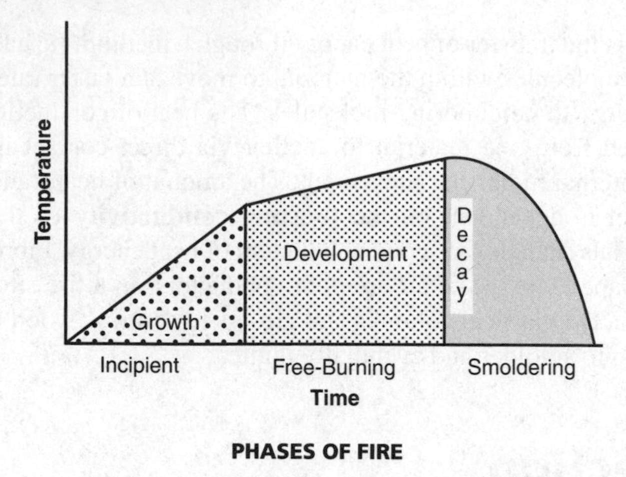


**PHASES OF FIRE**

There are three phases of fire: incipient (growth), free burning (fully developed), and smoldering (decay). Each phase has its own unique characteristics and dangers to firefighters and should be understood thoroughly to enhance safety during firefighting operations inside buildings and structures. These phases are part of the standard time/temperature curve, which helps in visualizing the heat energy and temperatures attained during a fire.

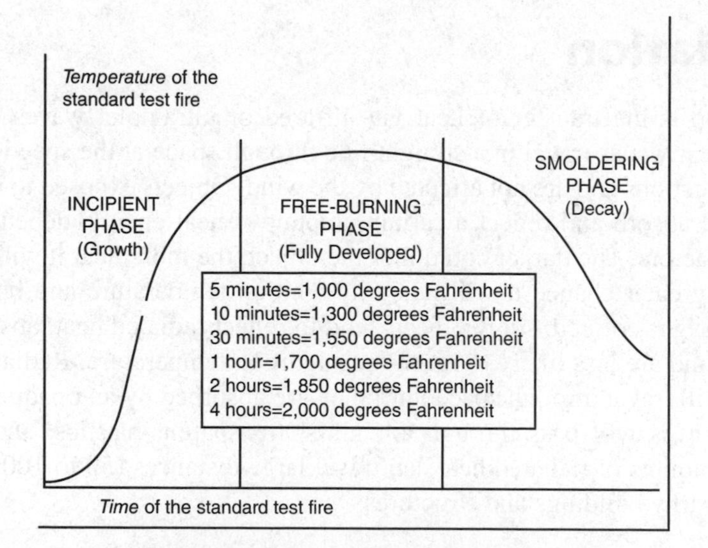


**THE STANDARD TIME/TEMPERATURE FIRE CURVE**

# Incipient (Growth) Phase

Most fires extinguished by firefighters are in this phase. In this phase, the fire is in the beginning, slow fuel combustion stage, with the oxygen content in the area still within the normal range (21 percent). There is limited heat being generated but high levels of smoke production and flammable carbon monoxide (CO) gas. Physical destruction from fire is limited to the immediate surrounding area. In certain situations, the introduction of fresh air by firefighters entering the area of fire can cause pent-up CO gas to react violently and explode (**backdraft**), leading to serious injury while increasing the intensity of the fire. Also during this phase, there is the possibility of fire gases reaching their ignition temperatures (**flashover**), causing the entire area's contents to become suddenly engulfed in fire, greatly increasing the temperature of the fire and leading to the next phase of fire, the free-burning phase.

## Free-Burning (Fully Developed) Phase

As fire spreads throughout an area, more heat and smoke are generated and travel in an upward direction toward the ceiling. During the free-burning phase, oxygen content in the area drops from 21 percent to approximately 15 percent, causing the volume of flames to eventually decrease, while smoke production continues to increase. When the oxygen level falls below 15 percent, flame generation ceases and the fire enters the next and last phase, the smoldering phase.

## Smoldering (Decay) Phase

During this phase, the oxygen content in the area is below 15 percent, causing the rate of heat production and active flaming to decrease rapidly. Combustibles in the room have been largely consumed by the fire and are no longer actively burning. These combustibles, however, are still emitting large amounts of smoke and flammable gases. If fresh air (oxygen) is introduced into the fire area at this time, a **backdraft** situation is possible, since the influx of oxygen will complete the fire triangle and cause reignition of the flammable gas mixture in the area.

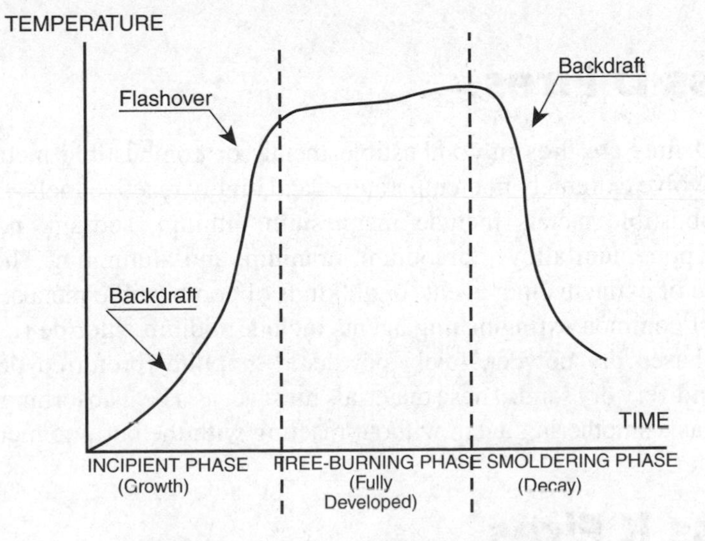


**BACKDRAFT AND FLASHOVER TIMELINE**

# CLASSIFICATION OF FIRE

There are five classifications of fire based on the type of fuel involved.

## Class A Fires

Class A fires include fires in ordinary combustibles (wood, wood products, paper, natural fibers, rubber, and plastics). Extinguishing fires in these types of materials requires water or foam or **clean agents** to absorb heat and smother the fire or dry chemical extinguishing agents (multipurpose) to inhibit the chemical chain reaction.

## Class B Fires

Class B fires include fires in animal-based, saturated fat cooking oils and greases, flammable and combustible liquids, and flammable gases. These fires need carbon dioxide to exclude air (oxygen), and dry chemical extinguishing agents, clean agents, or foam to inhibit the release of combustible vapors.

## Class C Fires

Class C fires are fires in live electrical equipment. These fires require an extinguishing agent that is nonconductive. Water mist (safe from electric shock), dry chemical extinguishing agents, carbon dioxide, and clean agents should be used on these types of fires.

**Note:** When electrical equipment is deenergized, extinguishing agents for Class A and Class B fires may be used.

## Class D Fires

Class D fires are fires in combustible metals or combustible metal alloys. They involve extremely hot temperatures and highly reactive fuels. Examples of combustible metals include magnesium, lithium, sodium, potassium, sodium potassium alloys, zirconium, uranium, and aluminum. There is no one type of extinguishing agent for all kinds of combustible metals. Some of the most common extinguishing agents include sodium chloride (table salt), copper-based dry powder, finely powdered graphite (preferred on lithium fires), and very dry sand. These materials must act as a heat absorbing medium as well as a smothering agent without reacting with the burning metal.

## Class K Fires

Class K fires are fires in unsaturated fat vegetable cooking oils used today with more efficient cooking appliances. These oils burn hotter than animal-based, saturated fat cooking oils and remain at high temperatures longer when used with new, thermally insulated cooking and frying equipment. They require wet chemical extinguishing agents.

# EXTINGUISHING AGENTS

## Water

Firefighters extinguish most fires using water. It is usually available in abundance at or near the fire. Water can be delivered onto the fire in a number of ways: hand lines stretched from the apparatus, hand lines stretched from a standpipe system located inside a building, sprinkler system, water mist system, master deluge nozzles, distributors, and so on.

Water extinguishes fire by **cooling** the material (absorption of heat), **smothering** (steam generation), **emulsification** (agitation of insoluble liquids to produce a vapor-inhibiting froth), and **dilution** (adding water to reduce the concentration of a burning soluble liquid and thereby raising its flash point). A review of the advantageous and disadvantageous properties of water follows.

### ADVANTAGEOUS PROPERTIES OF WATER

There are several advantages to water as an extinguishing agent. Some of the characteristics of water that make it advantageous in extinguishing fires are listed below.

- **Heavy, stable liquid** at ordinary temperatures.
- **High specific heat**—The heat capacity is given in terms of the mass of the substance in pounds and is the amount of heat required to raise the temperature of 1 pound of a substance 1°F. All solids, liquids, and gases have specific heats. Only two liquids have higher specific heats than water: ammonia and ether. One BTU is required to raise the temperature of 1 pound of water 1°F. To raise the temperature of 1 pound of water from 32°F to 212°F requires 180 BTU.
- **Latent heat of fusion—**Melting of 1 pound of ice into water at 32°F absorbs 143.4 BTU.
- **Latent heat of vaporization**—The latent heat of vaporization is the conversion of 1 pound of water into steam at a constant temperature with the absorption of 970.3 BTU.
- **Conversion to steam**—The conversion of liquid water to steam increases its volume approximately 1,600 times, which displaces an equal volume of air, thereby reducing the volume of oxygen available for the oxidation reaction.

### DISADVANTAGEOUS PROPERTIES OF WATER

Water also has disadvantages, some of which are listed below.

- It conducts electricity.
- It has low viscosity, which means it runs off smoldering material readily.
- It has high surface tension, so it has poor penetration qualities.

- It is transparent to radiated heat.
- It freezes at relatively high temperature.
- It displaces flammable liquids.
- It reacts violently with combustible metals.

## Foam

Mechanical foam used today is a mixture of air, water, and liquid foam concentrate (protein- or synthetic-based). The combination of these three ingredients produces a bubble blanket solution that will flow over and around combustible solids and liquids. Unlike water, foam is lighter than flammable liquids and will float on the surface of the liquid. Foam is used on both Class A and Class B fires. It will conduct electricity, however, and should not be used on energized electrical equipment. Foam extinguishes fires in four ways: cooling, smothering, suppressing vaporization, and separating the flames from the fuel surface, thereby reducing radiated heat feedback.

## Carbon Dioxide

Carbon dioxide ($CO_2$) is a relatively nonreactive gas that penetrates and spreads throughout the area of the fire. It is used primarily to extinguish Class B and Class C fires because it is nonconductive. $CO_2$ extinguishes fires by displacing the air (oxygen) content that is reacting with the fuel. It also has a cooling effect on the fire, which aids in lowering the temperature of the fuel below its ignition temperature.

## Dry Chemicals

The dry chemical ammonium phosphate extinguishing agent is used on Class A, Class B, and Class C fires, whereas the dry chemicals sodium bicarbonate or potassium bicarbonate are used to extinguish Class B and Class C fires only. Dry chemical agents are still being used to extinguish fires in animal-based, saturated fat cooking oils. These agents leave a residue that is nonflammable on the extinguished material, which reduces the likelihood of reignition. Dry chemical agents extinguish fires by cooling, smothering, and reducing radiated heat feedback, and most important, by inhibiting the chemical chain reaction of the oxidation process.

## Wet Chemicals

The wet chemical extinguishing agent contains a blend of potassium acetate, citrate, and water (40 to 60 percent). It is a low pH (acidic) agent that was originally developed for preengineered cooking equipment fire extinguishing systems for use on today's unsaturated fat vegetable cooking oils (Class K) that rapidly burn very hot and are difficult to extinguish. This agent is applied as a fine mist to help prevent grease splash and fire reflash while cooling the appliance.

## Clean Agents

Clean agents are gases, gas mixtures, and vaporized liquids that are replacements for halon gas extinguishing agents that are rapidly being phased out for being destructive to the Earth's ozone layer. These agents are nonconductive and noncorrosive, and they leave no residue and are therefore especially valuable in protecting telecommunication, electrical, and computerized equipment. Clean agents extinguish fire by interfering with the chemical chain reaction between oxygen and the fuel vapors, by absorbing heat, and by displacing the air (oxygen) content inside a room or space. Examples of clean agents include halocarbon gases, inert gases, powdered aerosols, and water mist. They are used to extinguish fires in Class A, Class B, and Class C materials.

## Dry Powder

Dry powder extinguishing agents are used to extinguish fires in combustible metals and combustible metal alloy (Class D) materials. They are generally in the form of a salt (sodium chloride) or powdered copper metal and graphite-based powder. Fire in a combustible metal causes the salt to cake and form a crust over the burning material. This transformation acts to smother the metal, thereby excluding oxygen and suppressing vaporization. Salts are preferred for the extinguishment of magnesium, uranium, and powdered aluminum, sodium, potassium, and sodium/potassium alloys. Powdered copper metal and graphite-based powder are preferred for fires involving lithium and lithium alloys.

# LABELING AND RATING OF PORTABLE FIRE EXTINGUISHERS

Fire extinguishers are labeled for quick identification of which classification of fire they will be effective on. The label shows both recommended and unacceptable uses for the extinguisher.

Also found on the label is the Underwriters Laboratories (UL) numerical rating. The system is divided into Class A and Class B ratings only. The rating system helps the user determine and compare the relative extinguishing effectiveness of portable fire extinguishers. A basic breakdown of the rating system is as follows.

The A rating is a water equivalent rating. The number 1 preceding the letter A is equal to 1.25 gallons of water. Numbers greater than 1 are multiplied by 1.25. For example, a 2A rating is equal to an extinguishing capacity of 2.5 gallons of water.

The B rating is an area of coverage equivalent rating. The number 1 preceding the letter B is equivalent to 1 square foot of coverage. Numbers greater than 1 are multiplied by 1. For example, a 20B rating is equivalent to 20 square feet of coverage.

**Note:** There is no numerical rating for Class C, Class D, and Class K fire extinguishers.

## REVIEW QUESTIONS

*Circle the letter of your choice.*

**1.** A rapid, self-sustaining oxidation process generating heat and light is commonly known as

(A) fuel
(B) oxygen
(C) fire
(D) pyrolysis

**2.** The percentage of oxygen in air at sea level is approximately

(A) 15%
(B) 21%
(C) 26%
(D) 28%

**3.** A rapid expansion of gases that have premixed prior to ignition is called

(A) an explosion
(B) an oxidizer
(C) an endothermic reaction
(D) spontaneous combustion

**4.** A detonation propagates at approximately what speed?

(A) Less than 200 feet per second
(B) Less than 600 feet per second
(C) More than 600 feet per second but less than 800 feet per second
(D) More than 1,000 feet per second

**5.** The primary flammable gas that ignites during a backdraft from the sudden influx of air into an oxygen-starved room is

(A) nitrogen
(B) carbon monoxide
(C) carbon dioxide
(D) hydrogen

**6.** A mechanical explosion caused by the weakening of a container shell due to heat via contact with fire is referred to as a

(A) BLEVE
(B) backdraft
(C) deflagration
(D) fire triangle

7. Spontaneous combustion is the result of what type of reaction?

(A) BLEVE
(B) Exothermic
(C) Mechanical
(D) Endothermic

8. An accurate statement describing the term known as pyrolysis is

(A) a physical explosion
(B) a rapid self-sustaining combustion process
(C) a decomposition reaction
(D) a container failure

9. The fire triangle is a model used to aid in the understanding of the three major elements required for ignition. Which of the following is not one of the elements of the fire triangle?

(A) Heat
(B) Fuel
(C) Oxidizer
(D) Chemical chain reaction

10. Another term used for heat as denoted in the fire triangle model is

(A) oxidizer
(B) chemical chain reaction
(C) thermal energy
(D) combustibles

11. The fire tetrahedron expands upon the one-dimensional fire triangle by adding a fourth element known as

(A) oxygen
(B) chemical chain reaction
(C) heat
(D) fuel

12. Which of the following is not a source of heat energy?

(A) Oxidation reaction
(B) Friction
(C) Fission of the atom
(D) Ignition temperature

13. Combustible matter commonly encountered by firefighters can be found in three states. Which of the following is not considered to be one of the three states of matter faced by firefighters on a daily basis?

(A) Plasma
(B) Liquid
(C) Gas
(D) Solid

14. Solids are materials with defined volume, size, and shape. Which of the following is an accurate example of a solid material?

(A) Kerosene
(B) Methane
(C) Cardboard
(D) Vinyl chloride

15. Which of the following is not considered to be a factor affecting the ignition and combustibility of wood and wood products?

(A) Thermal inertia
(B) Moisture content
(C) Physical form
(D) Type of weave

16. Wood having a low thermal inertia demonstrates what type of burning characteristic?

(A) It heats up and ignites readily.
(B) It heats up slowly.
(C) It is very difficult to ignite.
(D) None of the above.

17. Which of the following statements about a wood product having a large mass in relation to surface area compared to a similar wood product having less mass in relation to surface area is accurate?

(A) It will be less difficult to ignite the object with large mass.
(B) More heat energy will be required to ignite the object with large mass.
(C) Once the object with large mass is ignited, the rate of burning will be faster.
(D) All the statements above are accurate.

18. Wet wood is more difficult to ignite than dry wood. Wood will be very difficult to ignite when the moisture content rises above

(A) 8%
(B) 12%
(C) 15%
(D) Moisture content has no effect on combustibility.

19. The minimum temperature a material must be heated to for it to ignite and be self-sustaining without an external heat source is commonly referred to as a material's

(A) ignition temperature
(B) piloted-ignition temperature
(C) flash point
(D) specific gravity

20. Wood and wood products are classified as what type of fuel?

(A) Class A
(B) Class B
(C) Class C
(D) Class D

21. Plastics are common combustible solids encountered by firefighters. Which of the following is an accurate description concerning plastics and plastic materials?

(A) All plastics are petroleum based.
(B) All plastics are electrically conductive.
(C) Pyrolysis occurs less readily in plastics than in wood.
(D) Plastics are considered Class B materials.

22. Plastics can be divided into two categories: thermoplastics and thermosets. Which of the following statements concerning plastics is accurate?

(A) When subjected to the heat from a fire, thermoplastics decompose.
(B) Epoxies and alkyds are examples of thermoplastics.
(C) When subjected to the heat from a fire, thermosets melt and flow.
(D) Both thermoplastics and thermosets are formed by heat.

23. Textiles include clothing, bedding, and carpeting. Which of the following is an incorrect statement about textiles and textile fibers?

(A) Textiles can be divided into natural and synthetic fabrics.
(B) In general, all textile fibers are noncombustible.
(C) Natural fiber can be divided into plant and animal categories.
(D) Plant fiber will ignite more readily than animal fiber.

24. Which of the following selections is an example of plant fiber?

(A) Wool
(B) Mohair
(C) Camel hair
(D) Cotton

25. Which of the following statements about synthetic textile fibers is inaccurate?

(A) Examples of synthetic fibers are rayon, nylon, and polyester.
(B) Synthetic fiber is woven from artificial (plastic, metal, and glass) fiber.
(C) Synthetic fiber cannot be made flame retardant.
(D) Burning characteristics of synthetic fiber include decomposition and melting.

**26.** Combustible metals are solids that combine with oxygen to reach their ignition temperature and burn. A correct statement about combustible metals is that the metals

(A) undergo pyrolysis to produce combustible vapors when heated
(B) burn on their surface with no flaming combustion
(C) burn but do not produce an abundance of heat energy
(D) should be extinguished using water in small amounts

**27.** Combustible metals are classified as what type fuels?

(A) Class A
(B) Class C
(C) Class D
(D) Class K

**28.** Liquids are the second stage of matter. Which of the following is an incorrect statement about liquids that burn?

(A) Liquids have definite volume.
(B) Liquids have definite shape.
(C) Liquids that burn are classified as either combustible or flammable.
(D) Liquids can present other hazards (corrosiveness, toxicity) besides burning.

**29.** Gasoline is an example of a

(A) combustible liquid
(B) flammable liquid
(C) soluble liquid
(D) viscous liquid

**30.** Liquids that burn are classified as what type of fuels?

(A) Class B
(B) Class C
(C) Class D
(D) Class A

**31.** The minimum temperature of a liquid at which it emits vapors to form an ignitable mixture with air is known as its

(A) ignition temperature
(B) specific gravity
(C) boiling point
(D) flash point

**32.** A liquid is classified as combustible if its flash point is

(A) less than 100°F
(B) less than 70°F
(C) 100°F or more
(D) All of the above

**33.** Which of the following is an incorrect statement about the normal boiling point of a liquid?

(A) It is the temperature at which the liquid will liberate the most vapors.
(B) It is the temperature at which the vapor pressure of the liquid and atmospheric pressure are equal.
(C) The normal boiling point of a liquid is the temperature at which it will boil at sea level (14.7 psia).
(D) It is impossible to elevate the temperature of a liquid above its boiling point.

**34.** The specific gravity of a liquid is most accurately defined as

(A) the ratio of the weight of the liquid to the weight of an equal volume of water
(B) a measure of a liquid's flow through an opening in relation to time
(C) the percentage by weight of the liquid that will dissolve in water
(D) None of the above

**35.** The specific gravity of water is

(A) 0.8
(B) 1
(C) 1.4
(D) 14.7

**36.** Sulfuric acid, with a specific gravity of 1.8, will do what when mixed with water?

(A) It will sink.
(B) It will float.
(C) It will dissolve completely.
(D) None of the above

**37.** Asphalt is an example of what type of liquid?

(A) Soluble
(B) Flammable
(C) Viscous
(D) Alcohol

**38.** An important concept for firefighters to understand regarding gases and vapors emitted from liquids is vapor density. The term *vapor density* is most accurately stated as

(A) the relative density of the gas or vapor as compared to water
(B) the relative density of the gas or vapor as compared to air
(C) the percentage of gas or vapor that will mix with air
(D) the temperature required to cause the gas or vapor to rise

**39.** The vapor density of air is

(A) 0.55
(B) 2.1
(C) 1
(D) 1.5

**40.** Vapors heavier than air will have a vapor density of

(A) between 0.5 and 1
(B) less than 1
(C) more than 1
(D) 0.1

**41.** Gases that burn are classified as what type of fuels?

(A) Class A
(B) Class B
(C) Class D
(D) Class K

**42.** Helium is an example of what type of gas?

(A) Toxic
(B) Reactive
(C) Flammable
(D) Inert

**43.** Flammable range is defined as the ratio of gas or vapor in air that is between the upper and lower flammable limits. A gas or vapor above the upper flammable limit would

(A) be too rich to burn
(B) be too lean to burn
(C) ignite instantly
(D) None of the above

**44.** Which of the following is not considered a flammable gas?

(A) Acetylene
(B) Hydrogen
(C) Carbon dioxide
(D) Propane

**45.** A substance containing oxygen that will chemically react with fuel to start and/or feed a fire is the definition of a (an)

(A) toxic gas
(B) reactive gas
(C) oxidizer
(D) inert gas

**46.** An example of a reactive gas is

(A) nitrogen
(B) fluorine
(C) hydrogen cyanide
(D) argon

**47.** An oxygen-enriched atmosphere that will enhance the rate of burning will have what percent of oxygen in air?

(A) Less than 15%
(B) More than 21%
(C) Less than 21%
(D) Between 15% and 21%

**48.** During the chemical chain reaction of the combustion process, the heat directed back onto the burning material that aids in raising more fuel to its ignition temperature is called

(A) entrainment
(B) oxidation reaction
(C) latent heat of combustion
(D) radiated feedback

**49.** Heat is a measure of the quantity of energy contained within a material. When discussing heat units, the amount of heat energy required to raise the temperature of one pound of water (measured at 60°F at sea level) one degree Fahrenheit is called

(A) British thermal unit (BTU)
(B) joule
(C) calorie
(D) large calorie

**50.** The stored energy within combustible materials that is analyzed by fire protection engineers when designing and installing fire extinguishing systems is known as

(A) latent heat of vaporization
(B) latent heat of fusion
(C) latent heat of combustion
(D) none of the above

**51.** Wood and paper contain, within their structure, approximately how much heat energy (in BTU) per pound?

(A) 20,000–23,000
(B) 16,000–21,000
(C) 12,500–15,000
(D) 6,000–7,000

**52.** The amount of heat energy required to raise the temperature of one gram of water (measured at 15°C at sea level), one degree Celsius is called a

(A) British thermal unit (BTU)
(B) joule
(C) calorie
(D) none of the above

**53.** The temperature of an object is clearly and accurately stated as

(A) the total amount of molecular vibration inside a material
(B) a measure of how fast molecules are moving within a material
(C) the latent heat of combustion inside a material
(D) the amount of heat energy provided by one watt flowing for one second

**54.** On the Fahrenheit temperature scale, how many increments (degrees) are there between the temperature of melting ice and the boiling of water?

(A) 100
(B) 150
(C) 180
(D) 212

**55.** On the absolute temperature scales (Rankine and Kelvin), the zero point (absolute zero) represents a temperature at which

(A) water ceases to exist
(B) water boils
(C) all molecular movement ceases
(D) none of the above

**56.** Which of the following is not a means of heat transfer?

(A) Conduction
(B) Convection
(C) Direct flame contact
(D) Dilution

**57.** Conduction is most accurately defined as

(A) transfer of heat energy through a circulating medium (air)
(B) transfer of heat energy through a solid via molecular motion
(C) transfer of heat energy via infrared/ultraviolet rays
(D) none of the above

**58.** The amount and rate of heat transfer by conduction is dependent upon

(A) how readily hot air rises and travels up through the building
(B) moisture content in the air
(C) wind
(D) thermal conductivity of the material

**59.** Which item listed below is the best conductor of heat energy?

(A) Steel
(B) Wood
(C) Cloth
(D) Air

**60.** During firefighting operations, hot air expands and rises, as do the products of incomplete combustion. This behavior is what type of heat transfer method?

(A) Conduction
(B) Convection
(C) Direct flame contact
(D) Radiation

**61.** All things being equal, the darker and duller a solid object is, the more heat it will absorb and the greater chance it will reach its

(A) boiling point
(B) specific gravity
(C) thermal conductivity
(D) ignition temperature

**62.** Radiated heat waves will travel through space until they are absorbed by

(A) an opaque object
(B) air
(C) a transparent object
(D) water

**63.** Radiated heat (infrared/ultraviolet) rays travel through space at the speed of

(A) sound
(B) light
(C) gravity
(D) None of the above

**64.** Auto-exposure (flame traveling upward and outward from a window to a neighboring window above) is an example of what type of heat transfer method?

(A) Radiation
(B) Conduction
(C) Direct flame contact
(D) All of the above

**65.** There are three phases of fire. Which of the following is not considered to be one of the three phases?

(A) Incipient (growth) phase
(B) Backdraft
(C) Free-burning (fully developed) phase
(D) Smoldering (decay) phase

**66.** Most fires extinguished by firefighters are in what phase?

(A) Incipient
(B) Free-burning
(C) Smoldering
(D) Flashover

**67.** During what phase of a fire does the oxygen content in a room fall below 15 percent?

(A) Incipient
(B) Free-burning
(C) Backdraft
(D) Smoldering

**68.** What fire phenomenon can occur when fresh air is introduced into the fire area during the smoldering phase?

(A) Flashover
(B) Pyrolysis
(C) Backdraft
(D) None of the above

**69.** What is the missing component of the fire triangle that, when introduced, can cause a flashover situation in a room to occur?

(A) Fuel
(B) Heat
(C) Air (oxygen)
(D) Chemical chain reaction

**70.** What fuel (gas) listed below will not contribute to a backdraft situation?

(A) Oxygen
(B) Carbon monoxide
(C) Hydrogen
(D) Nitrogen

**71.** Flashover occurs during what phase of fire?

(A) Incipient (growth)
(B) Free-burning (fully developed)
(C) Smoldering (decay)
(D) All of the above

72. Fill in the blanks to answer the question. During the free-burning phase of fire, oxygen content in the fire area drops from ______% to below _____%.

(A) 15, 10
(B) 21, 15
(C) 10, 5
(D) 21, 10

73. There are five classifications of fire. Which of the following is not considered one of the five classes?

(A) Class A
(B) Class B
(C) Class K
(D) Class E

74. Fire is categorized alphabetically into Classes according to

(A) the time/temperature curve
(B) the type of fuel involved
(C) the phases of fire
(D) thermal conductivity

75. All of the materials listed below are considered Class A materials with the exception of

(A) magnesium
(B) paper
(C) wood
(D) plastics

76. All of the following extinguishing agents are used to put out Class A fires except

(A) clean agent
(B) foam
(C) carbon dioxide
(D) water

77. Which of the following is not considered to be a Class B material?

(A) Animal-based, saturated fat cooking oil
(B) Unsaturated fat vegetable cooking oil
(C) Fuel oil
(D) Hydrogen

78. Which of the following is a Class B extinguishing agent?

(A) Dry sand
(B) Wet chemical
(C) Dry chemical
(D) Copper-based dry powder

**79.** Fires in live electrical equipment are regarded as what classification of fire?

(A) Class A
(B) Class B
(C) Class C
(D) Class K

**80.** When live electrical equipment is deenergized, extinguishing agents for what class fires should be used?

(A) Class A and Class D fires
(B) Class B and Class K fires
(C) Class D and Class K fires
(D) Class A and Class B fires

**81.** Fires in combustible metals or combustible metal alloys are considered to be what classification of fire?

(A) Class A
(B) Class B
(C) Class C
(D) Class D

**82.** Select from the choices below the extinguishing agent that can be used to put out fire in certain Class D materials.

(A) Graphite
(B) Wet chemical
(C) Dry chemical
(D) Foam

**83.** Class D agents act in what ways to extinguish fires involving Class D materials?

(A) They deenergize the material and cool.
(B) They absorb heat and smother.
(C) They lower the oxygen level in the room to below 15 percent and deflect heat.
(D) All of the above

**84.** Fires involving unsaturated fat vegetable cooking oils are considered to be what classification of fire?

(A) Class B
(B) Class C
(C) Class D
(D) Class K

**85.** When comparing the burning characteristics of unsaturated fat (vegetable cooking oils) to animal-based, saturated fat cooking oils, vegetable oils

(A) burn slower and cooler than saturated fat oils
(B) ignite slower and burn cooler than saturated fat oils
(C) ignite faster and burn hotter than saturated fat oils
(D) have the same burning characteristics as saturated fat oils

**86.** Most fires are extinguished by firefighters using water. It can extinguish a fire in all of the following ways with the exception of

(A) cooling
(B) conducting electricity
(C) smothering
(D) emulsification

**87.** Adding water to reduce the concentration of a burning flammable liquid and thereby raising its flash point is an example of what type of extinguishing method?

(A) Dilution
(B) Cooling
(C) Smothering
(D) Emulsification

**88.** A principal advantage of water as an extinguishing agent is that it

(A) is transparent to radiated heat
(B) has high surface tension
(C) has low viscosity
(D) has high specific heat

**89.** To raise the temperature of 1 pound of water from 32°F to 212°F requires approximately how many BTU of energy?

(A) 1
(B) 32
(C) 212
(D) 180

**90.** The conversion of 1 pound of water into steam at a constant temperature with the absorption of approximately 970 BTU of energy is known as water's

(A) latent heat of fusion
(B) latent heat of vaporization
(C) displacement
(D) None of the above

**91.** The conversion of water from a liquid to steam is an important extinguishing property of water in that it

(A) decreases the volume of the water
(B) displaces oxygen in the air
(C) increases the volume of oxygen available for the oxidation process
(D) All of the above

**92.** A disadvantage of water concerning its extinguishing action is its

(A) latent heat of fusion
(B) heavy, stable liquid
(C) violent reaction with combustible metals
(D) None of the above

**93.** Mechanical foam extinguishing agent is a mixture of air, water, and liquid foam concentrate. It extinguishes fire using four actions. Which of the following is not considered one of the four actions?

(A) Deenergizing
(B) Smothering
(C) Cooling
(D) Reducing radiated heat feedback

**94.** Carbon dioxide is a relatively nonreactive gas that extinguishes fire primarily by

(A) suppressing vaporization and reducing radiated heat rays
(B) producing a bubble blanket solution
(C) displacing the air (oxygen)
(D) preventing grease splash

**95.** Dry chemical agents primarily extinguish fire by

(A) inhibiting the chemical chain reaction
(B) smothering
(C) cooling
(D) increasing radiated heat feedback

**96.** Wet chemical agent was originally developed for fire extinguishing systems protecting unsaturated fat (vegetable cooking oils). This agent extinguishes fire by

(A) inerting the atmosphere
(B) preventing grease splash and fire reflash
(C) deenergizing the equipment
(D) All of the above

**97.** Clean agents are gases, gas mixtures, and vaporized liquids that are replacements for halon fire extinguishing gases. They are especially valuable for protecting telecommunication, electrical, and computerized equipment for all but which of the following reasons?

(A) They are nonconductive.
(B) They are noncorrosive.
(C) They leave no residue.
(D) They have a low pH value.

**98.** Dry powder extinguishing agents put out fire in combustible metals and combustible metal alloys by

(A) reacting violently causing air (oxygen) displacement
(B) generating steam and hydrogen gas
(C) excluding oxygen and suppressing vaporization
(D) All of the above

**99.** Portable fire extinguishers are labeled for what purpose?

(A) To warn people not to use them around plants and animals
(B) To indicate what first aid measures to take if an agent is ingested
(C) To provide a water equivalent and square footage rating for Class A and Class B fires
(D) To provide washing instructions should the agent come in contact with clothing

**100.** Underwriters Laboratories (UL) provide a numerical rating for portable fire extinguishers to help determine and compare their relative effectiveness. There is no numerical rating for what class of fire materials?

(A) Class C, Class D, and Class K
(B) Class B, Class C, and Class D
(C) Class A, Class B, and Class C
(D) None of the above

# Answer Key

1. C
2. B
3. A
4. D
5. B
6. A
7. D
8. C
9. D
10. C
11. B
12. D
13. A
14. C
15. D
16. A
17. B
18. C
19. A
20. A
21. C
22. D
23. B
24. D
25. C
26. B
27. C
28. B
29. B
30. A
31. D
32. C
33. D
34. A
35. B
36. A
37. C
38. B
39. C
40. C
41. B
42. D
43. A
44. C
45. C
46. B
47. B
48. D
49. A
50. C
51. D
52. C
53. B
54. C
55. C
56. D
57. B
58. D
59. A
60. B
61. D
62. A
63. B
64. C
65. B
66. A
67. D
68. C
69. B
70. D
71. A
72. B
73. D
74. B
75. A
76. C
77. B
78. C
79. C
80. D
81. D
82. A
83. B
84. D
85. C
86. B
87. A
88. D
89. D
90. B
91. B
92. C
93. A
94. C
95. A
96. B
97. D
98. C
99. C
100. A

# ANSWER EXPLANATIONS

Following you will find explanations to specific questions that you might find helpful.

**1.** C Fire

**2.** B 21%

**3.** A An explosion

**4.** D More than 1,000 feet per second

**5.** B Carbon monoxide

**6.** A BLEVE (boiling liquid expanding vapor explosion)

**7.** D Endothermic reaction—a chemical reaction causing the absorption of heat
New substances formed by the chemical reactions contain more heat energy than prior to the reaction. An example is spontaneous combustion.

**8.** C A decomposition reaction

**9.** D Fire triangle: heat, fuel, oxidizer; chemical chain reaction is a component of the fire tetrahedron

**10.** C Thermal energy

**11.** B Chemical chain reaction

**12.** D Heat sources: oxidation reaction, friction, fission of the atom, sparks, arcs, the sun, compression of gases, electrical energy, hot surfaces. Ignition temperature is not a heat source.

**13.** A The three states of matter encountered by firefighters: liquid (gasoline), gas (methane), and solid (paper). Plasma is a hot ionized gas consisting of approximately equal numbers of positively charged ions and negatively charged electrons. The characteristics of plasmas are significantly different from those of ordinary neutral gases so that plasmas are considered a distinct "fourth state of matter."

**14.** C Cardboard. Kerosene is a liquid; methane is a gas; vinyl chloride is a gas.

**15.** D Type of weave (factor affecting the ignition and combustibility of textile products)

**16.** A Thermal inertia—resistance to heating, generally based on the specific gravity and density of the material, is known as its thermal inertia. Materials with a low thermal inertia (low specific gravity and density) will heat up and ignite more readily than materials with a high thermal inertia, specific gravity, and density.

**17.** B The greater the mass in relation to surface area, the more heat energy will be required to ignite it and the slower the rate of burning will be once ignited.
A more difficult; C slower; D incorrect statement

**18.** C 15%

**20.** A Class A fuels: wood, wood products, paper, cardboard, plastics, textiles
Class B fuels: combustible and flammable liquids, flammable gases
Class C fuels: energized electrical equipment and wiring
Class D fuels: combustible metals

**21.** C Pyrolysis doesn't occur as readily in plastics as in wood, and therefore plastics tend to have a higher ignition temperature than wood and wood products. A most; B can be electrically conductive or nonconductive (insulators); Class A

**22.** D A melt; B thermosets; C decompose

**23.** B Combustible

**24.** D Cotton
Wool, mohair, and camel hair are animal fibers.

**25.** C Can

**26.** B A do not undergo pyrolysis; C do produce an abundance of heat energy; D large amounts

**27.** C Class K fuels are unsaturated fat vegetable cooking oils used today with more efficient cooking appliances.

**28.** B A liquid has definite volume but takes the shape of the container it is being stored in.

**29.** B Flammable liquid with a flash point of less than 100°F

**30.** A Class B

**31.** D Flash point

**32.** C 100°F or more

**33.** D It is impossible to raise the temperature of a liquid above its boiling point, except if it is under pressure.

**34.** A B is the definition of viscosity of a liquid; C is the definition of solubility of a liquid.

**35.** B 1

**36.** A The specific gravity of a liquid is the ratio of the weight of the liquid to the weight of an equal volume of water. The specific gravity of water is 1. A liquid with a specific gravity less than water (gasoline, 0.8) will float on water, whereas a liquid with a specific gravity more than 1 (sulfuric acid, 1.8) will sink.

**37.** C The solubility of a liquid is the percentage by weight of the liquid that will dissolve in water. The solubility of a liquid ranges from negligible (less than one-tenth of 1 percent) to complete (100 percent).

**38.** B Vapor density is the relative density of the gas or vapor as compared to air. The vapor density of air is 1. A gas or vapor with a vapor density more than 1 (butane, 2.1) will be heavier than air and travel along the ground surface, where it may encounter an ignition source. A gas or vapor with a vapor density less than 1 (methane, 0.55) will rise and disperse readily into the air.

**39.** C 1

**40.** C More than 1

**41.** B A—A classification is for solids; C—D classification is for metals; D—K classification is for vegetable cooking oils, lards and fats

**42.** D An inert gas is a nonflammable gas that will not support combustion.

**43.** A Too rich to burn

**44.** C *Carbon dioxide* is a colorless, odorless, noncombustible *gas.*

**45.** C Oxidizer

**46.** B Fluorine; A inert, C toxic, D inert

**47.** B More than 21%

**48.** D Radiated feedback

**49.** A British thermal unit (BTU)
B joule: the amount of heat energy provided by 1 watt flowing for 1 second.
C calorie: the amount of heat energy required to raise the temperature of 1 gram of water (measured at 15°C at sea level) by 1°C.
D large calorie: the quantity of energy required to raise the temperature of 1 kg of water 1°C (more precisely, from 14.5 to 15.5°C); it is 1,000 times the value of the small calorie; used in measurements of the heat production of chemical reactions.

**50.** C Latent heat of combustion

**51.** D 6,000 to 7,000 BTUs per pound

**52.** C Calorie

**53.** B A measure of how fast molecules are moving within a material

**54.** C There are 180 increment degrees between the temperature of melting ice (32°) and the boiling of water (212°) on the Fahrenheit temperature scale.

**55.** C The zero point on the Rankine scalc is set at absolute zero, which is –457.6°F, the hypothetical point at which all molecular movement ceases. The Kelvin scale has its zero point set at absolute zero, which is –273.1°C.

**56.** D Dilution: adding water to reduce the concentration of a burning soluble liquid and thereby raising its flash point

**57.** B Conduction: transfer of heat energy through a solid via molecular motion
A Convection—transfer of heat energy through a circulating medium (air)
C Radiation—transfer of heat energy via infrared/ultraviolet rays

**58.** D A—relates to heat transfer via convection; B—incorrect; C—incorrect.

**59.** A Steel—dense materials (metals) are good conductors of heat energy. Fibrous materials (wood, paper, cloth) and air are poor conductors.

**60.** B Convection

**61.** D Solid objects that are dark and/or dull will absorb more radiated heat than light colored and/or shiny solid objects. This can lead to the solid object reaching its ignition temperature and bursting into flames.

**62.** A Radiated heat waves can pass through air, transparent objects and water.

**63.** B Speed of light

**64.** C Direct flame contact

**65.** B Three phases of fire: incipient (growth), free-burning (fully developed), smoldering (decay)

**66.** A Incipient (growth)

**67.** D Smoldering (decay)

**68.** C Backdraft; the influx of oxygen in the air will complete the fire triangle and cause reignition of the flammable gas mixture in the area.

**69.** B Heat; fire gases reaching their ignition temperatures, causing the entire area's contents to become suddenly engulfed in fire.

**70.** D Nitrogen is an inert gas and will not support combustion.

**71.** A During the incipient (growth) phase there is the possibility of fire gases reaching their ignition temperatures (flashover), causing the entire area's contents to become suddenly engulfed in fire, greatly increasing the temperature of the fire and leading to the next phase of fire, the free-burning (fully developed phase).

**72.** B 21%, 15%

**73.** D The five classifications of fire are: Class A—solid materials (wood, plastics, textiles); Class B—(liquids and gases); Class K—vegetable cooking oils, lards, fats; Class C—energized electrical equipment; Class D—combustible metals

**74.** B Type of fuel involved

**75.** A Magnesium is a combustible metal and, therefore, a Class D material.

**76.** C Carbon dioxide is used primarily to extinguish Class B and Class C fires.

**77.** B Unsaturated fat vegetable cooking oil is considered to be a Class K material.

**78.** C Dry chemical
A dry sand is a Class D agent; wet chemical is a Class K agent; copper-based dry powder is a Class D agent.

**79.** C

**80.** D A—Class D agent is used on combustible metals only; B—Class K agent is used on vegetable cooking oils, lards and fats; C—is incorrect (see explanations for answers A and B).

**81.** D

**82.** A B—wet chemical agent is used to extinguish fires in the kitchen (vegetable cooking oils, lards and fats); C—dry chemical agent is not effective on combustible metal fires; D—foam agent can cause combustible metal to worsen as the water content in the foam reacts with the metal.

**83.** B Class D agents absorb heat and smother.

**85.** C Unsaturated fat vegetable cooking oils ignite faster and burn hotter when compared to saturated fat cooking oils.

**86.** B Water conducting electricity is an electrocution/shock hazard to firefighters and not an extinguishment benefit.

**87.** A Dilution

A Cooling: absorption of heat; smothering: steam generation; emulsification: agitation of insoluble liquids to produce a vapor-inhibiting froth

**88.** D High specific heat—The heat capacity is given in terms of the mass of the substance in pounds and is the amount of heat required to raise the temperature of 1 pound of a substance 1°F. All solids, liquids, and gases have specific heats. Only two liquids have higher specific heats than water: ammonia and ether.

**89.** D One BTU is required to raise the temperature of 1 pound of water 1°F. To raise the temperature of 1 pound of water from 32°F to 212°F requires 180 BTU.

**90.** B Latent heat of vaporization

A Latent heat of fusion—melting of 1 pound of ice into water at 32°F absorbs 143.4 BTU.

C Displacement—of flammable liquid spills to spread over a greater area

**91.** B Displaces the oxygen in the air

A increases the volume of water approximately 1,600 times

C decreases the volume of oxygen available for the oxidation process

**92.** C Violent reaction with combustible metals

A advantage; B advantage

**93.** A Foam extinguishes fires in four ways: cooling, smothering, suppressing vaporization, and separating the flames from the fuel surface, thereby reducing radiated heat feedback.

**94.** C A—is incorrect; B—describes foam extinguishing agent; D—describes wet chemical extinguishing agent

**95.** A Dry chemical agents primarily extinguish fire by inhibiting the chemical chain reaction of the fire.

**96.** B A—describes the characteristics of carbon dioxide extinguishing agent; C—is incorrect; D—is incorrect.

**97.** D Question is seeking an incorrect answer. Answers A, B, and C are accurate descriptions of why clean agents are valuable as halon fire extinguishing gas replacements for protecting electrical and energized equipment. pH value is used to determine hydrogen content in aqueous liquids, not gases.

**98.** C A and B—are not compatible with extinguishment of the hazard. D—is incorrect

**99.** C A, B, and D are incorrect.

**100.** A B—Class B fire extinguishers have a numerical rating; C—Class A and Class B fire extinguishers have a numerical rating. D—is incorrect.

CHAPTER 10

# Hydraulics, Water Supply/Distribution, and Fire Protection Systems

## BASICS OF HYDRAULICS

Hydraulics, as it relates to the fire service, is the study of water pressure, flow, friction loss, and supply systems. It is also a branch of physics that studies the mechanical properties of water in motion and at rest. A satisfactory amount of water on a fire is essential for successful extinguishment. An inadequate quantity of water, however, will threaten fireground operations. The ability to find a sufficient water supply and use it effectively to control a fire remains one of the most basic functions of the fire service. A knowledge of hydraulics, therefore, can help ensure effective firefighting streams.

Water is a compound molecule formed when two hydrogen atoms (H) combine with one oxygen atom (O). Between 32ºF and 212ºF, water exists in a liquid state. Below 32ºF, water converts to a solid state of matter called ice. Above 212ºF, water converts into a gas called water vapor or steam.

### Definitions, Abbreviations, and Formulas

The following is a list of definitions and common abbreviations as they relate to fire service hydraulics:

**Attack hose**—designed to be used by trained firefighters to fight fires while inside a building.

**Back pressure**—generated by the weight of a column of water. Also known as head pressure. Figured at .433 pounds per square inch (psi) per foot of elevation.

**Cubic foot of water**—weighs 62.3 pounds and contains 7.48 gallons.

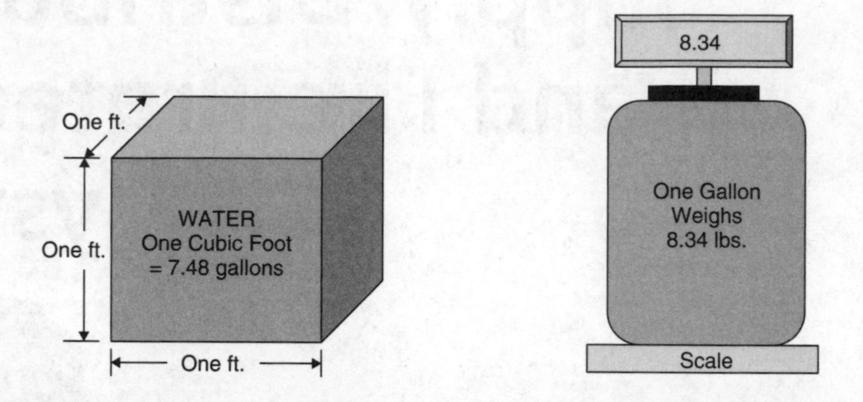


**Density**—weight (mass) of water per unit volume. $\rho$ (rho) = m/V. Density varies with the temperature of the water and is at its maximum at 39.2°F.

**Elevation pressure (EP)**—gained or lost due to elevation. When hose is stretched to an elevation higher or lower than the engine pump, elevation pressure must be calculated. As stated earlier, a column of water 1 foot high exerts a downward (head) pressure of .433 psi.

To find pressure when head is known: Pressure = .433 × H (Head)

The standard rule of thumb to determine head pressure when solving problems is:

0.5 POUNDS PER SQUARE INCH WILL LIFT WATER 1 FOOT

1.0 POUND PER SQUARE INCH WILL LIFT WATER 2 FEET

The reverse is also true:

0.5 POUNDS PER SQUARE INCH WILL BE GAINED FOR EACH 1 FOOT VERTICAL LOSS (drop) IN ELEVATION.

1.0 POUND PER SQUARE INCH WILL BE GAINED FOR EACH 2 FEET VERTICAL LOSS (drop) IN ELEVATION.

> **Example:** Looking at the picture on the next page, the ground level is 40 feet below the water level in the gravity water tank. To convert the elevation to pressure measured in psi, 1 foot of elevation change creates 0.433 psi of water pressure (40 ft. × 0.433 psi/ft. = 17.3 psi). Round this number up to 0.5 psi for fireground operations (40 ft. × .5 psi/ft. = 20 psi).

When supplying water to a sprinkler or standpipe system for a fire on the upper floor of a multistory building, for every 10 feet of height (or every story), a downward pressure of 5 psi is exerted. Pump pressure, therefore, must be increased accordingly to make up for this pressure loss.

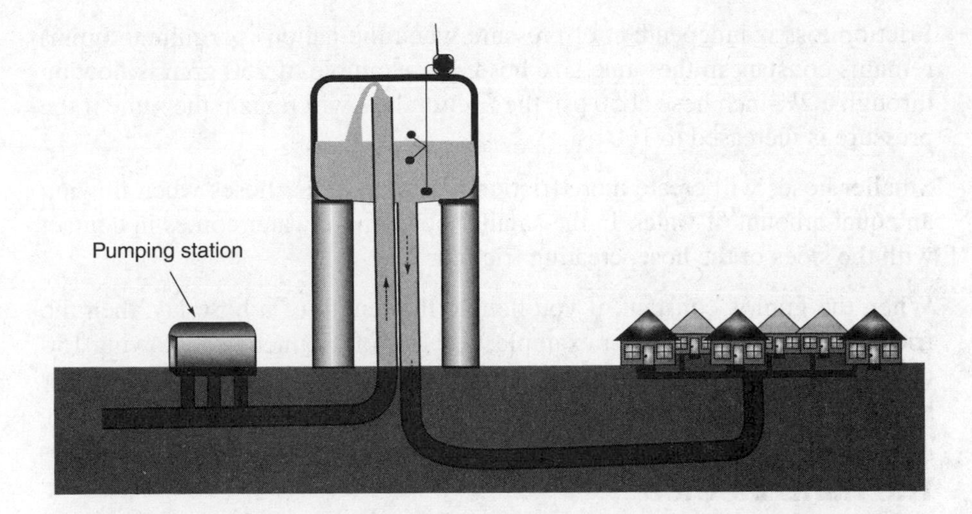


**Engine (pumper)**—fire apparatus consisting of a fire pump, water tank, and hose.

**Fire department connection (FDC)**—device that a pumper connects to in order to supply and augment the water flow in a standpipe and sprinkler fire protection systems. An FDC combines two hose lines into one and is also referred to as a siamese.

## FIRE DEPARTMENT CONNECTION

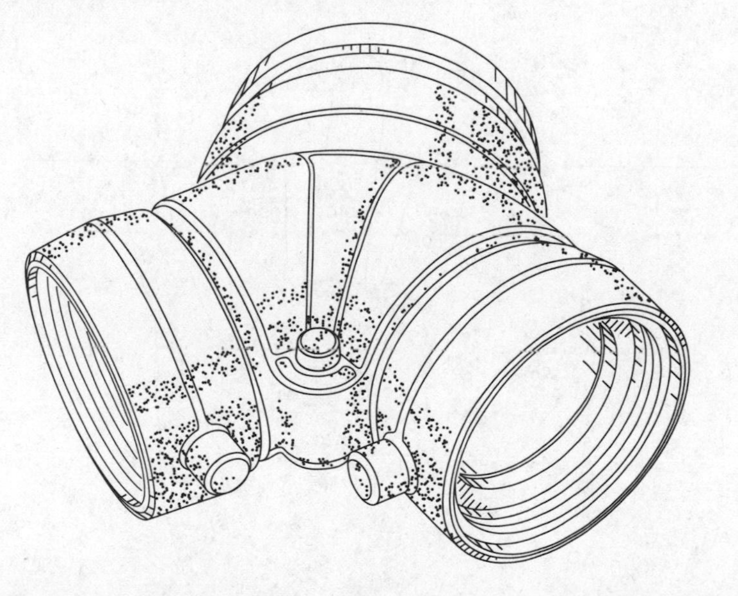

**Flow pressure**—created by the rate of flow of water coming from a discharge opening. Nozzle pressure, for example, is water discharged from a nozzle.

**Force (F)**—a simple measurement of weight, expressed in pounds. Force equals Pressure × Area (F = P × A)

**Friction loss (FL)**—pressure lost to overcome resistance while flowing water through a hose line or pipe. Causes of friction loss in a hose line include rough lining in the hose, sharp bends (kinks), couplings connecting lengths of hose improperly seated, protruding gaskets, change in hose size by adapters, and partially closed valves.

Friction loss is independent of pressure when the gallons per minute (gpm) remains constant in the same-size hose. For example, if 250 gpm is flowing through a 2½-inch hose at 50 psi, the friction loss will remain the same if the pressure is increased to 100 psi.

Smaller hoses will create more friction loss than larger hoses when flowing an equal amount of water. In the smaller hose, more water comes in contact with the sides of the hose, creating friction.

When the gpm is constant, if you double the length of a hose lay, then the friction loss will double. For example, 100 feet of 1¾-inch hose flowing 150 gpm has a 12 psi friction loss. Therefore, 200 feet of the same-size-diameter hose flowing 150 gpm will have a 24 psi friction loss.

## The Hand Method

One of the most widely used rules of thumb for determining friction loss is the hand method. This method works for flows from 100 to 500 gpm. It is used primarily for one of the most common attack hose lines: 2½-inch-diameter hose. It has also been modified to figure friction loss in another popular sized hose line: the 1¾-inch-diameter hose. Once the pump operator practices and becomes familiar with the hand method, it becomes very quick and easy to use. We will begin with the 2½-inch-diameter hose.

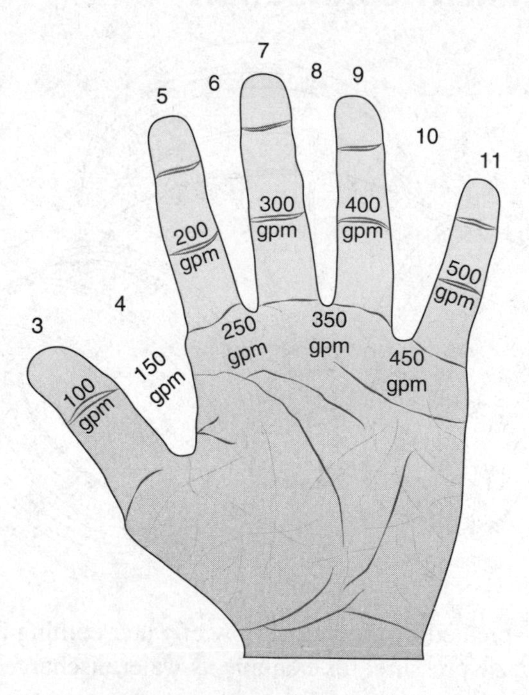


Starting with the thumb on the left hand, assign a gpm to the base of each finger. The thumb is 100, the index finger is 200, and so on. Between each pair of fingers are the half-hundred gpm (e.g., 150, 250). Move back to the thumb and assign odd numbers to the tips of each finger, starting with 3 (e.g., the thumb is 3, the index finger is 5). In between the fingertips are the even numbers. Divide the gpm flow by 100. Then multiply the quotient by the single-/double-digit number (at the fingertips) corresponding to the gpm flow. The product is the friction loss for each 100 feet of hose.

**Example 1:** What is the friction loss in 100 feet of 2½-inch-diameter hose flowing 200 gpm?

**Answer:** Divide 200 gpm by 100 = 2

$2 \times 5 = 10$ psi

**Example 2:** What is the friction loss in 400 feet of 2½-inch-diameter hose flowing 250 gpm?

**Answer:** Divide 250 gpm by 100 = 2.5

$2.5 \times 6 = 15$ psi (for each 100 feet of hose)

$15 \text{ psi} \times 4 = 60$ psi

The same method can be used for 1¾-inch-diameter hose with some modifications. This method is good only for flow rates between 100 and 200 gpm. Using this adapted method, the middle of each finger is labeled with the gpm and the tip of each finger is labeled 1 through 5, starting with the thumb. Take the flow in gpm and multiply the number at the tip of the finger corresponding to the gpm by 12. The product is the friction loss for each 100 feet of hose.

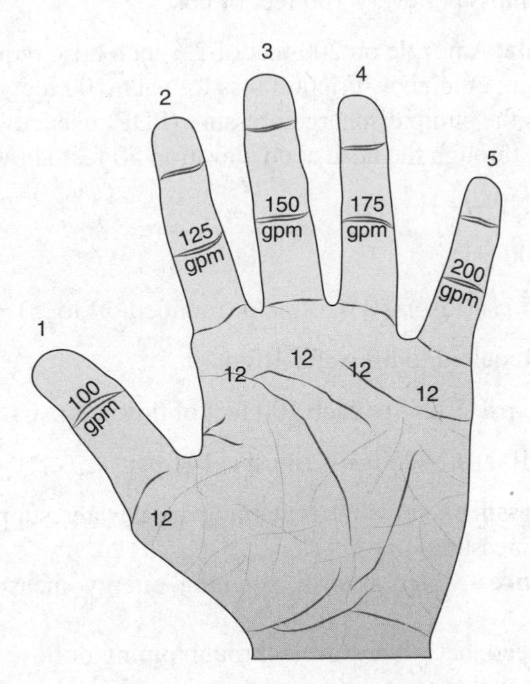


**Example 1:** Find the friction loss per 100 feet of 1¾-inch-diameter hose flowing 150 gpm.

**Answer:** 150 gpm corresponds to the number 3 at the tip of the finger

$3 \times 12$ (number found at the base of each finger) = 36 psi

**Example 2:** Find the friction loss in 300 feet of 1¾-inch-diameter hose flowing 175 gpm.

**Answer:** 175 gpm corresponds to the number 4 at the tip of the finger

$4 \times 12$ (number found at the base of each finger) = 48 psi

48 psi $\times$ 3 (300 feet of hose) = 144 psi

**Gallons per minute (gpm)**—a unit of volumetric flow rate.

**Large caliber stream**—a water flow of greater than 300 gpm.

**Nozzle pressure (NP)**—the water pressure in a fire hose measured at the nozzle.

**Pressure (P)**—force per unit area, expressed in pounds per square inch (psi); $P = F/A$

**Pump discharge pressure (PDP)**—is the amount of pressure in pounds per square inch (psi) indicated on the pressure gauge or any given orifice discharge gauge at the engine apparatus. PDP is the major controlling factor that changes nozzle pressure and volume of discharge.

PDP = NP +/– H + FL where:

PDP = Pump discharge pressure (discharge side of the pump)
NP = Nozzle pressure
+/– H = Head (elevation differential)
FL = Friction loss per every 100 feet of hose

**Example:** A nozzle on 200 feet of 2½-inch hose requires 100 psi for maximum efficiency. Friction loss for each 100 feet of hose is 8 psi. What is the pump discharge pressure (PDP) needed when 200 gpm is flowing through the hose at an elevation 50 feet above the pump?

PDP = NP +/– H + FL

NP = 100 psi

H = –25 psi (loss); 50 ft. $\times$ .433 (rounded off to .5) = –25 psi

0.5 psi required to lift water 1 foot

FL = 16 psi (8 psi for each 100 feet of hose), 8 psi + 8 psi =16 psi

PDP = 100 psi + 25 psi + 16 psi = 141 psi

**Residual pressure**—pressure remaining in a water supply when water is flowing; measured in psi.

**Static pressure**—water at rest; potential energy measured in psi (head pressure).

**Velocity**—speed that water travels through piping or hose line.

**Volume**—amount of space a three-dimensional object occupies. Measured in cubic units.

**Water hammer**—force created by the rapid deceleration of water, generally resulting from closing a hose nozzle or hydrant valve too quickly. The result is an energy surge transmitted in the direction opposite of water flow, often at many times the original pressure. Where large volumes of

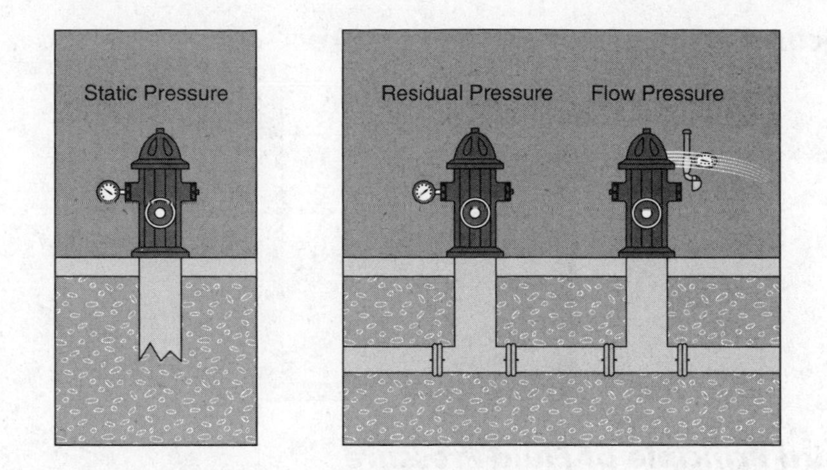


water are involved, such as in large-diameter hose layouts, water hammer can damage the fire pump, hose, and fire protection system piping.

## WATER HAMMER

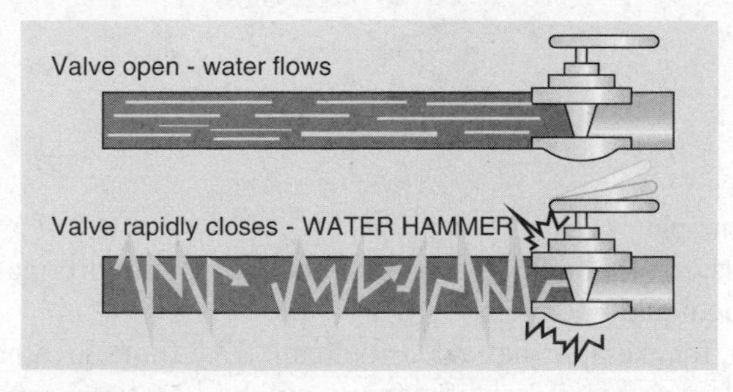


## Principles of Fluid Pressure

Water is acted on by the six principles of water under pressure. These principles are relevant to the study of fire service hydraulics.

A. First Principle
Fluid pressure is perpendicular to any surface on which it acts. If the pressure was not perpendicular, water would be in constant motion.

B. Second Principle
When a fluid is at rest, fluid pressure is the same in all directions.

C. Third Principle
Pressure applied to a confined fluid from without is transmitted equally in all directions. This principle denotes that water is incompressible.

D. Fourth Principle
The pressure at the base of a liquid in an open container is proportional to its depth.
One foot of water in an open container has a pressure of 0.433 psi. Two feet of water in an open container has a pressure of 0.866 psi. Three feet of water in an open container has a pressure of 1.302 psi.

### *Second Principle of Fluid Pressure*

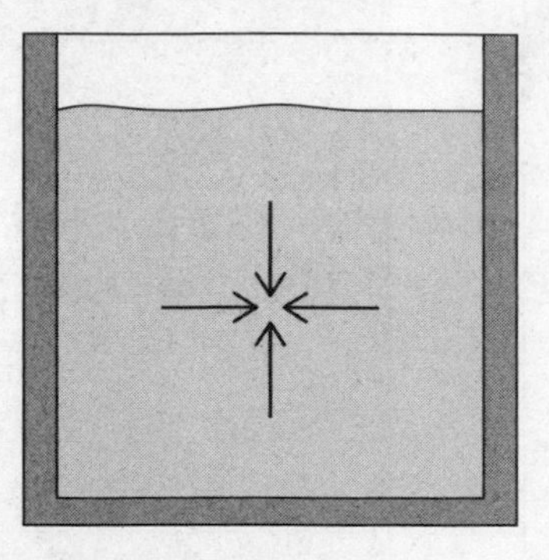

### *Third Principle of Fluid Pressure*

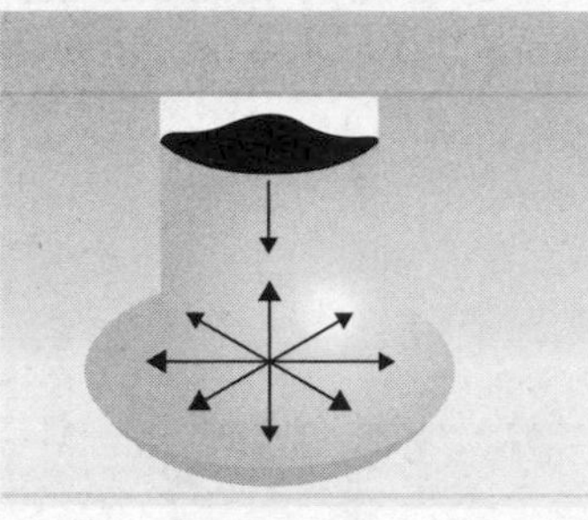

E. Fifth Principle
   The pressure of a liquid in an open container is proportional to the density of the liquid.
   Mercury, for example, has a density that is 13.6 times greater than water. Pressure created by 1 inch of mercury will be 13.6 times greater than the pressure created by 1 inch of water.
F. Sixth Principle
   Pressure only depends on the depth of the liquid, and is independent of the shape or volume of the container.

### *Sixth Principle of Fluid Pressure*

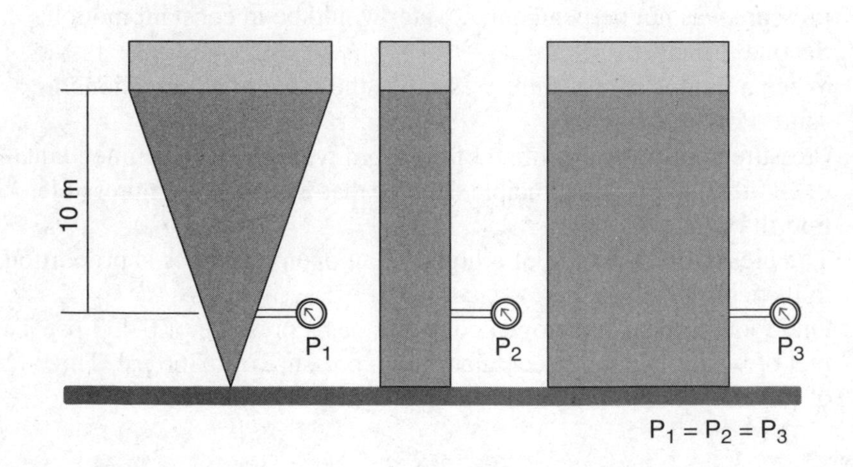

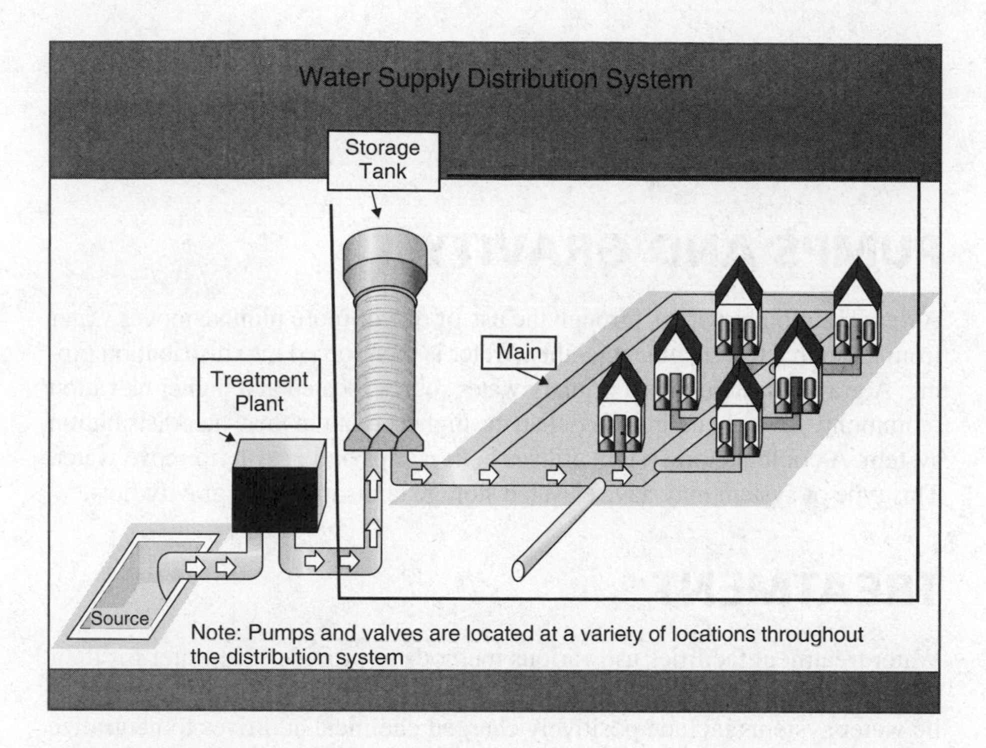


# HISTORY

Hydraulic principles relating to water supply and distribution date back 6,000 years. The Bible has numerous references to hydraulic problem solving pertaining to flood control and irrigation. Around 300 BC the Romans built large stone aqueducts to deliver water from upper elevations into the city of Rome. Remnants of this water supply system still exist today.

# COMPONENTS

A water supply and distribution system is engineered with hydraulic components that commonly include a source of water supply, pumping stations and/or elevation to move water, water treatment facilities, and a water distribution piping system to provide water to the consumer, as well as the fire service (fire hydrants and fire protection systems [sprinklers and standpipes]). Water supply and distribution systems are typically owned and maintained by municipalities or other public entities. Private water supply systems may also provide water under contract to a municipality. Storage and water delivery capacities should include maximum domestic (private) consumption combined with peak anticipated fire operational needs.

# SUPPLY

Natural sources, such as lakes and streams, are main sources of surface water. The water is withdrawn through pipe intakes extending from the shore into deep water. Intakes for large municipal supplies may consist of large conduits or tunnels. Manmade sources of water include towers, wells, and cisterns. Reservoirs are used to supplement the main source of the water supply and transmission system during peak demands. They also can provide water during a temporary failure of the supply system.

## PUMPS AND GRAVITY

A direct pumping system, through the use of one or more pumps, moves water from its source to a treatment facility. Water is then forced into distribution piping. A gravity system uses a primary water source located at a higher elevation (commonly several hundred feet) than the highest point in the water distribution system. A combination system utilizes both pumps and gravity to move water. This type of system may have elevated storage tanks to supply gravity flow.

## TREATMENT

Water treatment facilities use various methods to provide safe water for their communities. Today, the most common steps in water treatment used by public water systems include positively charged chemical additives to neutralize the negative charge of dirt and other dissolved particles in the water. The dirt then settles to the bottom of the water supply. The clear water on top will pass through filters to remove dust, parasites, bacteria, viruses, and chemicals. After the water has been filtered, a disinfectant (chlorine, for example) is added to eradicate any remaining parasites, bacteria, and viruses.

## PIPING

The distribution system consists of a series of pipes of different diameters. Adequate water delivery relies upon the carrying capacity of this piping. The most common type of pipe is as follows:

- **Cast iron**—most commonly used in water distribution systems mainly because they are relatively inexpensive, durable, and highly resistant to corrosion.
- **Steel**—used when pipes are subject to very high pressures and large-diameter pipe is required.
- **Galvanized iron (GI)**—used for water supply work into buildings.
- **Concrete**—unreinforced, reinforced, and prestressed large-diameter concrete pipe is generally used for major water supply works.
- **Plastic**—rigid polyvinyl chloride (PVC) pipes are used for water distribution piping. PVC is lightweight, inexpensive, and noncorrosive.

## FIRE HYDRANTS

There are two principal types of hydrants: dry barrel and wet barrel. Dry-barrel hydrants are pressurized and drained via a main valve located in the base of the hydrant. This type of hydrant does not have pressurized water up to its outlets. When the main valve is opened, the barrel is pressurized. When it is closed, the barrel drains. There are no valves at the outlets. The main valve is located below the frost line to protect the hydrant from freezing. Dry-barrel hydrants are especially suited for cold climates.

The wet-barrel hydrant has water in the barrel up to each of its outlet valves and is used exclusively in warm-weather areas of the country, where freezing is not an issue. Wet-barrel hydrants are simpler in construction than dry-barrel hydrants, and all their mechanical parts are above ground for easy accessibility. The outlet valves operate independently.

## Dry- and Wet-Barrel Hydrants



Fire hydrants are fitted on water distribution piping in urban and suburban areas where municipal water supply service is sufficient for firefighting use. The water supply design should strive to have all hydrants of the fire protection system within adequate distance to structures in need of protection from fire. Standard practice calls for the installation of hydrants every 500 ft.

Regularly scheduled inspections, testing, and maintenance of fire hydrants are essential to ensure optimal efficiency and operability. When water flows through pipes, its movement causes friction that results in a reduction of pressure. There is much less pressure loss in a water distribution system when fire hydrants are supplied from two or more directions. A fire hydrant that receives water from only one direction is known as a dead-end hydrant. Due to the limited supply to these type of hydrants, water may be insufficient during a major fire requiring large volumes of water.

## Gridiron Design System

Adequate and reliable distribution of water to hydrants is a product of sound engineering practices and an understanding of water distribution principles to support firefighting operations. Distribution systems include gridiron and branch design patterns. The most reliable means to provide water for firefighting is by designing redundancy into the system. This can be accomplished by connecting large-diameter transmission mains and smaller-diameter distribution mains together to form grids that allow water to flow from different directions. In the gridiron system, the piping is laid out in checkerboard fashion. Piping decreases in size as the distance increases from the source of supply.

Several advantages are gained by laying out water mains in this pattern. A grid with mains interconnecting at roadway intersections and other regular intervals will still allow water to be distributed through the system if a single section fails. The damaged section can be isolated, and the remainder of the system will still flow water. In addition, a gridiron system supplies water to fire hydrants from multiple directions. This has benefits during periods of peak fire flow demand. There will be less impact from friction loss in water mains because several mains will be sharing the supply. Discharges will, therefore, remain more stable when multiple hydrants are in use simultaneously. Street valves should be installed at every intersection for all mains tapping off from junctions. This design allows workers to isolate any single section of main to be taken out of service for repair without disrupting water service beyond the affected section of pipe.

*Gridiron System*

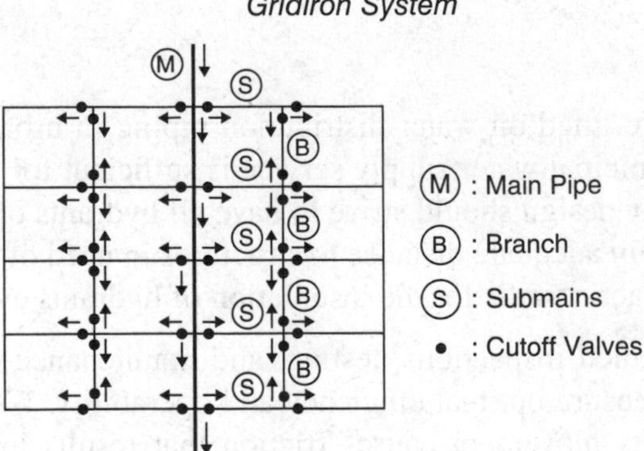

## Branch Design System

A branch system is a simple method of water distribution. Calculations are easy to formulate, and the required dimensions of the pipes are economical. This method of water distribution requires a comparatively small number of shutoff valves. Similar to the branching of a tree, a branch design pattern consists of a single main (trunk) line, reducing in size to submains and branches with increasing distance from the trunk's source of water supply. There is no water distribution to consumers directly from the trunk line. Submains are connected to the trunk line, and they are located along the main roads. Branches are connected to the submains, and they are situated along the streets. Service connections are provided to the consumers from the branches. Fire lines come off of the trunk line to feed fire hydrants and water-based fire protection systems.

Branch design is not commonly followed in modern waterworks practice. One reason for this is that water available for fighting fires will be limited because it is being supplied by only one water main. In addition, the pressure at the end of the line may become undesirably low as additional areas are connected to the water supply system. Other drawbacks include the fact that the area of the system receiving water from a pipe undergoing repair cannot be supplied until the work is finished. Branch systems also have a large number of dead ends in which water does not circulate and remains static.

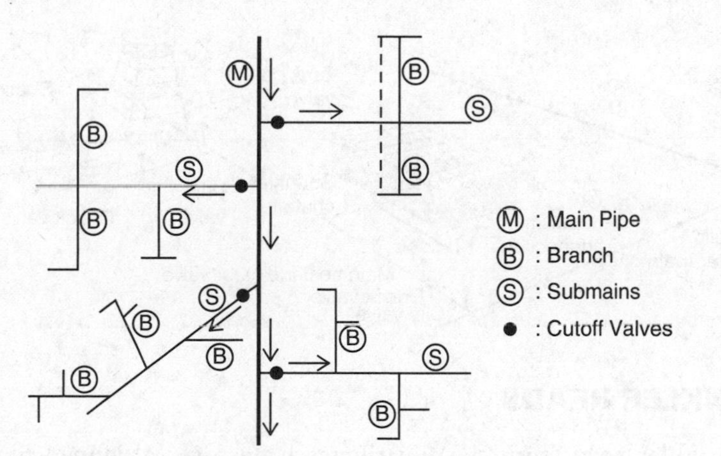


# FIRE PROTECTION SYSTEMS

## Sprinkler Systems

Fire sprinklers utilize water for direct application onto flames. This action cools the combustion process and prevents ignition of adjacent combustibles. Sprinkler systems are a series of water pipes supplied by a reliable water supply. At selected intervals along these pipes are independent, heat-activated sprinkler heads. Sprinkler heads are silent sentinels, always at the ready to deliver water on a fire when required.

During the beginning stage of a fire, heat output from burning combustibles is generally low and unable to cause sprinkler operation. As fire intensity increases, however, the sprinkler heads in the vicinity of the fire activate at a prescribed temperature. In many fire incidents, no more than two sprinkler heads are needed to control and suppress active flaming. In fast-growing fires, however, where flammable liquids are involved, more than a dozen sprinkler heads may be called into action.

Sprinkler systems are designed to deliver a satisfactory amount of water onto a fire hazard for control and/or extinguishment. Fire hazard is based on fuel load. Occupancy determines hazard classification:

**Light hazard occupancies**—have a relatively low amount of combustible materials present and the combustibility of the contents is low. These types of occupancies include schools, hospitals, and office buildings.

**Ordinary hazard occupancies**—laundries, automobile showrooms, and bakeries are just some of the occupancies grouped into the ordinary hazard classification. These occupancies hold a moderate amount of combustibles and the combustibility of the contents is moderate to high.

**Extra hazard occupancies**—include aircraft hangars, flammable liquid spraying areas, and plywood and particleboard manufacturing. Occupancies in this classification have large amounts of highly combustible materials.

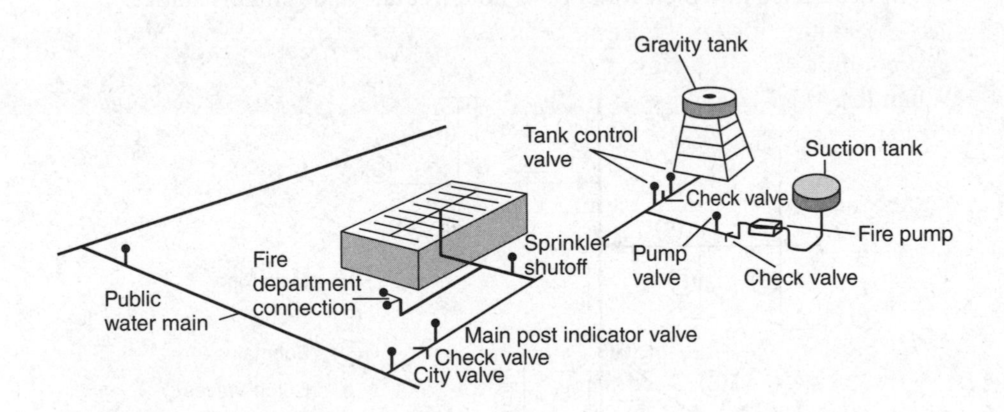


## SPRINKLER HEADS

The sprinkler head (sprinkler) distributes water over a defined fire hazard area. Each sprinkler operates by actuation of its own pre-engineered temperature linkage. The typical sprinkler consists of a frame, thermal operated linkage, cap, orifice, and deflector.

### Sprinkler Head Components

#### *Frame*

Provides the main structural component that holds the sprinkler together. The water supply pipe connects to the sprinkler at the base of the frame. The frame holds the thermal linkage and cap in place and supports the deflector during discharge. Standard finishes include brass, chrome, black, and white.

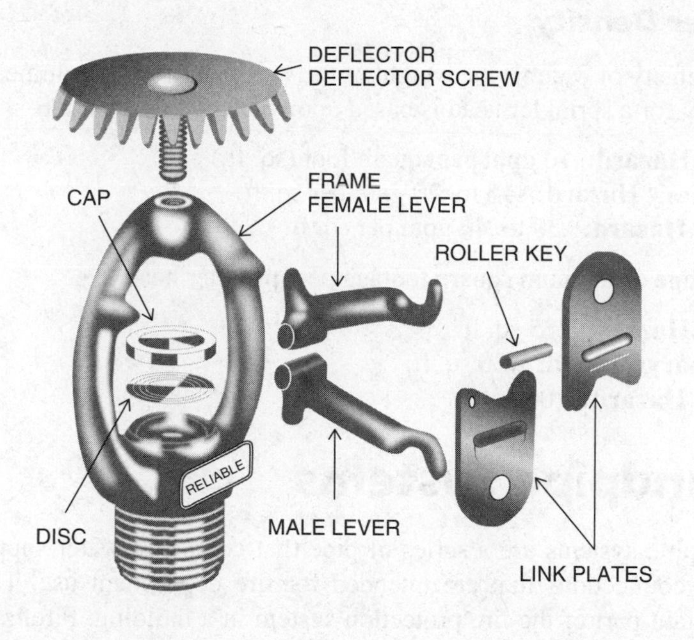


### *Thermal Linkage*

The thermal linkage is the component that controls water release. Under normal conditions the linkage holds the cap in place and prevents water flow. When the linkage is exposed to high heat, however, it weakens and releases the cap. Common linkage styles include soldered metal levers, frangible glass bulbs, and solder pellets.

### *Cap*

The cap is located over the sprinkler orifice. It provides a watertight seal. This element is held in place by the thermal linkage. Operation of the linkage causes the cap to drop from position and permit water flow. Caps are constructed of lightweight metal.

### *Orifice*

The machined opening at the base of the sprinkler frame is called the orifice. It is from this opening that water flows. Most orifice openings are ½-inch diameter. Extra hazard occupancies require larger diameter openings, whereas residential applications may have smaller bores.

### *Deflector*

The deflector is mounted on the frame opposite the orifice. The purpose of the deflector is to break up the water coming from the orifice in a more efficient distribution pattern. Common sprinkler mounting styles are upright (mounted above the pipe), pendant (mounted below the pipe), and sidewall (mounted in a lateral position from a wall).

### *Water Density*

The density of water is the weight of the water per its unit volume. The water density for a sprinkler head is based upon hazard classification:

**Light Hazard:** .10 gpm per square foot (sq. ft.)
**Ordinary Hazard:** .15 to .20 gpm per sq. ft.
**Extra Hazard:** .30 to .40 gpm per sq. ft.

Coverage (maximum square footage per sprinkler head)

**Light Hazard:** 225 sq. ft.
**Ordinary Hazard:** 130 sq. ft.
**Extra Hazard:** 100 sq. ft.

# Standpipe Systems

Standpipe systems are a series of pipe that connect a water supply to hose outlet connections that are intended for fire department use. They are an important part of the fire protection system in a building. Firefighters bring hoses inside the building and attach them to standpipe outlets located along the pipe throughout the structure. Water is fed into the piping system. The piping runs vertically (up and down) and horizontally (side to side) throughout the building. Risers (pipe running vertically) are normally installed inside staircase enclosures or in the hallways of the building. This piping system supplies water to every floor in the building.

The piping can be the sole supply for the standpipe system or include a combination system that supplies both a standpipe system and sprinkler system. The location of hose connections varies, but most commonly they are located in stairwells between floors. Standpipes are installed in very tall buildings (high-rise offices) as well as large area structures (malls).

## TYPES

Standpipes are broken down into types, based on the presence of water in the system. They may be wet (automatically charged with water) or dry (without water until manually supplied) by the fire department.

## WET STANDPIPE

Filled with pressurized water at all times. In contrast to dry standpipes, which can be used only by firefighters, wet standpipes can also be used by building occupants. A fire department connection can be used to augment the system's water supply.

## DRY STANDPIPE

Dry standpipes are not filled with water until needed in firefighting. Fire engines are used to supply this type of system with water through the fire department connection or hose outlet located on the lower floors of the building. Dry systems are installed in cold climates for buildings that are not heated.

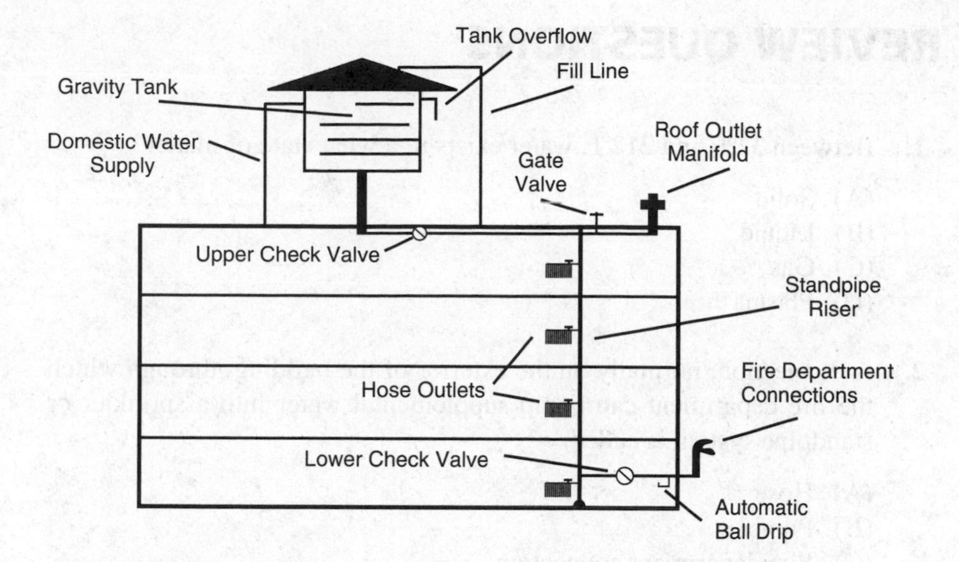


## CLASSIFICATIONS

Standpipe systems are classified depending on who is expected to use the system. The three classes are briefly described next.

Class I standpipes are designed for fire department use with 2½-inch-diameter hose and are intended for large-volume fire streams (250 gpm). Class I standpipes are typically not provided with preconnected fire hoses.

Class II standpipes are designed for trained occupant use with 1½-inch-diameter hoses. They are usually found in cabinets and are intended for small-volume fire streams (100 gpm). Class II standpipes often have pre-connected fire hoses.

Class III standpipes are designed for use by both firefighters and trained occupants. They are typically equipped with 1½-inch-diameter fire hose as well as 2½-inch fire hose.

# REVIEW QUESTIONS

1. Between 32ºF and 212ºF, water exists in a what state of matter?
   (A) Solid
   (B) Liquid
   (C) Gas
   (D) Plasma

2. A connection, normally on the exterior of the building, through which the fire department can pump supplemental water into a sprinkler or standpipe system is called:
   (A) Hose
   (B) Pumper
   (C) Fire department connection
   (D) Triplet valve

3. Five gallons of water weighs roughly how many pounds?
   (A) 45.6 pounds
   (B) 44.2 pounds
   (C) 41.7 pounds
   (D) 39.3 pounds

4. This hydraulic term is typically used to measure water flow rate or pump capacity.
   (A) Pounds per square inch (psi)
   (B) Gallons per minute (gpm)
   (C) Friction loss
   (D) Water density

5. Hoses designed to be used by trained firefighters to fight fires while inside a building are referred to as:
   (A) Attack hose
   (B) Supply hose
   (C) Large-caliber stream
   (D) Aggressive hose

6. Head pressure on a column of water 2.3 feet is approximately what?
   (A) .9 psi
   (B) 2.3 psi
   (C) 4.34 psi
   (D) .434 psi

7. The measurement of system pressure under water at rest (nonflow) conditions is called:

(A) Volume pressure
(B) Residual pressure
(C) Velocity pressure
(D) Static pressure

8. A water hammer is best described as what?

(A) Surge in pressure when a high-velocity flow of water is abruptly shut off
(B) Tool for opening stuck pipe valves
(C) Pressure exerted by flowing water in an open system
(D) Surge in pressure caused when nozzles on hose lines are shut down too slowly

9. The following chart contains information on three sizes of hose.

| Hose Diameter | Hose Length | Friction Loss per 50' Length | Maximum Working Pressure |
|---|---|---|---|
| 1¾-inch | 50' | 20 psi | 250 psi |
| 2½-inch | 50' | 5 psi | 250 psi |
| 3½-inch | 50' | 2.5 psi | 250 psi |

Reviewing the chart should cause you to select what statement as accurate?

(A) As the hose diameter increases, friction loss decreases by 10%
(B) The smaller the hose diameter, the lesser the friction loss
(C) The larger the hose diameter, the greater the friction loss
(D) The smaller the hose diameter, the greater the friction loss

10. A key characteristic of water that makes it useful in firefighting hydraulics is:

(A) Water is easily compressed
(B) Water does not leak readily out of a hose line
(C) Water boils at a high temperature
(D) Water is not easily compressed

11. Pressure applied to a confined liquid from without is transmitted equally in all directions. This principle demonstrates:

(A) Water is incompressible
(B) Pressure is directly related to height
(C) Pressure is proportional to density
(D) Pressure is independent of shape

12. Mercury has a density that is how many times greater than water?

(A) 10.8
(B) 11.9
(C) 12.1
(D) 13.6

13. One cubic foot of water has a density of 1g/mL. What is the density of one cubic foot of mercury?

(A) 4.77 g/mL
(B) 6.23 g/mL
(C) 13.7 g/mL
(D) 8.90 g/mL

14. Which one of these statements is accurate for when pressure is applied to water?

(A) Pressure can solidify water
(B) Pressure acts equally in all directions
(C) Pressure only acts vertically in a downward direction
(D) Pressure cannot act on water because water is incompressible

15. When a body is totally immersed in water, it is buoyed up by a force equal to:

(A) Weight of the body
(B) Weight of the water displaced by the body
(C) Both the weight of the body and the weight of displaced water
(D) Weight of the body divided by the weight of displaced water

16. A gravity tank flows water to distribution piping 60 feet below. What is the positive head pressure measured in psi (round 0.433 to .5 when doing your calculations)?

(A) 30 psi
(B) 40 psi
(C) 50 psi
(D) 60 psi

17. Using the hand method for friction loss in 2½-inch hose, what is the friction loss in 200 feet of hose flowing at 250 gpm?

(A) 15 psi
(B) 25 psi
(C) 30 psi
(D) 40 psi

**18.** Using the hand method for friction loss in 1¾-inch hose, what is the friction loss for 300 feet of hose flowing at 200 gpm?

(A) 200 psi
(B) 180 psi
(C) 150 psi
(D) 120 psi

**19.** A nozzle on 200 feet of 2½-inch hose requires 50 psi for maximum efficiency. Friction loss for each 100 feet of hose is 10 psi. What is the pump discharge pressure (PDP) needed when 250 gpm is flowing through the hose at an elevation 80 feet above the pump?

(A) 160 psi
(B) 150 psi
(C) 120 psi
(D) 110 psi

**20.** What civilization, around 300 BC, built large stone aqueducts to deliver water from upper elevations into the city?

(A) American Indians
(B) Chinese
(C) Romans
(D) Greeks

**21.** Which of the following choices is a manmade source of water supply?

(A) Lake
(B) River
(C) Stream
(D) Cistern

**22.** Cast iron pipe is the most commonly used water conveyor in distribution systems for all but which of the following reasons?

(A) Relatively inexpensive
(B) Lightweight
(C) Durable
(D) Highly resistant to corrosion

**23.** A combination water distribution system utilizes what methods to move water through piping?

(A) Treatment facilities and gravity
(B) Fire hydrants and sprinkler systems
(C) Pumps and gravity
(D) Water towers and reservoirs

24. Large-diameter concrete pipe is generally used for:

(A) Major water supply works
(B) Systems where very high pressures are used
(C) Lightweight, inexpensive, and noncorrosive systems
(D) Water supply work into buildings

25. A valve connection on a water supply distribution system with one or more outlets that is used to supply hose and fire department pumps with water is known as a:

(A) Standpipe
(B) Fire hydrant
(C) Gridiron system
(D) Water main

26. Select the inaccurate statement regarding dry-barrel fire hydrants:

(A) Pressurized and drained via a main valve
(B) Main valve is located in the base of the hydrant
(C) No valves at the outlets
(D) Especially suited for warm climates

27. Which characteristic of a wet-barrel hydrant is correct?

(A) More difficult to construct than dry-barrel hydrants
(B) All mechanical parts are below ground
(C) Mechanical parts are difficult to access
(D) Outlet valves operate independently

28. Fire hydrants are fitted on water distribution piping in urban and suburban areas where municipal water supply service is sufficient for firefighting use. The water supply design should strive to have all hydrants of the fire protection system within adequate distance to structures in need of protection from fire. Standard practice calls for the installation of hydrants every:

(A) 500 feet
(B) 700 feet
(C) 800 feet
(D) 1,000 feet

29. A fire hydrant that receives water from only one direction is known as a(an):

(A) Dead-end hydrant
(B) Lone wolf hydrant
(C) Isolated hydrant
(D) Remote hydrant

30. What is the most reliable way to provide water for firefighting in a distribution system?

(A) Bury the piping under sufficient pavement
(B) Design redundancy into the system
(C) Color-code fire hydrants to enhance recognition
(D) Restrict water mains to secondary roadways

31. How is piping laid out in a gridiron water distribution system?

(A) Tree branch fashion
(B) Clover leaf fashion
(C) Funnel fashion
(D) Checkerboard fashion

32. Which of the following is not correct pertaining to laying out interconnected water mains in a gridiron pattern?

(A) Supplies fire hydrants from multiple directions
(B) Damaged pipe can be isolated while remainder of system is still operational
(C) More impact from friction loss in water mains
(D) Discharges from multiple fire hydrants in use will remain more stable

33. Branch systems for water distribution are used for all but which of the following reasons?

(A) Simple design
(B) Hydraulic calculations are easy to formulate
(C) Required dimensions of the pipe are economical
(D) Limited number of dead-end mains

34. Select the accurate branch design pattern listed here:

(A) Single main, branches, submains
(B) Submains, single main, branches
(C) Single main, submains, branches
(D) Branches, single main, submains

35. Why is branch distribution design not followed in modern waterworks practice?

(A) Too much pipe required to build the system
(B) Water available for firefighting limited
(C) High pressure buildup in the trunk main
(D) Large number of shutoff valves needed

**36.** How are conventional sprinkler heads activated?

(A) Heat
(B) Water
(C) Smoke
(D) Light

**37.** Why are sprinkler heads called "silent sentinels"?

(A) Always ready to deliver water on a fire
(B) Don't require any maintenance or testing
(C) No sound is made upon activation
(D) Installed in the upright position

**38.** Sprinkler systems are designed to deliver a satisfactory amount of water onto a fire hazard for control and/or extinguishment. What is the fire hazard based on?

(A) Height of the building
(B) Size of the building
(C) Fuel load
(D) Water supply

**39.** Occupancy determines hazard classification. Occupancies that hold a moderate amount of combustibles and the combustibility of the contents is moderate to high are examples of what hazard classification?

(A) Light
(B) Ordinary
(C) Extra
(D) Heavy

**40.** Schools, hospitals, and office buildings are examples of what occupancy classification?

(A) Ordinary
(B) Moderate
(C) Extra
(D) Light

**41.** All but which of the following is a component of a typical sprinkler head?

(A) Frame
(B) Reflector
(C) Cap
(D) Deflector

**42.** Thermal linkage is the sprinkler head component that controls water release. Choose an incorrect statement regarding this component:

(A) Under fire conditions, it holds the cap in place
(B) It prevents water flow under normal conditions
(C) When exposed to high heat, it weakens
(D) Common linkage styles include metal levers and frangible glass bulbs

**43.** Most sprinkler head orifice openings are how large in diameter?

(A) ¼ inch
(B) ½ inch
(C) ¾ inch
(D) 1 inch

**44.** An inaccurate statement concerning a sprinkler head deflector can be found in which choice?

(A) Mounted on the frame
(B) Purpose is to minimize the amount of water coming from the orifice
(C) Provides a more efficient water distribution pattern
(D) Common mounting styles include upright, pendant, and sidewall

**45.** The density of water is the weight of the water per its unit volume. What is the water density for a sprinkler head protecting an extra hazard occupancy?

(A) .5 to .10 gpm per square foot
(B) .10 to .15 gpm per square foot
(C) .20 to .30 gpm per square foot
(D) .30 to .40 gpm per square foot

**46.** What is the maximum square footage per sprinkler head coverage for an ordinary hazard occupancy?

(A) 310 sq. ft.
(B) 225 sq. ft.
(C) 130 sq. ft.
(D) 100 sq. ft.

**47.** Standpipe systems are an important part of the fire protection system in a building. Select the incorrect statement concerning standpipes from the choices provided:

(A) Series of pipe that connects a water supply to hose outlet connections
(B) Firefighters bring hoses inside the building and attach them to standpipe outlets
(C) Risers run horizontally throughout the protected building
(D) Piping system supplies water to every floor in the building

**48.** A correct point of information regarding a combination standpipe system can be found in which choice?

(A) Supplies both a standpipe system and a sprinkler system
(B) Supplies two different type occupancies
(C) Cannot be fed water through a fire department connection
(D) Are only installed in very tall buildings

**49.** Standpipes are broken down into two types, based on the presence of water in the system. They may be wet or dry. Which statement is not correct?

(A) Wet standpipes are filled with pressurized water at all times
(B) Wet standpipes can be used by firefighters
(C) Dry standpipes can be used by building occupants
(D) Dry standpipes are not filled with water until needed in firefighting

**50.** Standpipe systems are classified depending on who is expected to use the system. Which classification description is correct?

(A) Class I standpipes are designed for large-volume fire streams
(B) Class II standpipes are intended for small-volume fire streams
(C) Class III standpipes are designed only for firefighters
(D) Class I standpipes are typically not provided with preconnected fire hoses

# Answer Key

**1.** B
**2.** C
**3.** C
**4.** B
**5.** A
**6.** A
**7.** D
**8.** A
**9.** D
**10.** D
**11.** A
**12.** D
**13.** C
**14.** B
**15.** B
**16.** A
**17.** C
**18.** B
**19.** D
**20.** C
**21.** D
**22.** B
**23.** C
**24.** A
**25.** B
**26.** D
**27.** D
**28.** A
**29.** A
**30.** B
**31.** D
**32.** C
**33.** D
**34.** C
**35.** B
**36.** A
**37.** A
**38.** C
**39.** B
**40.** D
**41.** B
**42.** A
**43.** B
**44.** B
**45.** D
**46.** C
**47.** C
**48.** A
**49.** C
**50.** C

## ANSWER EXPLANATIONS

Following you will find explanations to specific questions that you might find helpful.

**3.** C 1 gallon of water weighs 8.34 pounds
8. 34 pounds × 5 = 41.7 pounds

**6.** A 1 foot of head = .434
.434 + .434 = .868 psi or approximately .9 psi

**8.** A C in a closed system
D too quickly

**9.** D A 5 psi is 25% of 20 psi
2.5 psi is 50% of 5 psi
B greater the friction loss
C less friction loss

**16.** A Pressure = .433 (rounded off to .5 psi) × H (Head)
P = 0.5 psi/ft × 60 ft. = 30 psi

**17.** C Divide 250 gpm by 100 = 2.5
2.5 × 6 (from hand method for 2½-inch hose) = 15 psi (for each 100 feet of hose)
15 psi × 2 (200 feet of hose) = 30 psi

**18.** B 200 gpm corresponds to the number 5 (hand method for 1¾-inch hose)
5 × 12 (hand method for 1¾-inch hose) = 60 psi (per 100 feet of hose)
60 psi × 3 = 180 psi

**19.** D PDP = NP +/− H + FL
NP = 50 psi
H = −40 psi (loss); 80 ft. × .433 (rounded off to .5) = −40 psi
0.5 psi required to lift water 1 foot
FL = 20 psi (10 psi for each 100 feet of hose), 10 psi + 10 psi = 20 psi
PDP = 50 psi + 40 psi + 20 psi = 110 psi

**24.** A B steel
C plastic
D galvanized iron

**26.** D cold climates

**27.** D A simpler
B above ground
C easy

**32.** C less

**42.** A normal

**44.** B break up the water coming from the orifice

**47.** C vertically

**49.** C cannot

**50.** C and trained occupants

# CHAPTER 11

# Principles of Mechanics

## Your Goals for This Chapter

- Learn basic physical science definitions and formulas
- Explore the functions and uses of simple machines
- Understand the mechanical advantage of simple machines
- Test your mechanics knowledge with review questions

Archimedes (c. 287–212 BC), the great Sicilian inventor, physicist, mathematician, and engineer, is credited with saying, "Give me a place to stand and I will move the earth." He understood the concept of using simple machines to gain a mechanical advantage to move or lift heavy objects with less force. He invented a simple machine (Archimedes' screw) operated by manually turning a screw-shaped surface inside a hollow tube to raise water up from low-lying areas into irrigation ditches.

The first simple machines were probably wooden levers used to move large boulders and sharp rocks used as wedges to scrape off the skins of dead animals. Around 3000 BC, logs were used as rollers to move heavy objects. This concept developed into the wheel and axle simple machine. Some 200 years later, the builders of Stonehenge used levers, rollers, and pulleys. Soon afterward, inclined planes, ramps, and rollers were used to build the Great Pyramids of Egypt. Progress continued with the development of the screw principle by Archimedes in the third century BC. Early civilizations throughout the world continued to develop and invent simple machines to construct their buildings and facilitate work. Progress was steady throughout the next 20 centuries—with the invention of the spinning wheel, compass, printing press, telescope, mechanical clock, and engine—but it was given a big leap forward by improvements to the steam engine by James Watt of England in the mid-eighteenth century. That ushered in the Industrial Revolution and the machine age for modern man.

This chapter reviews some key mechanical principles to help you visualize and understand how simple machines work—and how to solve commonly asked questions in firefighter written exams.

## PHYSICAL SCIENCE DEFINITIONS AND FORMULAS

**Force** is generally thought of as a push, pull, dropping, stretching, or squeezing of an object that results in a change in the shape and/or a change in the motion of the object. Examples of forces include gravity, magnetism, and electricity.

Force is measured in the SI (System International), or metric system, in units called **Newtons** (N), for the great English mathematician and physicist Isaac Newton (1642–1727). In the British system of measurement, force is measured in foot × pounds.

$$\text{Force } (F) = \text{Mass } (M) \times \text{Acceleration } (A)$$

**Mass** is the quantity of matter of an object. In the metric system, the unit of measurement of mass is the kilogram (kg).

**Weight** is a force originating when a mass is acted on by gravity. Weight (W) is the product of an object's mass (m) and the acceleration of gravity (g) at the location of the object, or W = mg. The units of weight in the SI are Newtons. Weight is also measured by the gram in the metric system and by the ounce or pound in the British system.

**Velocity** is a measure of how fast an object is moving in a given direction. In the metric system, velocity is measured in meters per second (m/s). In the British system, the measurement is in miles per hour (mph).

**Momentum** is the term used when mass has a velocity. Its unit of measure has no metric or British name.

$$\text{Momentum} = \text{Mass} \times \text{Meters/Second}$$

**Acceleration** is the rate of change of velocity. The metric unit of measure for acceleration is meters per second per second (m/s/s). The British system measures acceleration in feet per second per second (ft/s/s).

**Length** is measured in meters (m) in the metric system. In the British system, length is measured in feet (ft).

**Newton** is a measure of the amount of force required to accelerate a mass of one kilogram (kg) at a rate of one meter (m) per second (s) per second (s).

$$1 \text{ N} = 1 \text{ kg} \times \text{m/s/s}$$

**Friction** is a force that reduces the motion of objects. It occurs when two objects rub against each other with heat as a by-product. Friction is reduced by polishing a surface and by using lubricants (oils and greases) and rollers to allow objects to move more easily.

**Work** is the use of force to cause motion. It is the transfer of energy through motion. Energy (work) can never be destroyed; it can only be transferred.

For work to take place, a force must be exerted through a distance. The amount of work performed depends on the amount of force that is exerted and the distance over which the force is applied.

Work (W) = Force (F) × Distance (D)

Mechanical work is measured by the joule (J), named after the nineteenth-century English physicist James Prescott Joule. (One joule is equal to 1 Newton multiplied by 1 meter.)

$1\ J = 1\ N \times 1\ m$

The British system measures mechanical work in foot-pounds (ft-lb).

**Torque** is a twisting force that occurs when the force is not applied to the object's center of mass.

**Power** is the rate at which work is performed. It is derived by measuring work per unit of time. The metric system unit used to measure power is the **Watt** (W). One watt is equal to 1 joule per second.

$1\ W = 1\ J/s$

1,000 Watts = 1 Kilowatt (KW)

The British unit of measurement for power is the horsepower (HP).

1 HP = 550 ft-lb/s = 746 watts

# SIMPLE MACHINES—DEFINITION AND USES

A machine is any device that applies mechanical energy at a given point and delivers it in a more efficient form at another point. There are many kinds of machines with varying capabilities and functions. Specifically, machines are used:

- to transform energy from one type to another (steam turbine)
- to transfer energy (automobile drive train)
- to increase force (pry bar)
- to multiply speed (bicycle gears)
- to change the direction of force (pulleys)
- to reduce friction (rollers)

Machines may be powered by motors, engines, or simply human effort. Prior to the age of motors and engines, animals were used to assist workers in moving and lifting heavy objects.

Simple machines were invented and used to overcome resistive forces and enable workers to get the job done. Firefighters have been using simple machines to perform the work they do since ancient times. Some of the simple machines they use daily include the axe, pry bar, hook, hammer, pliers, Vise-Grip pliers, shovel, crowbar, chisel, screwdriver, wheelbarrow, wedge, chock, pulley, block and tackle, hydraulic spreading and cutting tools, jacks, bulldozer, backhoe, and many, many more.

Simple machines are normally used when the amount of force required cannot be applied without the aid of a machine. They are also used to change the direction of effort (force) when the direction of the force to be applied is not the desired one, as in a pulley. An important point to understand is that most simple machines do not save energy; they simply allow the force required to do the work to be distributed over a longer distance or to be increased over a shorter distance. They provide a gain in effort (force) or a gain in distance, but not both. Common simple machines are the **inclined plane**, **wedge**, **screw**, **lever**, **wheel and axle**, **pulley**, and **gear and belt drive**.

More complicated (**complex**) machines are basically combinations of two or more simple machines. Machines that transform heat energy into mechanical energy are known as engines (steam engine and internal combustion engine). Electric motors change electrical energy into mechanical energy.

## What Is Mechanical Advantage?

As stated above, simple machines are used to reduce the amount of force needed to perform a given task, such as moving heavy objects or lifting a load. By definition, a machine is a device that provides a mechanical advantage (MA). However, there is a trade-off when simple machines are used: you must apply this force over a greater distance than if the load were moved directly. When trading off effort for distance, the advantage gained of increasing our effort force is called a mechanical advantage. The mechanical advantage of a machine is the factor by which a machine multiplies the force or effort being exerted on it. The mechanical advantage is a ratio of a load to the applied force (MA = load/applied force). A mechanical advantage greater than 1 is considered good. The greater the mechanical advantage, the less force or effort is required to accomplish the task.

## How to Calculate Mechanical Advantage

Basic equations for finding the mechanical advantage of simple machines are:

$$\text{MA} = \frac{\text{resistance force (load)}}{\text{effort (force)}}$$

$$\text{MA} = \frac{\text{effort (force) distance}}{\text{resistance force (load) distance}}$$

# TYPES OF SIMPLE MACHINES AND THEIR MECHANICAL ADVANTAGE

Simple machines are derived from either the inclined plane or the lever. Seven common simple machines, with illustrations and examples, are discussed below.

# The Inclined Plane

The inclined plane—a slanted surface raised at one end—is a simple machine that does not move. It is used to lift heavy loads by providing the worker with a mechanical advantage. The inclined plane provides for less effort, not for less work, to lift thc object. Getting heavy boxes or barrels onto a loading dock is much easier if you slide the objects up a ramp rather than lift them up. The trade-off is the greater distance to travel. Stairs and ramps provide examples of inclined planes.

The mathematical formula for the mechanical advantage of an inclined plane is the length of the inclined plane—the effort (force) distance—divided by the height—the resistance force (load) distance. The length of the inclined plane can never be less than the height; therefore, the MA can never be less than 1.

$$MA = \frac{\text{length of the slope}}{\text{height of the slope}}$$

$$MA = \frac{\text{effort (force) distance}}{\text{resistance force (load) distance}}$$

**Example:** A ramp 20 feet in length is 5 feet in height. What is the mechanical advantage of this simple machine?

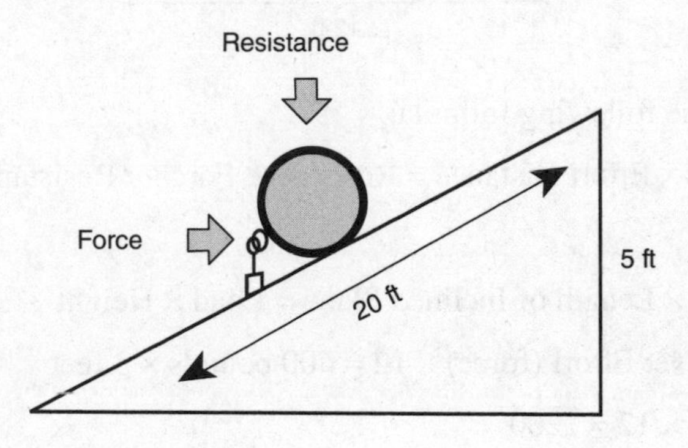


$$MA = \frac{\text{length of the slope}}{\text{height of the slope}}$$

$$MA = \frac{20 \text{ feet}}{5 \text{ feet}}$$

$$MA = 4$$

The trade-off is that the length of the slope (effort distance) is 4 times greater than the resistance force (load) distance, or height of the slope.

**Example:** Select the inclined plane diagram below that gives the best mechanical advantage in lifting a heavy barrel to the height of the platform.

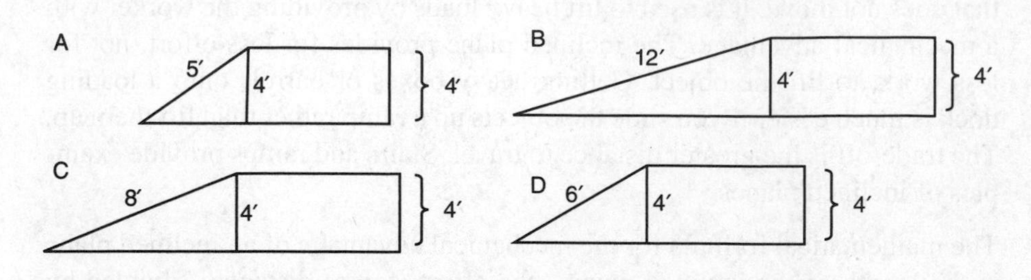


The correct choice is answer B. Using the MA formula for the four inclined planes shown reveals that choice B has the greatest MA, 3, compared to 1.25 for A, 2 for C, and 1.5 for D. However, B would require the longest effort (force) distance, 12 feet.

**Example:** Determine the amount of force, or effort, required to move a 400-pound load to a height of 3 feet, using a 12-foot ramp.

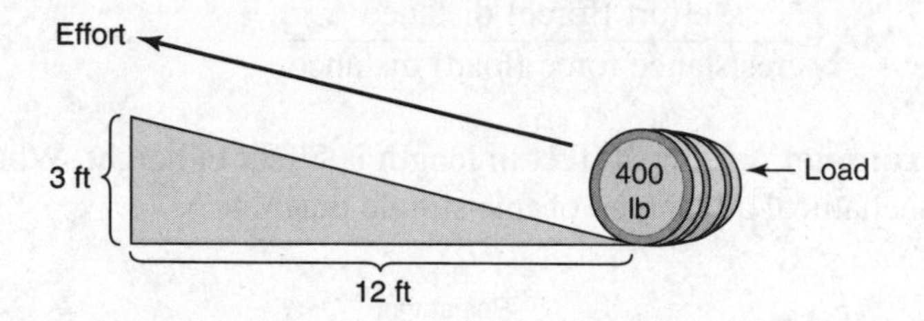


Use the following formula:

Effort × Effort Distance = Resistance Force × Resistance Distance

OR

Force × Length of Inclined Plane = Load × Height

To solve: Effort (force) × 12 = 400 pounds × 3 feet

$\text{Effort} \times 12 = 1200$

$\text{Effort} = \dfrac{1200}{12}$

Effort = 100 pounds

The inclined plane with an MA of 4 allows a 400-pound object to be raised 3 feet with a 100-pound force.

## The Wedge

A wedge is an inclined plane that tapers to a sharp edge. It is used to increase force. **Double wedges** are made up of two inclined planes and are used to split (e.g., fireplace logs), fasten, or cut. However, a wedge can also be one sloping surface (**single wedge**) such as that used by firefighters as a chock or doorstop.

When used for cutting, the longer and thinner the wedge, the less effort (force) is required to overcome resistance. You can enhance the mechanical advantage of an axe, a type of double wedge, by sharpening it. Effort (force) is applied to the thicker edge of the wedge and is transferred to the thinner end. The wedge is also used to change the direction of force. When force is applied downward on a double wedge, it will push out in two directions, helping to push things apart at right angles. Unlike the inclined plane, the wedge moves. For example, when a wedge is used to split a log, it is the log that remains in place while the wedge moves through it.

Many common objects are double wedges. Nails, for example, are wedges used to fasten objects together. The tip of a slotted screwdriver is a simple wedge. Knives, axe heads, chisels, and scissors are sharpened double wedges used for cutting.

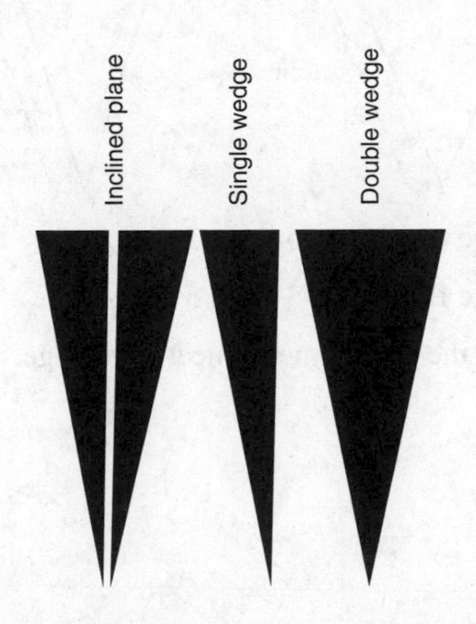


To ascertain the mechanical advantage of a **single wedge** when it is used as a chock, simply use the same formula as that for the inclined plane, as shown below:

$$\text{MA} = \frac{\text{length of the slope}}{\text{height of the slope}}$$

To ascertain the mechanical advantage for a **double wedge** use the formula:

$$\text{MA} = \frac{\text{effort (force) distance}}{\text{resistance force (load) distance}}$$

**Example:** What is the mechanical advantage of a single wedge with a length of slope of 18 cm and height (thickness) of 6 cm being used to chock open a door by firefighters stretching hose line?

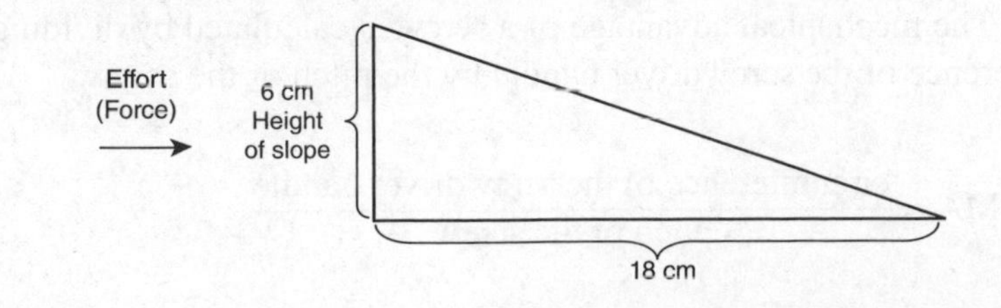

$$MA = \frac{\text{length of the slope}}{\text{height of the slope}}$$

$$MA = \frac{18\text{ cm}}{6\text{ cm}} = 3$$

**Example:** Which double wedge, A or B, being used to split logs, has the greater mechanical advantage?

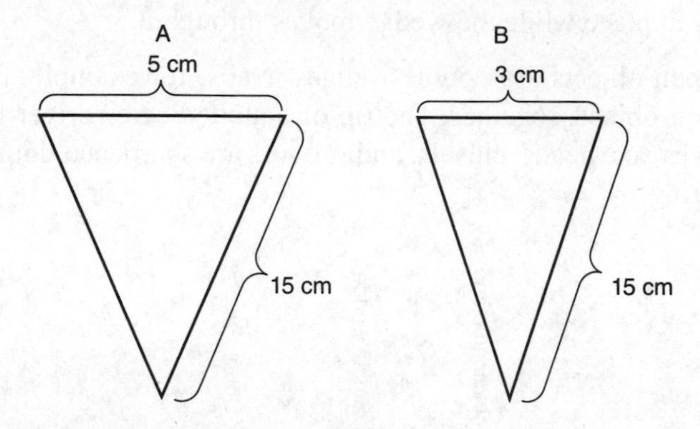


MA for double wedge A = 15 cm/5 cm = 3

MA for double wedge B = 15 cm/3 cm = 5

Double wedge B has the greater mechanical advantage.

## The Screw

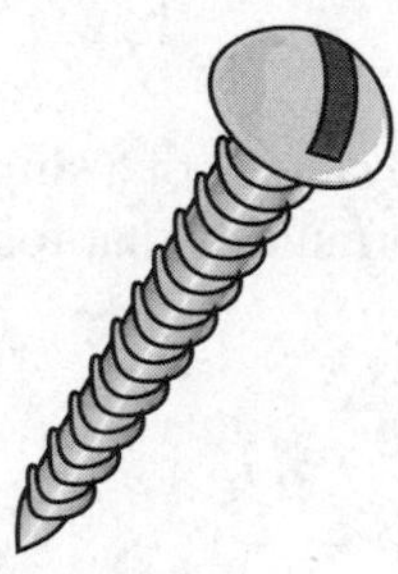

A screw is a simple machine similar to an inclined plane or a wedge, but in a screw the incline wraps around a shaft. A screw converts rotational motion to linear motion. The **pitch** of a screw is the distance between its threads; this is the distance the screw will advance during one complete rotation. Screw jacks are operated by turning a lead screw. They are commonly used to lift heavy weights such as large vehicles.

To find the mechanical advantage that a screw provides, you have to know the pitch. The mechanical advantage of a screw is calculated by dividing the circumference of the screwdriver handle by the pitch of the screw.

$$MA = \frac{\text{circumference of the screwdriver handle}}{\text{pitch of the screw}}$$

OR

$$\text{MA} = \frac{\text{effort (force) distance}}{\text{resistance force (load) distance}}$$

The circumference of the screwdriver handle can be considered the effort distance and the pitch of the screw can be considered the resistance force.

**Example:** A screw with 10 threads per inch is being turned by a screwdriver that has a handle with a radius (*r*) of 1 inch. What is the mechanical advantage?

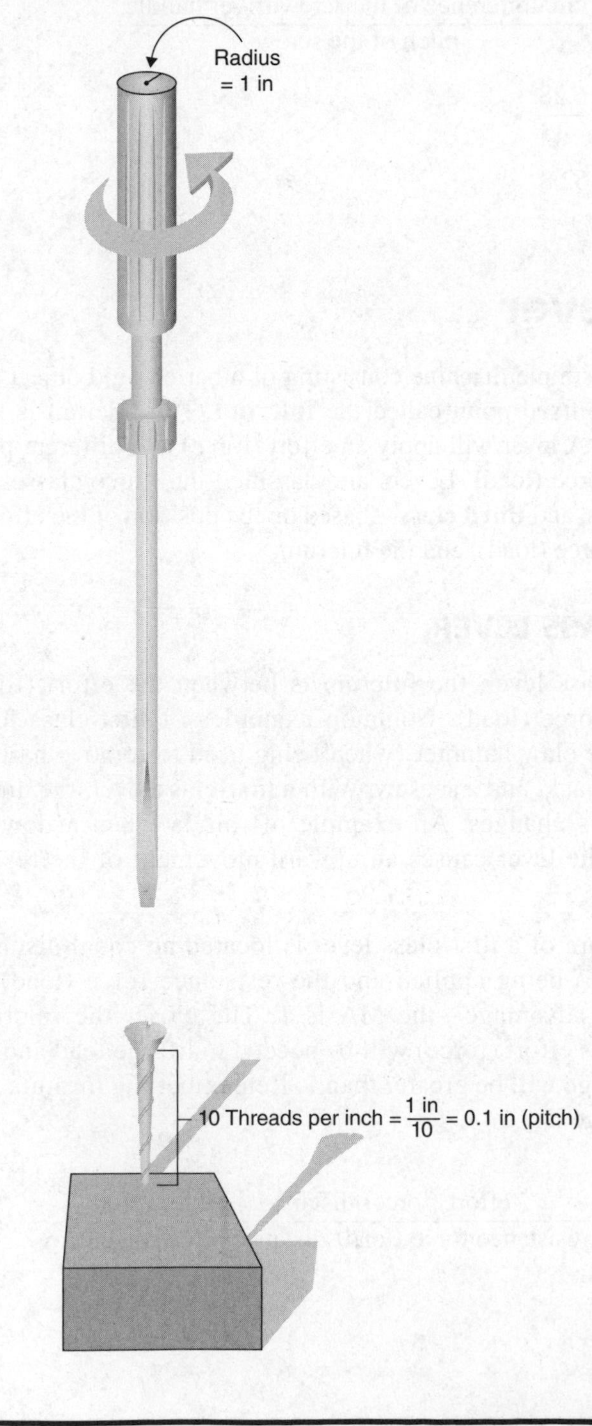

To determine the mechanical advantage, first calculate the circumference of the handle of the screwdriver.

$C$ (circumference) $= 2\pi r$

$C = 2(3.14)(1 \text{ inch})$

$C = 6.28$ inches

When using a screwdriver, the mechanical advantage of a screw is calculated as follows:

$$\text{MA} = \frac{\text{circumference of the screwdriver handle}}{\text{pitch of the screw}}$$

$$\text{MA} = \frac{6.28}{.1}$$

$$\text{MA} = 62.8$$

# The Lever

A lever is a simple machine consisting of a bar or rigid object that is free to turn about a fixed point called the **fulcrum**. The fulcrum is known as the **pivot point**. A lever will apply an effort (force) at a different point from the resistance force (load). Levers are classified into three classes—**first class**, **second class**, and **third class**—based on the position of the effort (force), the resistance force (load), and the fulcrum.

## FIRST-CLASS LEVER

In a first-class lever, the fulcrum is between the effort (force) and the resistance force (load). Common examples of first-class levers are the crowbar, the claw hammer (when being used to remove nails), pliers, tin snips, a car jack, and a seesaw. With a first-class lever, the direction of the force always changes. An example of this is when a downward effort (force) on the lever causes an upward movement of the resistance force (load).

If the fulcrum of a first-class lever is located an equal distance from the effort (force) being applied and the resistance force (load), there is no mechanical advantage—the MA is 1. The closer the fulcrum is to the load, the less effort (force) will be needed to lift the load and the mechanical advantage will be greater than 1. Remember the formula for mechanical advantage is:

$$\text{MA} = \frac{\text{effort (force) distance} = \text{effort arm}}{\text{resistance force (load) distance} = \text{resistance arm}}$$

**Example:** In the illustration shown below, the load and mechanical advantage are known, and the effort needed can be calculated:

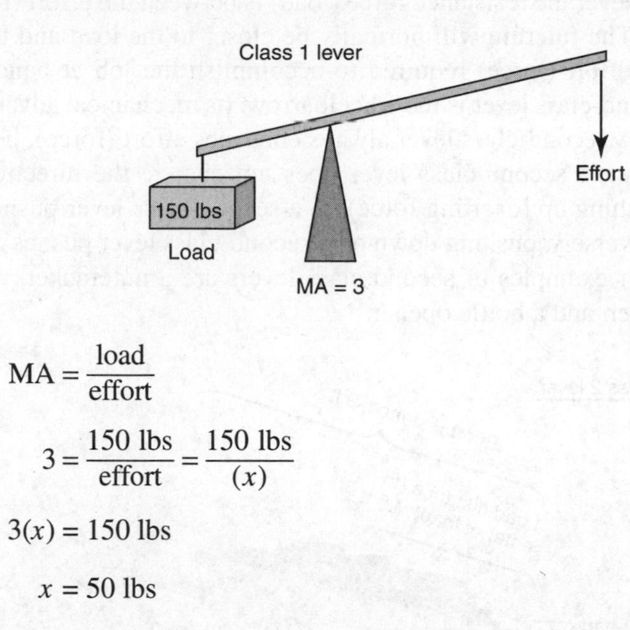


$$\text{MA} = \frac{\text{load}}{\text{effort}}$$

$$3 = \frac{150 \text{ lbs}}{\text{effort}} = \frac{150 \text{ lbs}}{(x)}$$

$$3(x) = 150 \text{ lbs}$$

$$x = 50 \text{ lbs}$$

The trade-off, however, is that the effort (force) will have to move (down) a greater distance and the load will move (rise up) a smaller distance.

Conversely, if the fulcrum is moved closer to the effort (force) being applied, greater effort (force) will be required to lift the resistance force (load) but the effort (force) will have to move a shorter distance and the resistance force (load) will move (rise up) a greater distance.

**Note:** Another mechanical principle formula that can be used to calculate the value of lever systems is the following:

Force × Effort Distance = Load × Resistance Distance

**Example:** A 300-pound weight is placed on the end of a plank 3 feet from a fulcrum. How much effort (force) would a firefighter have to exert on the opposite end of the plank at 9 feet from the fulcrum to obtain equilibrium?

Force × Effort Distance = Load × Resistance Distance

$(x) \times 9 \text{ ft} = 300 \text{ lbs} \times 3 \text{ ft}$

Divide both sides by 9 ft to isolate the variable.

$$\frac{(x) \times 9 \text{ ft}}{9 \text{ ft}} = \frac{300 \text{ lbs} \times 3 \text{ ft}}{9 \text{ ft}}$$

$$(x) = \frac{300 \text{ lbs}}{3} = 100 \text{ lbs}$$

$$\text{MA} = 3$$

## SECOND-CLASS LEVER

In a second-class lever the resistance force (load) is between the effort (force) and the fulcrum. The fulcrum will normally be closer to the load and therefore reduces the effort (force) required to accomplish the job at hand. An example of a second-class lever is the wheelbarrow. Its mechanical advantage is greater than 1. A second-class lever always enhances effort (force). Unlike the first-class lever, a second-class lever does not change the direction of effort (force). Pushing up (exerting force) on a second-class lever pushes up on the (load); conversely, pushing down on a second-class lever pushes down on the load. Other examples of second-class levers are a nutcracker, a bellows, a paper cutter, and a bottle opener.

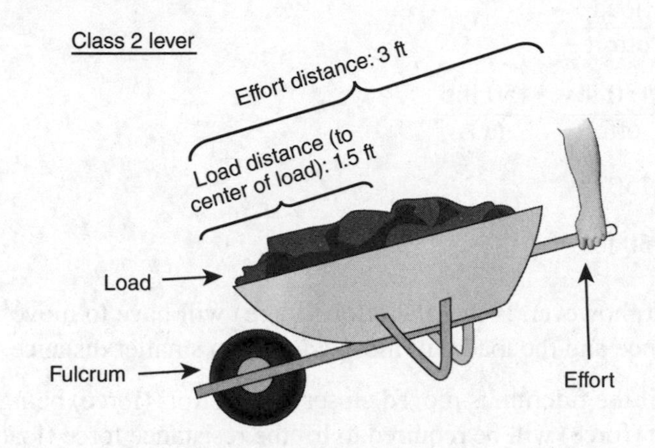


**Example:** A wheelbarrow filled with building materials weighing 100 pounds that is 1.5 feet from the wheel would require how much force to lift it off the ground using handles that are 3 feet from the load? What is the mechanical advantage?

Force × Effort Distance = Load × Resistance Distance

$(x) \times 3 \text{ ft} = 100 \text{ lbs} \times 1.5 \text{ ft}$

Divide both sides by 3 ft to isolate the variable.

$$(x) = \frac{100 \text{ lbs} \times 3 \text{ ft (36 in)}}{3 \text{ ft}} = \frac{100 \text{ lbs} \times 1.5 \text{ ft (18 in)}}{3 \text{ ft}}$$

$$(x) = \frac{100 \text{ lbs} \times 1.5 \text{ ft (18 in)}}{3 \text{ ft(36 in)}}$$

$$(x) = \frac{100 \text{ lbs}}{2}$$

$$x = 50 \text{ lbs}$$

If effort is 50 pounds to move a load of 100 pounds, the MA = 2.

## THIRD-CLASS LEVER

A third-class lever has the effort (force) between the fulcrum and the resistance force (load). In a third-class lever, the effort (force) required to lift the load is actually increased and, therefore, the mechanical advantage is less than 1. The trade-off, however, is an increase in speed and distance of travel of the load. An example of a third-class lever is a shovel. The worker's hands supply the effort (force) while the elbows act as a fulcrum. The load (soil, sand, coal) is moved at the end of the shovel. As in the second-class lever, the dirction of effort (force) does not change. Examples of third-class levers are a pitchfork, tweezers, a hoe, tongs, and a broom.

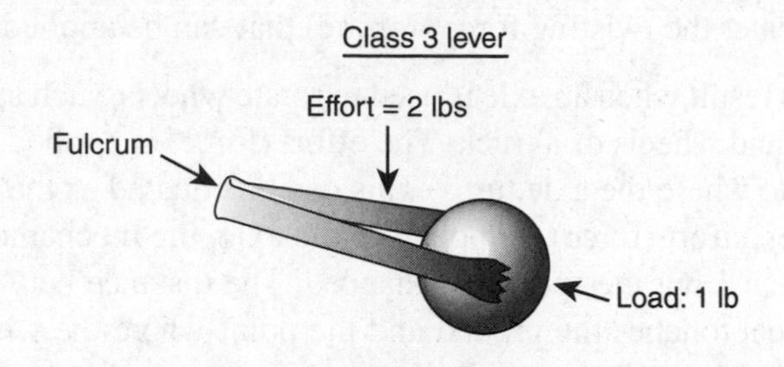


**Example:** A pair of tongs is being used by fire marshals at a fire scene to pick up a 1-pound circular ball to be used as evidence. Two pounds of effort force is required to lift the object. The fingers being used to squeeze the tongs are 4 inches from the object and 2 inches from the fulcrum of the tongs. What is the mechanical advantage of the tongs?

$$MA = \frac{\text{resistance force (load)}}{\text{effort (force)}} = \frac{1\,\text{lb}}{2\,\text{lbs}} = \frac{1}{2}$$

OR

$$MA = \frac{\text{effort (force) distance}}{\text{resistance force (load) distance}} = \frac{2}{4} = \frac{1}{2}$$

# The Wheel and Axle

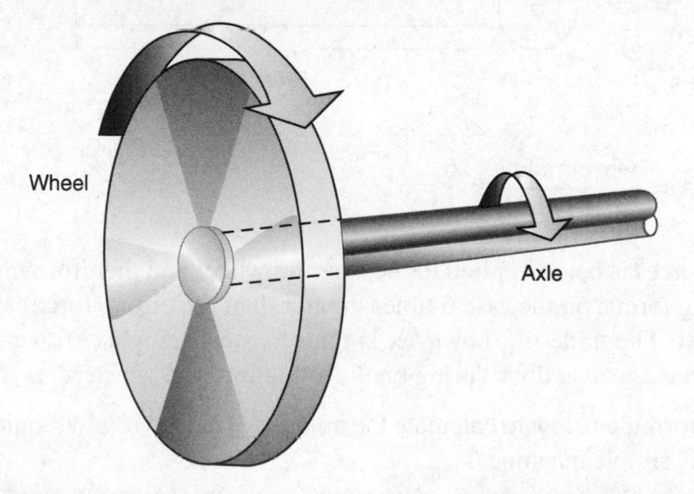


THE WHEEL AND AXLE IS A WHEEL CONNECTED TO A RIGID POLE.

The wheel and axle is another type of simple machine that moves objects across distances. Wheels help move objects along the ground by decreasing the amount of friction between what is being moved and the surface. The work of this simple machine can result from the larger wheel being utilized to turn a smaller axle wheel. [Example: The steering wheel and shaft of a car enhances effort (force) with the trade-off, once again, of having to apply effort (force) over a greater distance.] Movement is created as the steering wheel turns, thereby applying a rotating force to its cylindrical axle post. The bigger the wheel, the greater the twisting force (torque) that can be applied to the axle.

Work can also result when an axle is used to rotate wheels, such as the example of a rear axle and wheels of a truck. The effort (force) is applied to the axle at a point close to where the axle turns. This can be equated as the effort (force) distance. When effort (force) is applied to the axle, the mechanical advantage will be less than 1 but the speed is enhanced. The distance between the point where the wheel touches the ground and the point where the wheel turns can be called the resistance force (load) distance. These two distances are equal to the radius of the axle and the radius of the wheel, respectively.

To calculate the mechanical advantage of a wheel and axle assembly, divide the radius of the wheel by the radius of the axle.

> **Example:** What is the mechanical advantage provided by a car's steering wheel assembly when the radius of the steering wheel is 6 inches and the radius of the axle is 1 inch?

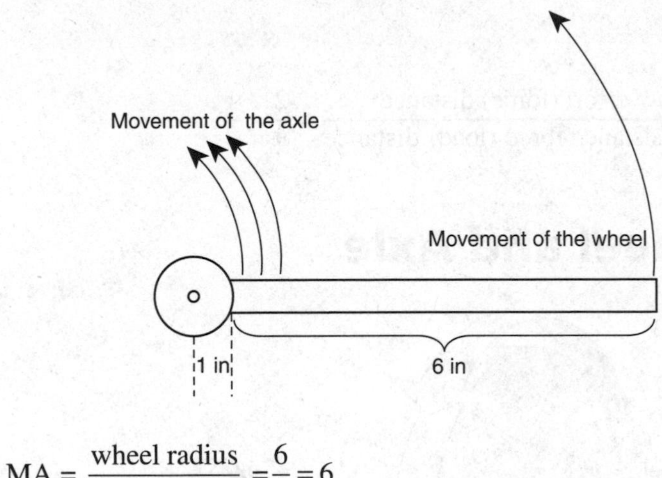


$$MA = \frac{\text{wheel radius}}{\text{axle radius}} = \frac{6}{1} = 6$$

Effort (force) is being applied to the steering wheel and therefore multiplied, providing torque on the axle 6 times greater than the effort (force) applied to the wheel. The trade-off, however, is that the steering wheel travels 6 times farther than the axle does during one full rotation.

Use the formula below to calculate the amount of effort (force) required when using this simple machine.

$$\text{Effort} \times \text{Circumference} = \text{Resistance Force} \times \text{Circumference}$$

$$(\text{Force}) \times (\text{Wheel}) = (\text{Load}) \times (\text{Axle})$$

**Example:** In the drawing below of a well crank (windlass), the handle is attached to a 2-inch radius axle. The turning circumference of the crank is 16 inches. How much effort (force) is required to lift a bucket of water weighing 40 pounds?

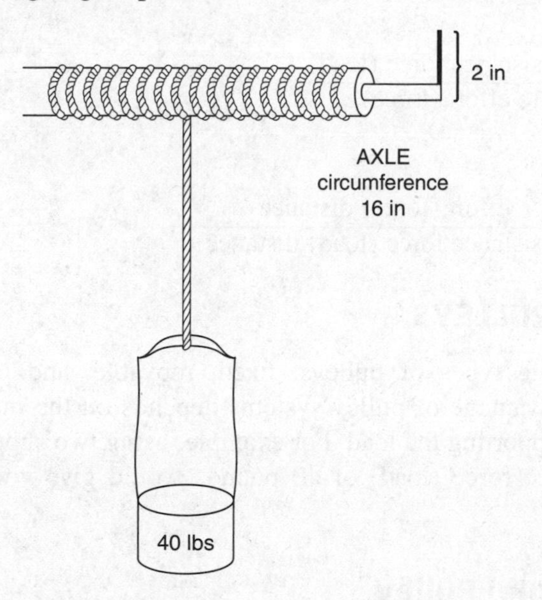


$$\underset{\text{(Crank Wheel)}}{\text{Effort (Force)} \times \text{Circumference}} = \underset{\text{(Load)}}{\text{Resistance Force}} \times \underset{\text{(Axle)}}{\text{Circumference}}$$

$$\text{Effort (Force)} \times 16 \text{ inches} = 40 \text{ pounds} \times \underset{\text{(Axle)}}{\text{Circumference}}$$

Find: Circumference of Axle

$C = 2 \times \pi \times \text{radius}$

$C = 2 \times 3.14 \times 2 \text{ in}$

$C = 4 \text{ inches} \times 3.14 = 12.56 \text{ in}$

$\text{Effort (Force)} \times 16 \text{ in} = 40 \text{ lbs} \times 12.56 \text{ in}$

$$\text{Effort (Force)} = \frac{40 \text{ lbs} \times 12.56 \text{ in}}{16 \text{ in}}$$

$$\text{Effort (Force)} = \frac{502.4 \text{ lbs}}{16} = 31.4 \text{ lbs}$$

# The Pulley

A pulley can be considered as a circular lever. It is a wheel with a grooved rim and axle with a rope, belt, or chain attached to it in order to change the direction of the pull and lift a load. The effort (force) distance is the radius of the pulley (length from the axle to the side of the rope being pulled). The resistance force (load) distance is the radius of the pulley from the axle to the load-carrying side of the rope. Pulleys are used to lift heavy loads and can be found in block and tackles, cranes, hydraulic systems, and chain hoists.

They change the direction of effort (force), making it easier to lift the object or they enhance the effort (force).

Mechanical advantage for pulley systems can be found using the following formulas:

$$MA = \frac{\text{resistance force (load)}}{\text{effort (force)}}$$

OR

$$MA = \frac{\text{effort (force) distance}}{\text{resistance force (load) distance}}$$

## TYPES OF PULLEYS

There are three types of pulleys: fixed, movable, and compound. The mechanical advantage of pulley systems depends on the number of ropes, chains, etc. supporting the load. For example, using two supporting ropes to lift a resistance force (load) of 40 pounds would give you a mechanical advantage of 2.

### Fixed (Single) Pulley

A fixed (single) pulley is attached to a stationary object like a wall or ceiling. It acts as a first-class lever having the fulcrum at the axis and the rope acting as the bar. Fixed (single) pulleys only change effort (force) direction (you can pull down on the rope to lift the load instead of pushing up on it). They do not enhance the effort (force). Effort (force) distance equals resistance force (load) distance and, therefore, each foot of pull on the rope will lift the load 1 foot. It provides no mechanical advantage (MA = 1).

> **Example:** The fixed (single) pulley has a resistance force (load) at one end of the rope (see diagram below). The other end must have effort (force) applied downward to raise the load. The effort (force) is equal to the load in this pulley system and there is no mechanical advantage, with the MA equal to 1.

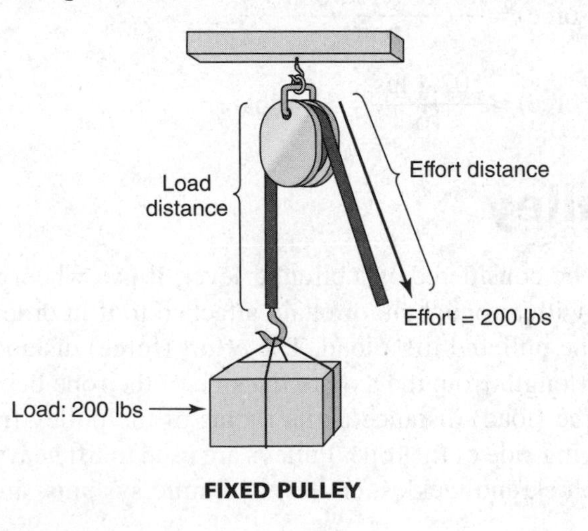


**FIXED PULLEY**

## Movable Pulley

A movable pulley moves up and down with the effort (force). It acts as a second-class lever having the resistance force (load) between the fulcrum and the effort (force). Unlike the fixed (single) pulley, it cannot change the direction of the effort (force). Movable pulleys, however, enhance effort (force). Their mechanical advantage is greater than 1. The trade-off is that the effort (force) distance is greater than the resistance force (load) distance.

The movable pulley (see diagram below) has the resistance force (load) supported by both the rope ends [the rope end attached to the upper bar and the rope end to be pulled effort (force) in the upward direction]. The two upward tensions are equal and opposite in direction to the load. The mechanical advantage is 2.

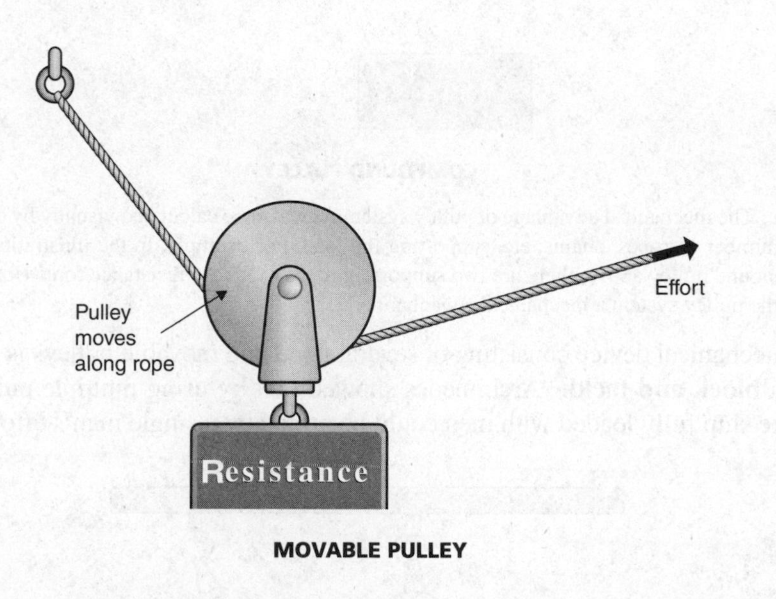


MOVABLE PULLEY

## Compound Pulley

A compound pulley utilizes both a fixed (single) pulley and a movable pulley. Compound pulleys provide both a change in the direction of the effort (force) as well as dramatically decreasing the effort (force) required to lift the resistance force (load). The mechanical advantage of this type of pulley is 2. The effort (force) distance, however, like with the movable pulley, will be greater than the resistance force (load) distance.

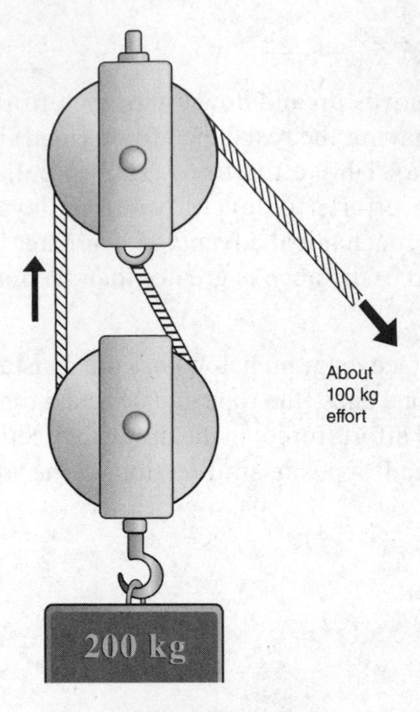


**COMPOUND PULLEY**

**Note:** The mechanical advantage of pulley systems can also be calculated visually by counting the number of ropes, chains, etc. supporting the load. For example, in the illustration of the compound pulley above, there are two supporting ropes to lift the resistance force (load), giving the pulley system a mechanical advantage of 2.

A mechanical device consisting of several fixed and movable pulleys is known as a **block and tackle**. Archimedes showed that by using multiple pulleys, a large ship fully loaded with men could be pulled by a single man's effort.

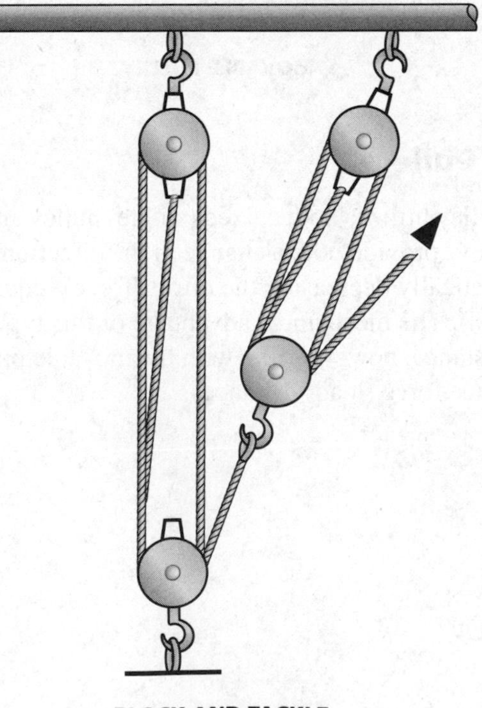

**BLOCK AND TACKLE**

### EFFICIENCY OF A PULLEY SYSTEM

To calculate the **efficiency** of a pulley system, first determine the mechanical advantage. Next, determine the **velocity ratio** by dividing the distance moved by effort (force) by the distance moved by the resistance force (load). Finally, divide the mechanical advantage by the velocity ratio and multiply this number by 100 percent.

**Example:** A pulley system can lift an object weighing 50 N with an effort (force) of 10 N. The input distance is 5 m and the output distance is 0.5 m. What is the efficiency of the pulley system?

$$\text{MA} = \frac{\text{resistance force (load)}}{\text{effort (force)}} = \frac{50\text{ N}}{10\text{ N}} = \mathbf{5}$$

$$\text{Velocity ratio} = \frac{\text{effort (force) distance}}{\text{load distance}} = \frac{5\text{ m}}{0.5\text{ m}} = \mathbf{10}$$

$$\text{Efficiency} = \frac{5}{10} \times 100\% = 50\%$$

# Gear and Belt Drive Systems

Gears are toothed wheels that are meshed together to transmit a twisting force (torque) and motion. They are usually attached to a shaft and can be considered a rotating lever. Utilizing leverage principles, gears can enhance or inhibit effort (force) or change effort (force) direction. Gears are either turned by a shaft or they turn the shaft. A large gear can apply more twisting force on a shaft to which it is attached than a smaller gear.

Gears that have straight teeth (perpendicular to their facing) and that mesh together in the same plane with axles parallel are known as **spur** gears. Spur gears provide an important way of transmitting a positive motion between two shafts. They give a smooth and uniform drive.

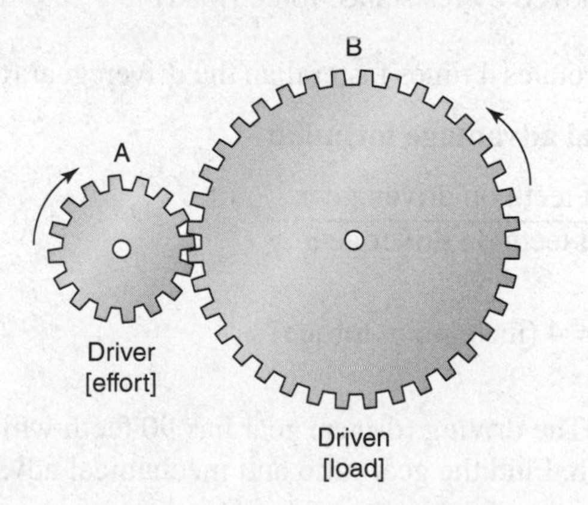


In the diagram, one gear, labeled the **driver** (also known as the driving gear) is turned by a motor. As it turns, it turns the other gear, known as the **driven** gear.

A basic rule concerning gears states that each gear in a series of gears reverses the direction of rotation of the previous gear.

Another basic rule of gears is that when you have a pair of meshing gears and the smaller gear with less pitch diameter (number of teeth) is the driver, torque output will be enhanced, with the trade-off being a decrease in speed of rotation. Conversely, when a larger gear with greater pitch diameter is the driver, torque output will be inhibited, but the trade-off is the speed of rotation will be enhanced.

**Example:** The driving (driver) gear has 9 teeth while the driven gear has 36 teeth. Find the gear ratio and mechanical advantage (torque) of this gear system.

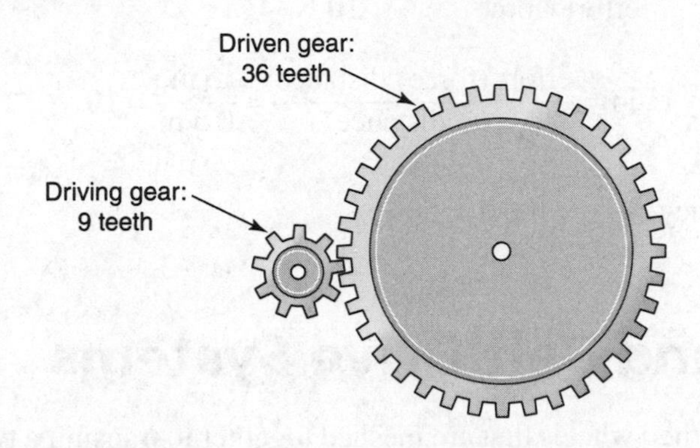


**Note:** When computing **gear ratio**, always compare the larger gear rotating once to the smaller gear regardless of whether it is a driver or driven gear.

**Gear ratio formula:**

Input movement (Driver):Output movement (Driven)

4:1

OR

$$\frac{\text{distance moved by effort (force)}}{\text{distance moved by resistance force (load)}} = \frac{4}{1}$$

The driver gear rotates 4 times faster than the driven gear (decrease in speed).

**Mechanical advantage formula:**

$$MA = \frac{\text{\# of teeth on driven gear}}{\text{\# of teeth on driver gear}}$$

$$MA = \frac{36}{9} = 4 \text{ (increase in torque)}$$

**Example:** The driving (driver) gear has 90 teeth while the driven gear has 15 teeth. Find the gear ratio and mechanical advantage (torque) of this gear system.

**Gear ratio formula:**

Input movement (Driver):Output movement (Driven)

1:6

OR

$$\frac{\text{distance moved by effort (force)}}{\text{distance moved by resistance force (load)}} = \frac{1}{6}$$

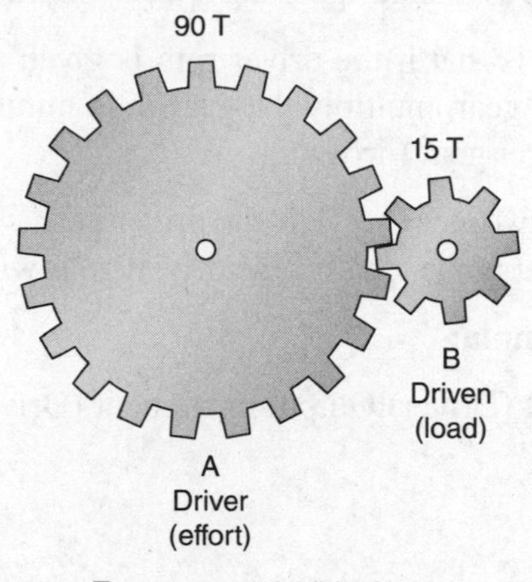


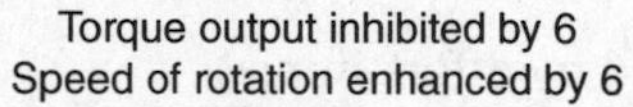


The driver gear rotates 6 times slower than the driven gear (increase in speed).

**Mechanical advantage formula:**

$$MA = \frac{\text{\# of teeth on driven gear}}{\text{\# of teeth on driver gear}}$$

$$MA = \frac{15}{90} = \frac{1}{6} \text{ (decrease in torque)}$$

## REVOLUTIONS PER MINUTE

As a general rule, if the number of revolutions per minute (rpm) of the driver gear is given and the driver gear is smaller than the driven gear, divide the gear ratio number of the driver gear into the rpm of the smaller gear (driver).

**Example:** In a two-gear system, the driver gear with 25 teeth revolves at 60 rpm; what is the rpm for the driven gear with 75 teeth?

**Gear ratio formula:**

Input movement (Driver):Output movement (Driven)

3:1

OR

$$\frac{\text{distance moved by effort (force)}}{\text{distance moved by resistance force (load)}} = \frac{3}{1}$$

The driver gear rotates 3 times faster than the driven gear (decrease in speed).

Divide the gear ratio number of the driver into the rpm for the driver gear.

(60 rpm/3) = 20 rpm (agrees with driver gear rotating 3 times faster than the driven gear)

This adheres to the general rule of gears, which states that when a smaller gear drives a larger gear the speed should decrease.

Another general rule is that if the driver rpm is given and the driver gear is larger than the driven gear, multiply the gear ratio number of the driven gear by the rpm of the larger gear (driver).

**Example:** In a two-gear system, the driver gear with 60 teeth revolves at 120 rpm; what is the rpm for the driven gear with 30 teeth?

**Gear ratio formula:**

Input movement (Driver):Output movement (Driven)

1:2

OR

$$\frac{\text{distance moved by effort (force)}}{\text{distance moved by resistance force (load)}} = \frac{1}{2}$$

The driver gear rotates 2 times slower than the driven gear (increase in speed).

Multiply the gear ratio number of the driven gear by the driver gear's rpm.

$2 \times 120$ rpm = 240 rpm (agrees with driver gear 2 times slower than the driven gear)

This adheres to the general rule of gears, which states that when a larger gear drives a smaller gear the speed should increase.

## GEAR TRAINS

When solving questions using more than two gears (**gear trains**), concentrate on two gears at a time. Give each gear a letter designation. Focus on the driver gear and the direction in which it is rotating (clockwise or counterclockwise). Gears on either end of the driver gear will rotate in the opposite direction. These gears in turn will cause the gears abutting them to rotate in opposite directions. A gear train component used to allow adjacent gears to rotate in a desired direction is known as an **idler** gear.

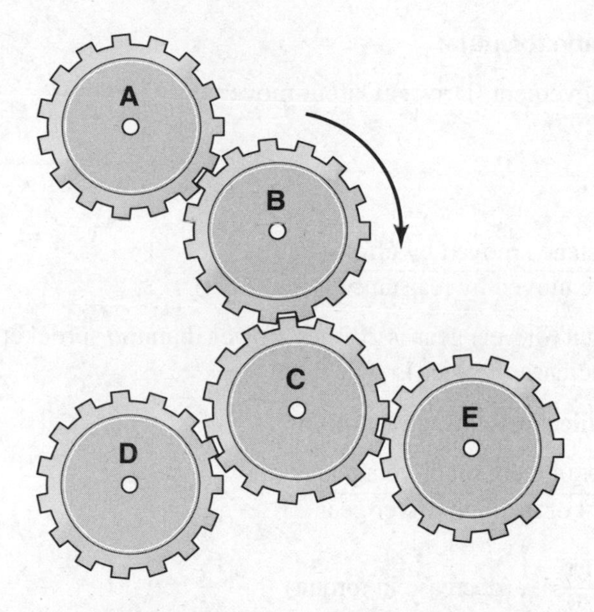


In the illustration shown, the driver (gear B) rotates in a clockwise direction. Gears A and C, abutting gear B, rotate in a counterclockwise direction. Gears D and E, which are abutting gear C, rotate in a clockwise direction.

## PEDAL AND SPROCKET GEARS

Special gears with elongated teeth are connected with a chain. These **pedal and sprocket gears** are found quite commonly in bicycles. They operate the same way as gears except that the direction of the rotating gears is not reversed. When the large pedal gear toward the front of the bicycle revolves, the chain pulls around the sprocket gear wheel at the rear in the same direction. Gear ratios and mechanical advantage are calculated via the same formulas used for gear systems.

> **Example:** If the pedal (driver) gear with 30 teeth in the illustration below revolves once, how many times will the sprocket (driven) gear with 15 teeth revolve? Find the gear ratio as well as the mechanical advantage of this gear system.

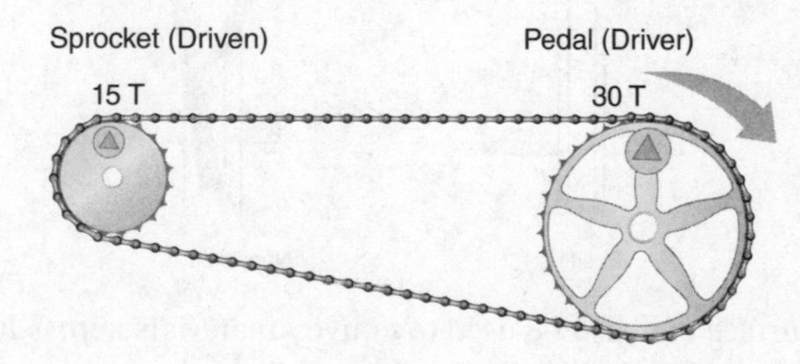

**Gear ratio formula:**

Input movement (Driver):Output movement (Driven)

1:2

OR

$$\frac{\text{distance moved by effort (force)}}{\text{distance moved by resistance force (load)}} = \frac{1}{2}$$

The pedal (driver) gear is 2 times slower than the sprocket (driven) gear (increase in speed).

**Mechanical advantage formula:**

$$MA = \frac{\text{\# of teeth on driven gear}}{\text{\# of teeth on driver gear}}$$

$$MA = \frac{15}{30} = \frac{1}{2} \text{(decrease in torque)}$$

## BELT DRIVE

In machinery, a pair of pulley wheels attached to parallel shafts and connected by a flexible band of flat leather, rubber, or similar material is known as a **belt drive**. Like gear trains, a belt drive can be used to increase or reduce the speed and mechanical advantage (torque) of the pulley wheels they are attached to and modify rotational motion from one shaft to the other. Unlike gears, however, belt drives rotate in the same direction.

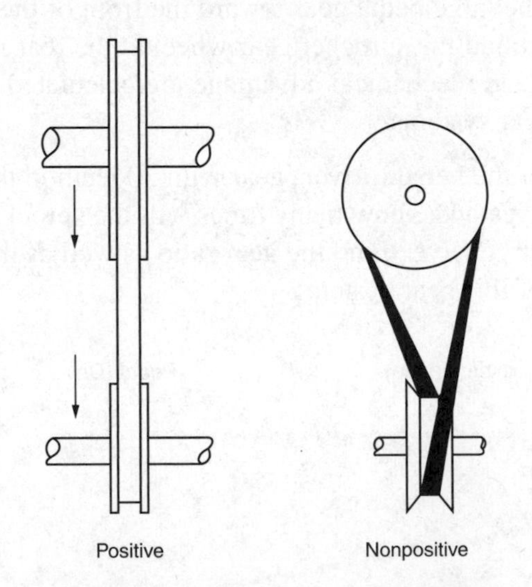


A belt's top surface can also be used to convey materials across it, such as on a conveyor belt. Belts are installed under tension in order to create friction, which allows the belt to grip and turn the pulley wheels. Substantial tension also keeps the belt from slipping off the pulley wheels when rotating. Belt drives are better for greater distances with smaller forces since they are not directly joined.

There are two types of belt drives, positive and nonpositive. Positive belt drives (gear belts) consist of belts with teeth that mesh with pulley wheels that also have teeth. This type of belt drive does not allow the belt to slip. They are used in applications requiring a higher horsepower or torque capacity. A nonpositive belt drive system utilizes smooth belts and pulley wheels. The design of a nonpositive belt is in the shape of a V. These belts require less tension than do flat belts because they have more surface area in contact with the pulley wheels. Nonpositive belt drives are useful for connecting shafts that are in close proximity to one another. The V-belt found inside of an automobile engine is an example of the nonpositive type. Crossed (twisted) belts cause shafts to rotate in opposite directions.

The mechanical principles and applications that apply to gears, in general, also apply to belt drives. To calculate the mechanical advantage of a belt drive system, instead of counting the teeth as you would when working with gears, divide the diameter of the driven pulley wheel by the diameter of the driver pulley wheel.

Whenever the driven pulley wheel diameter is larger than the driver pulley wheel diameter, you will obtain a mechanical advantage.

**Example:** A belt drive system has a 9-inch diameter driven pulley wheel and a 3-inch diameter driver pulley wheel. What is the mechanical advantage of the system?

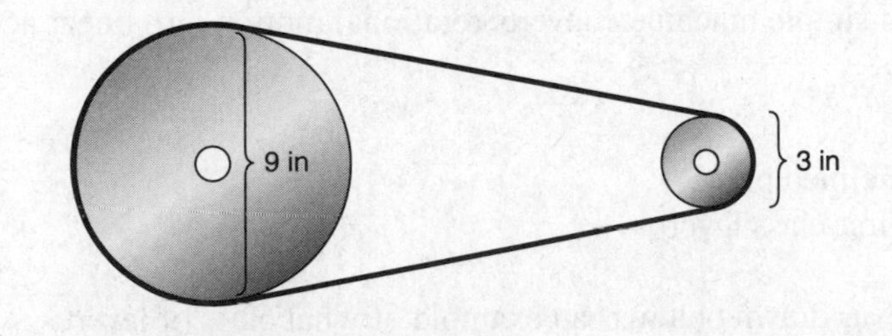


$$\text{MA} = \frac{\text{driven pulley wheel diameter}}{\text{driver pulley wheel diameter}}$$

$$\text{MA} = \frac{9 \text{ in}}{3 \text{ in}} = 3$$

To calculate the speed of the belt drive pulley system above, which has a driver pulley speed of 1,200 rpm, use the following formula:

$$\underset{\text{(rpm)}}{\text{Speed (Driver)}} \times \text{Diameter Driver} = \underset{\text{(rpm)}}{\text{Speed (Driven)}} \times \text{Diameter (Driven)}$$

$$1{,}200 \text{ rpm} \times 3 \text{ in} = (x) \times 9 \text{ in}$$

$$\frac{1{,}200 \text{ rpm} \times 3 \text{ in}}{9 \text{ in}} = x$$

$$x = \frac{1{,}200 \text{rpm}}{3} = 400 \text{rpm}$$

## REVIEW QUESTIONS

*Circle the letter of your choice.*

**1.** A wheelbarrow is what type of simple machine?

(A) Wheel and axle
(B) First-class lever
(C) Second-class lever
(D) Third-class lever

**2.** A shovel is an example of what classification of lever?

(A) First class
(B) Second class
(C) Third class
(D) A shovel is not considered a lever.

**3.** Which of the following simple machines is a class 1 lever?

(A) Crowbar
(B) Wheelbarrow
(C) Shovel
(D) None of the above

**4.** Which simple machine converts rotational motion into linear action?

(A) Wedge
(B) Screw
(C) Inclined plane
(D) First-class lever

**5.** The lever drawn below is an example of what class of lever?

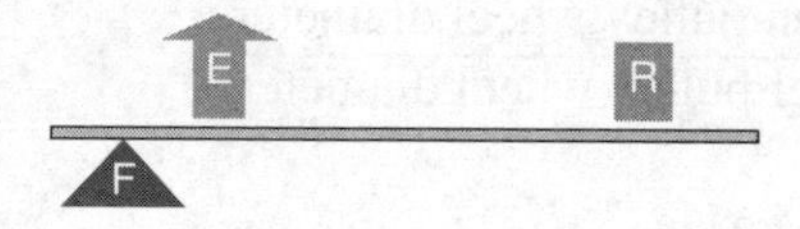


(A) First class
(B) Second class
(C) Third class
(D) All of the above

**6.** A metal bar is being used as a first-class lever to lift a heavy object weighing 2,500 N. The effort (force) being applied to the metal bar is 250 N. What is the mechanical advantage of the lever?

(A) 2,500
(B) 250
(C) 500
(D) 10

7. What is the mechanical advantage when 500 N of effort (force) is used on a bicycle pedal creating 50 N of force to move the bicycle in a forward direction?

(A) 0.1
(B) 10
(C) 100
(D) 1000

8. If the force to raise a flag up a flagpole using a pulley system is 100 N and the resistance force (load) is also 100 N, what is the mechanical advantage of the pulley system?

(A) 0
(B) 1
(C) 2
(D) 100

9. A fixed point on a lever is called the

(A) effort (force)
(B) load
(C) fulcrum
(D) mechanical advantage

10. The load that works against the effort (force) of the simple machine is called the

(A) input
(B) lever arm
(C) fulcrum
(D) resistance force

11. What is the mechanical advantage of a class 1 lever system when the length of the lever between the effort (force) being applied and a stone being used as a fulcrum is 1 meter and the length of the lever from the fulcrum to the load being lifted is 0.1 meter?

(A) 1
(B) 2
(C) 10
(D) 100

12. What type of simple machine is being used when you split a log with an axe?

(A) Third-class lever
(B) Wedge
(C) Screw
(D) None of the above

**13.** Which of the following is true of third-class levers?

(A) Effort (force) is multiplied.
(B) The distance the object is moved is increased.
(C) Resistance force (load) is between the effort (force) and fulcrum.
(D) All of the above

**14.** Which of the following is NOT true of an inclined plane?

(A) It is a simple machine with a sloped surface.
(B) It will not change the amount of work necessary to complete a task.
(C) It will not reduce the effort (force).
(D) It requires much less effort to push a heavy object up an incline than to lift it straight up.

**15.** A crank handle simple machine assembly, shown in the diagram, is being used to lift a heavy object. The axle circumference is 6 inches. The crank handle has a turning radius of 8 inches. How much resistance force (load) can be lifted if 200 pounds of effort (force) is applied to the crank handle?

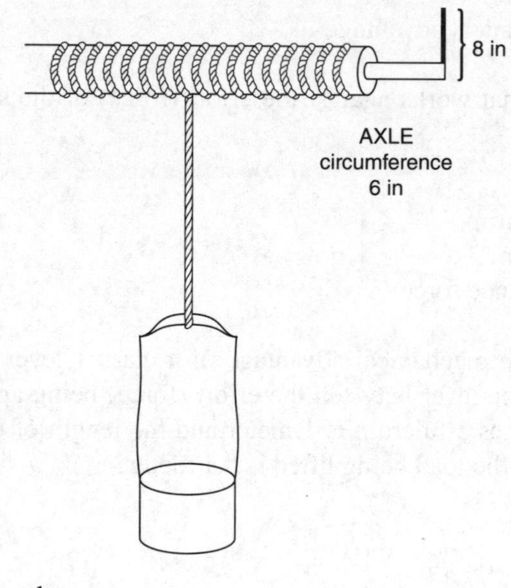


(A) 1,675 pounds
(B) 1,800 pounds
(C) 1,200 pounds
(D) 2,200 pounds

**16.** Which of the following is true regarding the gear train assembly shown in the diagram?

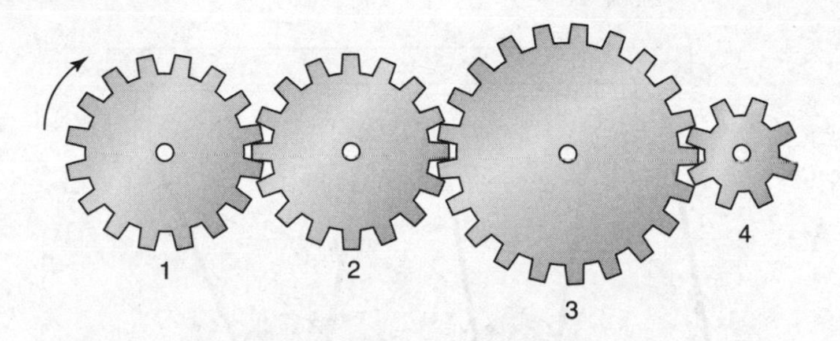


(A) All gears are rotating clockwise.
(B) Gear 1 and gear 3 are both rotating clockwise.
(C) Gears 2, 3, and 4 are all rotating counterclockwise.
(D) Gears 1, 2, and 4 are all rotating clockwise.

**17.** A pulley is used to raise a heavy weight. Effort (force) of 200 N is needed to lift a resistance force (load) of 1,600 (N). The mechanical advantage of this pulley system is

(A) 2
(B) 4
(C) 6
(D) 8

**18.** If the resistance force (load) is 10,000 N, what would be the effort (force) if a wedge with a mechanical advantage of 0.75 is being utilized to lift the corner of a house from its foundation?

(A) 13,333
(B) 7,555
(C) 10,000
(D) 15,757

**19.** The mechanical advantage of the block and tackle shown is

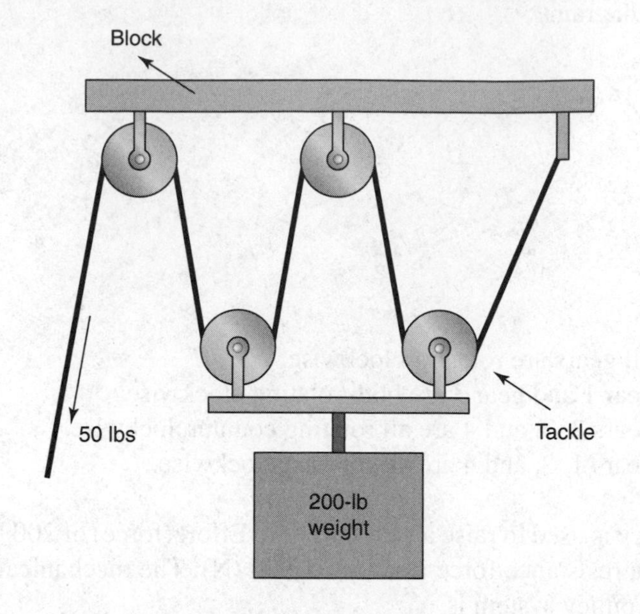


(A) 2
(B) 4
(C) 5
(D) 6

**20.** In a gear train (multiple gears of different sizes), each gear travels

(A) at the same speed as the one next to it
(B) always in the same direction as the one next to it
(C) faster than the one next to it
(D) in the direction opposite to the one in which the gear next to it is traveling

**21.** A wheel and axle device consisting of a small cylinder (axle) having a crank handle used to apply force to a rope or cable while winding it around the cylinder is called a

(A) fulcrum
(B) screw
(C) windlass
(D) third-class lever system

**22.** A pulley system lifts a 200-N weight with an effort (force) of 40 N. The pulley rope where the effort (force) is being applied moves 6 m while the load is lifted 1 m. The efficiency of the pulley system is most nearly

(A) 78.8%
(B) 56.7%
(C) 83.3%
(D) None of the above

**23.** The circumference of a screw multiplied by the number of times it is turned is equal to its effort (force) distance. A screw with a circumference of 15 mm is turned 3 times causing it to enter into a piece of wood a distance of 5 mm. What is the mechanical advantage of the screw?

(A) 3
(B) 6
(C) 7
(D) 9

**24.** The figure shows a belt drive system with two pulley wheels being rotated by a crossover belt. Which of the following is true regarding this type of belt drive system?

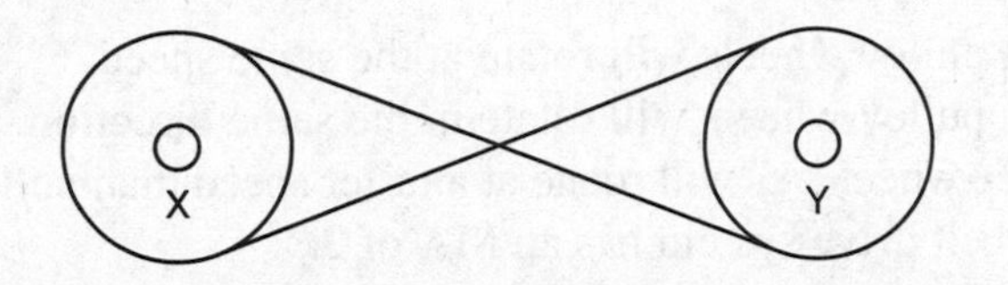


(A) Pulley wheel (X) rotates in the same direction as pulley wheel (Y).
(B) Pulley wheel (X) rotates faster than pulley wheel (Y).
(C) Pulley wheel (X) rotates in the opposite direction from pulley wheel (Y).
(D) Pulley wheel (X) rotates slower than pulley wheel (Y).

**25.** What is the mathematical formula for work?

(A) Work = Force × Distance
(B) Work = Velocity × Distance
(C) Work = Force × Time
(D) None of the above

**26.** In physical science, the product of mass × acceleration is called

(A) velocity
(B) force
(C) momentum
(D) mechanical advantage

**27.** Which of the following is true regarding the belt drive system shown?

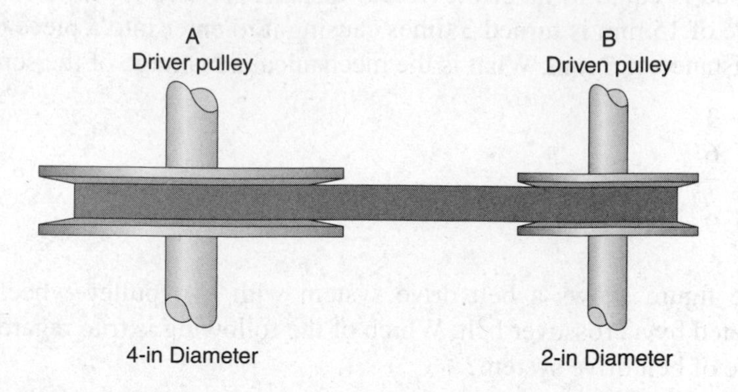


(A) Both pulley wheels will rotate at the same speed.
(B) Both pulley wheels will rotate in the same direction.
(C) Pulley wheel "A" will rotate at a faster speed than pulley wheel "B."
(D) The belt drive system has an MA of 2.

**28.** How much weight can a jackscrew with a pitch of ¼ inch and a handle 16 inches long lift when a force of 4 pounds is applied to the handle?

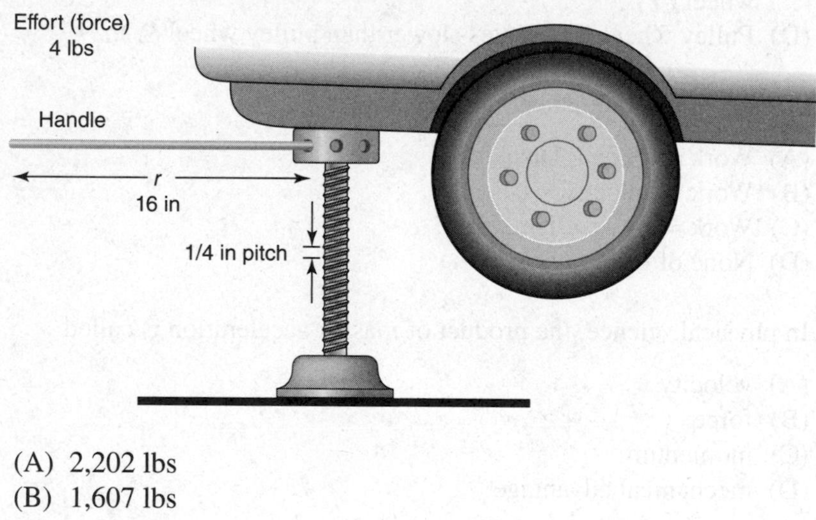


(A) 2,202 lbs
(B) 1,607 lbs
(C) 4,011 lbs
(D) 3,084 lbs

**29.** Which inclined plane has the greatest mechanical advantage and therefore requires the least amount of effort to raise a barrel up onto the platform?

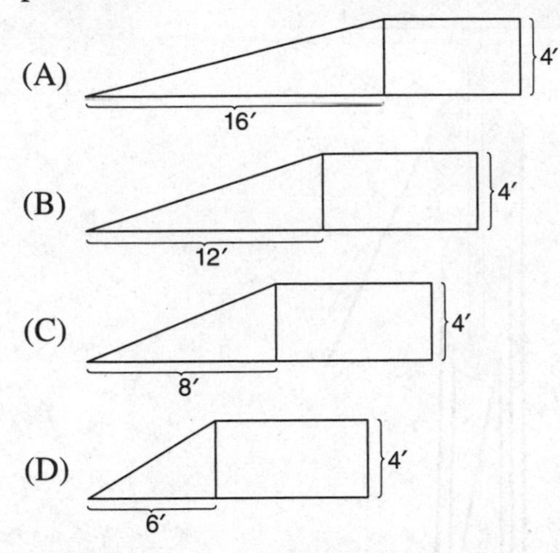


**30.** What is the mechanical advantage of the inclined plane shown?

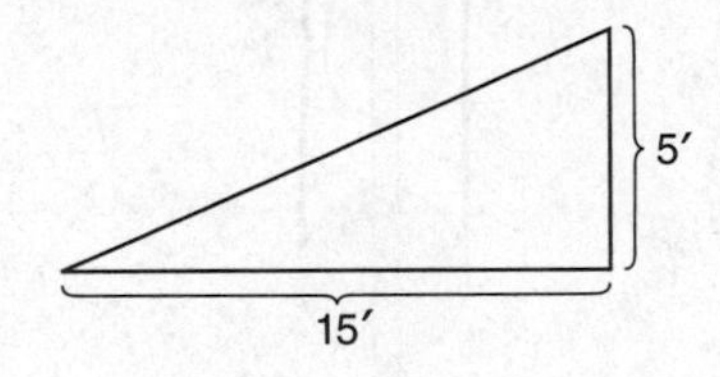


(A) 0
(B) 1
(C) 2
(D) 3

**31.** What force should be applied to the handle of the claw hammer shown in order to lift the nail out of the block of wood safely?

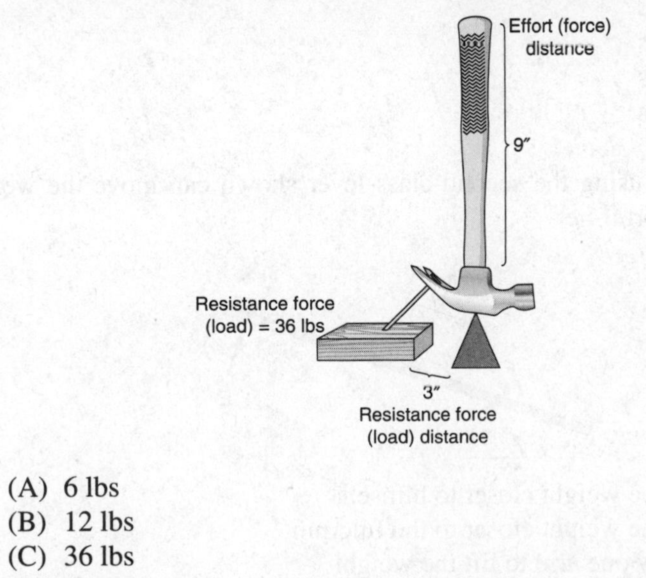


(A) 6 lbs
(B) 12 lbs
(C) 36 lbs
(D) 72 lbs

**32.** How much weight attached to the block and tackle shown will a pull equal to 200 pounds applied to line #1 be able to lift?

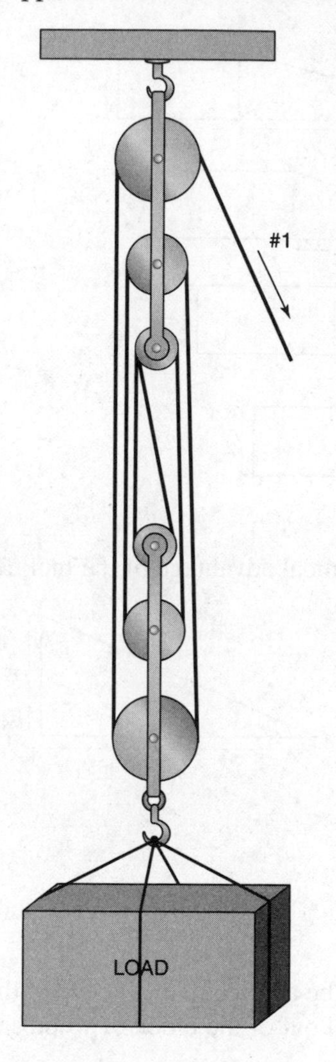


(A) 1,200 lbs
(B) 1,400 lbs
(C) 1,740 lbs
(D) 1,960 lbs

**33.** A firefighter using the second-class lever shown can move the weight with less effort if he

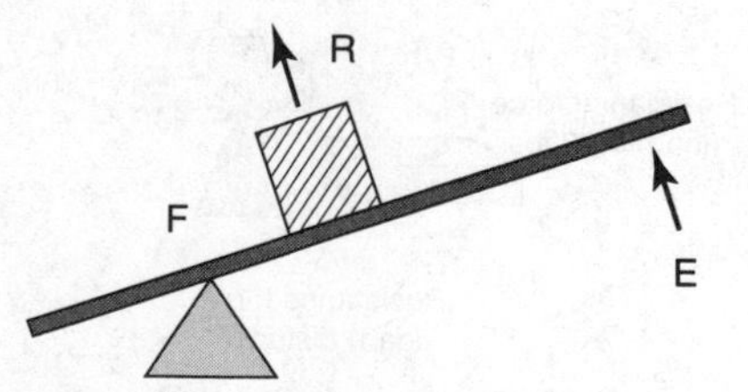


(A) moves the weight closer to himself
(B) moves the weight closer to the fulcrum
(C) uses only one arm to lift the weight
(D) moves the fulcrum farther away from the weight

**34.** If the weight attached to the pulley system shown is to be lifted up 2 feet, how much rope must be pulled downward?

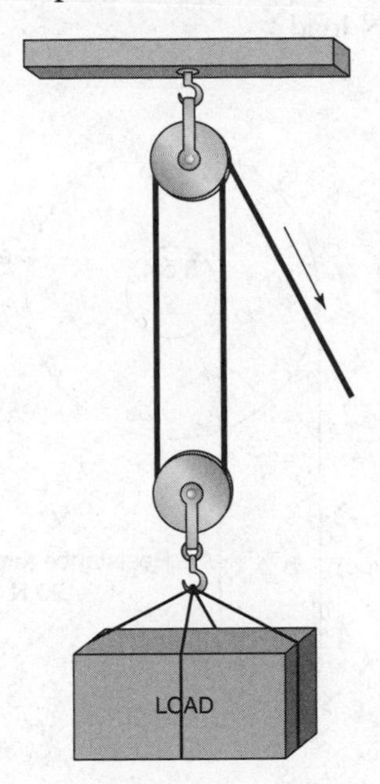


(A) 2 ft
(B) 3 ft
(C) 4 ft
(D) 6 ft

**35.** A firefighter using Vise-Grip pliers to back out a keyway cylinder from a door during a through-the-lock forcible entry demonstration would require less effort to complete the task if she

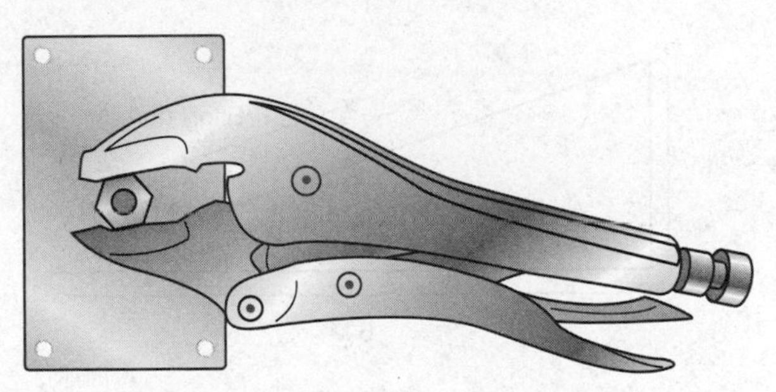

(A) squeezed tighter on the pliers
(B) placed her hand farther down the handle of the pliers
(C) moved her hand closer to the keyway cylinder
(D) turned the pliers upside down

**36.** The wheel and axle assembly shown has a wheel with a circumference of 30 cm and an axle circumference of 5 cm. How much effort would be required to lift a 90-N load?

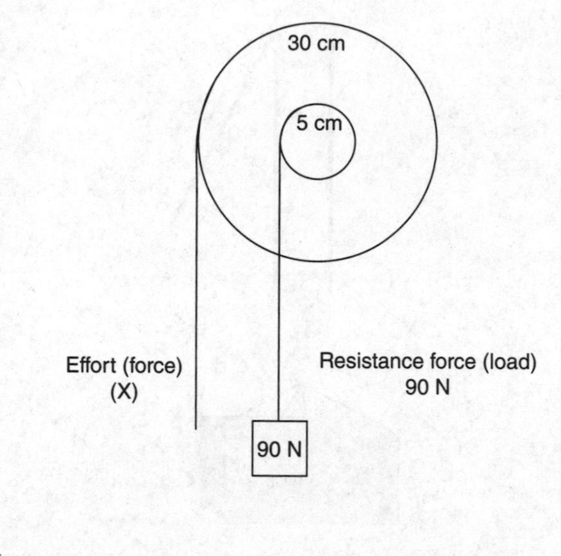


(A) 5 N
(B) 10 N
(C) 15 N
(D) 30 N

**37.** How much force would be required to roll a 15-N ball up a 6-meter-long inclined plane (ramp) to a platform that is 3 meters in height?

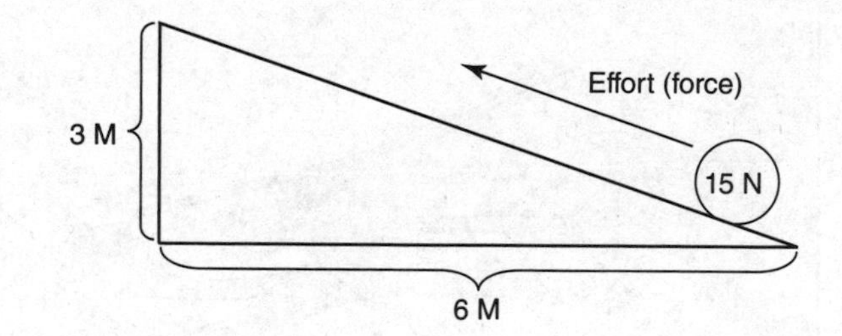


(A) 7.5 N
(B) 5.5 N
(C) 15 N
(D) 3.5 N

**38.** The pulley system shown below has an MA of 4 (four supporting ropes). An effort (force) of 100 N can lift an object of how much weight?

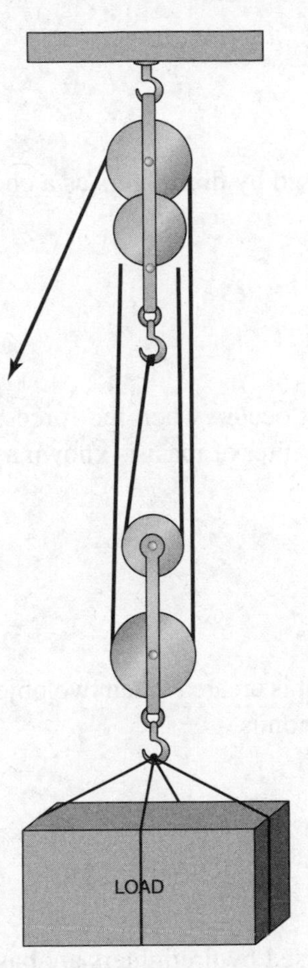


(A) 25 N
(B) 100 N
(C) 200 N
(D) 400 N

**39.** In ancient times, an inventor created a simple machine whose motion made it easier to transport water from low-lying areas to upper areas for irrigation purposes. The device was called

(A) Achilles' wedge
(B) Archimedes' screw
(C) Archimedes' lever
(D) Archimedes' pulley

**40.** A bicycle utilizes a chain to connect its gears. This chain has links that fit into the teeth of the gears. A gear that has teeth that fits into the links of a chain is called a

(A) driver gear
(B) sprocket gear
(C) bevel gear
(D) None of the above

**41.** The distance a screw will travel during one full rotation is called the

(A) thread
(B) circumference
(C) pitch
(D) radius

**42.** A simple machine used by firefighters as a chock or doorstop is the

(A) ramp
(B) axle
(C) windlass
(D) wedge

**43.** A twisting force that occurs when the force being applied does not go through an object's center of mass is known as

(A) joule
(B) watt
(C) torque
(D) momentum

**44.** Friction is a force that is created when two objects rub against each other. A by-product of friction is

(A) lubricant
(B) oil
(C) water
(D) heat

**45.** Complex machines used by firefighters are basically

(A) two or more simple machines used together
(B) less efficient than simple machines
(C) easier to operate than simple machines
(D) bigger than simple machines

**46.** The tip of a slotted screwdriver is known as a

(A) screw
(B) gear
(C) wheel and axle
(D) simple wedge

**47.** Unzipping a jacket is an example of what type of simple machine?

(A) Screw
(B) Wedge
(C) Gears
(D) None of the above

**48.** You can enhance the mechanical advantage of an axe by

(A) sharpening the blade
(B) blunting the blade
(C) cutting off the handle
(D) oiling the blade

**49.** A gear train component used to allow adjacent gears to rotate in a desired direction is called an

(A) driver gear
(B) driven gear
(C) idler gear
(D) train gear

**50.** A nail is an example of what type of simple machine?

(A) Gears
(B) Screw
(C) Wedge
(D) Pulley

# Answer Key

1. C
2. C
3. A
4. B
5. C
6. D
7. A
8. B
9. C
10. D
11. C
12. B
13. B
14. C
15. A
16. B
17. D
18. A
19. B
20. D
21. C
22. C
23. D
24. C
25. A
26. B
27. B
28. B
29. A
30. D
31. B
32. A
33. B
34. C
35. B
36. C
37. A
38. D
39. B
40. B
41. C
42. D
43. C
44. D
45. A
46. D
47. B
48. A
49. C
50. C

# ANSWER EXPLANATIONS

Following you will find explanations to specific questions that you might find helpful.

1. C Second-class lever; in a second-class lever, the resistance force (load) is between the effort (force) and the fulcrum.
2. C Third-class lever; a third-class lever has the effort (force) between the fulcrum and the resistance force (load).
3. A Crowbar; in a first-class lever, the fulcrum is between the effort (force) and the resistance force (load).
4. B A screw is a simple machine similar to an inclined plane or a wedge, but in a screw the incline wraps around a shaft. A screw converts rotational motion into linear motion.
5. C Third-class lever; in a third-class lever, the effort (force) required to lift the load is actually increased and, therefore, the mechanical advantage is less than 1. The trade-off, however, is an increase in speed and distance of travel of the load.
6. D MA = resistance force (load)/effort (force); MA = 2,500N/250N = MA = 10
7. A A bicycle is a first-class lever. MA = resistance (force)/effort (force) = 50 N / 500 N = 0.1
8. B The effort (force) is equal to the resistive (force) (load) in this pulley system, and there is no mechanical advantage, with the MA equal to 1. MA = effort (force)/resistance (force) = 100 N / 100 N = 1
9. C Fulcrum
10. D Resistance force
11. C MA = effort (force) distance/resistance (force) distance = 1/0.1 = 10
12. B Wedge
13. B In a third-class lever, the effort (force) required to lift the load is actually increased and, therefore, the mechanical advantage is less than 1. The trade-off, however, is an increase in speed and distance of travel of the load.
14. C The inclined plane provides for less effort, not for less work, to lift the object. Getting heavy boxes or barrels onto a loading dock is much easier if you slide the objects up a ramp rather than lift them up. The trade-off is the greater distance to travel.
15. A effort (force) × circumference (crank wheel) = resistance force (load) × circumference (axle)
    Solve for the circumference of the crank wheel: $C = 2 \cdot \pi \cdot 8 = 2 \cdot 3.14 \cdot 8$ in = 50.24 in
    200 pounds · 50.24 in = *x* [resistance force (load)] · 6 in
    200 pounds · 50.24 in/6 in = *x* [resistance force (load)]
    1,675 pounds = resistance force (load)

**16.** B A basic rule concerning gears states that each gear in a series of gears reverses the direction of rotation of the previous gear.

**17.** D MA = resistance force (load)/effort (force) = 1600 N/200 N = 8

**18.** A MA = resistance force (load)/effort (force) = 10,000 N/0.75 = 13,333

**19.** B The mechanical advantage of the block and tackle pulley system shown can be calculated visually by counting the number of ropes supporting the load. There are four supporting ropes to lift the resistance force (load), giving the pulley system a mechanical advantage of 4.

**20.** D A basic rule concerning gears states that each gear in a series of gears reverses the direction of rotation of the previous gear.

**21.** C Windlass or well crank

**22.** C MA = resistance force (load)/effort (force) = 200N/40N = 5

Velocity ratio = effort (force) distance/load distance = 6 m/1 m = 6

Efficiency = 5/6 × 100% = 83.3%

**23.** D The circumference of a screw multiplied by the number of times it is turned is equal to its effort (force) distance.

MA = effort (force) distance/resistance force (load) distance

(15 mm) · 3 = 45 mm/5 mm = 9

**24.** C Crossed (twisted) belts cause shafts to rotate in opposite directions.

**25.** A Work (W) = Force (F) × Distance (D)

**26.** B Force (F) = Mass (M) × Acceleration (A)

**27.** B Unlike gears, belt drives rotate in the same direction.

**28.** B Jackscrew handle 16 in long. Circumference = 2 · π · r = 2 · 3.14 · 16 = 100.48 in

100.48/.25 (pitch) = 401.92 · 4 lbs. effort (force) = 1,607 lbs.

**29.** A MA = length of slope/height of slope: 16/4 = 4
12/4 = 3; 8/4 = 2; 6/4 = 1.5

**30.** D MA = length of slope/height of slope: 15/5 = 3

**31.** B MA = effort (force) distance/resistance force (load) distance = 9 in/3 in = 3 resistance force (load) = 36 lbs. With MA = 3, effort (force) required to pull nail out of block is 36 lbs./3 = 12 lbs.

**32.** A The mechanical advantage of pulley systems can be calculated visually by counting the number of ropes supporting the load. The illustration of the block and tackle pulley system shows six supporting ropes to lift the resistance force (load), giving the pulley system a mechanical advantage of 6. Therefore, an effort (force) of 200 pounds applied to line #1 will lift 1,200 pounds.

**33.** B The MA of a second-class lever is greater than 1. If the load is closer to the fulcrum, then more force has to be applied and there is a greater MA. MA = effort (force) distance/resistance force (load) distance.

Increasing the effort (force) distance and decreasing the resistance force (load) distance will increase the MA.

**34.** C As you increase the number of pulleys, you also increase the distance you have to pull the rope. If you use two pulleys, it takes half the effort to lift something, but you have to pull the rope twice as far. Weight is moved 2 ft upward while the rope is pulled 4 ft. downward.

**35.** B The pliers is a first-class lever. The ratio of the forces is inversely proportional to the distances from the pivot point where both the load and fulcrum are located. It would therefore require less effort to complete the task if the firefighter placed her hand farther down the handle of the pliers to increase the effort (force) distance.

MA = effort (force) distance/resistance force (load) distance

**36.** C effort (force) · circumference (wheel) = resistance force (load) · circumference (axle)

X · 30 cm = 90 N · 5 cm

X = 90 N · 5 cm/30 cm = 90 N/6 = 15 N

**37.** A effort (force) · effort (force) distance = resistance force (load) · resistance (force) distance

effort (force) · 6 m = 15 N · 3 m

effort (force) + 15 N · 3 m/6 m = 15 N/2 = 7.5 N

**38.** D MA = 4; effort (force) of 100 N can therefore lift 100 N · 4 = 400 N

**39.** B Archimedes' screw

**40.** B Sprocket gear

**41.** C Pitch

**42.** D Wedge

**43.** C Torque

**44.** D Heat

**45.** A Two or more simple machines used together; example: flat head axe and Halligan tool used for forcible entry

**46.** D Simple wedge

**47.** B Wedge

**48.** A When used for cutting, the longer and thinner the wedge, the less effort (force) is required to overcome resistance. You can enhance the mechanical advantage of an axe, a type of double wedge, by sharpening it. Effort (force) is applied to the thicker edge of the wedge and is transferred to the thinner end.

**49.** C Idler gear

**50.** C Wedge

CHAPTER 12

# Tools of the Trade

> **Your Goals for This Chapter**
> - Learn to recognize basic tools
> - Learn to recognize firefighting tools and know their uses
> - Test your tool knowledge with review questions

This chapter includes a brief discussion of basic hand tools, power tools, and measuring tools, but the main emphasis is on tools specific to firefighting, used not only on the fireground but also in the firehouse during maintenance operations. This chapter and Chapter 11, "Principles of Mechanics," provide the reader with a comprehensive review of tools and how and why they are used.

## BASIC HOME TOOLS

The following is a basic review of the types of tools that candidates preparing for the firefighter exam should be familiar with.

### Hand Tools

#### HAMMERS

A **nail hammer** is among the most widely used hand tools. It is used for striking and removing nails. Hammers are made in two general patterns: **straight claw** (ripping) and **curved claw**. Handles are made of wood, steel, or fiberglass. A hammer blow should always be struck squarely with the striking face of the hammer parallel to the object being struck. When striking another tool (chisel, wedge, hand punch, etc.), the hammer's striking face should have a diameter larger than the struck face of the other tool.

In addition to the basic straight and curved claw hammers, there are hammers designed for specialized purposes, such as light-duty **tack hammers** designed for driving small nails; round ball-shaped **ball-peen hammers** for riveting and shaping unhardened metal; **soft-face hammers**, or **mallets**, made from wood or rubber and designed for delivering blows to objects that would mar if struck with a metal hammer, and **drywall hammers** designed to dimple drywall prior to nailing.

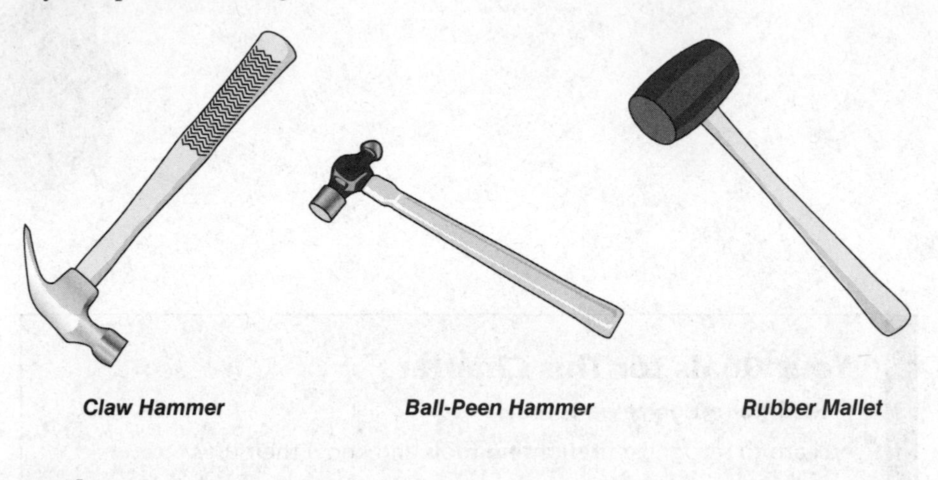


## Nails

Nails are made from wire. They range in length from 2 d (1 inch) to 60 d (6 inches). The abbreviation "d" is used for the term *penny* and is derived from the first letter of the Roman coin denarius. The common 16 d penny nail is 3½ inches long and is used to fasten structural building elements together. Other types of nails include concrete, drywall, finishing, ring, roofing, shingle, and spiral nails.

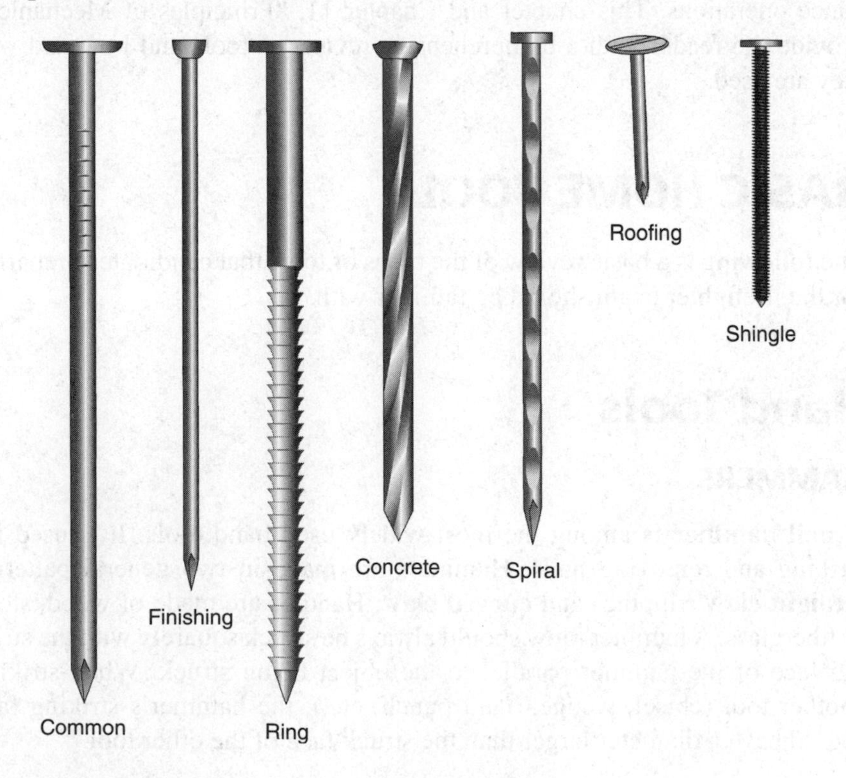

## CHISELS

Various types of chisels are used for cutting, shaping, and trimming different materials. A cold chisel made from steel is used for cutting and shaping metals such as cast iron, bronze, and copper. A wood chisel is designed for rough work on wooden materials. A masonry chisel is used with a hand drilling hammer to score or trim brick or block.

*Wood Chisel*

## PLIERS

Pliers are hand tools used to grip, turn, pull, or crimp a large variety of objects. Pliers direct the power of the handgrip into a precision grip. The long handles in relation to the nose of the pliers act as levers, enhancing the force in the hand's grip to the object being acted upon.

The variety of pliers today exceeds most, if not all, other types of hand tools. **Linesman pliers** with a side cutting feature bend lightweight metal and sever wire. **Wire stripping (electrician) pliers** sever and remove insulation on electrical wire without damaging the wire. **Long-nose** or **needle-nose pliers** are used to grip and shape lightweight metal; the slim head design facilitates crimping wires in confined, narrow spaces. **Locking pliers**, also called **LockJaw** or **Vise-Grip pliers**, are basically a handheld vise that allow for the purchase on an object to be locked in and tightened prior to applying force. They are used to firmly grip lightweight metal or remove round door-lock cylinders.

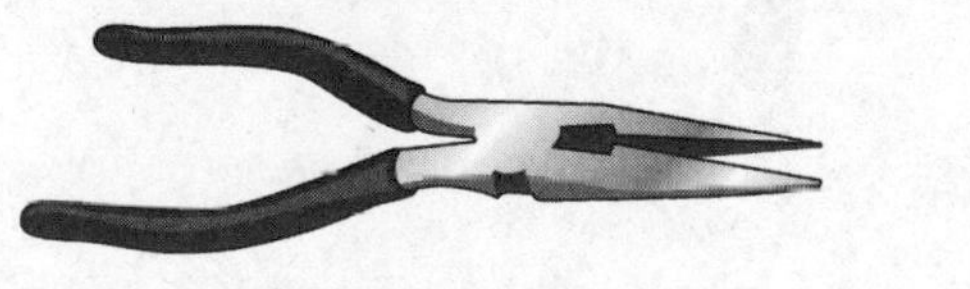

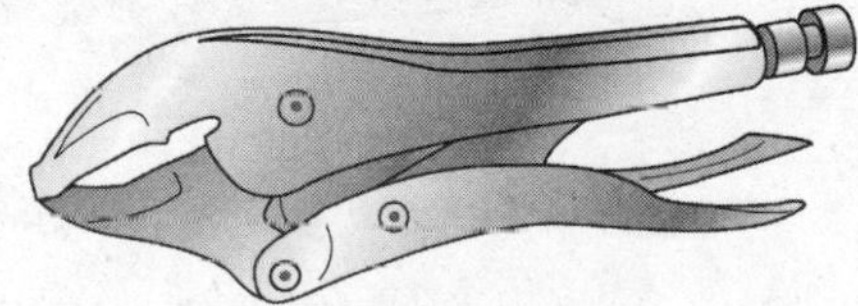

*Needle-Nose Pliers*

*Locking Pliers*

## WRENCHES

Wrenches are used for tightening and loosening nuts, bolts, pipes, and many other objects that are hard to turn. The tool works as a lever. The mouth of the wrench is used to grip the object to be turned. The wrench is pulled at a right angle to the axis of the lever action and the turned object. Wrenches can be nonadjustable or adjustable to fit better around objects of various sizes that need to be turned.

An **open-end wrench** is a one-piece wrench with a smooth, U-shaped opening(s). It is designed to grip two opposite faces of a bolt or nut from the side. This is advantageous in areas that are difficult to access or are obstructed.

A **box-end wrench** is a one-piece wrench with recessed, grooved, enclosed opening(s). It is often designed double-ended with different-sized box ends. The enclosed opening grips all the faces of the bolt or nut, providing more torque than the open-end wrench without slipping or stripping the bolt or nut. A **combination wrench** is a double-ended tool with one end open and the other end enclosed. Both ends generally fit the same size bolt or nut.

An **offset wrench** is designed to provide access to obstructed bolts and nuts in recessed areas. It allows for hand clearance when turning an object flush with a work surface. An **adjustable (crescent) wrench** is an open-end wrench with smooth, adjustable jaws used to turn bolts or nuts. A **pipe** or **Stillson wrench** is an adjustable wrench having serrated jaws for gripping soft iron pipe and pipe fittings. A **hex key (Allen) wrench** is an L-shaped, six-sided wrench used to turn machined setscrews or bolt heads designed with a hexagonal recess. And, finally, the well-known **monkey wrench** (named for its inventor, Charles Moncky—I kid you not!) is an old-type adjustable wrench with smooth jaws that is used for turning bolts or nuts.

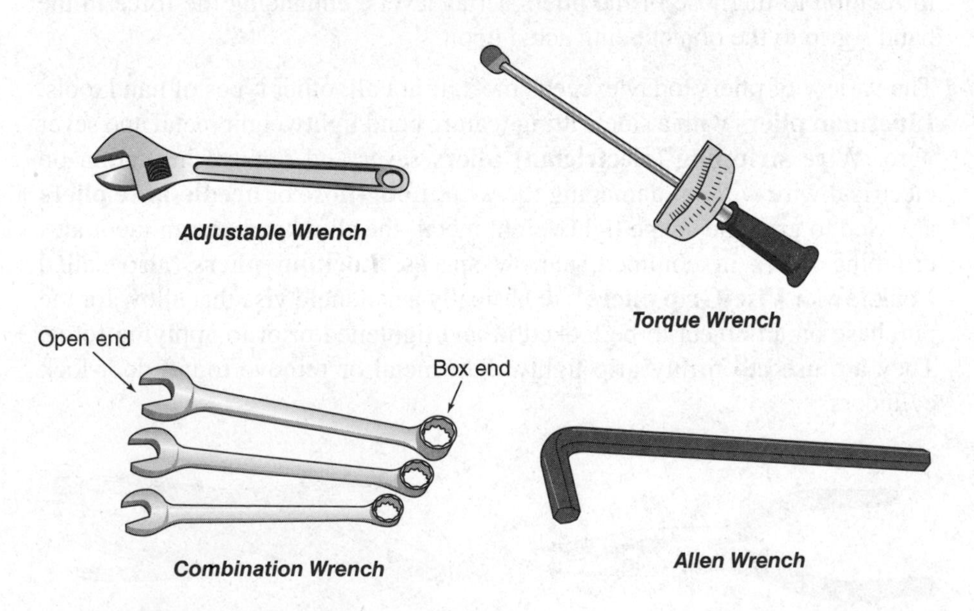


A **ratchet box**, designed for turning bolts or nuts, has a mechanism that eliminates the need to readjust the wrench during the return stroke. A **socket with ratchet handle** is a hollow cylinder (socket) that fits over a bolt or nut head, used in conjunction with a drive tool (ratchet handle). Sockets are generally sold in sets of various sizes.

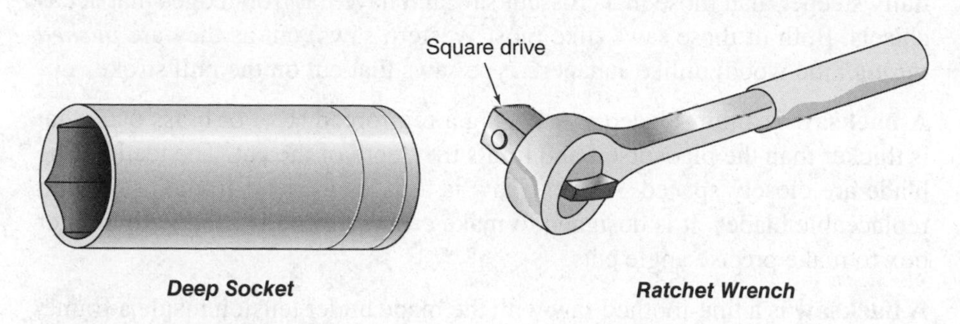


*Deep Socket* *Ratchet Wrench*

## SCREWDRIVERS

The simple handheld screwdriver is designed to tighten or loosen and remove screws. It consists of a tip or head at the end of an axial shaft that is encased inside a cylindrical handle. The handle allows the shaft to be rotated, thereby applying torque at the tip. Screwdrivers are made in a wide variety of sizes to match different screw sizes. Screwdriver heads come in many types, the most common of which are mentioned below.

A **slot-head screwdriver** is a flat-bladed screwdriver that fits a single slot screw. A **Phillips head screwdriver** is a cross-headed screwdriver with rounded corners. It is designed to slip off the screw when under high torque to prevent overtightening. A **hex head (Allen) screwdriver** has a six-sided head and is used as an alternative tool to the hex key wrench. An **offset screwdriver** is used to access hard-to-reach and obstructed screws where a straight shaft screwdriver is inappropriate. A **ratchet screwdriver** is designed for high-speed turning using a ratchet handle.

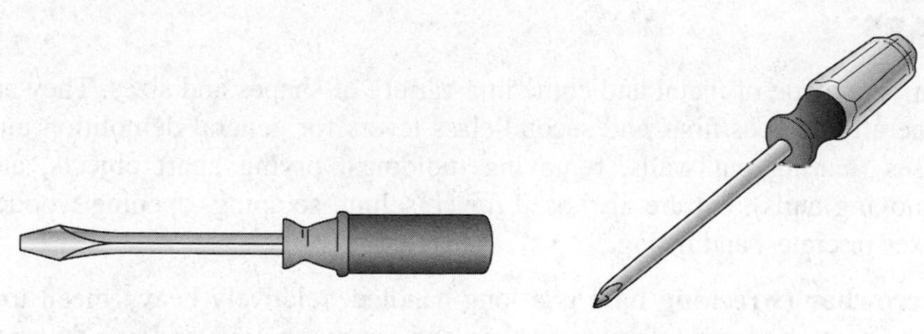

*Standard Screwdriver* *Phillips Screwdriver*

## SAWS

The key component of the saw is a **blade** with a cutting edge. Other components include the **heel**, the end closest to the handle; the **toe**, the end farthest from the handle; the **front**, or **bottom** edge; and **back**, or **top** edge.

There are many different types of saws, designed for specific uses. The most common saws are mentioned below.

A **crosscut saw** is designed for making cuts in lumber perpendicular (at a right angle) to the grain. The cutting edge of the blade is beveled, allowing the blade to act like a knife edge and slice through the wood. A **ripsaw** is designed for cutting lumber parallel to the grain. The saw teeth are substantially steeper than those in a crosscut saw and have flat front edges that act as chisels. Both of these saws (like most Western saws) cut as they are *pushed through* the wood; unlike Japanese-type saws that cut on the pull stroke.

A **backsaw** is a thin-bladed saw having a reinforced steel or brass back that is thicker than the blade itself and limits the depth of the cut. The teeth of the blade are closely spaced. A **miter saw** is a back or metal-framed saw with replaceable blades. It is designed to make crosscuts and is used with a miter box to make precise angle cuts.

A **hacksaw** is a fine-toothed saw with the blade under tension inside a frame. It is designed to cut metal. Finally, a **coping (jigsaw)** is designed to cut intricate shapes in wood. It has a thin blade tensioned inside a metal frame.

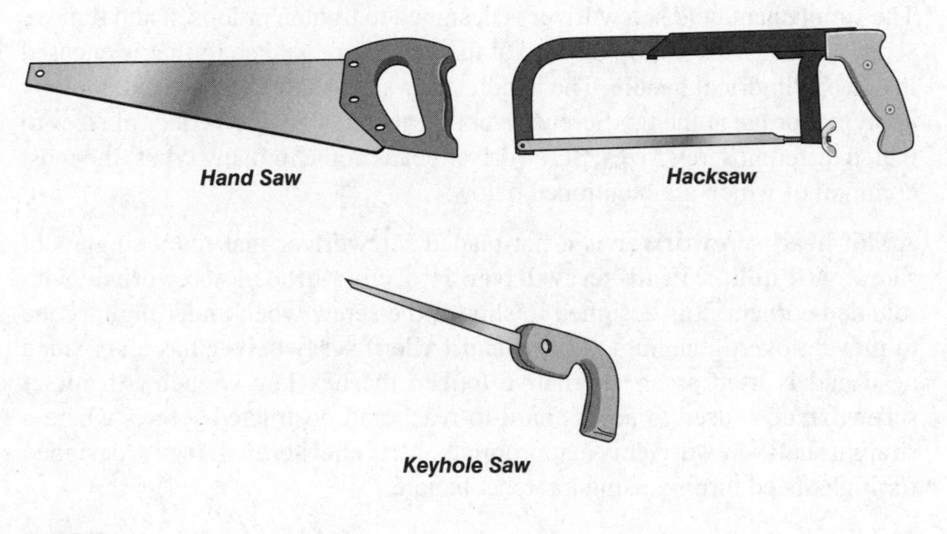


## BARS

Bars are made of metal and come in a variety of shapes and sizes. They are generally used as first- and second-class levers for general demolition purposes (tearing out walls, removing moldings, prying apart objects, and removing nails), but are also used for chiseling, scraping, opening wooden boxes or crates, and lifting.

A **crowbar (wrecking bar)** is a long-handled, relatively heavy metal tool with one curved end (claw) for prying and removing nails and one flattened end.

A **pry bar** is a long-handled metal tool, generally smaller and lighter than the crowbar. It has one flattened end and one tapered, pointed end and is used for prying, lifting, and removing nails. A combination pry bar and chisel tool, an offset **rip bar**, is similar in appearance to the crowbar. It has a small claw end and wide chisel end in an offset pattern for difficult-to-reach objects. A **utility pry bar**, a small, lightweight metal tool having beveled cutting edges at both ends, is designed for pulling nails, prying light objects, and scraping. A **cat's paw** is a small-handled pry bar with one flattened end and one rounded, curved end for prying and removing nails.

A **nail puller** is a metal bar designed to easily slide under the head or into the shank of an embedded nail for easy removal. The other end is a striking head for a hammer.

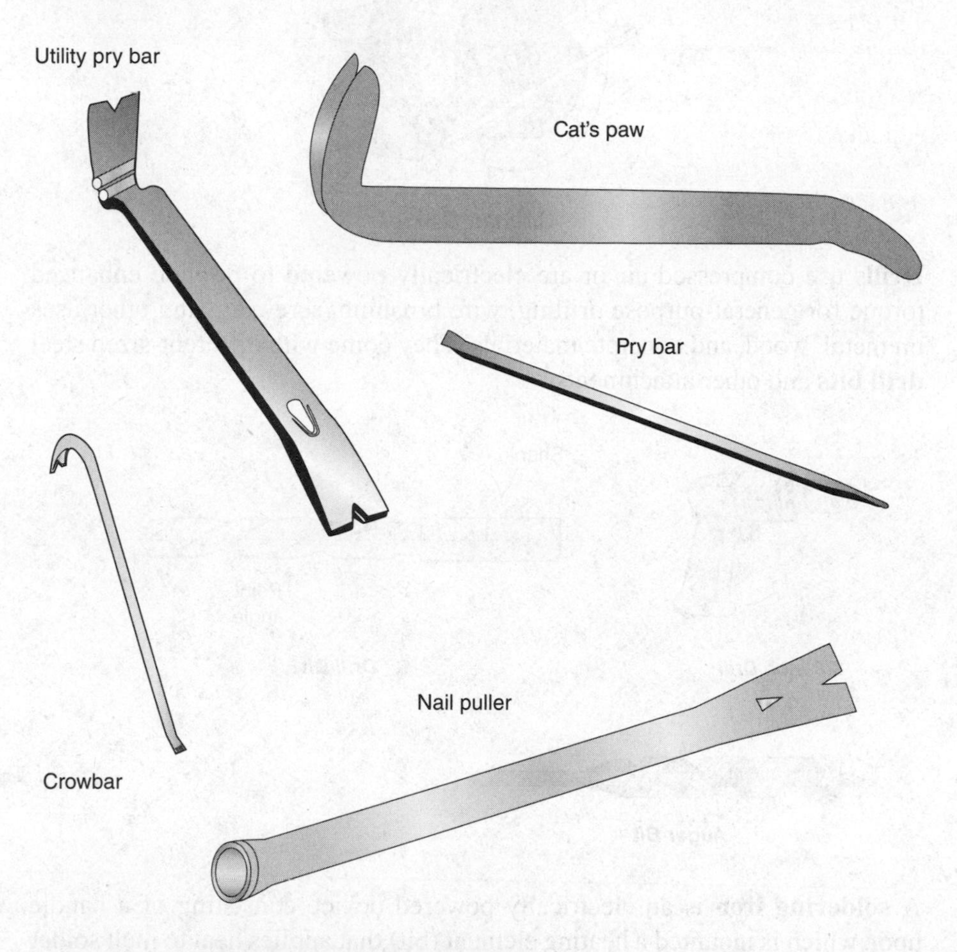


## OTHER COMMON HAND TOOLS

Other common hand tools include **hand punches** designed to mark metal and other surfaces softer than the punch itself and to align holes and drive or remove pins and rivets; **drift pins** used for aligning holes in metal; and **star drills** used, along with a **hand drilling hammer**, to drill holes in masonry.

# Power Tools

Power tools are basically hand tools that use electrical, pneumatic, or fuel power. Many common hand tools discussed above are now also available as power tools.

Different types of saws are available as power tools. A **circular saw** with a rotary blade is used for cutting wood; a **reciprocating saw**, with a thin, straight blade, is used for cutting metal, pipe, and wood; a lightweight **jig**, or **saber saw**, is used to cut custom designs in soft wood and light metal; and a **chainsaw** is used for logging, tree trimming, and harvesting firewood.

*Circular Saw*

**Drills** use compressed air or are electrically powered to provide enhanced torque for general-purpose drilling, wire brushing, screwing, and other uses on metal, wood, and concrete materials. They come with different-sized steel **drill bits** and other attachments.

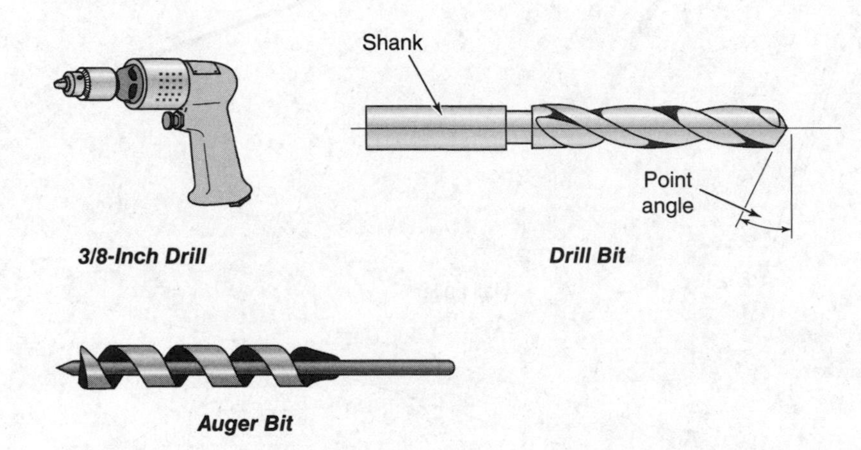


*3/8-Inch Drill*

*Drill Bit*

*Auger Bit*

A **soldering iron** is an electrically powered device consisting of a handle upon which is mounted a heating element (bit) that applies heat to melt solder (lead and tin alloy) for joining metal components together. It is used in electrical work to connect wires and on circuit boards.

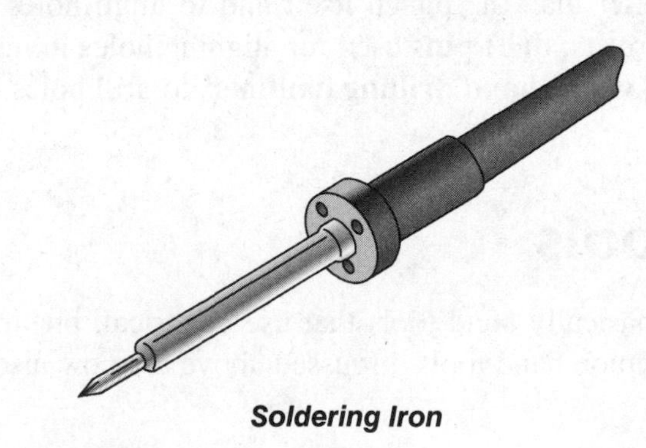

*Soldering Iron*

A **sander** uses compressed air or electric power to sand, polish, and otherwise provide prepainting preparation on both wood and metal.

A **nail gun** is a compressed air tool powered by electricity or a battery that provides fast, consistent nail penetration into lumber.

Other power tools include the **impact socket wrench** used in automotive work; a **bench grinder** used to polish and sharpen tools; an **air chisel** used to cut lightweight metal, masonry, and wood; and a **spur point** used for drilling holes for dowels in woodworking.

## Measuring Tools

Common measuring devices are the **tape measure**, **level**, and **chalk line tool**, which is a chalk-covered cord inside a metal casing that when stretched taut between two points and snapped leaves a straight chalk line on a surface. **Calipers**, devices used to measure the thickness and internal and external size of an object, are also used; some, known as **vernier calipers**, provide accuracy to one ten-thousandth of an inch.

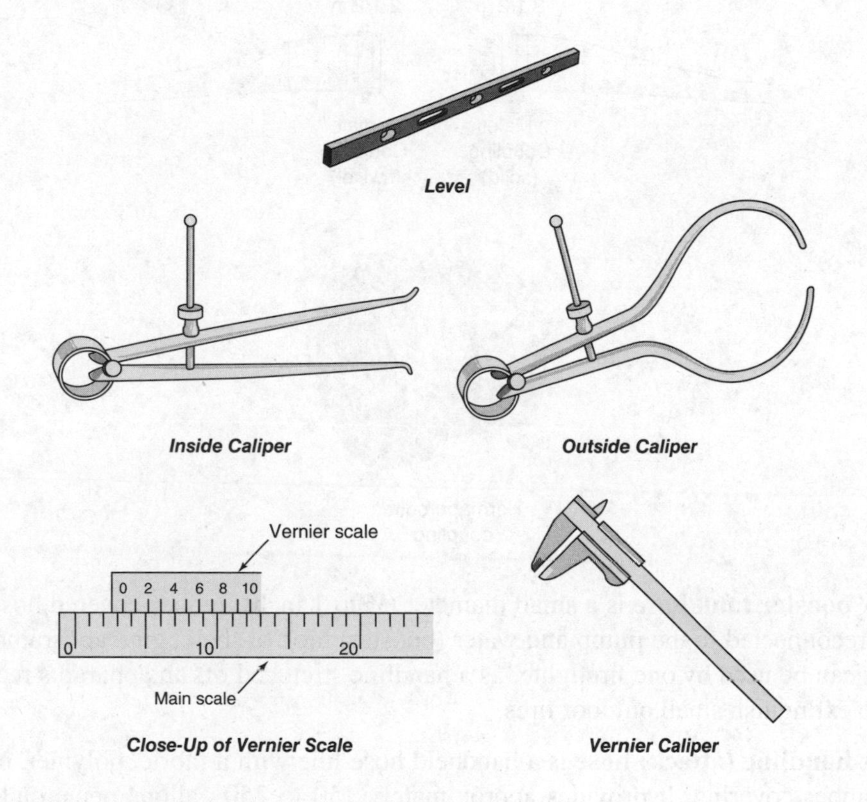


*Level*

*Inside Caliper*

*Outside Caliper*

*Close-Up of Vernier Scale*

*Vernier Caliper*

# FIREFIGHTING TOOLS

This section focuses on tools and equipment used by firefighters on the fireground. The tools and equipment may be grouped into two categories: engine company and ladder company tools and equipment.

## Engine Company Tools

The engine apparatus contains an abundance of hoses, nozzles, tips, fittings, and appliances needed for firefighting operations.

## HOSES

Hoses are classified by inside diameter and construction material. A length of hose is approximately 50 feet, with a male coupling (connector) on one end and a female coupling (connector) on the other end. Hose couplings can be standard type (male/female threaded) or hermaphrodite type (identical, nonthreaded).

A **male coupling** has a nonswiveling (solid) connection whose threads are located externally. A **female coupling** has a swiveling connection whose threads are located internally. A **hermaphrodite coupling** is a quick-connect coupling having no threads; identical connections are found at both ends of this type of hose. Generally, connections are made on hermaphrodite couplings using a half-turn rotational action.

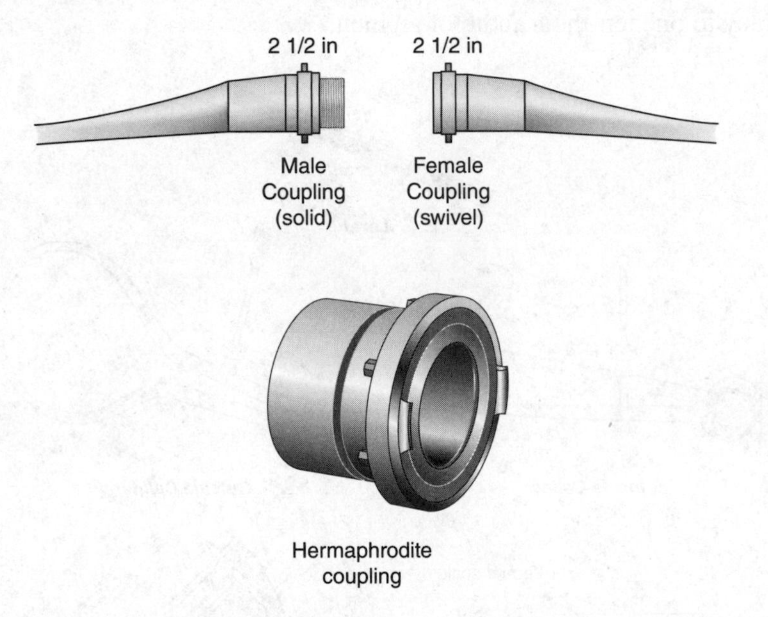


A **booster tank hose** is a small diameter (½ to 1 inch), rubber-covered hose preconnected to the pump and water (booster) tank of the engine apparatus. It can be used by one firefighter as a handline stretched off an apparatus reel to extinguish small outdoor fires.

A **handline (attack) hose** is a handheld hose line with a fabric, polymer, or rubber covering. It provides approximately 150 to 250 gallons per minute (GPM) and generally requires more than two firefighters to stretch and operate as an interior attack hose line. It is stretched off the back or side hose bed of the apparatus to extinguish fires inside buildings during interior structural firefighting operations.

A **supply (large diameter) hose** is a hose line generally 3 to 6 inches in diameter having a covering of fabric, polymer, or rubber. It is used to provide water at greater than 250 GPM to fire apparatus, large caliber stream appliances, and building fire extinguishing systems.

## NOZZLES AND TIPS

Nozzles and tips are designed to provide a fire stream of water or foam extinguishing agent. **Nozzles** are attached to handlines and have shutoff mechanisms designed to close, open, and regulate hose streams. **Tips** are also attached to handlines but do not have shutoffs and are placed on nozzles and fire extinguishing appliances to provide a fire stream in various patterns (straight, narrow, wide, fog). A smooth solid bore, straight stream tip is, for example, designed to produce a compact, penetrating stream with little breakup; a **fog nozzle** provides a wide or narrow patterned water stream composed of fine droplets; and a **foam nozzle** mixes water, foam concentrate, and air to produce a foam extinguishing agent stream.

## FITTINGS

Fittings are devices used in conjunction with hose line couplings to solve hose connection problems. An **increaser** has larger-sized male coupling threads than female threads, whereas a **reducer** has larger-sized female coupling threads than male threads. A **double male connection** is a fitting with male coupling threads on both ends, used for connecting two female couplings together; a **double female connection**, a fitting with female coupling threads on both ends, is used for connecting two male couplings together.

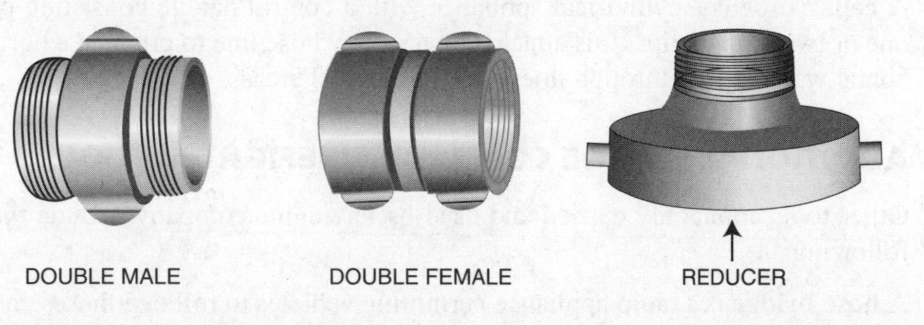


A **National Standard adapter** is a fitting permitting National Standard appliances, fittings, and hoses to be used with local fire department equipment. It has local fire department threads at the female end and the National Standard thread at the male end.

## APPLIANCES

Appliances are connected to hose lines or hydrants and are used to control, augment, divide, and discharge water streams or fire extinguishing agent. A **ball valve** is a flow control appliance used to open and close the flow of water in a supply hose line. They are in the open position (water flowing) when the handle is in line with the hose and closed when the handle is at a right angle to the hose. A **gate valve** is used to open and close the flow of water from a hydrant. It has a baffle that moves up and down through the turning of a handle.

A **wye** is an appliance that divides a supply hose line entering through its female inlet into two equal hose lines of smaller diameter out its male outlets. It may or may not have control handles (gates) for its male outlets.

A **water thief** is an appliance that divides a supply hose into three hose lines, one of the three being of larger diameter than the other two.

A **Siamese** is an appliance used to augment a supply hose line or building fire extinguishing equipment. It consists of two or more female inlets and one male outlet.

A **hose jacket** is a metal or leather encasing appliance used to reduce leakage from cut, damaged, or improperly coupled hose lines.

**Master (deluge) nozzles** are appliances (generally straight stream) with smooth solid bore tip(s) used to direct a large (300+) GPM water stream. They are connected to a Siamese in order to ensure an adequate water supply.

**Monitor nozzles** are used from the ground position. **Deck pipes** are mounted atop the engine company's apparatus. **Ladder pipes** are attached to the lead ladder section of a ladder company's aerial apparatus.

A **distributor** is an appliance with swiveling/rotating outlet heads attached to a supply hose line. It is used to produce a circular, spray pattern of water in hard-to-access areas (basements, cellars, piers). It is equipped with handles so it can be supported when suspended through the opening it is operating into.

A **cellar pipe** is a cylindrical appliance with a control handle consisting of one or two straight tips. It is attached to a supply hose line to produce a horizontal water stream through flooring into sublevel areas.

### ADDITIONAL ENGINE COMPANY FIREFIGHTER TOOLS

Other tools commonly carried and used by an engine company include the following:

A **hose bridge** is a ramp appliance permitting vehicles to roll over hoses and couplings without damaging them.

A **hydrant wrench** is a tool (box wrench or adjustable wrench) used to open the valve of a fire hydrant.

A **hose spanner** is a rigid metal tool used to loosen and tighten hose line couplings.

A **hose roller hoist** is a curved metal device with two or more rollers designed to fit over a windowsill or the edge of a roof. It is used in conjunction with rope to safely raise or lower hose and equipment.

A **hose strap** is a short length of rope with an eye-loop at one end and a metal hook at the other end. It is wrapped around a hose and tied off to a banister or railing to support the weight of hose couplings when hose lines are stretched vertically up stairwells and fire escapes and charged with water.

## Ladder Company Tools

The ladder company apparatus also has a wide array of tools to aid the firefighter in completing firefighting tasks. Some of these tools include ladders, ropes, forcible entry tools, ventilation fans, and tarps.

## PORTABLE LADDERS

Hand-operated (ground) portable ladders are tools used by firefighters to reach areas that cannot be accessed by normal means. They come in a variety of lengths, from less than 10 feet to more than 50 feet.

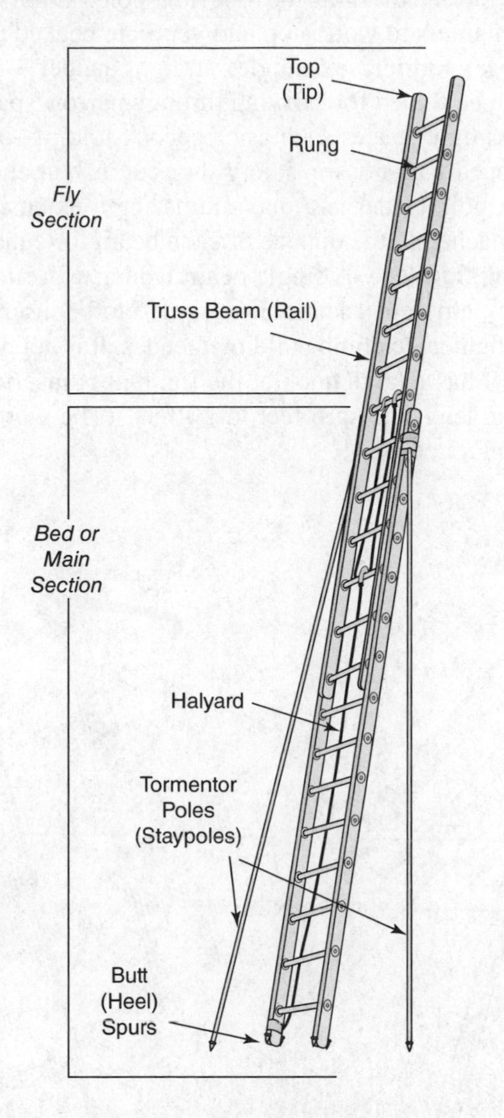


Ladders commonly used by firefighters are listed below.

**Single (straight) ladder**—A ladder that is nonadjustable in length and consists of only one section, a single (straight) ladder generally ranges in length from 12 to 24 feet and can be placed into operation by one firefighter acting alone. A single straight ladder is used for quick access to the roofs and windows of two-story structures.

**Extension ladder**—A ladder that is adjustable in length and generally consists of one or more movable sections beyond the base, an extension ladder uses a rope (halyard) and pulley system for extension and retraction. It ranges in length from 24 to 50 feet and is usually placed in operation by two or more firefighters.

**Combination ladder**—A ladder designed to be used three ways—single, extension, and A-frame; a combination ladder ranges in length from 8 to 14 feet.

**Roof ladder**—A single, straight ladder with folding hooks attached to the beams at the top end that grab the roof ridgepole. A roof ladder is designed to allow firefighters to work atop and ventilate peaked roofs.

**Folding (suitcase) ladder**—A single, straight ladder with hinged rungs allowing it to be folded for carrying through narrow spaces like hallways and to access tight spaces such as closets. A folding, or suitcase, ladder comes equipped with nonslip safety shoes at its butt end.

**Pole ladder**—A pole ladder is a maximum-length extension ladder having stay poles attached to the outside of each beam for enhanced stability.

**Pompier (scaling) ladder**—A single beam ladder with rungs extending out on both sides, a pompier ladder is designed to be inserted into windows to allow firefighters to climb building facades. It is not designed to rest on the ground. It has a large hook at the top that is inserted into a window and ranges in length from 6 feet to 20 feet. (The word *pompier* means firefighter in French.)

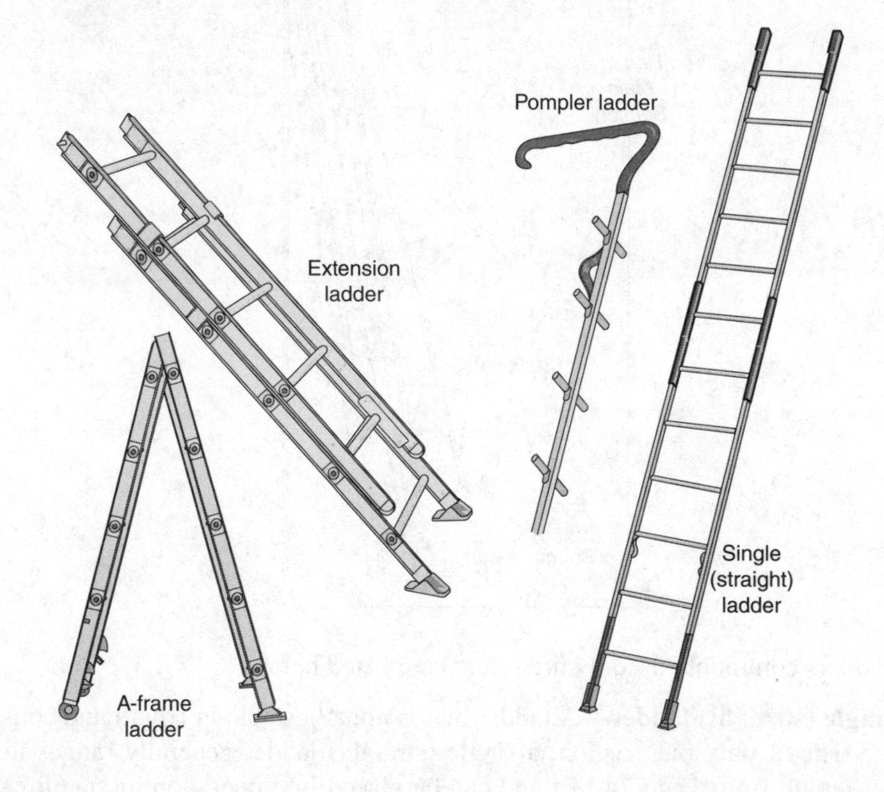


## ROPES

Ropes are used in a variety of ways by firefighters. They can be employed to hoist, secure, and lower tools, hoses, and appliances and for search-and-rescue purposes.

The part of a rope that is used in tying knots is known as the **working end**. The long part not used when tying knots is the **standing end**.

## Types of Ropes

Ropes are made of several types of materials, each with its characteristic properties.

**Nylon rope**—The strongest rope in common use, a nylon rope is excellent in absorbing shock loads when used for rescue and lifting loads and has superior abrasion, oil, and chemical resistance properties.

**Polyester rope**—A strong rope that stretches very little, a polyester rope doesn't have the shock-absorbing capability of nylon rope.

**Polypropylene rope**—A lightweight rope, a polypropylene rope is the only rope used in the fire service that floats, and it is therefore used in marine firefighting operations. It is not as strong as nylon or polyester.

**Natural fiber (Manila)**—A type of rope that holds knots firmly and stretches very little, a natural fiber rope is subject to deterioration by chemicals and mildew.

## Uses of Ropes

Ropes are used for numerous purposes in firefighting operations.

**Lifesaving rope**—Approximately 150 feet long, a lifesaving rope is used to lower firefighters and victims from dangerous positions during fire operations.

**Personal rope**—Approximately 50 feet long, a personal rope is used by firefighters to lower themselves from dangerous positions.

**Search rope**—Approximately 200 feet long, a search rope is used by firefighters as a retrieval lifeline and to perform searches safely in large, complex areas. It provides a fixed anchor point to keep firefighters from becoming disoriented during searches.

**Utility rope**—Approximately 40 feet long, the utility rope is a multipurpose rope used for securing tools and equipment. It is not used for lifesaving purposes.

## KNOTS

Making knots in rope entails tying the parts of one or more ropes together. Knots are used by firefighters to perform many tasks, such as temporarily securing an object with a **hitch** or **single/simple knot**. Knots are also used to make a rope longer by tying two ropes together (bend knot). A rope can also be bent to form two parallel sides (**bight**) or made into a **loop** by crossing the sides of a bight.

There are many different types of knots, including the following:

**Overhand (thumb) knot**—The simplest of all knots, an overhand knot is used as a safety hitch or binder to ensure a knot will not loosen.

**Half-hitch**—A half-hitch knot is used to hoist and lower tools and equipment.

**Clove hitch**—Formed by making two half-hitches and secured with an overhand knot (binder), a clove hitch is also used for hoisting and lowering tools and equipment.

**Rolling hitch**—A rolling hitch is a series of half-hitches used to secure hose lines that have been hoisted along the exterior of the building. They are generally secured with a binder.

**Becket bend**—Also known as a sheet bend, a becket bend knot is particularly suited for tying two ropes of unequal diameter together.

**Square knot**—A square knot is used to tie two ropes of equal diameter together.

**Bowline**—A bowline knot will not slip or tighten under tension and is used for hoisting and lowering portable ladders.

**Bowline on a bight (rescue knot)**—Used to lower and lift firefighters and victims from roofs, windows, and confined spaces during rescue operations. A bowline is used to support the body, while the bight prevents the rope from tightening and slipping. It is used in conjunction with a slippery hitch to secure the working end of the rope during lowering and lifting rescue operations.

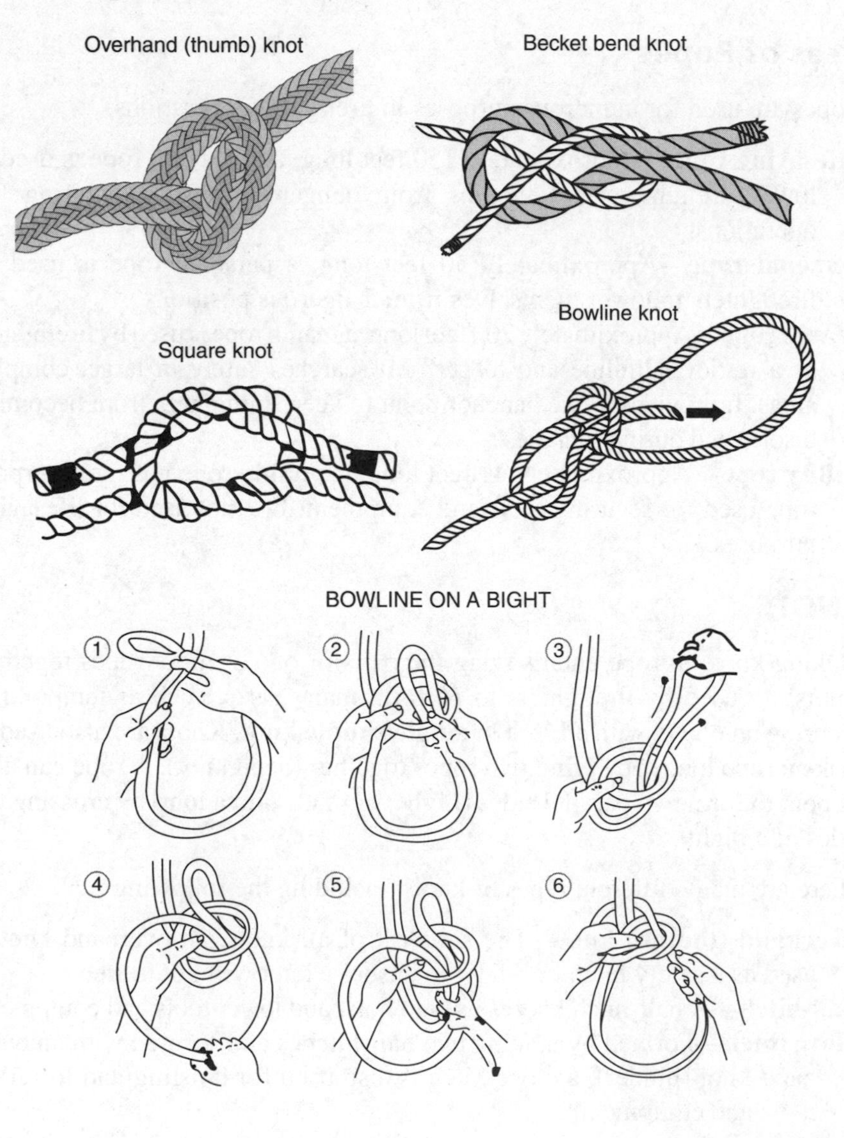

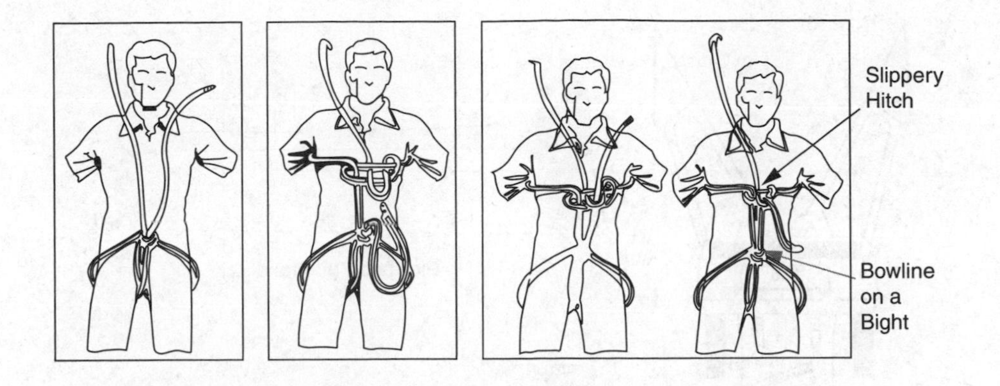


## Forcible Entry Tools

Forcible entry tools are designed to break and pry locks and locking devices to gain entry into buildings and fenced-in areas during fire and emergency operations. These tools are also used for forcibly opening and entering doors, windows, and roof openings, as well as for striking, pulling, prying, and removing building material components (roofing, plaster, drywall, wood lath, brick, masonry, tile) in search of hidden fire.

**Halligan tool**—Originally designed by Deputy Chief Hugh Halligan of the Fire Department of New York in the 1940s, the Halligan tool is a metal bar tool that has on one end a spike (for ripping and prying) and an adz (for chopping, prying, and cutting) at right angles to each other and at the opposite end a fork (for prying and cutting). Traditionally it is nested with the axe to form a firefighter's set of forcible entry "irons." It is used for forcing open doors, windows, and locks.

**Flat head axe**—A flat head axe consists of a 6 to 8 pound steel head attached to a long wooden or fiberglass handle. The head has a flat poll side designed for striking as well as receiving hammer blows and an axe blade for cutting. It is "married" to the Halligan tool for forcible entry work or used separately for cutting floorboards, roof boards, and windowsills.

**Pick head (fire) axe**—A pick head (fire) axe has a 6- to 8-pound pound head attached to a long wooden handle. The head has a square, pointed pick on one end and an axe blade on the opposite end. The pick end is designed for enhanced penetration through and prying of floor and roof boards.

**Kelly tool**—One of the original firefighting forcible entry tools, the Kelly tool, which was designed by FDNY Captain John Kelly of Ladder Co. 163, is a metal bar tool with a fork end and an adz end that protrudes into a hammerhead for striking and receiving hammer blows. It is designed for lock breaking.

**Claw tool**—Another original firefighting forcible entry tool, the claw tool is a metal bar tool having a hook (claw) at one end and a fork at the other.

**T-N-T (Denver) tool**—A multipurpose, forcible entry hand tool. The T-N-T, or Denver, tool is designed to perform the work of an axe, pry bar, ram, pike pole, and sledgehammer. It is used on automobiles and inside buildings.

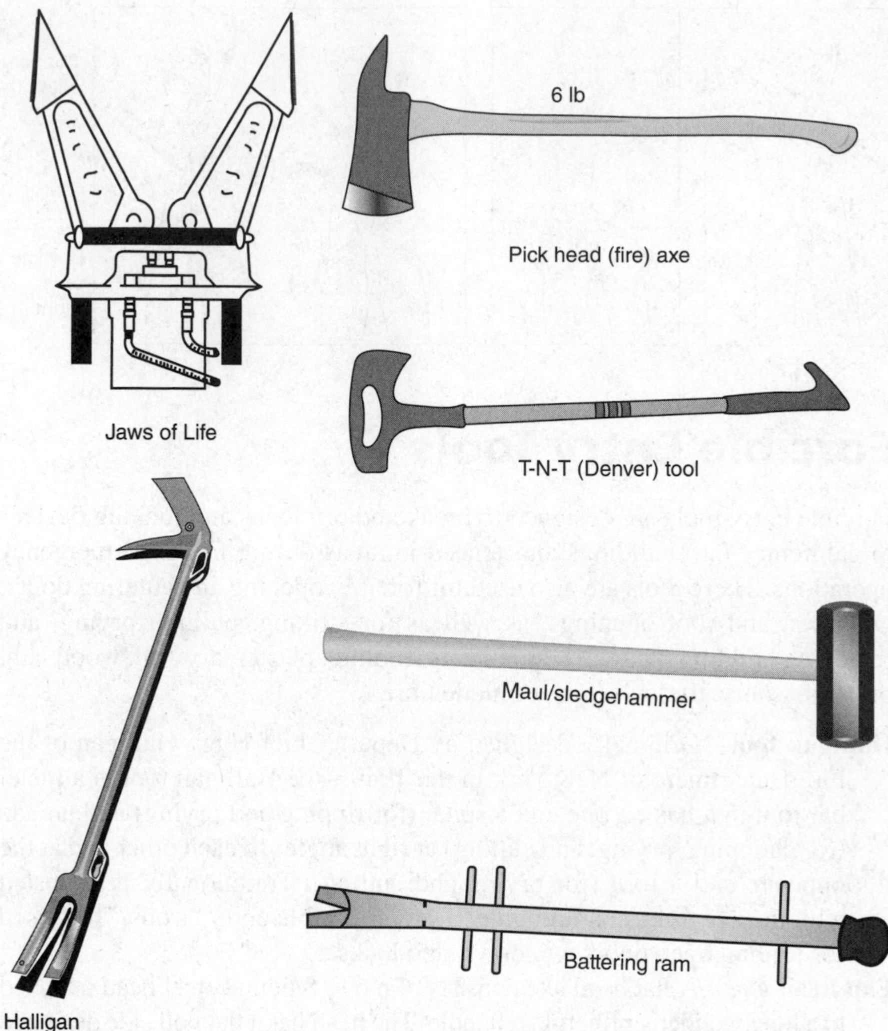


**Maul/sledgehammer**—A heavy, long-handled hammer having a head with two flat striking surfaces, the sledgehammer is used for forcible entry and breaking up walls and flooring consisting of drywall, concrete, tile, brick, and masonry.

**Battering ram**—A battering ram is a heavy metal bar designed to be used by two or four firefighters for pushing in heavy doors and breaching walls.

**K-tool**—A square (3 inches by 3 inches) steel block with a sharp-edged K-shaped notch on one side. The K-tool is designed to be slipped over cylinder locks to remove them. The other side of the tool has a U-shaped flange for prying when used in conjunction with metal bar tools.

**Automatic center punch**—A metal, hand-sized spring-loaded pointed tool, an automatic center punch is used by firefighters to break tempered glass in vehicles during extrication emergencies.

**Shove knife**—A shove knife is a rigid, flat metal blade used by firefighters to retract door spring latches during forcible entry.

**Hydraulic spreader (Jaws of Life)**—A mechanical levering tool powered by a hydraulic pump engine, the Jaws of Life is used for forcible entry and spreading of car components (doors, hood) to permit extrication of trapped victims.

**Hydra-Ram**—A lightweight hydraulically operated forcible entry tool consisting of a control handle and working end piston, a Hydra-Ram is used primarily by firefighters to force inward- and outward-opening doors.

**Pneumatic air bags**—Bags constructed of neoprene rubber reinforced with steel that are inflated using compressed air cylinders, pneumatic air bags are designed to move and lift heavy loads.

**Power saw**—A power saw is a portable, gas-fueled tool. A power saw uses circular blades to cut wood, lightweight metal, and masonry.

**Oxyacetylene cutting torch**—Using a mixture of oxygen and acetylene, this burning tool is used by firefighters to cut locks, heavy iron bars, and metal plating.

**Poles (hooks)**—Poles are long-handled tools primarily designed for opening up (pulling) ceilings and walls made of drywall (Sheetrock) and wood lath in the search for hidden fire. They are also utilized by firefighters to vent windows and roof openings (skylights, scuttle covers) as well as to pull up roof coverings and roof boards. Poles come in a variety of lengths from less than 5 feet to more than 15 feet.

**Pike pole (hook)**—A long (6 to 15 feet), wooden-handled striking and pulling tool used on windows, ceilings, and partition walls. The steel head of the tool has a short hook for pulling and a pointed tip for striking and penetrating through plaster, drywall, and wood lath. There are several types of pike poles, or hooks, including a **closet hook**, which is a short (under 5 feet), wooden **D-handled** pike pole for use in small, tight spaces; a **multihook** with a pointed, penetrating metal head and two flared adz end hooks for a greater pulling surface and easier removal of large areas of material in comparison to the pike pole; and a **Sheetrock hook** with a pointed metal head consisting of a 4-inch, toothed curved hook for a larger contact point and pulling surface. It is specifically designed to facilitate the pulling and removing of Sheetrock, lath, and plaster.

## Ventilation Equipment

**Positive pressure ventilation (PPV) fans**—Electric or fuel-driven fans primarily designed to provide forced, uncontaminated air into a room or building to displace the by-products of fire (smoke and toxic gases).

**Smoke ejectors**—Fans primarily designed to eject smoke and toxic gases from an area. They have some use in drawing fresh air into an area but do not move as much air as positive pressure fans.

## Salvage Equipment

Salvage operations require specialized equipment. The following are some of the more common types of tools or equipment used.

**Salvage covers** are generally used indoors to protect furniture, equipment, and valuables from water damage, but they may also be used as a chute to funnel water out windows and down staircases. The tarpaulins are made of canvas, waterproof-treated drop-cloth material, or plastic, with reinforced edges and grommets for hanging.

**Plastic sheeting** is used to cover openings on roofs and windows to protect the interior of the structure from adverse weather conditions.

**Dewatering pumps** are electrical, engine-driven, and hydraulic-powered pumps used in conjunction with a hose line to lift and remove water from below-grade areas.

**Water vacuums** are designed to suction off water from floors and carpeting. These devices consist of a vacuum nozzle and catch tank worn on the back of the firefighter.

## Safety Equipment and Clothing

To protect their health and lives, firefighters use a variety of safety equipment while performing their duties. Some of this equipment is discussed briefly below.

**Self-contained breathing apparatus (SCBA)**—Firefighters wear self-contained breathing apparatus during interior structural firefighting operations to protect against breathing toxic fumes and smoke. An SCBA is also used inside confined spaces where insufficient oxygen or poisonous vapors are present. The SCBA supplies compressed, breathable air to the wearer and includes an air cylinder, high and low pressure hoses, a regulator, and a facepiece. It may or may not have an integral alarm device that operates automatically should the firefighter become disabled.

**Personal alert safety system (PASS) device**—A personal alert safety system is an alarm device that emits a signal when a firefighter is disabled, lost, or otherwise in distress. It can be activated both manually and automatically should the firefighter stop moving for an extended period of time.

**Respirator**—A respirator is a full-face or half-face respiratory protection device that protects the user by filtering out dust particles, organic vapors, and acid gases. The device has replaceable cartridges. The term is also used for disposable dust and paint masks, which lack filters and do not provide protection from organic vapors and acid gas. Air-purifying respirators (APRs) remove contaminants by moving ambient air through a filter. APRs can have a full or partial facepiece.

**Bunker (turnout) gear**—The structural firefighting ensemble (coat and pants) that provides flame, thermal, and mechanical (cuts, abrasions) protection is known as bunker gear. The term may also refer to the entire firefighting ensemble, including helmet, hood, and boots. The gear is made from fire-resistant fiber, synthetics, and polymer materials, such as Nomex, Kevlar, and polybenzimidazole (PBI), and generally has high-visibility striping material.

**Reflective vest**—A reflective vest is a high-visibility garment worn over firefighting clothing. It is used by firefighters when working with and around cranes and heavy construction equipment and trucks during technical rescue and building collapse operations.

**Fire helmet**—Constructed of fiberglass, plastic, or leather, the helmet firefighters wear provides thermal and impact head protection during firefighting operations. A long curved rear brim keeps cinders and hot runoff water off the firefighter's neck. The helmet also provides thermal ear protection (ear flaps) and eye protection (eye shields), and it has a chinstrap to secure the helmet to the head.

**Hood**—Made from fire-resistant material similar to that used for manufacturing bunker gear, the hood is worn over the head to protect the areas of the head and face not covered by the helmet and facepiece of the SCBA.

**Fire boots**—Structural firefighting footwear, or fire boots, made from rubber and/or leather, provide thermal (Nomex/Kevlar), puncture, and impact (steel sole/shank/toe) protection of the foot, ankle, and lower leg. Some boots are designed in lengths to be pulled up over the knees for enhanced thermal protection of the lower extremities.

**Additional protective gear**—Includes earmuffs/earplugs, safety glasses/goggles, kneepads, safety shoes, and gloves, all made of material designed to protect the firefighter. Firefighters working in confined spaces and in technical rescue work wear a hard hat.

# Miscellaneous Firefighting Tools

Firefighters also use **flashlights**, **pocketknives**, and **portable fire extinguishers** in their duties. A **thermal imaging camera**, a handheld heat-detecting tool that uses infrared rays to find hidden fire and disabled victims inside a building under smoky firefighting conditions, is another invaluable tool. A **multigas detector**, a metering device that detects and measures oxygen level, as well as toxic (hydrogen cyanide) and flammable (carbon monoxide) gases, is also a valuable tool.

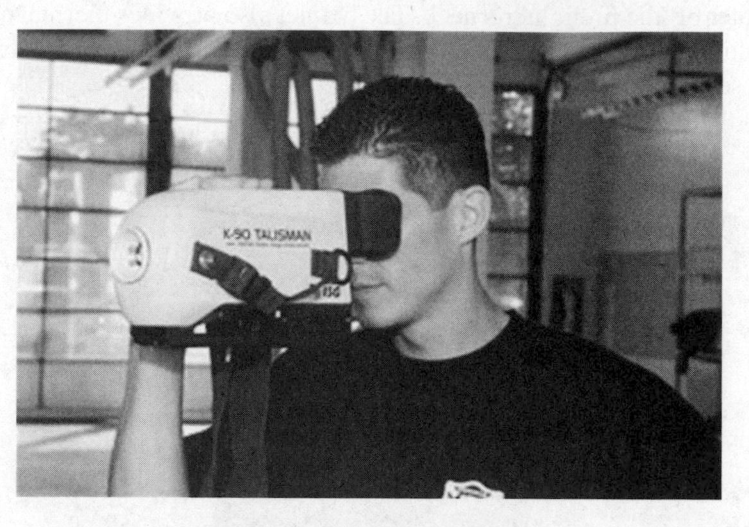


THERMAL IMAGING CAMERA

## Specialty Tools

Firefighters at times need to employ tools that are generally used in specific trades. For example, they may use circuit meters, multimeters, and voltage continuity testers to measure a household's electrical condition (current flow, voltage, leakage, and resistance). In maintaining the fire station, they may use painting, woodworking, masonry, glazer, and plumbing tools. Basic automotive tools are used when maintaining the fire apparatus.

## Wildland Firefighting Tools and Equipment

Wildland firefighting tools and equipment include buckets, hoes, rakes, shovels, swatters, and brooms. It is beyond the scope of this book to discuss all of these tools in detail, but a short list of some of the most common pieces of equipment used in wildland firefighting follows.

**Bambi bucket**—A collapsible bucket slung below a helicopter that is dipped into a source of water (lake, river, or reservoir) and emptied onto the fire.

**Drip torch**—A handheld container with a fuel fount, burner arm, and igniter, used to ignite a prescribed burn by dripping flaming liquid fuel (diesel fuel and gasoline) on foliage and brush.

**Combination (Combi) tool**—A military entrenching tool with a long handle developed for firefighting. Serves as a light-duty shovel and scraper.

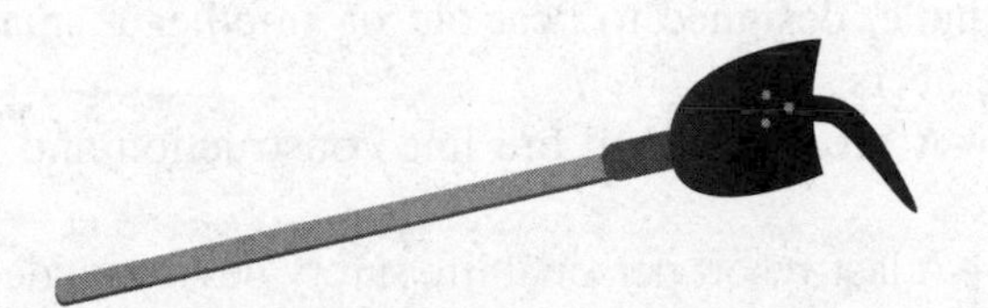

**Pulaski tool**—A wooden-handled, steel-headed tool with an axe blade at one end and an adz blade on the other end, used for chopping and trenching. Invented by USFS ranger Ed Pulaski in 1911.

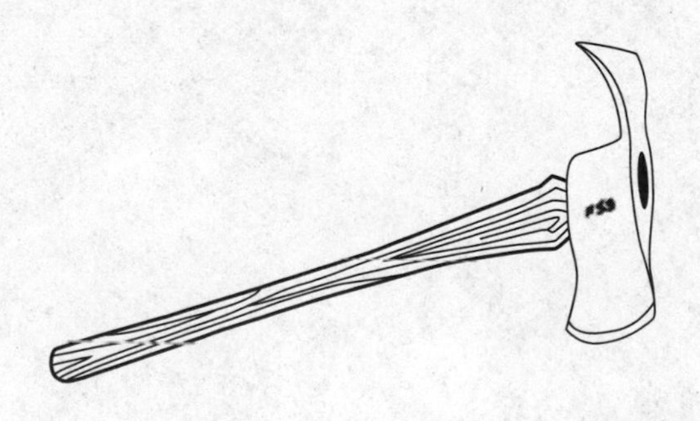

**McLeod tool**—A long, wooden-handled tool that is a combination rake and hoe, with one serrated edge for raking and one sharpened edge for cutting and hoeing. Designed by Sierra National Forest ranger Malcolm McLeod in 1905.

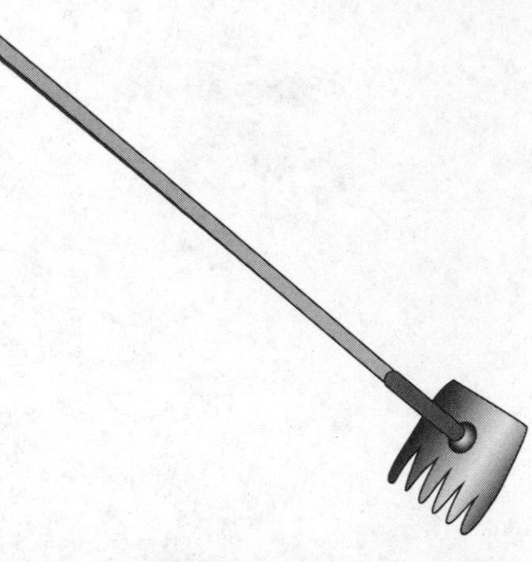

**Knapsack (Indian) pump**—A backpack-mounted water tank or bladder bag (5 gallons) equipped with a high-pressure, double-action squirt pump used like a portable fire extinguisher for small brush fires.

**Round point shovel**—A long, wooden-handled tool with a narrow, pointed blade designed for digging out burning roots and logs and burying smoldering fires with dirt.

**Fire rake**—A long, wooden-handled rake with steel teeth, used for raking fire lines and cutting under brush and foliage.

**Fire swatter (flapper)**—A flexible, square-shaped rubber flap connected to a long handle, designed to beat out or smother a small ground fire or burning embers.

**Fire broom**—A broom used in fire line construction and in patrolling fire breaks.

**Fire shelter**—A last-resort personal life safety device made of several layers of aluminum foil, silica cloth, and fiberglass that reflect most of a fire's radiant heat and can be deployed quickly.

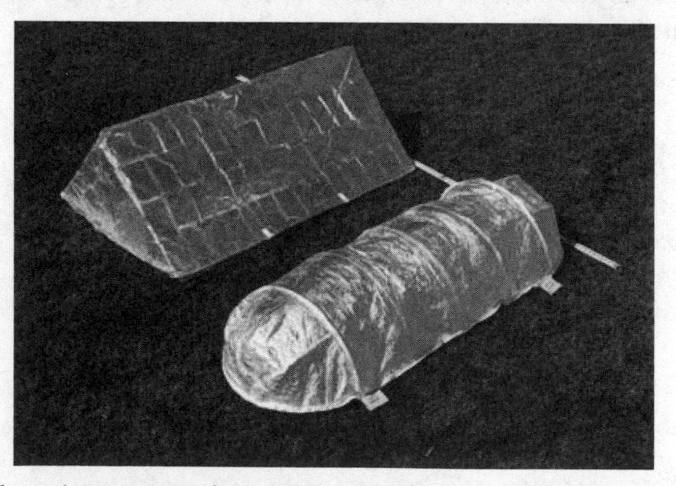

Fire shelters (next generation to the right). New shelter offers more protection against flame and radiant heat than the standard type (left).

# REVIEW QUESTIONS

*Circle the letter of your choice.*

**1.** The type of hammer used primarily for striking and removing nails is called a

(A) curved claw hammer
(B) ball-peen hammer
(C) tack hammer
(D) hand drilling hammer

**2.** Which of the following is a tool that is made from wire and is commonly used to fasten structural building elements together?

(A) A drift pin
(B) A wood splitting wedge
(C) A star drill
(D) A nail

**3.** A chisel used for cutting, shaping, and removing metals such as cast and wrought iron, steel, and bronze is called a

(A) wood chisel
(B) cold chisel
(C) nail set
(D) masonry chisel

**4.** Pliers with a high-leverage gripping action that are designed to bend lightweight metal and have a side cutting feature to sever wires are called

(A) wire stripping (electrician) pliers
(B) long-nose or needle-nose pliers
(C) linesman pliers
(D) LockJaw (Vise-Grip) pliers

**5.** Pliers with a long, slim head designed to facilitate crimping wires in confined, narrow spaces are called

(A) diagonal cutting pliers
(B) groove-joint (Channellock) pliers
(C) end cutting pliers
(D) long-nose or needle-nose pliers

**6.** Pliers primarily designed to sever and remove insulation on electrical wire without damaging the wire are called

(A) wire stripping (electrician) pliers
(B) long-nose or needle-nose pliers
(C) linesman pliers
(D) diagonal cutting pliers

7. Pliers that can be locked in and tightened prior to applying force are called

(A) wire stripping (electrician) pliers
(B) groove-joint (Channellock) pliers
(C) LockJaw or Vise-Grip pliers
(D) slip-joint pliers

8. A one-piece wrench with smooth U-shaped opening(s) that is designed to grip two opposite faces of a bolt or nut is called a (an)

(A) box end wrench
(B) monkey wrench
(C) open-end wrench
(D) combination wrench

9. A wrench that is constructed with a mechanism that eliminates the need to readjust its position during the return stroke is called a (an)

(A) adjustable (crescent) wrench
(B) offset wrench
(C) socket with screwdriver handle (nut driver)
(D) ratchet box wrench

10. An L-shaped, six-sided wrench used to turn machined setscrews or bolt heads that are designed with a hexagonal recess is a

(A) hex key (Allen) wrench
(B) Stillson wrench
(C) socket with screwdriver handle (nut driver)
(D) torque wrench

11. Flat-bladed screwdrivers are called

(A) Phillips head screwdrivers
(B) Pozidriv head screwdrivers
(C) slot-head screwdrivers
(D) Frearson head screwdrivers

12. Cross-headed screwdrivers with rounded corners, designed to slip off a screw when under high torque to prevent overtightening, are called

(A) slot-head screwdrivers
(B) Pozidriv head screwdrivers
(C) Frearson head screwdrivers
(D) Phillips head screwdrivers

13. A screwdriver used to access hard-to-reach and obstructed screws where a straight shaft screwdriver is inappropriate is termed

(A) unorthodox
(B) offset
(C) irregular
(D) cylindrical

14. A type of saw designed with a blade for cutting lumber perpendicular to the grain as it is being pushed through the wood is called a

(A) ripsaw
(B) miter saw
(C) crosscut saw
(D) hacksaw

15. A type of saw made to cut lumber parallel to the grain with a flat front edge that acts as a chisel is called a

(A) backsaw
(B) crosscut saw
(C) coping (jigsaw)
(D) ripsaw

16. A thin-bladed saw with closely spaced teeth having a reinforced steel back that is thicker than the blade itself is called a

(A) backsaw
(B) bucksaw
(C) bow saw
(D) chainsaw

17. A type of saw that is designed primarily to make crosscuts and precise angle cuts on lumber for picture framing and for ceiling/wall molding joint connections is called a

(A) bucksaw
(B) miter saw
(C) coping (jigsaw)
(D) ripsaw

18. A long-handled, heavy metal demolition tool with one curved end (claw) for prying and removing nails and one flattened end is called a

(A) pry bar
(B) pinch (Jimmy) bar
(C) crowbar (wrecking bar)
(D) nail puller

19. A long-handled tool with one flattened end and one tapered, pointed end that is designed for prying, lifting, and removing nails is called a

(A) nail puller
(B) cat's paw
(C) crowbar (wrecking bar)
(D) pry bar

**20.** A small-handled, lightweight pry bar having one rounded, curved end designed for prying and removing nails and one flattened end is called a

(A) cat's paw
(B) pinch (Jimmy) bar
(C) offset rip
(D) utility pry bar

**21.** A handheld power tool that is an electrically powered rotary saw with round blades designed for cutting wood, lumber, paneling, and plywood is a

(A) reciprocating saw
(B) chainsaw
(C) circular saw
(D) jigsaw or saber saw

**22.** An electric or battery-powered saw that uses thin, straight blades to simulate the back and forth motion of a handsaw and that is designed for cutting metal, pipe, and wood is a

(A) circular saw
(B) reciprocating saw
(C) chainsaw
(D) None of the above

**23.** A measuring tool using a bubble in a vial is a

(A) vernier caliper
(B) level
(C) tape measure
(D) chalk line tool

**24.** In general, a length of hose is approximately

(A) 15 feet
(B) 20 feet
(C) 25 feet
(D) 50 feet

**25.** A quick connect coupling having no threads in which connections are made up using a half-turn rotational action is called a

(A) female coupling
(B) male coupling
(C) hermaphrodite coupling
(D) None of the above

**26.** A small-diameter, rubber-covered hose that is preconnected to the pump and water tank of the engine apparatus and can be used by one firefighter as a handline stretched off the apparatus hose reel is called a

(A) supply hose
(B) handline (attack) hose
(C) booster tank hose
(D) none of the above

**27.** A type of hose that is stretched off the back or side hose bed of the engine apparatus to extinguish fires inside buildings during interior structural firefighting operations is called a

(A) handline (attack) hose
(B) supply hose
(C) booster tank hose
(D) hydrant hose

**28.** A large-diameter hose line used to provide water at greater than 250 GPM to fire apparatus, large-caliber stream appliances, and building fire extinguishing systems is called a

(A) booster hose
(B) hydrant hose
(C) handline (attack) hose
(D) supply hose

**29.** A device with a shutoff mechanism commonly attached to handlines that is designed to close, open, and regulate hose streams is a

(A) tip
(B) deck pipe
(C) fitting
(D) nozzle

**30.** A device designed to produce a wide or narrow patterned water stream composed of fine droplets is called a

(A) smooth solid bore, straight stream tip
(B) fog nozzle
(C) foam nozzle
(D) play pipe

**31.** The type of fitting that has larger female coupling threads than male threads is called a (an)

(A) reducer
(B) increaser
(C) double female connection
(D) double male connection

**32.** A fitting used for connecting two male couplings together is called a (an)

(A) double male connection
(B) double female connection
(C) increaser
(D) None of the above

**33.** Ball valves regulate the flow of water in supply hose lines. They are in the closed position when their control handle is

(A) at a right angle (perpendicular) to the hose
(B) in line (parallel) to the hose
(C) lifted up and over the hose outlet
(D) None of the above

**34.** A device that divides a supply hose line entering through its female inlet into two equal hose lines of smaller diameter out its male outlets is called a

(A) Siamese
(B) gate valve
(C) wye
(D) hose jacket

**35.** An appliance that is used to augment a supply hose line or building fire extinguishing equipment and that consists of two or more female inlets and one male outlet is called a

(A) water thief
(B) wye
(C) Siamese
(D) cellar pipe

**36.** Which of the following is not considered to be a large caliber stream (greater than 300 GPM) appliance?

(A) National Standard adapter
(B) Deck pipe
(C) Ladder pipe
(D) Monitor nozzle

**37.** Which piece of equipment has swiveling/rotating outlet heads, is attached to a supply hose line, and produces a circular, spray pattern of water in hard-to-access places?

(A) A cellar pipe
(B) A deck pipe
(C) A ladder pipe
(D) A distributor

**38.** An adjustable ladder generally consisting of one or more movable sections beyond the base that uses a halyard and pulley system for raising and lowering is called a (an)

(A) single (straight) ladder
(B) folding (suitcase) ladder
(C) roof ladder
(D) extension ladder

**39.** A single beam ladder with rungs extending out on both sides and a large hook at the top for inserting into windows is called a

(A) folding (suitcase) ladder
(B) pompier (scaling) ladder
(C) combination ladder
(D) pole ladder

**40.** Single (straight) ladders generally range in length from

(A) 12 to 24 feet
(B) 6 to 20 feet
(C) 10 to 50 feet
(D) 8 to 14 feet

**41.** A ladder designed for carrying through narrow hallways and used in hard-to-enter spaces (closets) is called a

(A) pole ladder
(B) pompier ladder
(C) suitcase ladder
(D) roof ladder

**42.** The strongest type of rope in common use among firefighters is made of

(A) polyester
(B) natural fiber (Manila)
(C) polypropylene
(D) nylon

**43.** A rope approximately 150 feet in length that is used to lower firefighters and victims from dangerous positions during fire operations is called a

(A) lifesaving rope
(B) utility rope
(C) search rope
(D) personal rope

**44.** A rope approximately 200 feet in length that is utilized by firefighters as a retrieval lifeline is a

(A) utility rope
(B) search rope
(C) personal rope
(D) lifesaving rope

**45.** A knot used by firefighters as a temporary way to secure an object is called a

(A) bend
(B) bight
(C) loop
(D) hitch

**46.** Tying two ropes together at their ends to make one long rope is called a

(A) hitch
(B) bend
(C) bight
(D) loop

**47.** The simplest of all knots that is used as a safety hitch or binder to ensure a knot will not loosen is called a

(A) clove hitch
(B) square knot
(C) overhand (thumb) knot
(D) bowline

**48.** Which of the following is also referred to as the firefighter's rescue knot and is frequently used to lower firefighters and victims from dangerous positions on the fireground?

(A) A bowline on a bight
(B) A square knot
(C) A half-hitch
(D) An overhand (thumb) knot

**49.** A heavy, long-handled hammer having a head with two flat striking surfaces that is used for breaking up walls and flooring made of drywall, concrete, tile, brick, and masonry is called a

(A) Kelly tool
(B) battering ram
(C) maul
(D) pick head (fire) axe

**50.** A multipurpose tool designed to perform the work of an axe, pry bar, ram, pike pole, and sledgehammer is called a

(A) K-tool
(B) T-N-T (Denver) tool
(C) Jaws of Life
(D) Hydra-Ram

**51.** A spring-loaded pointed metal tool used by firefighters to break tempered glass in vehicles during extrication emergencies is the

(A) automatic center punch
(B) K-tool
(C) shove knife
(D) pneumatic air bag

**52.** A firefighter's tool that is commonly referred to as the "Jaws of Life" and is designed for forcing entry into vehicles to facilitate victim removal is called a

(A) pneumatic air bag
(B) oxyacetylene cutting torch
(C) power saw
(D) hydraulic spreader

**53.** A lightweight forcible entry tool consisting of a control handle and working end piston that is used by firefighters primarily to force open doors is a

(A) claw tool
(B) Kelly tool
(C) Hydra-Ram
(D) sledgehammer

**54.** Poles (hooks) are long-handled tools primarily designed for all of the following firefighter tasks EXCEPT

(A) venting windows
(B) opening up ceilings
(C) pulling roof boards
(D) forcing padlocks

**55.** A long, wooden-handled pole having a steel head with a short hook and a pointed tip is a

(A) pike pole
(B) multihook
(C) Sheetrock hook
(D) closet hook

**56.** Positive pressure ventilation (PPV) fans are primarily designed to

(A) draw fresh air into an area by ejecting smoke
(B) scrub toxic gases out of smoke
(C) provide forced air to displace smoke
(D) limit the amount of oxygen required by the fire

**57.** Salvage covers are used by firefighters for all but which of the following tasks?

(A) Protecting furniture from water damage
(B) Suctioning off water from carpeting
(C) Funneling water out windows
(D) Covering valuable electrical equipment

**58.** The thermal imaging camera is used by firefighters primarily for

(A) extinguishment of small fires in their incipient phase
(B) filming the exterior of the fire building for training purposes
(C) finding hidden fire and disabled victims inside buildings
(D) none of the above

**59.** An alarm device that emits a distress signal when a firefighter is disabled or lost is commonly referred to by the acronym

(A) SCBA
(B) SCUBA
(C) PASS
(D) SOS

**60.** The piece of equipment that is worn by firefighters during interior, structural firefighting operations and is designed to protect the wearer against breathing toxic fumes and smoke by providing compressed breathable air is referred to by which acronym?

(A) PASS
(B) SCUBA
(C) SCBA
(D) T-N-T

# Answer Key

| | | |
|---|---|---|
| **1.** A | **21.** C | **41.** C |
| **2.** D | **22.** B | **42.** D |
| **3.** B | **23.** B | **43.** A |
| **4.** C | **24.** D | **44.** B |
| **5.** D | **25.** C | **45.** D |
| **6.** A | **26.** C | **46.** B |
| **7.** C | **27.** A | **47.** C |
| **8.** C | **28.** D | **48.** A |
| **9.** D | **29.** D | **49.** C |
| **10.** A | **30.** B | **50.** B |
| **11.** C | **31.** A | **51.** A |
| **12.** D | **32.** B | **52.** D |
| **13.** B | **33.** A | **53.** C |
| **14.** C | **34.** C | **54.** D |
| **15.** D | **35.** C | **55.** A |
| **16.** A | **36.** A | **56.** C |
| **17.** B | **37.** D | **57.** B |
| **18.** C | **38.** D | **58.** C |
| **19.** D | **39.** B | **59.** C |
| **20.** A | **40.** A | **60.** C |

## ANSWER EXPLANATIONS

Following you will find explanations to specific questions that you might find helpful.

**1.** A Light-duty **tack hammers** are designed for driving small nails; round, ball-shaped **ball-peen hammers** are used for riveting and shaping unhardened metal; **hand drilling hammers** are used to drill holes in masonry.

**2.** D A **drift pin** is a tool used in metalworking for enlarging holes or for aligning holes prior to bolting or riveting metal parts together; a **wood splitting wedge** is a tool you use to split harder woods that are too tough for a traditional axe; a **star drill** has a star-shaped point that is used for making holes in stones or masonry.

**3.** B A **cold chisel**, made from steel, is used for cutting and shaping metals such as cast iron, bronze, and copper; a **wood chisel** is designed for rough work on wooden materials; a **masonry chisel** is used with a hand drilling hammer to score or trim brick or block; a **nailset** is a small metal tool that looks much like an icepick used for driving finish nails at or below the surface of wood.

**4.** C **Wire stripping (electrician) pliers** sever and remove insulation on electrical wire without damaging the wire; **long-nose or needle-nose pliers** are used to grip and shape lightweight metal; the slim head design facilitates crimping wires in confined, narrow spaces; **linesman pliers** with a side-cutting feature bend lightweight metal and sever wire; locking pliers, also called **LockJaw or Vise-Grip pliers** are basically a handheld vise that allow for the purchase on an object to be locked in and tightened prior to applying force. They are used to firmly grip lightweight metal or remove round door-lock cylinders.

**5.** D **Diagonal cutting pliers** are designed to cut wire; they cut by indenting and wedging the wire apart; **groove-joint (Channellock) pliers** are slip-joint pliers that have serrated jaws generally set 45 to 60 degrees from the handles; **end-cutting pliers**, also called end nippers, are useful for a wide range of cutting applications, including cutting electrical and piano wire, glass, wood, and concrete.

**8.** C A **box-end wrench** is a one-piece wrench with recessed, grooved, enclosed opening(s); it is often designed double-ended with different-sized box ends. The enclosed opening grips all the faces of the bolt or nut, providing more torque than the open-end wrench but without slipping or stripping the bolt or nut; the **monkey wrench** is an old-type adjustable wrench with smooth jaws that is used for turning bolts or nuts; a **combination wrench** is a double-ended tool with one end open and the other end enclosed—both ends generally fit the same size bolt or nut.

**9.** D An **adjustable (crescent) wrench** is an open-end wrench with smooth, adjustable jaws used to turn bolts or nuts; an **offset wrench** is designed to provide access to obstructed bolts and nuts in recessed areas—it allows for hand clearance when turning an object flush with a work surface; a **socket with a screwdriver handle (nut driver)** is a hollow cylinder (socket) that fits over a nut head designed to loosen or tighten the nut.

**10.** A A **Stillson wrench** is an adjustable wrench with serrated jaws for gripping soft iron pipe and pipe fittings. A **torque wrench** is a tool used to apply precisely a specific torque to a fastener such as a nut or bolt.

**11.** C A **Phillips head screwdriver** is a cross-headed screwdriver with rounded corners. It is designed to slip off the screw when under high torque to prevent overtightening, which increases the likelihood of material damage and distortion to both screw head and driver. **Pozidriv head screwdrivers** are also cross-headed. They are designed such that they cannot jump out if overtorqued. **Frearson head screwdrivers** are cross-headed screwdrivers that have a sharper head than the Phillips head screwdriver.

**14.** C A **crosscut saw** is designed for making cuts in lumber perpendicular (at a right angle) to the grain; the cutting edge of the blade is beveled, allowing the blade to act like a knife edge and slice through the wood; a **hacksaw** is a fine-toothed saw with the blade under tension inside a frame; it is designed to cut metal.

**15.** D A **coping (jigsaw)** is designed to cut intricate shapes in wood. It has a thin blade tensioned inside a metal frame.

**16.** A A **bucksaw** is a manual frame saw generally used with a sawbuck to cut logs or firewood to length; a **bow saw** is a metal-framed crosscut saw in the shape of a bow with a coarse wide blade; it can cut through dry wood and lumber and is designed for demanding applications and tough environments of construction sites. A **chainsaw** is a portable, mechanical saw that cuts with a set of teeth attached to a rotating chain that runs along a guide bar. It is used in activities such as tree felling, limbing, bucking, pruning, cutting firebreaks in wildland fire suppression, and harvesting of firewood.

**18.** C A **pinch (Jimmy) bar** is a metal hand tool featuring two work ends (one pointed and one flat) that are ideal for prying, positioning, lifting, and aligning applications that need greater leverage. A **nail puller** is a metal bar designed to easily slide under the head or into the shank of an embedded nail for easy removal. The other end is a striking head for a hammer.

**20.** A A combination pry bar and chisel tool, an **offset rip bar**, is similar in appearance to the crowbar; it has a small claw end and wide chisel end in an offset pattern for difficult-to-reach objects. A **utility pry bar** is a small, lightweight metal tool with beveled cutting edges at both ends; it is designed for pulling nails, prying light objects, and scraping.

**23.** B Calipers are devices used to measure the thickness and internal and external size of an object. **Vernier calipers** provide accuracy to one ten-thousandth of an inch. A **tape measure** is a flexible ruler used to measure distance. A **chalk-line tool** is a measuring device consisting of a chalk-covered cord inside a metal casing that when stretched taut between two points and snapped leaves a straight chalk line on a surface.

**25.** C A **female coupling** has a swiveling connection whose threads are located internally. A **male coupling** has a nonswiveling (solid) connection whose threads are located externally.

**26.** C A **booster tank hose** is a small-diameter (½ to 1 inch), rubber-covered hose preconnected to the pump and water (booster) tank of the engine apparatus.

**27.** A **Hydrant hose** is used to connect the fire pumper inlet with a pressurized hydrant.

**29.** D **Tips** are attached to handlines but do not have shutoffs and are placed on nozzles; **deck pipes** are nozzles mounted atop the engine company's apparatus; **fittings** are devices used in conjunction with hose line couplings to solve hose connection problems.

**30.** B A **smooth solid bore, straight stream tip** is designed to produce a compact, penetrating stream with little breakup; a **foam nozzle** mixes water, foam concentrate, and air to produce a foam extinguishing agent stream; a **play pipe** is an elongated nozzle with a set of handles on each side for better control and efficiency

**31.** A An **increaser** has larger-sized male coupling threads than female threads; a **double female connection** is a fitting with female coupling threads on both ends; it is used for connecting two male couplings together; a **double male connection** is a fitting with male coupling threads on both ends used for connecting two female couplings together.

**34.** C A **gate valve** is used to open and close the flow of water from a hydrant; a **hose jacket** is a metal or leather encasing appliance used to reduce leakage from cut, damaged, or improperly coupled hose lines.

**35.** C A **water thief** is an appliance that divides a supply hose into three hose lines, one of the three being of larger diameter than the other two; a **cellar pipe** is a cylindrical appliance with a control handle consisting of one or two straight tips; it is attached to a supply hose line to produce a horizontal water stream through flooring into sublevel areas.

**36.** A A **National Standard adapter** is a fitting permitting National Standard appliances, fittings, and hoses to be used with local fire department equipment. It has local fire department threads at the female end and the National Standard thread at the male end; **deck pipes** are mounted atop the engine company's apparatus; **ladder pipes** are attached to the lead ladder section of a ladder company's aerial apparatus; **monitor nozzles** are used from the ground position.

**38.** D A **single (straight) ladder** is a ladder that is nonadjustable in length and consists of only one section; a **roof ladder** is a single, straight ladder with folding hooks attached to the beams at the top end that grab the roof ridgepole. A roof ladder is designed to allow firefighters to work atop and ventilate peaked roofs.

**39.** B A **combination ladder** is a ladder designed to be used three ways—single, extension, and A-frame; a combination ladder ranges in length from 8 to 14 feet; the **pole ladder** is a maximum-length extension ladder with stay poles attached to the outside of each beam for enhanced stability.

**43.** A A **personal rope** is approximately 50 feet long; it is used by firefighters to lower themselves from dangerous positions; a **utility rope** is approximately 40 feet long and is a multipurpose rope used for securing tools and equipment. It is not used for lifesaving purposes.

**46.** B A rope can also be bent to form two parallel sides (**bight**) or made into a **loop** by crossing the sides of a bight.

**47.** C A **clove hitch** is formed by making two half-hitches and secured with an overhand knot (binder); a clove hitch is also used for hoisting and lowering tools and equipment. A square knot is used to tie two ropes of equal diameter together. A bowline knot will not slip or tighten under tension and is used for hoisting and lowering portable ladders.

**48.** A A **half-hitch knot** is used to hoist and lower tools and equipment.

**49.** C The **Kelly tool** is one of the original firefighting forcible entry tools; it is a metal bar tool with a fork end and an adz end that protrudes into a hammerhead for striking and receiving hammer blows, designed for lock breaking. A **battering ram** is a heavy metal bar designed to be used by two or four firefighters for pushing in heavy doors and breaching walls. A **pick head (fire) axe** has a 6- to 8-pound head attached to a long wooden handle. The head has a pointed pick on one end and an axe blade on the opposite end. The pick end is designed for enhanced penetration through and prying of floor and roof boards.

**50.** B A **K-tool** is a square (3 inches by 3 inches) steel block with a sharp-edged K-shaped notch on one side. The K-tool is designed to be slipped over cylinder locks to remove them. The other side of the tool has a U-shaped flange for prying when used in conjunction with metal bar tools. The hydraulic spreader (Jaws of Life) is a mechanical levering tool powered by a hydraulic pump engine; the Jaws of Life is used for forcible entry and spreading of car components (doors, hood) to permit extrication of trapped victims.

**51.** A A **shove knife** is a rigid, flat metal blade used by firefighters to retract door spring latches during forcible entry. **Pneumatic air bags** are constructed of neoprene rubber reinforced with steel that are inflated using compressed air cylinders; pneumatic air bags are designed to move and lift heavy loads.

**52.** D The **oxyacetylene cutting torch** uses a mixture of oxygen and acetylene; this burning tool is used by firefighters to cut locks, heavy iron bars, and metal plating. A **power saw** is a portable, gas-fueled too;, it uses circular blades to cut wood, lightweight metal, and masonry.

**53.** C The **claw tool** is an original firefighting forcible entry tool; it is a metal bar tool with a hook (claw) at one end and a fork at the other. The **sledgehammer (maul)** is a heavy, long-handled hammer with a head with two flat striking surfaces; it is used for forcible entry and breaking up walls and flooring consisting of drywall, concrete, tile, brick, and masonry.

**55.** A A **multihook** has a pointed, penetrating metal head and two flared adz end hooks for a greater pulling surface and easier removal of large areas of material in comparison to the pike pole; the **Sheetrock hook** has a pointed metal head consisting of a 4-inch, toothed, curved hook for a larger contact point and pulling surface. It is specifically designed to facilitate the pulling and removing of Sheetrock, lath, and plaster; the **closet hook** is a short (under 5 feet), wooden D-handled pike pole for use in small, tight spaces.

**59.** C **Personal Alert Safety System**

**60.** C **Self-Contained Breathing Apparatus**

CHAPTER 13

# Emergency Medical Care

> **Your Goals for This Chapter**
> - Review some basics of human anatomy
> - Find out how to perform initial patient assessment
> - Learn about cardiopulmonary resuscitation (CPR) and other emergency medical techniques
> - Find out about emergency removal and rescue techniques

Modern firefighters are trained in the assessment and treatment of life-threatening conditions sustained by the public they serve. They are responsible for stabilizing the scene of an accident or emergency situation and performing basic first aid procedures, as well as assisting prehospital providers. Patient assessment and basic medical treatment is essential to the role of the firefighter. Upon the arrival of the ambulance, firefighters will transmit valuable information to emergency medical technicians (EMTs) and paramedics concerning the patient's condition, chief complaints, and any medical assistance that was administered prior to their arrival. Today's firefighter is truly a multifaceted, lifesaving professional.

## MEDICAL DIRECTIONAL COMMUNICATION

The human body is divided into areas with which firefighters should be very familiar. A working knowledge of these will aid the first responder to isolate the part of the body that is injured or that is giving the victim the most pain. It will facilitate providing treatment and communicating information to incoming medical personnel. The diagram shows the major areas of the human body.

The fire department is often the first agency to arrive at incidents where people are injured. Firefighters therefore should know the basic anatomy or structure of the body in order to communicate medical information to medical professionals (EMS) who will arrive subsequently.

| **Medical Directional Terms** | |
|---|---|
| **Direction** | **Meaning** |
| Anterior | Toward the front of the body |
| Posterior | Toward the back of the body |
| Superior | Toward the head |
| Inferior | Toward the feet |
| Proximal | Toward the trunk of the body |
| Distal | Away from the trunk of the body |
| Medial | Toward the midline (imaginary line drawn down the center of the body) |
| Lateral | Away from the midline of the body |
| Right | The right side of the patient's body |
| Left | The left side of the patient's body |

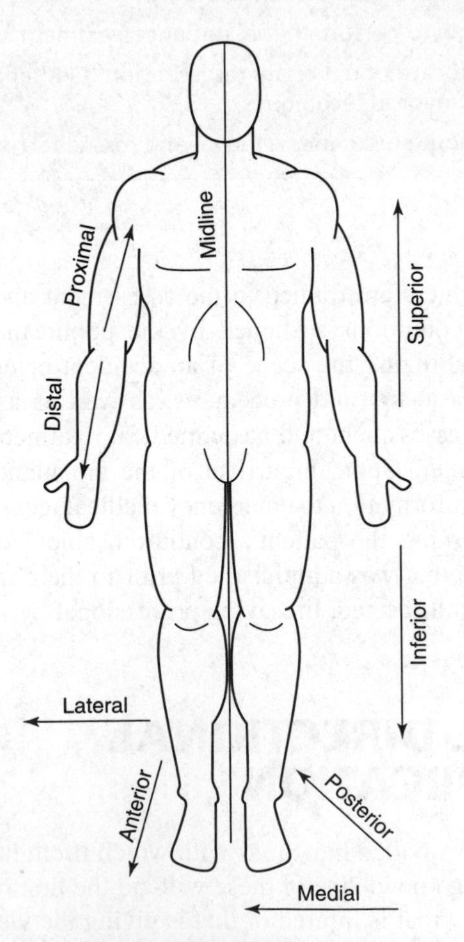


**MEDICAL DIRECTIONAL TERMS**

# INITIAL PATIENT ASSESSMENT

When firefighters arrive at the scene of an accident or medical emergency, they must perform an initial assessment of the patient's status. A general overview of the patient entails evaluating the scene to determine whether you are dealing with a **trauma** situation (injury sustained from a violent impact or sudden force) or a **medical** condition. If you are responding to the scene of an auto accident, you can assume that the patient(s) will be suffering from trauma injuries. Conversely, arriving at a patient's home and noticing pills or drug paraphernalia on the kitchen table would most likely indicate you have a person with a medical or drug problem. Other general conditions to consider are the patient's age and gender. Look for clues concerning the condition of the victim when you arrive on the scene. Is the person bleeding? Does the victim have contusions or abrasions that commonly result from trauma? Are there obvious deformities to the extremities, shoulders, or pelvis area that could indicate that the victim is suffering from a dislocation or bone fracture? Palpate or feel the patient's body to search for irregularities (swelling, depressions, or tenderness). Also, utilize your sense of smell to ascertain if the victim has been drinking alcoholic beverages or has been poisoned. The first responder should also check the victim's **vital signs** (respirations, pulse, temperature, pupils, and blood pressure) to obtain an accurate assessment of the patient's condition.

The **CUPS** status assessment criteria may be used by first responders to determine the seriousness of the patient's injury or illness. Victims considered in **critical** condition are receiving cardiopulmonary resuscitation (CPR), in respiratory arrest, or requiring and receiving life-sustaining ventilation/circulation support. **Unstable** victims are unresponsive to external stimuli or responsive but unable to obey commands. They also may exhibit difficulty breathing. Indications that a patient is going into shock (pale skin/rapid breathing), uncontrolled bleeding, and severe pain in any area of the body would warrant being classified as **potentially unstable**. People with minor injuries and illnesses would be determined to be in **stable** condition.

## Vital Signs

Assessing the **vital** signs of a victim is an essential way to monitor the functions of the body and discover abnormalities. Evaluating the information gleaned from a patient's vital signs dictates the kind of treatment the firefighter will administer while awaiting the arrival of medical professionals.

### RESPIRATION

Measure the **rate** of breathing for 30 seconds. Count the number of breaths for this time frame and multiply by two to get the number of respirations per minute. The normal rate of breathing for adults is between 12 and 20 breaths per minute. Infants and children have a faster normal rate of breathing (20 to 30 breaths per minute). Observe the chest and abdomen for abnormal **depth** (deep or shallow) and **pattern** (irregular sequence) of breathing. Abnormalities can indicate brain trauma or difficulty breathing. Listen for any unusual sounds (wheezing, snoring) while the patient is breathing for indications of asthma, bronchitis, or airway obstruction.

## PULSE

The pulse is the expansion and contraction of the body's arteries as blood flows away from the heart. It is an indicator of the function of the patient's circulatory system. It is measured by palpating (touching) an artery located close to the skin. Common arteries used by first responders to evaluate pulse rate, rhythm, and quality are the carotid (in an unresponsive adult or child), brachial (in a child or infant), radial (in an adult or child), and femoral (in an unresponsive child). Measure the pulse rate by counting the number of impulses or beats for 30 seconds. Multiply by two the number of beats for this time frame to ascertain the pulse rate per minute. The normal pulse range for adults and adolescents is approximately 60 to 100 beats per minute. Younger people tend to have higher normal pulse rates (80 to 130), while newborns have normal pulse rates from 120 to 160. The rhythm of a pulse is either regular or irregular, which directly reflects the functioning of the heart. A pulse's quality can be normal, bounding (thumping), or vaguely detectable (thready). A bounding pulse could mean the victim has high blood pressure, while a thready pulse rate may indicate cardiac arrest or shock.

## TEMPERATURE

Use the posterior (back) surface of your hand on the patient's forehead or side of the face to assess general body temperature. Also note the relative moisture on the skin. Findings can range from hot and dry, which can indicate heat stroke, to cold skin (hypothermia).

## PUPILS

The pupil is the center, dark part of the eye. It normally changes diameter in relation to the amount of light to which it is exposed, constricting (shrinking) under bright light and dilating (expanding) in the absence of light. Constricted pupils in a dark area or dilated pupils in a well-lighted room can indicate that the victim is under the influence of drugs or has been poisoned. Pupils of unequal size are associated with head injuries, stroke, or damage to the eye, as are nonreactive pupils that do not react to a beam of light directed into the eye of the victim by a first responder's **penlight**.

## BLOOD PRESSURE

Blood pressure is a measure of the force the blood exerts on the walls of the arteries. It provides a functional status indicator of the heart and blood vessels. A victim's blood pressure is commonly recorded using the auscultation method (a **stethoscope** and **blood pressure cuff**, referred to as a sphygmomanometer). Blood pressure is denoted by two numbers in fractional form (the systolic blood pressure number over the diastolic blood pressure number).

The systolic number is the pressure when the heart is contracting (pumping) blood through the blood vessels. The diastolic number is the pressure in the blood vessels when the heart is relaxed and not pumping blood. A first responder can use the auscultation method by placing the blood pressure cuff around the patient's brachial artery just above the elbow. The brachial artery is palpated and then the stethoscope is positioned over the brachial pulse point. Next, air is pumped into the cuff using its squeeze bulb until the gauge on the cuff stops rising and falling, usually between 150 and 200 mmHg (millimeters of mercury). Listen for brachial pulse sounds through the stethoscope. At 150 to 200 mmHg, brachial pulse sounds tend to disappear. This is the time to slowly release air from the cuff and begin listening for audible brachial pulse sounds. Read the gauge when the pulses are first heard; this is the systolic blood pressure number. Read the gauge again when the pulse sounds stop; this is the diastolic blood pressure number. Normal blood pressure for adults varies with age. It is generally in the range of from 100–150 over 60–90. The normal blood pressure range for children is from 100–130 over 60–80. High blood pressure can indicate hypertension and is associated with cardiovascular disease, kidney malfunction, stroke, and drug use. Low blood pressure can indicate hypotension and may be the result of shock.

**Normal Respirations, Pulse, and Blood Pressure**

| Age | Respiratory Rate (breaths/minute) | Pulse Rate (beats/minute) | Blood Pressure (mmHg) |
|---|---|---|---|
| Infant | 20–30 | 80–140 | 84–106/56–70 |
| 2–6 years | 20–30 | 80–120 | 98–112/64–70 |
| 6–13 years | 18–30 | (60–80)–100 | 104–124/64–80 |
| 13–16 years | 12–20 | 60–100 | 118–132/70–82 |
| Adult | 12–20 | 60–100 | 100–150/60–90 |

## Patient Responsiveness

Another important procedure is to evaluate how alert the patient is to exterior stimulus. Is the patient conscious, semiconscious, or unconscious? If the patient is conscious, is he or she in an oriented or disoriented state? Does the patient know her name, where she is, and what happened? Can the patient communicate to you his or her chief complaint? If the victim is semiconscious, what type of stimulus (verbal or touch) gets a response? An unresponsive person is considered unconscious. Tapping the victim's shoulders or chest, rubbing the breast bone (sternum), or pinching the muscles of the neck are common physical ways to assess responsiveness. Your ability to gather valuable medical information from the patient will depend on the response status of the individual you encounter.

# PATIENT CARE

Emergency medical care must first focus on establishing the patient's airway, breathing, and circulation.

## Airway

The patient's airway can initially be assessed by responsiveness. In general, a person who can readily communicate to you his or her chief complaint can be considered to have a clear airway. A patient having trouble talking and/or breathing can be assessed as having a partially obstructed airway. Encourage the patient to cough in an attempt to dislodge the object. If the object is not expelled, the person should be transported via EMS to the nearest medical facility.

In unresponsive adults and children where there is no sign of trauma, firefighters should have them placed on their back (supine) and open the airway performing **the head tilt–chin lift** technique. It is applied from the side of the victim by placing one hand on the patient's forehead to tilt the head back while lifting the jaw forward with your fingers.

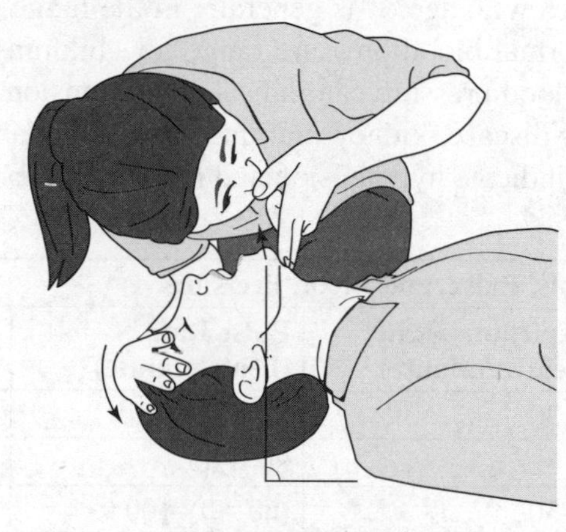

HEAD TILT–CHIN LIFT

An airway obstructed by fluid should be cleared using a **suction unit** with attached catheter that is placed inside the victim's mouth. Mucus, vomit, and fluid are cleared from the mouth using a negative pressure through the catheter. An **oropharyngeal airway** (OPA) device is used by first responders to maintain an open airway on an unresponsive patient once it has been established. The OPA is a hard plastic, C-shaped device positioned into the mouth of the patient using a twisting (rotational) action to move the tongue forward. Sizes vary to fit infants through adults.

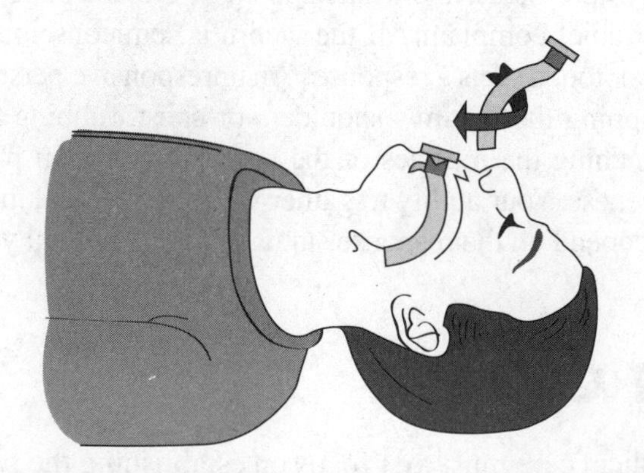

OROPHARYNGEAL AIRWAY (OPA) DEVICE INSERTION TECHNIQUE

For conscious adults and children with a completely obstructed (solid foreign object) airway, showing signs of choking (inability to speak, breathe, or cough), position yourself behind the victim, support the victim with one arm around the waist or shoulders and deliver five **back blows** with the heel of your hand. Your hand should land right between the shoulder blades.

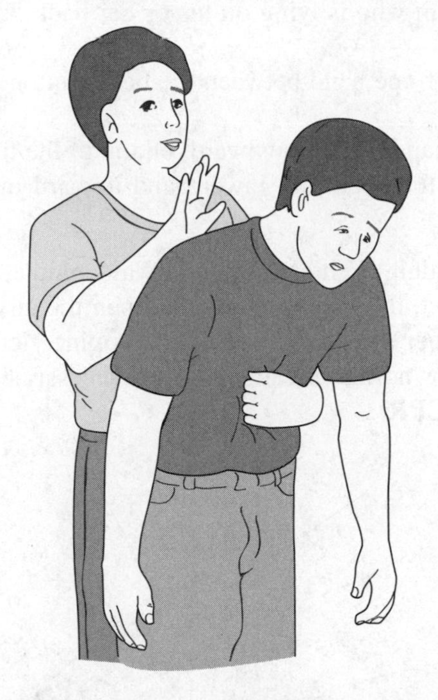

**BACK BLOWS ADMINISTERED ON A CONSCIOUS ADULT CHOKING VICTIM**

If the back blows do not dislodge the foreign object, remain in your position behind the victim and place your fist with the thumb-side just above the navel and the second hand atop the first. Apply five quick inward and upward **abdominal thrusts**. Continue sets of five back blows and five abdominal thrusts until the object is dislodged, the victim can cough or breathe, or the person becomes unconscious.

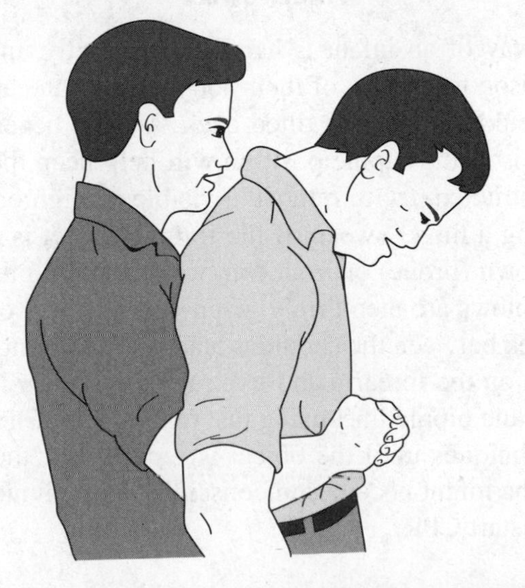

**ABDOMINAL THRUST POSITION FOR A STANDING CONSCIOUS ADULT**

## ABDOMINAL THRUSTS - LYING DOWN POSITION

An alternative way to perform the abdominal thrusts when a choking victim is largely built and discovered in the lying-down position is to follow these steps:

Facing the victim who is lying on his or her back, kneel down near the person's thighs.
Place the heel of one hand between the belly button and rib cage of the victim.
Put your other hand, palm downward, on top of the first hand.
Press down on the heel with inward and upward movements until the blockage clears.

In an unconscious adult or child with an airway completely obstructed by a visible foreign object, the firefighter should open the airway using the head tilt–chin lift and **finger sweep** the mouth on a supine victim in an attempt to remove it. Check for normal breathing. If unsuccessful, start cardiopulmonary resuscitation **(CPR)**.

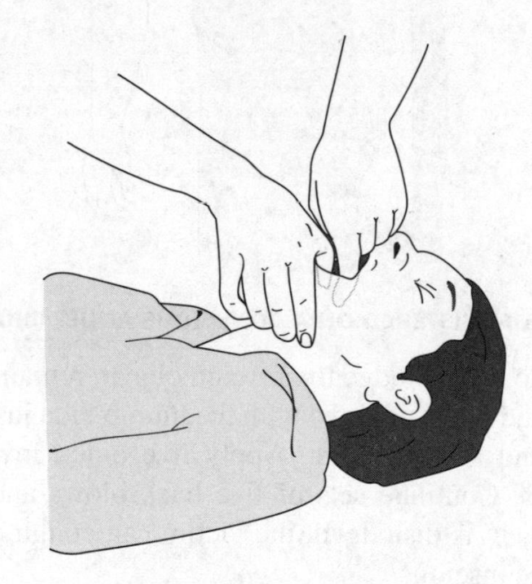

**FINGER SWEEP**

Opening the airway of an infant is handled differently. Infants have larger heads in comparison to the rest of their body which can cause the airway to be closed off should the head be tilted back. A hand beneath the shoulders while the victim is in the supine position will help keep the airway open. If the airway is obstructed, try to remove a visible foreign object from a conscious infant using a finger sweep. If the foreign object is not visible, place the infant face down (prone) on your arm while cradling the face with your hand. Five back blows are then firmly given with the heel of one hand to the middle of the back between the shoulder blades. The infant is then shifted to a supine position on the forearm and five rapid, downward chest thrusts are given, just below the nipple line, using just two fingers. Check for breathing. Repeat these techniques until the object is removed or the infant becomes unconscious. If the infant becomes unconscious, stop giving back blows and chest thrusts and start CPR.

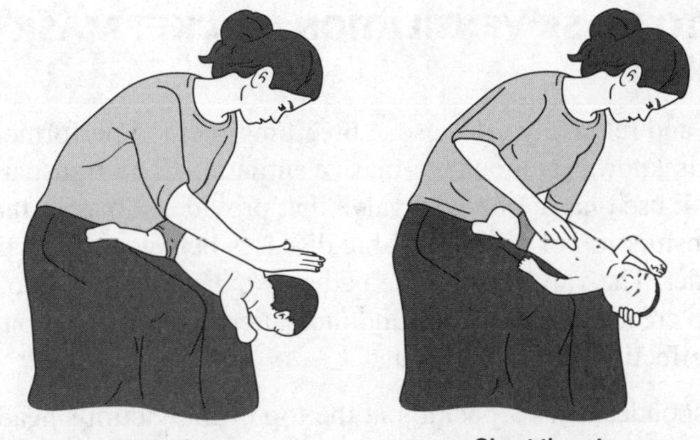


**BACK SLAPS AND CHEST THRUSTS ON CONSCIOUS INFANT (CHOKING VICTIM)**

To remove a visible foreign object from the airway of an unconscious infant, place the baby on your forearm face up with your hand supporting the back of the head. Use the head tilt–chin lift to open the airway and look for an obstruction. If visible, remove using finger sweep. Check for breathing. If the victim is not breathing, start CPR.

# Breathing

The use of the **look, listen, and feel** method is a simple way to determine the patient's breathing status. Look to see if the person is struggling to breathe. Is the patient breathing very quickly or slowly? When encountering an unconscious person, check to see if the chest of the victim is rising and falling. Check the color of the skin. Blue-gray color around the mouth and the end of the fingers is usually attributed to a lack of oxygen to the body and respiratory distress. Listen to the victim. Put your ear next to the victim's nose and mouth and listen for the sounds of breathing if the person is unconscious. If the person is conscious, is he or she able to communicate difficulty in breathing? Is the patient experiencing loud, labored breathing? In an unconscious victim, feel for breathing by placing your hand on the chest to check for movement. Place the side of your head next to the victim's mouth and feel for any air movement against your cheek. A victim who is not breathing requires the administration of positive pressure ventilation (PPV) methods or **rescue breathing**.

## SINGLE- AND DUAL-CYLINDER OXYGEN UNIT

First responders frequently utilize a combination fixed-flow **inhalator** (for patients who are able to breathe with difficulty) and mouth-to-mask rescue breathing **resuscitator oxygen unit** (for patients who are not breathing) to provide supplemental oxygen to victims having difficulty breathing or forced ventilation to those who are not breathing. Cylinders typically provide 20 to 60 minutes of oxygen to the patient depending on how the unit is being used.

## MOUTH-TO-MASK VENTILATION (POCKET MASK RESCUE BREATHING)

A common and relatively safe rescue breathing method performed today by firefighters is known as mouth-to-mask ventilation. The transparent plastic **pocket mask** used has a one-way valve that provides a barrier that can prevent the transmission of communicable diseases between the victim and the first responder. The valve diverts the patient's exhalations. It also allows the firefighter to create a tight seal around the patient's nose and mouth in order to perform effective rescue breathing.

The first responder takes a position at the top of the victim's head as shown in the diagram, maintains an open airway, and places the pocket mask over the victim's nose and mouth. The firefighter then places his or her mouth over the mouthpiece of the pocket mask. The firefighter breathes slowly into the mouthpiece, causing the victim's chest to rise visibly. The firefighter then allows the victim to exhale by removing his or her mouth from the mouthpiece.

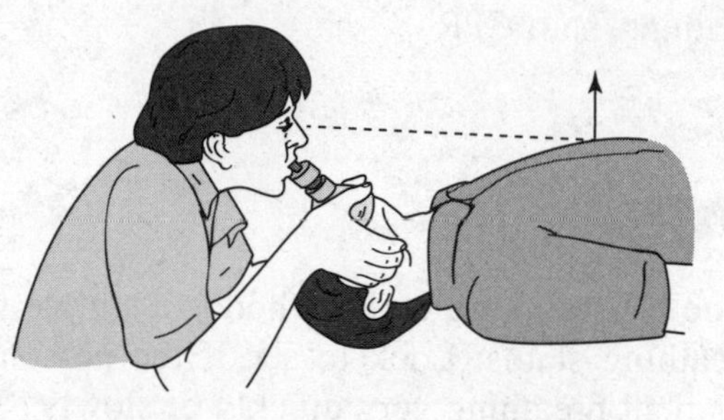

POCKET MASK WITH ONE-WAY VALVE

## BAG-VALVE-MASK DEVICE

Another rescue breathing method utilizes a **bag-valve-mask device**. The bag-valve-mask is made of soft, collapsible plastic and can be used to provide air forcibly to the victim with or without **supplementary oxygen** attached. All of these rescue breathing devices and equipment are used in conjunction with an OPA.

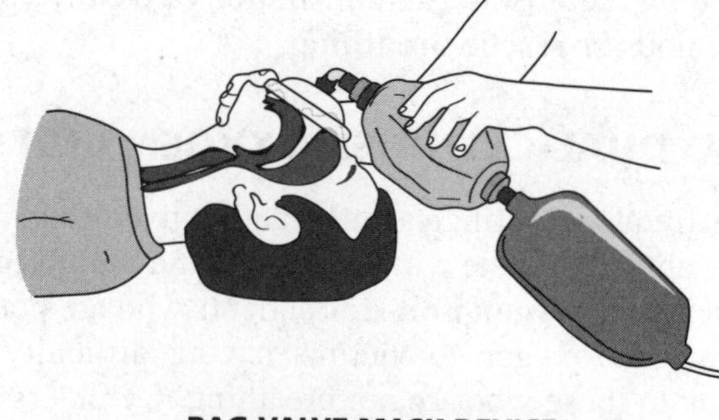

BAG-VALVE-MASK DEVICE

## RESCUE BREATHING

Rescue breathing is also known as artificial respiration. First responders should administer rescue breathing when a victim stops breathing and becomes unconscious. Administering rescue breathing greatly increases the probability of recovery without permanent brain damage from the lack of oxygen. Position the victim on his or her back and tilt the head back slightly (if there is no obvious head or neck injury) to open the airway. Check for possible obstructions in the airway and that the victim is, in fact, not breathing before performing rescue breathing. Pinch the victim's nose shut gently to negate ventilation leakage and lift the chin prior to giving rescue breaths. Take a normal breath before giving mouth-to-mask ventilation while checking to see if the victim's chest rises.

Blowing slowly but firmly, give one breath every five seconds for an adult. Wait five seconds, then repeat the breath and check the victim's pulse. If there is no pulse, you will have to perform cardiopulmonary resuscitation (CPR). Otherwise, keep breathing into the victim every five seconds until he or she is able to breathe unassisted. For children and infants, give two slow breaths into the victim's mouth or mouth and nose (infant). Deliver rescue breaths every three seconds for a minute. After one minute, stop and check for pulse and breathing for approximately five to ten seconds. If there is no pulse, you will have to perform CPR.

| **Rate of Rescue Breathing** | | |
|---|---|---|
| Adult | Child | Infant |
| 12 breaths/min | 20 breaths/min | 20 breaths/min |
| 1 breath every 5 seconds | 1 breath every 3 seconds | 1 breath every 3 seconds |

# Circulation

Assess the patient's circulation by checking for bleeding, pulse rate, and skin color. Analyze the scene to see if the patient is bleeding heavily (hemorrhaging). Uncontrollable bleeding is potentially life threatening. Is there a pool of blood on the ground? Are there bloodstains on the victim's body, hair, or clothing? Check the back of the victim for the possibility of hidden blood. The loss of large amounts of blood can also cause the victim to go into shock. A shock victim may exhibit pale skin color. The patient's skin color also reflects the status of the circulatory system. Bluish (cyanotic) skin indicates a restriction of oxygen to the living tissues and organs of the body.

The pulse rate is commonly taken at one of the four arteries listed in the table "CPR Summary" and shown in figures later in this section, dependent on whether the victim is responsive or unresponsive and his or her age. Determine if the pulse rate is regular or irregular, fast or slow, weak or strong. A patient without a pulse requires the first responder to start CPR.

## CARDIOPULMONARY RESUSCITATION (CPR)

CPR should be administered by first responders on victims of cardiac arrest (lacking a pulse and not breathing). It is used to promote blood flow (compressions) and provide **oxygen** (ventilation) to the body's heart and brain. When blood stops circulating, oxygen cannot be transported to the body's vital organs and tissues. Without oxygen for four to six minutes, brain cells begin to die.

Ensure that unresponsive adults and children with no sign of trauma are positioned supine on the ground. Use the head tilt–chin lift technique to open airway. Rescuers place themselves on their knees at opposite sides of the victim to administer CPR.

TWO-RESCUER CPR ON ADULT VICTIM

### Compressions-Airway-Breathing (C-A-B) for CPR

The American Heart Association (AHA) now recommends that chest compressions be the first step for rescuers to revive victims of sudden cardiac arrest. Therefore, the acronym C-A-B for compressions-airway-breathing is used when performing CPR. During the first few minutes of cardiac arrest, victims will have oxygen still remaining in their lungs and bloodstream. All victims in cardiac arrest, however, will require chest compressions to pump oxygenated blood to the brain and heart. AHA research has shown that rescuers who started CPR with opening the airway took much longer (30 seconds) to begin chest compressions than rescuers who began CPR with chest compressions.

C-A-B entails starting compressions within 10 seconds of recognition of a cardiac arrest. Compressions are at a rate of at least 100 per minute with a depth of at least two inches for adults, approximately two inches for children, and about one and a half inches for infants. Administering compressions requires complete chest recoil, allowing the heart to refill with blood between compressions. Effective ventilation is indicated when the victim's chest rises.

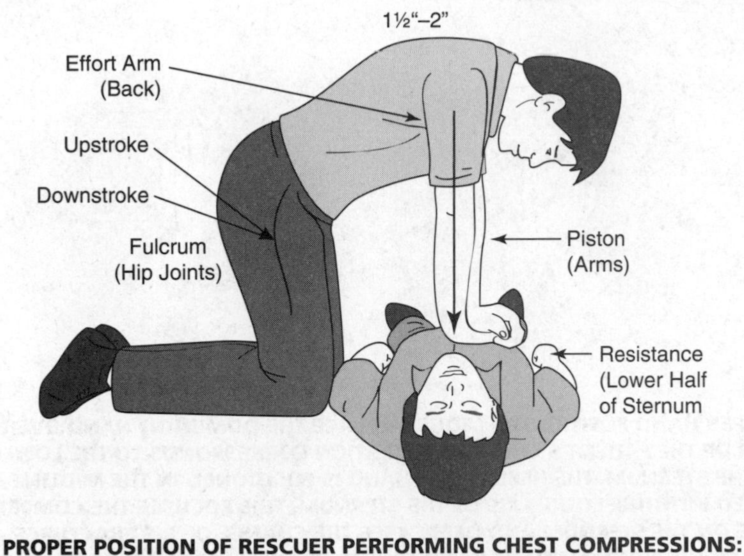


**PROPER POSITION OF RESCUER PERFORMING CHEST COMPRESSIONS: SHOULDERS DIRECTLY OVER VICTIM'S STERNUM; ELBOWS LOCKED**

## Adult CPR (One or Two Rescuers)

Adult CPR techniques are performed on victims from eight years old and up. Generally two first responders administer CPR to an adult. Occasionally, CPR is performed by a lone rescuer. If the victim is not breathing, administer 30 chest compressions at a rate of 100 compressions per minute, followed by ventilation. Give the victim a breath big enough to make the chest rise. Repeat ventilation once more when the chest falls. Repeat chest compressions and ventilation for approximately two minutes (five cycles of 30 compressions and two ventilations). After two minutes, check the victim for breathing. If the victim is still not breathing, continue CPR. The pulse check for adults and children is the carotid artery (neck). In two-rescuer CPR, rotate the rescuer performing compressions at regular intervals.

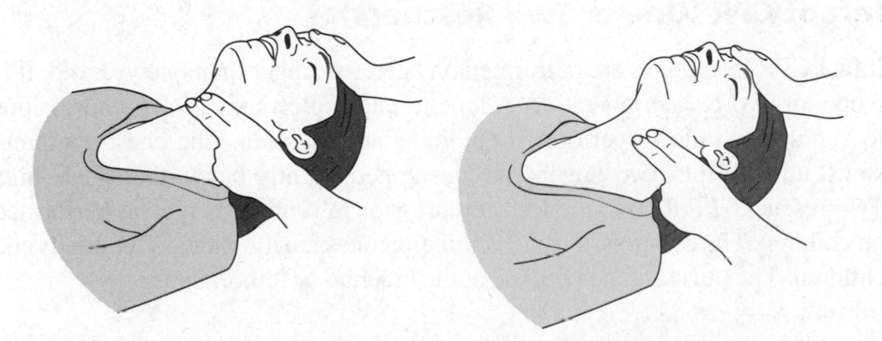

**LOCATING THE CAROTID PULSE ON AN ADULT VICTIM**

**Note:** Chest compressions are given with the heel of one hand positioned in the middle of the chest between the nipples. Place the heel of the other hand on top and interlace your fingers. Elbows are locked as compressions are given directly over the sternum. Once the chest is compressed, it should be allowed to come back up to its original position.

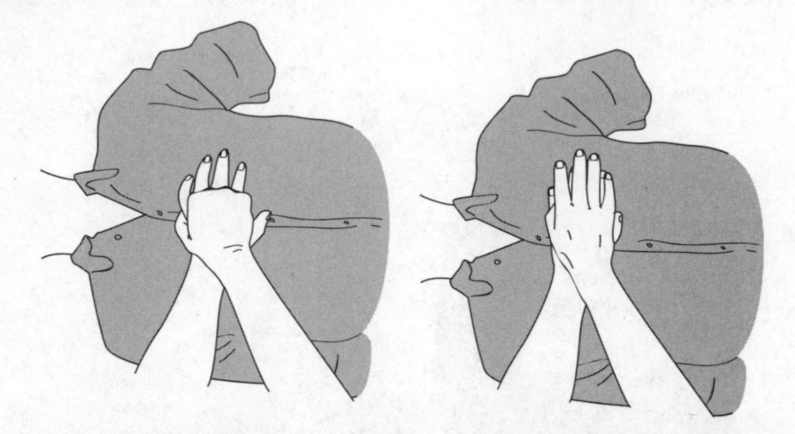

**PROPER HAND POSITIONING (ADULT) – PLACE THE DOMINANT HAND OVER THE CENTER OF THE PATIENT'S CHEST. THIS POSITION CORRESPONDS TO THE LOWER HALF OF THE STERNUM. THE HEEL OF THE HAND IS POSITIONED IN THE MIDLINE AND ALIGNED WITH THE LONG AXIS OF THE STERNUM. THIS FOCUSES THE COMPRESSIVE FORCE ON THE STERNUM AND DECREASES THE CHANCE OF RIB FRACTURES. NEXT, PLACE THE NONDOMINANT HAND ON TOP OF THE FIRST HAND SO THAT BOTH HANDS ARE OVERLAPPED AND PARALLEL. THE FINGERS SHOULD BE ELEVATED OFF THE PATIENT'S RIBS TO MINIMIZE COMPRESSIVE FORCE OVER THE RIBS.**

## Child CPR (One or Two Rescuers)

Child CPR techniques are administered on victims one to eight years of age. Similar to adults, it is generally performed by two rescuers. A lone rescuer would follow the same 30:2 compression to ventilation ratio and rate as stated for an adult. For two rescuers performing CPR, one rescuer gives 15 chest compressions followed by two breaths by the second rescuer. The compression and ventilation rates are the same as for adults. When performing chest compressions on a child, proper hand placement is even more important than with adults. Place two fingers at the sternum (bottom of the rib cage where the lower ribs meet) and then place the heel of your other hand directly on top of your fingers. The victim's pulse can be checked at the brachial or femoral artery.

## Infant CPR (One or Two Rescuers)

Infant CPR techniques are performed on babies younger than one year old. It is a one- or two-rescuer operation. A lone rescuer follows the 30:2 compressions to ventilation ratio as performed on adults and children. The chest compressions, utilizing just two fingers, are positioned slightly below the nipple line. Two-rescuer CPR follows the 15:2 compression to ventilation ratio as performed on children. The compression and ventilation rates are the same as for adults and children. The pulse can be checked at the brachial or femoral artery.

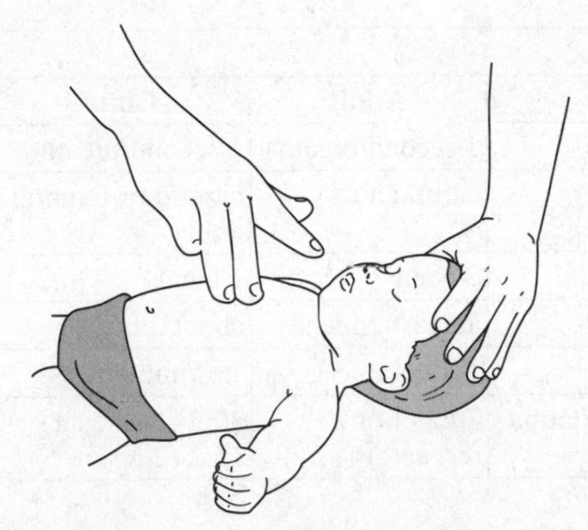

**CHEST COMPRESSIONS (INFANT) – PLACE THREE FINGERS IN THE CENTER OF THE INFANT'S CHEST WITH THE TOP FINGER ON AN IMAGINARY LINE BETWEEN THE INFANT'S NIPPLES. RAISE THE TOP FINGER UP AND COMPRESS WITH THE BOTTOM TWO FINGERS.**

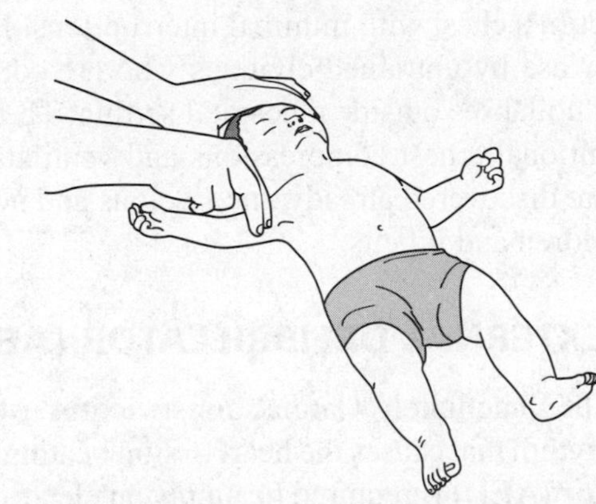

**PULSE CHECK FOR INFANT AT THE BRACHIAL ARTERY (ARM)**

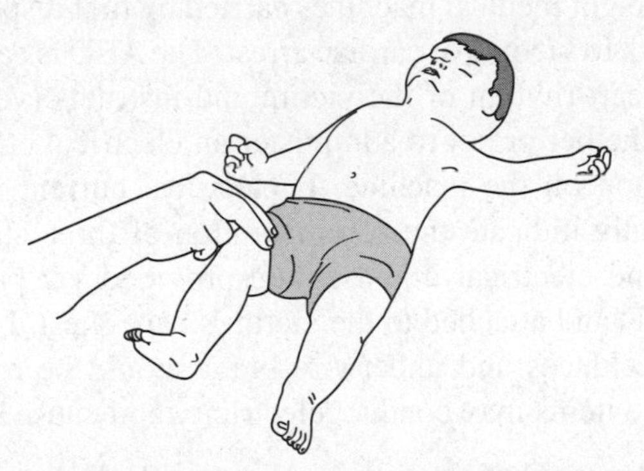

**PULSE CHECK FOR INFANT AT THE FEMORAL ARTERY (LEG)**

| CPR Summary | | | |
|---|---|---|---|
| | **Adult** | **Child** | **Infant** |
| Breaths | 1 second/breath | 1 second/breath | 1 second/breath |
| Pulse | carotid artery | carotid or femoral artery | brachial or femoral artery |
| Compression Method | 2 hands | 1 hand | 2 fingers |
| Compression Depth | at least 2 inches | about 2 inches | about 1½ inches |
| Compression Rate | 100/minute | 100/minute | 100/minute |
| Compression/Ventilation Ratio | 30:2 (1 or 2 rescuers) | 30:2 (1 rescuer) 15:2 (2 rescuers) | 30:2 (1 rescuer) 15:2 (2 rescuers) |

### Hands-Only CPR for Adults

Hands-only CPR is conducted without ventilation. The practitioner begins hands-only CPR by providing chest compressions by pushing hard and fast in the center of the victim's chest with minimal interruptions. Hands-only CPR is recommended for use by untrained civilians who are confronted with an adult who suddenly collapses outside a hospital setting. The AHA, however, recommends conventional (chest compressions and ventilation) CPR for all adult victims who are discovered already unconscious and not breathing normally as well as children and infants.

## AUTOMATED EXTERNAL DEFIBRILLATOR (AED)

CPR cannot sustain life indefinitely. Cardiac arrest victims usually suffer from an abnormal heart rhythm that causes the heart to stop beating. **An automated external defibrillator (AED)** is required to supply an electrical current to the heart **(defibrillation)** in order for it to start to beat and regain a stable rhythm.

AEDs are lightweight medical machines carried by first responders to provide an electrical shock to victims of cardiac arrest. The AED is easy to use because it analyzes the heart rhythm of the victim and instructs (voice prompts) the first responder whether or not to administer an electrical current by pressing the SHOCK button on the machine. If electrical current is warranted, the AED will generally indicate the administration of three stacked SHOCKS. Heart analysis and electrical impulses are provided via pads that are connected to the AED and attached to the victim's bare chest. If the chest is wet, dry it. Metal (necklaces and underwire bras) should be removed from the victim since these items may conduct electricity and cause burns.

Safety also dictates that before analysis and the administration of a shock all rescuers stay clear of the victim. Additionally, before using an AED, check for accumulations of water in and around the area you will be working in. If necessary, move the victim to a dry location. Checking for breathing and pulse, providing CPR, and reusing the AED are based on the victim's recovery status subsequent to the initial use of the machine.

Pediatric AEDs, designed to apply a reduced electrical charge through small pads, are employed by rescuers for attempted defibrillation of children one to eight years of age. If, however, a pediatric AED is not available, rescuers should use a standard AED. For infants less than one year of age, a manual defibrillator (generally found in hospitals) is preferred.

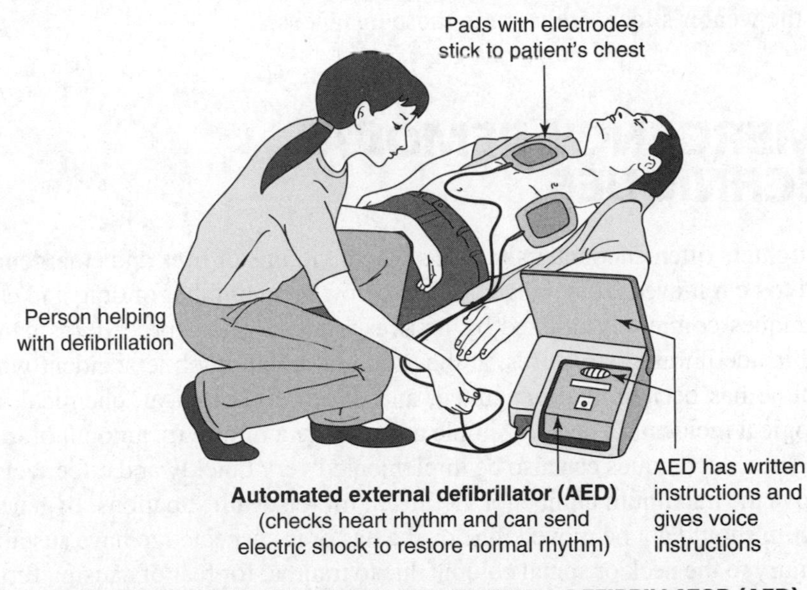


**TYPICAL SETUP USING AN AUTOMATED EXTERNAL DEFIBRILLATOR (AED). THE AED HAS STEP-BY-STEP INSTRUCTIONS AND VOICE PROMPTS THAT ENABLE A RESCUER TO CORRECTLY USE THE MACHINE.**

## EMERGENCY MEDICAL PERSONAL PROTECTION EQUIPMENT (PPE)

To prevent the spread of disease, it is imperative that firefighters carry and utilize adequate medical personal protection equipment (PPE) when administering first aid and CPR. The level of protection worn should correlate with the potential threat of infection and disease. The following is a short list of essential PPE that all firefighters generally wear when providing patient care.

**Gloves**—In general, gloves (made from latex) are worn by firefighters as standard practice during emergency medical incidents. However, if the firefighter or patient has allergies to this material, vinyl or other type synthetic material gloves are substituted. The patient should be asked if he or she has a latex allergy prior to care. A new pair of gloves must be worn for each patient treated.

**Eye protection** is worn to inhibit bodily fluids from entering the rescuer's eyes as well as to provide a certain measure of impact resistance. Safety glasses or goggles are commonly used for this purpose. First responders who wear prescription glasses may require some modification to their protective eyewear.

**Gowns**—Disposable outer garments to protect exposed skin and uniforms from bodily fluid contamination and blood-borne pathogens should be available to firefighters during certain emergency medical care situations. Gowns are of special value when treating trauma patients.

**Masks**—To protect the firefighter from **airborne pathogens** and bodily fluids, surgical-type masks are used. If a patient is suspected of having a communicable respiratory disease, however, greater protection will be needed. Respirators, which require fit testing to ensure a proper seal on the wearer's face, are worn in these instances.

# EMERGENCY REMOVAL TECHNIQUES

Firefighters often encounter situations where victims of fires and emergencies need to be removed from danger. Listed below are a number of drag and carry techniques commonly utilized by first responders. These procedures can be used inside burning structures, at the scene of an automobile accident where gasoline has been spilled or ignited, and during radiological, chemical, and biological incidents. They are simple and require a minimum amount of training. These techniques can also be implemented very quickly and effectively to remove the maximum number of victims in life-or-death situations. In general, a victim should not be moved if he or she has or is suspected to have sustained an injury to the neck or spinal column due to trauma, for fear of causing further injury and paralysis. If time allows, spinal immobilization techniques using a **cervical collar** (head and neck) and **short board**, **immobilization vest**, or **long backboard** (spine) should be employed prior to movement.

## Rolls

A victim is usually moved from a prone to a supine position for reasons ranging from medical evaluation to transport. Rolls are generally performed by more than two rescuers.

### LOG ROLL

To perform a log roll to place a backboard under a supine victim when a **neck or spine injury** is suspected, the lead rescuer is repositioned from the side of the victim to the top of the victim's head. Three other rescuers line up on their knees along the same side of the victim (see figure). These three rescuers reach across the prone victim to the far side of the body. The lead rescuer provides **manual head stabilization** by placing his or her hands on either side of the victim's head. The next rescuer holds the victim's arms and shoulders. The third rescuer holds the trunk and thigh area of the victim. The end rescuer holds the victim's feet and pelvis. The victim is rolled on the order of the lead rescuer in a single movement toward the rescuers to a side position, and the backboard is placed beside the victim on the side opposite the rescuers. The victim is then rolled in a single movement back to a supine position on the backboard upon the order of the lead rescuer. The victim is strapped to the backboard before being moved.

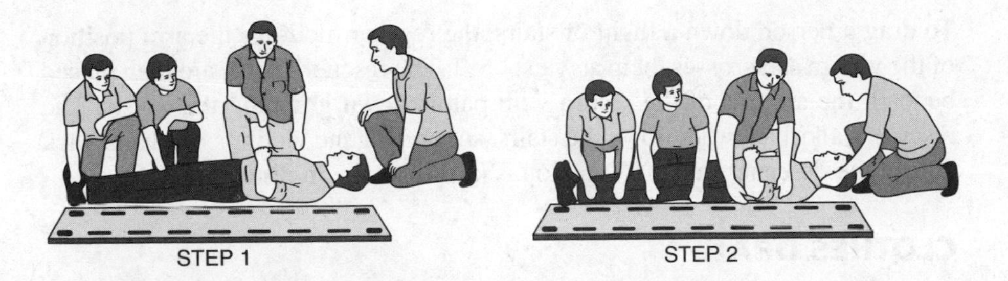


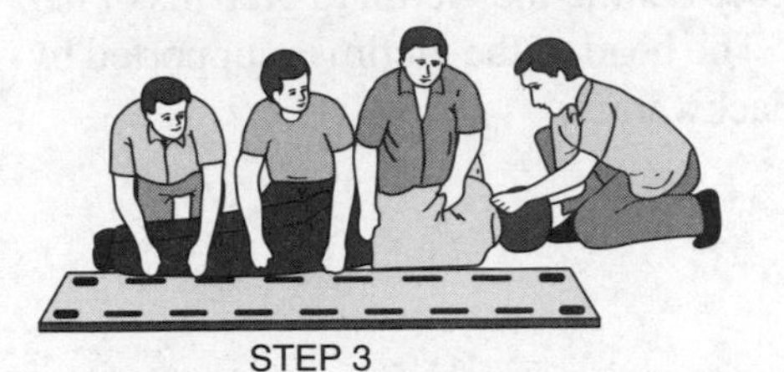


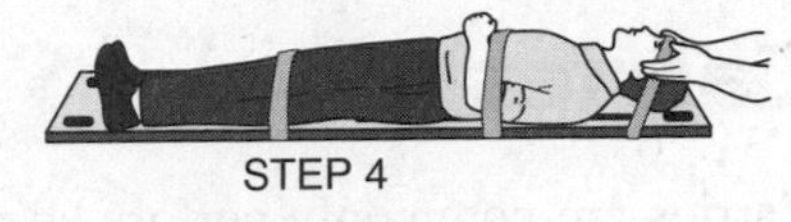


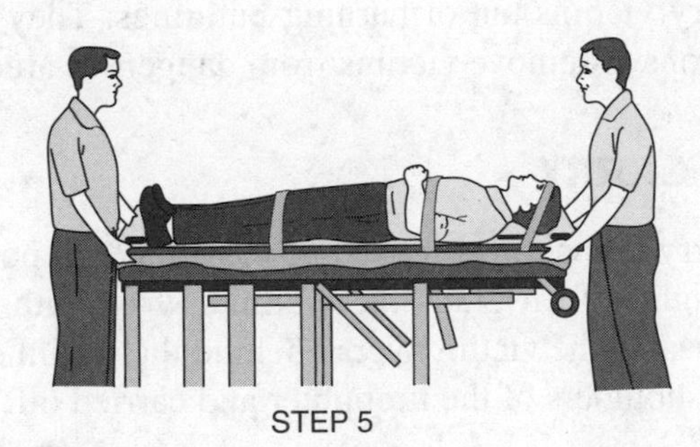


**LOG-ROLL METHOD OF MOVING A NECK OR SPINE INJURY VICTIM**

# Drags

Drags are normally performed by a lone rescuer. A second rescuer, however, can be employed to roll the victim to a supine position and to take the weight off the lower trunk as the victim is being removed.

## FIREFIGHTER'S DRAG

A firefighter's drag is performed during structural firefighting operations to remove a victim from a smoky or high heat environment. The rescuer instructs the supine victim to drape his or her arms onto the back of the rescuer's neck while the rescuer is straddling the victim on hands and knees. The rescuer crawls forward on hands and knees with the victim. This drag method allows the rescuer to move the victim low to the floor, thereby protecting him or her from high heat conditions and the products of combustion.

## SHOULDER DRAG

In a shoulder drag, the rescuer goes behind the victim and raises up the victim's head and shoulders, using the arms to under-hook the upper trunk. The victim's arms are at the side. The rescuer's arms are positioned beneath the armpits of the victim with palms up as the rescuer walks backward dragging the victim.

To drag a person down a flight of stairs, the rescuer modifies the arm position of the victim and crosses them at waist level. The rescuer's arms are then placed beneath the armpits of the victim with palms down grabbing the wrists. The rescuer walks backward down the stairs, supporting the victim's head and back during the descent. This modification is known as the **incline drag**.

### CLOTHES DRAG

To perform a clothes drag, the rescuer goes behind the victim to grab his or her clothes at the neck collar/shoulder area. The head of the victim is supported by the arms of the rescuer while moving backward.

## Carries

Carries are commonly performed on the fireground to remove unconscious and nonambulatory victims out of burning buildings. They are also used in emergency situations to remove victims from dangerous areas.

### FIREFIGHTER CARRY

In a firefighter carry the victim is raised up to a standing position facing the rescuer. The firefighter then grabs the victim's wrist with one hand while using the arm closest to the victim to grab behind the victim's leg. The victim is raised onto the shoulders of the firefighter and carried off.

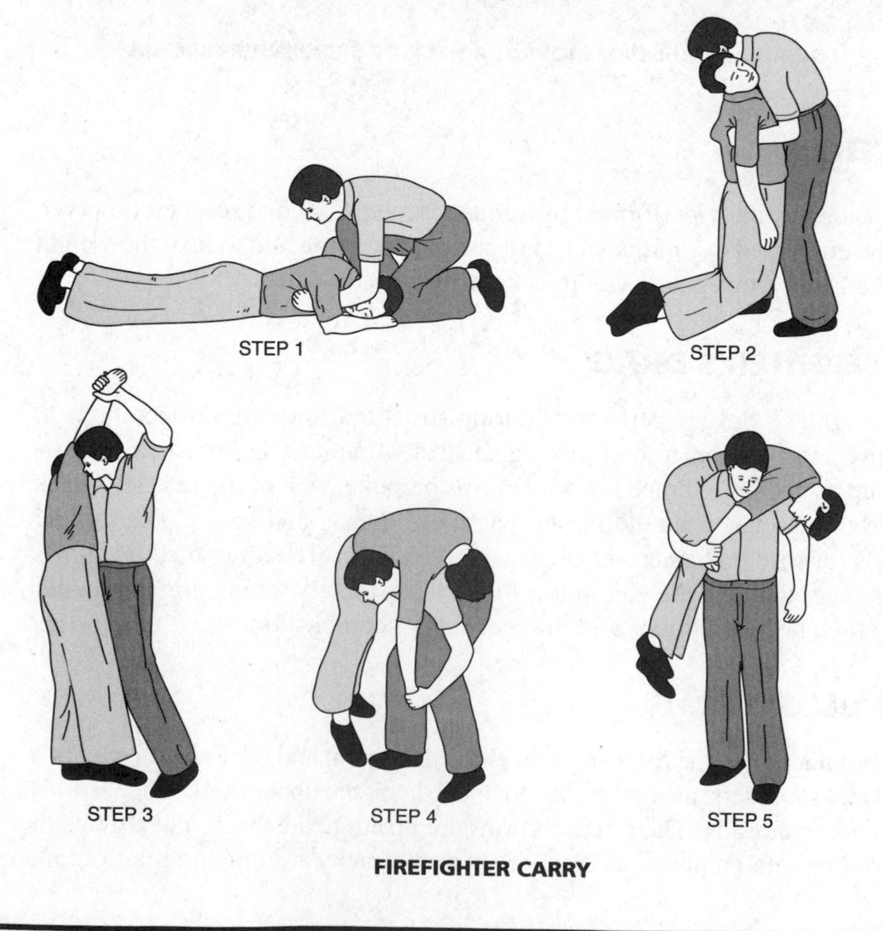


**FIREFIGHTER CARRY**

### PACK STRAP CARRY

The victim is raised to a standing position and his or her arms are positioned around the firefighter's neck as the rescuer rotates 180 degrees to carry the victim off. The firefighter leans slightly forward while walking to support the person's weight and raise the victim off the ground.

### CRADLE-IN-ARMS CARRY

The victim is in a supine or sitting position on the ground. If supine, place one arm behind the victim's back (just below the shoulders) and raise him or her to a sitting position. The firefighter's other arm is placed under the victim's knees. While keeping your back straight, raise the victim to approximately waist height and remove to safety.

## Transport Equipment

**Portable stretcher**—A simple transport device commonly consisting of two wooden poles and a canvas attachment.

**Scoop stretcher**—Composed of two pieces of light metal that can be separated and placed on either side of the victim to be transported and then reattached when properly positioned.

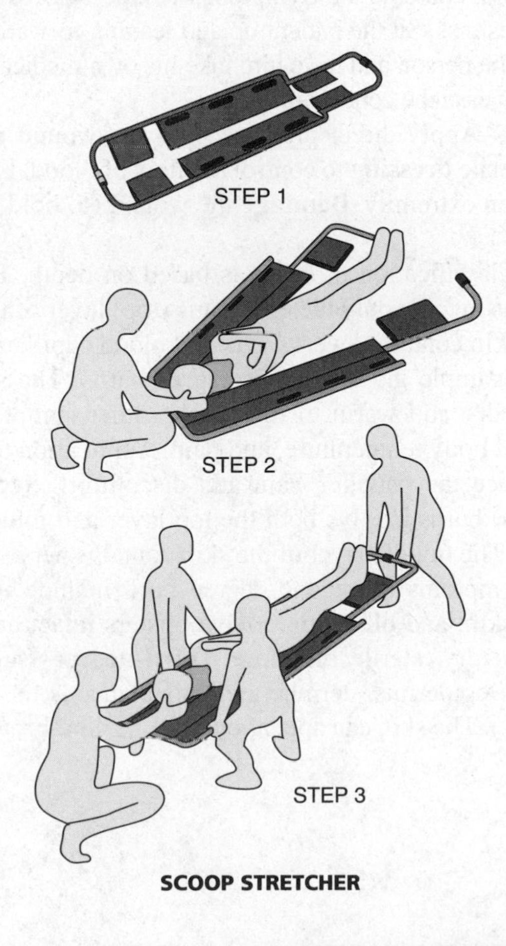


SCOOP STRETCHER

**Basket stretcher**—Made from plastic, fiberglass, or lightweight metal, this type of device is used to remove victims vertically from confined spaces.

**Wheeled stretcher**—Provides easy transfer of a victim to, into, and out of an ambulance.

**Stair chair**—Used to transport victims down stairs.

**Long backboard**—Used for victims who are lying down or standing up and to immobilize victims with neck and spine injuries prior to transporting.

# COMMON INJURIES, ILLNESSES, SYMPTOMS, AND TREATMENT

Below is a brief listing of some of the more common first-aid, trauma, and medical emergency incidents encountered by firefighters during a typical tour of duty. Signs and symptoms of the injury are noted where applicable. Basic stabilization and treatment procedures are provided to give the reader a fundamental knowledge of emergency medical care techniques.

**Asthma**—Difficulty breathing as a result of muscle spasms, which are triggered by an allergic reaction to pollen, dust, smoke, animal fur, medications, certain foods, mold, and cold air. Symptoms include wheezing, coughing, anxiousness, or distress. Sit the patient up and leaning forward to facilitate breathing. Calm the person and have him take his own medication (aerosol spray or puffer) to ease the condition.

**Bleeding (external)**—Apply direct pressure onto the wound using a protected hand or **sterile dressing** to control the flow of blood. Elevate, if the wound involves an extremity. **Bandage** the wound (to hold the dressing in place).

**Burns (thermal)**—Classification of burns is based on depth. **First-degree** (superficial) burns involve only the epidermis (top) layer of the skin. The top layer of the skin contains sweat ducts and blood capillaries. Sunburn is the standard example given for first-degree burns. The skin appears reddened and is dry and warm to the touch. Other symptoms include swelling, elevated body temperature, and pain. Application of cool water will help to reduce the patient's pain and discomfort. **Second-degree** (partial-thickness) burns involve both the top layer and middle (dermis) layer of the skin. The middle layer of the skin contains nerve endings and hair follicles. Symptoms of second-degree burn include intense pain, pink or reddish skin, and blistering. Leave blisters intact and cover the burn area with a dry, sterile dressing. **Third-degree** (full-thickness) burns involve the epidermis, dermis, and subcutaneous tissue (fat cells and blood vessels). The skin can appear charred, yellow-brown, dark red, or white. The patient feels no pain because the nerve cells of the skin are destroyed. Flush the damaged area with water and remove smoldering clothing. Cover the burn area with a dry, sterile dressing. For burns to the hands or feet, separate the fingers or toes with dressings. Burns that involve muscle and bone are sometimes referred to as **fourth-degree** burns. To determine the percentage of body surface that the burn covers, use the **rule of nines**. The rule of nines accounts for approximately 100 percent of the total body surface.

The head (9%), chest (9%), abdomen area (9%), upper back (9%), lower back (9%), each arm—anterior and posterior (9% × 2 = 18%), each leg—anterior (9% × 2 = 18%), and each leg—posterior (9% × 2 = 18%) account for 99 percent of an adult's body surface. The groin area (calculated as 1%) raises the total to 100%.

For children the rule of nines is slightly modified. The head counts for 18%. The calculations for the chest (9%), abdomen area (9%), upper back (9%), lower back (9%), and arms—anterior and posterior (9% × 2 = 18%) remain the same as for adults. The legs, anterior and posterior, however, account for 13.5% × 2 = 27% . The estimate for the groin area (1%) of a child is similar to adults.

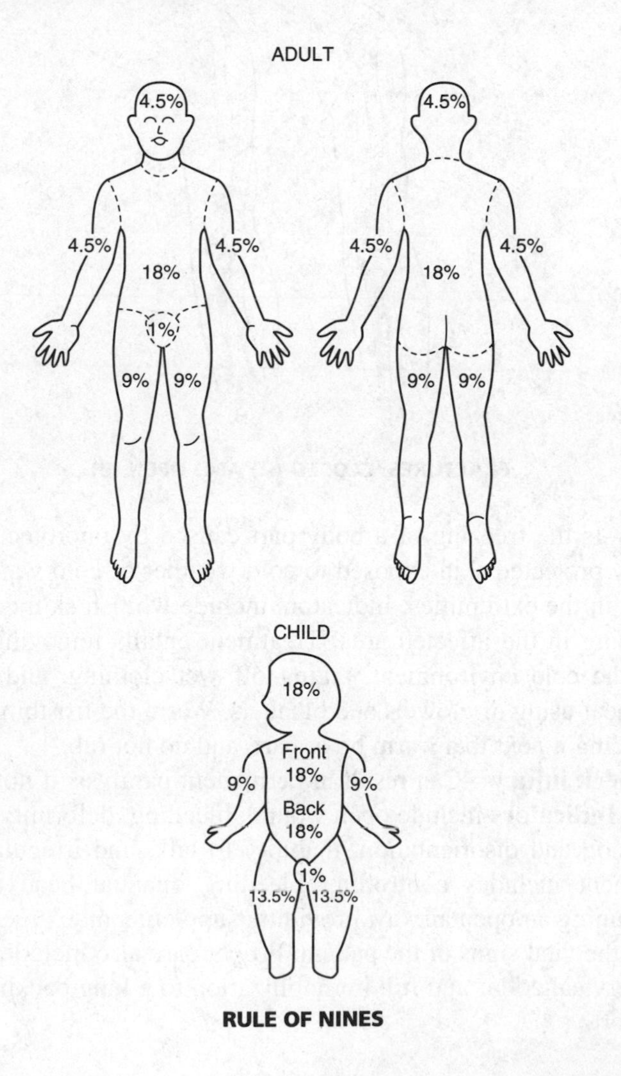


RULE OF NINES

**Chest injury**—Indicators include difficulty breathing and chest pain. Treat the patient by assisting with breathing and provide supplemental oxygen as needed.

**Contusion (bruise)**—Apply cold or an ice pack to inhibit swelling and elevate the injured part.

**Eye wound**—To protect against eye movement and further injury, bandage both eyes of the patient, even if only one eye is injured, because the eyes move in unison.

**Fracture (broken bone)**—Classified as either an **open** (skin is broken) or **closed** (skin is not broken) fracture. Indicators include open wound, deformity, pain with movement, tenderness, and swelling. Initially the first responder should cut away clothing to expose the injury using **trauma shears**. Also check for **dislocation** (separation of the bone from its joint). Treatment involves stabilizing the area above and below the site of injury. Apply a dressing to open wounds, administer cold or an ice pack to inhibit swelling, and splint the injury as required. A **splint** is a device used to immobilize the area around the broken bone(s) and joint. In general, bones should be splinted in the position found. Splints can be either rigid (wood, padded cardboard) or soft and flexible (foam, air).

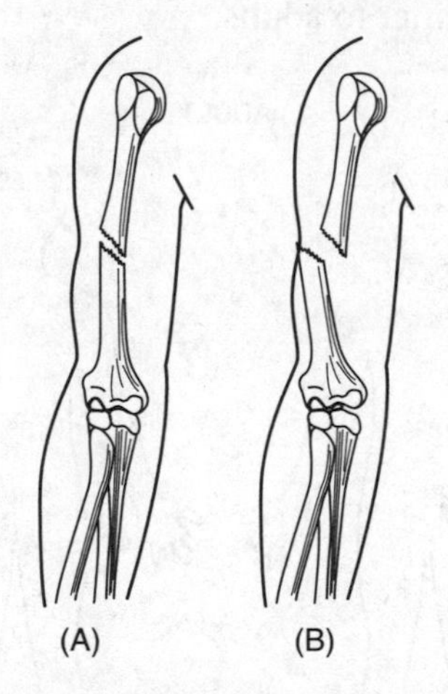


**FRACTURES: CLOSED (A) AND OPEN (B)**

**Frostbite**—Is the freezing of a body part caused by unprotected or inadequately protected skin exposed to cold weather or cold water. It usually occurs in the extremities. Indicators include whitish skin color and loss of feeling in the affected area. Treatment entails removing the patient from the cold environment, taking off wet clothing, and maintaining body heat using dry towels and blankets. Warm the frostbitten body part by placing it next to a warm body part, and do not rub.

**Head or neck injury**—Can result in permanent paralysis if not treated correctly. **Indicators** include open wound, bleeding, deformity or swelling, confusion and disorientation, inability to talk, and irregular breathing. Treatment includes controlling bleeding, manual head stabilization, maintaining an open airway, providing supplemental oxygen, and monitoring the vital signs of the patient. Proper care also includes application of a cervical collar and full immobilization to a long backboard prior to transport.

**Heat exhaustion**—Caused by the excessive loss of fluids from the body due to overwork or overexercising. Signs and symptoms include sweating profusely, cold and moist skin, rapid/shallow breathing, and dizziness. The first responder should remove the patient to a cool environment and check vital signs. Remove tight-fitting clothing and allow the victim to drink fluids (water). Reduce body temperature by wetting the skin with cool water and apply ice packs to the patient's neck, back, armpits, and groin to reduce body temperature. If left untreated, heat exhaustion can progress to heat stroke, which is a more serious condition.

**Heat stroke**—A condition where the body cannot cool itself. If not treated quickly and properly, it can cause brain damage and be life threatening. Indicators of heat stroke are inconsistent breathing (deep/shallow) and pulse (rapid strong/rapid weak) patterns; dry, hot skin; dilated pupils; and seizures. If possible, move the victim to a cool/shady area or into an air-conditioned environment and remove any unnecessary clothing. Cool the victim off by wetting the skin. Apply ice packs to the patient's neck, back, armpits, and groin to reduce body temperature.

**Hypoglycemia**—Abnormally low blood sugar. A diabetic health problem (the patient has taken insulin and not eaten or the patient has not taken insulin as required) whose major symptom is an altered mental state (combativeness, hostility, agitation) due to the brain's need for glucose. Other signs include rapid heart rate; fruity odor on the breath; cold, pale, and moist skin; and dilated pupils. Treat for shock. Additionally, maintain and monitor an open airway and administer supplemental oxygen, when required. Check vital signs and provide sugar or glucose solution orally to conscious patients.

**Hyperglycemia (high blood sugar)**—Occurs when there is enough glucose in the body but not enough insulin being produced in the pancreas. This condition is encountered less commonly than hypoglycemia. Symptoms include rapid breathing, weak/rapid pulse, and intense thirst. Administer fluids if the person is conscious. Treat for shock and continually monitor the victim's vital signs.

**Hypothermia**—Is low body temperature, generally below 95°F (normal adult body temperature is 98.6°F). Hypothermia is commonly caused when victims are exposed to cold weather, snow, or cold water for long periods of time. Indications include cold skin temperature, decreased mental and motor functioning, and muscle stiffness. Treatment should include removal from the cold environment, taking off wet clothing, and maintaining body heat.

**Impaled wound**—Impaled objects include knives, sticks, metal bars, arrows, and fencing. Expose the object to perform a medical examination. Do not remove the object from the wound unless it is impairing the victim's breathing. Control bleeding and secure the object with bulky dressings so it does not move.

**Nosebleed**—Have the patient sit down and lean forward to keep blood from flowing back into the throat. Pinch the nostrils together.

**Open wounds**—Treatment for abrasions (scrapes), lacerations (cuts), puncture (stab) wounds, and avulsions (skin tears) involve controlling the bleeding and applying a sterile dressing and bandage.

**Poisoning**—Common causes include medicines, carbon monoxide, household cleaning fluids, paints, insecticides, illegal drugs, chemicals, and food. Some symptoms are headache, weakness, dizziness, confusion, profuse sweating, chemical-smelling breath, burns around the mouth, unusual odors, nausea, diarrhea, eye irritation, muscle twitching/convulsions, difficulty in breathing, rapid pulse, pinpoint pupils, and unconsciousness. Treatment may entail removing the victim to a clean air environment; flushing with water; inducing vomiting; monitoring airway, breathing, and circulation; checking vital signs; administering rescue breathing or CPR; and contacting the local Poison Control Center for advice.

**Seizures**—May involve twitching, jerking, and violent contracting of the muscles (convulsions). Seizures can have a variety of causes, including a medical condition, fever, infection, poisoning, and head injury. Protect the patient from objects in the immediate area that can cause injury. Vomiting may occur and the first responder must therefore monitor and ensure an open airway. Provide oxygen to the patient as necessary. Seizures may be for a short (several minutes) or long duration. Do not restrain the victim or put anything in the person's mouth, which could cause further injury. Prolonged seizures can be life threatening.

**Shock**—A condition that is a result of inadequate blood circulation in the body and thereby a reduction of oxygen to living cells. It is indicated by anxiety and restlessness in the patient. The patient's skin may be pale, cool, and moist. The patient will also exhibit an increased pulse and respiratory rate. Place the individual in a supine position with legs slightly elevated to facilitate blood flow from the lower extremities back to the heart. Loosen restrictive clothing. Keep the victim immobile and cover with a blanket to keep warm. Do not administer fluids. Check and maintain the **ABCs** of patient care.

**Smoke inhalation**—The most common cause of death in fires. Symptoms include burning of the eyes, nose, and throat and difficulty breathing. Treatment entails removing the victim to a clean and uncontaminated environment. Follow the ABCs for patient care and treat for shock. Begin CPR, if needed.

**Spinal column injury**—Can result in permanent paralysis if not treated correctly. This injury is usually associated with trauma. Signs and symptoms include confused mental status, loss of feeling and movement in the extremities, "pins and needles" sensation in extremities, neck/back pain, and loss of bladder or bowel control. The victim should be fully immobilized utilizing a cervical collar and long backboard prior to transport. Attention should be given to maintaining an open airway.

**Sprains (damaged ligament) and strains (damaged muscle)**—Apply cold or an ice pack to the injured area to control swelling and pain. Splint the injury if necessary.

**Stroke**—Caused by the disruption of blood to the brain. Symptoms include partial face drooping or paralysis, slurred speech, dilated pupils, nausea/vomiting, and blurred vision. Treatment entails monitoring and maintaining an open airway, rescue breathing, and CPR if required.

## REVIEW QUESTIONS

*Circle the letter of your choice.*

1. An injured auto accident victim who is displaying signs and symptoms of shock would be assessed according to CUPS criteria by first responders as

   (A) stable
   (B) unstable
   (C) potentially unstable
   (D) critical

2. The **normal anatomical position** for a person is most accurately described in which of the following statements?

   (A) Sitting up facing you with arms at the side and palms facing you
   (B) Prone with arms at the side and palms facing away from you
   (C) Supine with arms crossed and palms facing you
   (D) Standing upright facing you with arms at the side and palms facing you

**Directions:** Match the body directional term in column A with its correct corresponding meaning in column B. Write the letter of your choice in the space provided.

| Column A | Column B |
|---|---|
| ____ **3.** Lateral | (A) Toward the midline of the body |
| ____ **4.** Anterior | (B) To the left side of the patient's body |
| ____ **5.** Medial | (C) Away from the trunk of the body |
| ____ **6.** Proximal | (D) Toward the feet |
| ____ **7.** Superior | (E) Toward the front of the body |
| ____ **8.** Left | (F) Away from the midline of the body |
| ____ **9.** Inferior | (G) Toward the head |
| ____**10.** Right | (H) Toward the back of the body |
| ____**11.** Distal | (I) To the right side of the patient's body |
| ____**12.** Posterior | (J) Toward the trunk of the body |

13. An injury sustained from a violent impact or sudden force is called

   (A) a medical condition
   (B) trauma
   (C) depression
   (D) initial patient assessment

**14.** A first responder should check a patient's vital signs to help monitor the individual's health condition and discover abnormalities. All but which of the following is included in the evaluation of a victim's vital signs?

(A) Pulse rate
(B) Respiration rate
(C) Blood pressure
(D) CPR

**15.** The normal rate of breathing for adults is

(A) between 12 and 20 breaths per minute
(B) between 20 and 30 breaths per minute
(C) between 30 and 40 breaths per minute
(D) None of the above

**16.** An infant's normal rate of breathing is

(A) between 8 and 12 breaths per minute
(B) between 12 and 20 breaths per minute
(C) between 20 and 30 breaths per minute
(D) between 30 and 40 breaths per minute

**17.** The pulse is an indicator of the functioning of the victim's

(A) central nervous system
(B) respiratory system
(C) body temperature
(D) circulatory system

**18.** The normal pulse range for adults and adolescents is approximately

(A) 120 to 160 beats per minute
(B) 100 to 130 beats per minute
(C) 60 to 100 beats per minute
(D) 30 to 60 beats per minute

**19.** The artery commonly used by first responders to check the pulse rate of a responsive infant is the

(A) radial artery
(B) brachial artery
(C) femoral artery
(D) carotid artery

**20.** The carotid artery is normally used to check the pulse rate of a (an)

(A) unresponsive adult or child
(B) responsive infant
(C) responsive adult or child
(D) unresponsive infant

**21.** The pulse rate of a victim is commonly determined by first responders in which manner listed below?

(A) Counting the number of beats for three minutes and dividing by two
(B) Counting the number of beats for one minute and multiplying by four
(C) Counting the number of beats for half a minute and multiplying by two
(D) Counting the number of beats for two minutes and dividing by four

**22.** A bounding (thumping) pulse would most likely indicate the patient is suffering from

(A) high blood pressure
(B) cardiac arrest
(C) shock
(D) None of the above

**23.** What procedure is usually performed by the first responder to assess a patient's general body temperature?

(A) Inserting a finger into the victim's mouth under the tongue
(B) Grabbing the wrist of the victim
(C) Examining the pupils of the patient
(D) Placing the back of the hand on the patient's forehead

**24.** Cold and clammy skin with profuse sweating can be indicative of what condition listed below?

(A) Heat cramps
(B) Heat exhaustion
(C) Heat stroke
(D) None of the above

**25.** Nonreactive pupils that do not react to a beam of light from the penlight of the first responder directed into the eye of the victim are indicative of all of the following with the exception of

(A) damage to the eye
(B) stroke
(C) head injury
(D) the victim is wearing contact lenses

**26.** Blood pressure is the measure of the force the blood exerts on the walls of the arteries. It provides a functional status of

(A) the lungs
(B) the circulatory system
(C) the central nervous system
(D) the musculoskeletal system

**27.** A victim's blood pressure is commonly recorded using a

(A) bag-valve-mask device
(B) watch
(C) stethoscope
(D) none of the above

**28.** The proper position for a blood pressure cuff when taking a reading on a patient is

(A) around the brachial artery
(B) around the radial artery
(C) around the carotid artery
(D) around the ankle

**29.** Blood pressure is denoted by two numbers in fractional form. Select the correct answer that accurately denotes what these two numbers are called and what they represent.

(A) The systolic number is the denominator and represents the pressure in the blood vessels when the heart is relaxed.
(B) The diastolic number is the numerator and represents the pressure when the heart is contracting and pumping blood through the blood vessels.
(C) The systolic number is the numerator and represents the pressure when the heart is contracting and pumping blood through the blood vessels.
(D) None of the above

**30.** Normal blood pressure for adults is generally within what range listed below?

(A) 150–170/120
(B) 150/200
(C) 80–130/40–60
(D) 100–150/60–90

**31.** High blood pressure is associated with all but which of the following?

(A) Hypotension
(B) Cardiovascular disease
(C) Kidney malfunction
(D) Drug use

**32.** Select the choice that is not considered to be a common physical way to assess patient responsiveness.

(A) Pinching the muscles of the neck
(B) Rubbing the breast bone
(C) Pulling the hair
(D) Tapping the chest/shoulders

**33.** The ABCs of patient care represent all of the choices listed below with the exception of

(A) airway
(B) anatomy
(C) breathing
(D) circulation

**34.** To open the airway on a victim with a solid object lodged inside his or her mouth the first responder should utilize what correct procedure on the patient?

(A) Insert a suction unit catheter into the mouth of the victim
(B) Use the head tilt–chin lift technique
(C) Insert an OPA device into the victim's mouth
(D) None of the above

**35.** The medical technique used by first responders to dislodge a foreign object from the throat of a conscious victim is known as a

(A) finger sweep
(B) side straddle
(C) head tilt–chin lift
(D) abdominal thrust

**36.** Firefighters should follow what foreign object removal procedure stated below for an unconscious adult with an airway completely obstructed?

(A) Place the adult face up and apply chest compressions using two fingers.
(B) Position the victim face down and perform abdominal thrusts.
(C) Place the victim face up and perform a finger sweep.
(D) Position the victim face down and give five strong back blows.

**37.** Abdominal thrusts are best delivered in what manner noted below?

(A) Inward and upward with hands just below the victim's navel
(B) Inward and upward with two fingers on top of the navel
(C) Inward and upward with hands just above the victim's navel
(D) None of the above

**38.** Select the procedure that is NOT correct to remove a foreign object from the airway of a conscious infant.

(A) Place the infant face down on your arm.
(B) Deliver five back blows with the heel of one hand.
(C) Shift the infant to a supine position on your arm.
(D) Apply chest compressions with the heel of one hand.

**39.** Which of the following is NOT one of the simple ways of determining a patient's breathing status?

(A) Administer positive pressure ventilation
(B) Listen for sounds of breathing
(C) Look to see if the victim is struggling to breathe
(D) Feel for breathing by placing your hand on an unconscious victim's chest

**40.** A common and relatively safe rescue breathing method used today by firefighters is known as

(A) mouth-to-mouth resuscitation
(B) mouth-to-mask ventilation
(C) mouth-to-stoma ventilation
(D) none of the above

**41.** The rescue breathing ratio for an adult victim is

(A) three breaths every five seconds
(B) one breath every three seconds
(C) one breath every five seconds
(D) none of the above

**42.** A firefighter performing rescue breathing on a child or infant should administer one breath every

(A) two seconds
(B) one second
(C) three seconds
(D) five seconds

**43.** Assessing a patient's circulation includes all but which of the following checks?

(A) Airway
(B) Pulse rate
(C) Bleeding
(D) Skin color

**44.** A patient exhibiting a cyanotic skin color will look

(A) pink
(B) yellow
(C) blue
(D) orange

**45.** A patient without a pulse requires the first responder to start

(A) abdominal thrusts
(B) CPR
(C) finger sweeps
(D) None of the above

**46.** CPR should be administered by first responders on victims of

(A) eye injury
(B) hemorrhaging
(C) cardiac arrest
(D) high blood pressure

**47.** CPR, if performed correctly, will provide the victim with

(A) controlled bleeding and a reduction of pain
(B) oxygen and enhanced blood flow
(C) needed fluids and electrolytes
(D) None of the above

**48.** Without oxygen, brain cells begin to die in

(A) one to two minutes
(B) two to four minutes
(C) four to six minutes
(D) eight to twelve minutes

**49.** While performing CPR on an adult, first responders will normally check the pulse at the

(A) brachial artery
(B) radial artery
(C) femoral artery
(D) carotid artery

**50.** During two-rescuer CPR on an adult victim, the ratio of chest compressions to ventilation is

(A) 30 compressions to 2 ventilations
(B) 2 compressions to 15 ventilations
(C) 15 compressions to 2 ventilations
(D) 1 compression to 5 ventilations

**51.** Adult CPR techniques are performed on patients aged

(A) one year old and up
(B) five years old and up
(C) eight years old and up
(D) 15 years old and up

**52.** Chest compressions administered during CPR, regardless of the age of the victim, are given at the rate of

(A) 100 per minute
(B) 80 per minute
(C) 60 per minute
(D) 40 per minute

**53.** Select the INCORRECT point of information concerning two-rescuer CPR on a child victim.

(A) Child CPR techniques are performed on victims aged one to eight years.
(B) One rescuer gives 15 chest compressions followed by two ventilations given by the second rescuer.
(C) Compressions are performed directly over the sternum using two fingers.
(D) Effective ventilation is indicated when the victim's chest rises.

**54.** An INCORRECT statement concerning infant CPR technique performed by a lone rescuer is

(A) it is performed on babies younger than one year old
(B) CPR can also be performed on infants by two rescuers
(C) the ratio of chest compressions to ventilation is 15 to 2
(D) chest compressions are given utilizing just two fingers

**55.** The benefit of an automated external defibrillator (AED) is best described in

(A) it maintains an open airway
(B) it supplies an electric current to the heart
(C) it provides positive pressure ventilation
(D) None of the above

**56.** Pediatric AEDs are employed by rescuers for attempted defibrillation of children within what age bracket listed below?

(A) Children from ages one to eight
(B) Children from ages eight to 12
(C) Children older than 12 years of age
(D) All of the above

**57.** The pads connected to the AED get attached to

(A) the head of the victim
(B) the hands of the rescuers
(C) the feet of the victim
(D) the bare chest of the victim

**58.** When using the AED, safety dictates that before analysis and the administration of a shock all rescuers

(A) touch the victim
(B) interlock arms
(C) stay clear of the victim
(D) remove metal objects from their pockets

**59.** In general, a victim should not be moved if he or she has or is suspected to have sustained an injury to the neck or spinal column due to trauma for fear of

(A) loss of blood
(B) disorientation
(C) seizures
(D) paralysis

**60.** The device commonly used to immobilize the head and neck of a trauma injury victim is called a (an)

(A) cervical collar
(B) short backboard
(C) immobilization vest
(D) stair chair

**61.** Rolls are used on victims by first responders for all but which of the following reasons?

(A) To move the victim from a prone to a supine position
(B) For medical evaluation
(C) For transport
(D) They can be performed easily by a lone rescuer.

**62.** The end rescuer at the lower legs performing a log roll on a victim grabs what parts of the victim's anatomy?

(A) The arms and shoulders
(B) The feet and pelvis
(C) The trunk and thigh
(D) None of the above

**63.** The lead rescuer during a log roll of a victim with a suspected neck injury should perform what technique during the procedure?

(A) Lifting of the arms and shoulders of the victim
(B) Manual head stabilization
(C) Head tilt–chin lift technique
(D) Rescue breathing

**64.** A drag technique requiring crawling on hands and knees, performed during structural firefighting operations to remove a victim safely from a smoke-filled environment, is called

(A) a shoulder drag
(B) a clothes drag
(C) a firefighter's drag
(D) a duck walk drag

**65.** What area of the body is grabbed during the clothes drag technique of victim removal?

(A) The collar/shoulder area
(B) The pant leg area
(C) The belt/pelvis area
(D) The ankle area

**66.** A drag technique requiring the rescuer to under-hook the upper trunk of the victim while walking backward is known as the

(A) clothes drag
(B) firefighter's drag
(C) shoulder drag
(D) blanket drag

**67.** The carry technique that requires the rescuer to grab the victim's wrist and leg and lift the victim onto the shoulders is known as the

(A) pack strap carry
(B) firefighter carry
(C) piggyback carry
(D) None of the above

**68.** A carry that requires the rescuer to position the victim's arms around his or her neck as the rescuer leans forward while walking to support the victim's weight is called

(A) a firefighter carry
(B) a cradle carry
(C) a pack strap carry
(D) None of the above

**69.** A stretcher used to remove a victim vertically from a confined space is referred to as a

(A) basket stretcher
(B) scoop stretcher
(C) wheeled stretcher
(D) portable stretcher

**70.** A stretcher composed of two pieces of light metal that can be separated and placed on either side of the victim to be moved and then reattached is called a

(A) basket stretcher
(B) portable stretcher
(C) stair chair
(D) scoop stretcher

**71.** A device used to immobilize a victim who is lying down or standing up with a suspected spinal chord injury prior to transport is known as a

(A) stair chair
(B) scoop stretcher
(C) portable stretcher
(D) long backboard

**72.** Choose the item that is NOT an example of personal protective equipment (PPE) for a medical emergency.

(A) Utility knife
(B) HEPA mask
(C) Latex gloves
(D) Gown

**73.** It is essential for first responders to wear PPE during medical emergencies to protect against

(A) head trauma injury
(B) heat exhaustion
(C) infection and disease
(D) burns from direct flame contact

**74.** Which of the following is NOT a proper way to control the flow of blood from a laceration of the leg?

(A) Splint the wound.
(B) Apply direct pressure with a gloved hand.
(C) Place a dressing over the wound.
(D) Bandage the dressing.

**Directions:** Match the burn classification in column A with the correct description found in column B. Write the letter of your choice in the space provided.

| Column A | Column B |
|---|---|
| ____ **75.** First-degree burn | (A) Involves muscle and bone |
| ____ **76.** Second-degree burn | (B) Involves only the top layer of the skin |
| ____ **77.** Third-degree burn | (C) Symptoms include intense pain, pink skin, and blisters |
| ____ **78.** Fourth-degree burn | (D) Known as a full-thickness burn |

**79.** Proper treatment for a second-degree burn is to

(A) scrape off blisters with a knife and apply a dry, sterile dressing
(B) apply petroleum jelly ointment to blisters and don't cover
(C) leave blisters intact and cover with dry, sterile dressing
(D) None of the above

**80.** A standard example used to depict a first-degree burn is

(A) blistering skin from contact with boiling water
(B) sunburn
(C) charring of skin from direct flame contact
(D) All of the above

**81.** The middle layer of the skin is known as the

(A) dermis layer
(B) epidermis layer
(C) subcutaneous layer
(D) muscle and bone layer

**82.** A third-degree burn can demonstrate which of the following skin characteristics?

(A) A charred appearance
(B) A yellow-brown color
(C) A dark red or white color
(D) All of the above

**83.** For burns to the hands or feet, separate the fingers or toes with

(A) splints
(B) bandages
(C) dry, sterile dressing
(D) ice packs

**84.** To determine the percentage of body surface that a burn covers on a victim, first responders use

(A) a stethoscope
(B) a thermometer
(C) the rule of nines
(D) None of the above

**85.** The head of an adult accounts for what percentage of the body when determining the extent of a burn using the rule of nines?

(A) 18%
(B) 9%
(C) 13.5%
(D) 1%

**86.** An adult victim with burns covering the entire chest and abdominal area would be estimated using the rule of nines to have what percentage of the body burned?

(A) 18%
(B) 9%
(C) 27%
(D) 36%

**87.** A child having burned both legs (anterior and posterior) would be estimated by first responders as having what percentage of the body burned using the rule of nines?

(A) 36%
(B) 18%
(C) 9%
(D) 27%

**88.** A child or adult having burns on both arms (anterior and posterior) would be determined by the rule of nines to have what percentage of the body affected?

(A) 9%
(B) 18%
(C) 27%
(D) 36%

**89.** A cerebral vascular accident (stroke) is commonly caused by

(A) thermal burn
(B) nosebleed
(C) disruption of blood to the brain
(D) inability of the pancreas to produce insulin

**90.** All of the following are indicators of a closed fracture with the exception of

(A) an open wound
(B) a deformity
(C) pain with movement
(D) tenderness

**91.** An example of a soft splint is a

(A) wood splint
(B) padded cardboard splint
(C) air splint
(D) All of the above

**92.** Which of the following is NOT a correct treatment procedure for a frostbite victim?

(A) Removing the patient from the cold environment
(B) Taking off wet clothing
(C) Maintaining the body heat of the victim using dry towels and blankets
(D) Rubbing thc affcctcd area to stimulate blood circulation

**93.** What immobilization device is essential to apply on a victim suffering from a traumatic head and neck injury?

(A) A short board
(B) A cervical collar
(C) A rigid splint
(D) A vest-type device

**94.** Which of the following is not indicative of heat stroke?

(A) Inconsistent breathing and pulse patterns
(B) Cold and clammy skin
(C) Dilated pupils
(D) Seizures

**95.** A diabetic health problem caused by abnormally low blood sugar resulting in an altered mental state on the part of the victim is called

(A) hypoglycemia
(B) hyperglycemia
(C) hypothermia
(D) None of the above

**96.** Normal adult body temperature is

(A) below 95 degrees Fahrenheit
(B) 96.5 degrees Fahrenheit
(C) 97.8 degrees Fahrenheit
(D) 98.6 degrees Fahrenheit

**97.** Hypothermia is commonly caused when victims

(A) have too much insulin
(B) ingest too much sugar
(C) are exposed to cold weather
(D) are suffering from a head or neck injury

**98.** An INCORRECT treatment measure to take when encountering a patient suffering a seizure with convulsions is to

(A) protect the patient from objects in the immediate area that can injure him or her
(B) monitor and ensure an open airway
(C) place a tongue depressor in the patient's mouth
(D) provide supplemental oxygen as necessary

**99.** An INCORRECT measure in the administration of medical care to a patient in shock is

(A) place the victim in a prone position with legs slightly elevated
(B) keep the person warm with blankets
(C) check and maintain the ABCs of patient care
(D) provide supplemental oxygen if required

**100.** A spinal column injury patient discovered inside a playground was IMPROPERLY treated by first responders if

(A) a cervical collar was applied to immobilize the head and neck
(B) the patient was transported to the hospital prior to immobilization
(C) attention was given to maintaining an open airway
(D) a long backboard was utilized to immobilize the spine

# Answer Key

1. C
2. D
3. F
4. E
5. A
6. J
7. G
8. B
9. D
10. I
11. C
12. H
13. B
14. D
15. A
16. C
17. D
18. C
19. B
20. A
21. C
22. A
23. D
24. B
25. D
26. B
27. C
28. A
29. C
30. D
31. A
32. C
33. B
34. B
35. D
36. C
37. C
38. D
39. A
40. B
41. C
42. C
43. A
44. C
45. B
46. C
47. B
48. C
49. D
50. A
51. C
52. A
53. C
54. C
55. B
56. A
57. D
58. C
59. D
60. A
61. D
62. B
63. B
64. C
65. A
66. C
67. B
68. C
69. A
70. D
71. D
72. A
73. C
74. A
75. B
76. C
77. D
78. A
79. C
80. B
81. A
82. D
83. C
84. C
85. B
86. A
87. D
88. B
89. C
90. A
91. C
92. D
93. B
94. B
95. A
96. D
97. C
98. C
99. A
100. B

# ANSWER EXPLANATIONS

Following you will find explanations to specific questions that you might find helpful.

**29.** C Blood pressure is denoted by two numbers in fractional form (the systolic blood pressure number over the diastolic blood pressure number). The systolic number is the pressure when the heart is contracting (pumping) blood through the blood vessels. The diastolic number is the pressure in the blood vessels when the heart is relaxed and not pumping blood.

**31.** A Low blood pressure can indicate hypotension and may be the result of shock.

**33.** B The ABCs of patient care represent Airway, Breathing, and Circulation.

**35.** D Abdominal thrust: position yourself behind the victim, place your fist with the thumb-side just above the navel and the second hand atop the first. Apply five quick inward and upward abdominal thrusts.

**38.** D The infant is then shifted to a supine position on the forearm and five rapid, downward chest thrusts are given, just below the nipple line, using just two fingers.

**45.** B Cardiopulmonary resuscitation (CPR)

**46.** C CPR should be administered by first responders on victims of cardiac arrest (lacking a pulse and not breathing).

**53.** C Place two fingers at the sternum (bottom of the rib cage where the lower ribs meet) and then place the heel of your other hand directly on top of your fingers.

**54.** C Infant CPR techniques are performed on babies younger than one year old. It is a one- or two-rescuer operation. A lone rescuer follows the 30:2 compressions to ventilation ratio as performed on adults and children.

**61.** D To perform a log roll to place a backboard under a supine victim when a neck or spine injury is suspected, the lead rescuer is repositioned from the side of the victim to the top of the victim's head. Three other rescuers line up on their knees along the same side of the victim.

**81.** A Skin has three layers: the epidermis, the outermost layer of skin, provides a waterproof barrier and creates our skin tone; the dermis, beneath the epidermis, contains tough connective tissue, hair follicles, and sweat glands; the deeper subcutaneous tissue (hypodermis) is made of fat and connective tissue.

**84.** C To determine the percentage of body surface that the burn covers, use the rule of nines. The rule of nines accounts for approximately 100 percent of the total body surface. The head (9%), chest (9%), abdomen area (9%), upper back (9%), lower back (9%), each arm anterior and posterior (9% × 2 = 18%), each leg anterior (9% × 2 = 18%), and each leg posterior (9% × 2 = 18%) account for 99 percent of an adult's body surface. The groin area (calculated as 1%) raises the total to 100 percent. For children the rule of nines is slightly modified. The head counts for 18 percent. The calculations for the chest (9%), abdomen area (9%), upper back (9%), lower back (9%), and arms anterior and posterior (9% × 2 = 18%) remain the same as for adults. The legs, anterior and posterior, however, account for (13.5% × 2 = 27%). The estimate for the groin area (1%) of a child is similar to adults.

**91.** C Splints can be either rigid (wood, padded cardboard) or soft and flexible (foam, air).

**94.** B Indicators of heat stroke are inconsistent breathing (deep/shallow) and pulse (rapid strong/rapid weak) patterns; dry, hot skin; dilated pupils; and seizures. Signs and symptoms of heat exhaustion include sweating profusely, cold and moist skin, rapid/shallow breathing, and dizziness.

**98.** C Do not place anything in the person's mouth, which could cause further injury.

**99.** A Place the individual in a supine position with legs slightly elevated to facilitate blood flow from the lower extremities back to the heart.

CHAPTER 14

# On-the-Job Scenarios

## Your Goals for This Chapter

- Learn how to answer on-the-job scenario questions
- Test yourself with on-the-job scenario review questions

The questions in this section of the exam put you in the position of a firefighter. Scenarios typically include firefighting operations, firehouse duties, and fire prevention inspections, dealing with fellow firefighters and supervisory officers, and interactions with the public. The questions are designed to assess your judgment, reasoning, common sense, moral character, and problem-solving capabilities. For the most part, prior knowledge of fire department rules and regulations is not required to answer these kinds of questions correctly. Questions may involve factual information and ask for the most accurate answer. Often you may be presented with a problem that requires a common-sense solution. You may be asked to pick the best solution, or you may be asked why a particular solution is best. To choose the correct answer, use deductive and inductive reasoning, logical thinking, validity, and sound judgment.

## GENERAL TIPS ON ANSWERING SCENARIO-TYPE QUESTIONS

### Best Practices to Follow

- Read the question carefully to be sure that you fully understand the scenario presented and information provided.
- Read all four suggested answers and eliminate those that are obviously wrong.

- Reread the question and underline key words such as *same/opposite*; *always/never*; *best/worst*; *most often/least often*; and so on.
- Choose the appropriate answer and test it against the information given in the question.
- Consider the desirable and undesirable consequences of the answer you have selected.

## Answers to Avoid

- Any answer that endangers you and others unnecessarily
- Disobeying an order from a supervisor or superior officer
- Taking action without considering other people's feelings
- Overlooking violations of the law
- Misusing the public trust
- Violating the rules and regulations of the fire department
- Acting outside the scope and authority of your rank and position
- Doing something solely for your own gratification
- Taking no action and avoiding your responsibility as a civil servant
- Acting unlawfully
- Circumventing authority
- Discourteous actions
- "Passing the buck" to your supervisor
- Knowingly conveying wrong information to firefighters and the public

## On-the-Job Scenario (Example 1)

An example of on-the-job scenario questions will have you as a probationary firefighter being told by your company officer that you will be responsible for cleaning and polishing the firehouse poles each day tour that you work. You know it is a dirty job, requiring lots of elbow grease to perform properly. You also recognize that none of the other firefighters in your unit ever clean the poles. What should your thought process be concerning this issue?

### REASONING

Your reasoning in this scenario ought to be that if your company officer gives you an assignment, it should be performed readily and cheerfully without any negative feelings toward the officer and other members of your company. Remember, as a probationary firefighter you are the lowest person on the totem pole. Commonly, you will be sought out by supervisors and senior members to take on such tasks. There are many chores performed in the fire-house that firefighters carry out throughout their career. Some are not very pleasant (making beds, cleaning the bathrooms, mopping floors, reloading apparatus hose beds, polishing sliding poles, laundry, taking out the garbage, etc.). You must be patient—in time as you get some years under your belt, assignments will be more interesting and challenging. Do not make an issue over this assignment even if you feel it is unfair. Every occupation has chores and responsibilities that are undesirable. In the fire service be grateful that you are a firefighter given the opportunity to serve the public. Try to understand that as a new firefighter you should expect to be asked or ordered to do more than other members who have more time on the job than you. It is only fitting that a new firefighter be given extra duties. Doing the bulk of the work in and around the firehouse is part of the initiation process. You should also want to do more than your fair share to demonstrate your desire to fit in. Firefighters before you when they were probationary firefighters were also asked to do more.

## On-the-Job Scenario (Example 2)

Another on-the-job scenario may have you, as a new recruit, once again performing building inspection with your unit at a hardware store. During the implementation of your duties you discover a rear exit blocked by hand trucks, portable ladders, and paint cans. You bring this infraction of the Fire Code to the attention of the owner and are told the rear door is never used by customers. He also tells you that your company has been inspecting the store for many years and never have they given him a violation. Moreover, the owner states that he gives a price break to the firefighters on all his hardware items. What would be your next course of action?

### REASONING

Issue a violation order to the owner, subsequent to informing your officer of the situation. Your officer may be able to assist you in writing the correct information on the form. You have discovered a violation of the law and a potential safety hazard during a fire or emergency. It is important that you explain to the owner that even though the rear exit is not the primary means of egress, it is necessary for you to issue a violation order to clear the area in front of and around the exit. If the owner removes the blockage in your presence, you can rescind the order (with your officer's approval) yet hold it in abeyance to serve as a warning to the owner to keep the doorway accessible. It should be made clear that what has or has not been done previously by your unit does not excuse you from doing your duty as a public servant. No matter what other members of your company may say to the contrary, your actions are correct and in the best interest of the community, which you have sworn to protect when you became a firefighter.

## Sample On-the-Job Scenario Questions

The two sample on-the-job scenario questions next describe situations related to a firefighter's duties and daily experiences. The goal of the questions is to represent a variety of real-life circumstances. The questions will ask you what you would do or what you think about the situation. For these types of questions, there can be more than one correct answer and perspective. Written exam questions that put you in the role of a firefighter, however, should be answered with the viewpoint of a civil servant. Remember, your words and actions to people outside the fire service are extremely important. They have an impact on the entire fire department, either positively or negatively. After reading a scenario, think about the ramifications of using impolite speech or demonstrating rude behavior. Answer the questions in a way that will instill pride in your superiors and your department.

## Question 1

You are a probationary firefighter manning the house watch at the front of the firehouse. Often, peddlers with small goods knock on the door seeking to sell their merchandise. You find this situation disturbing your concentration in listening to the department radio for situational awareness. What is the best action you should take to help alleviate this problem?

(A) Don't answer the door when they knock.
(B) Open the firehouse door and tell the peddler to move along in a loud and angry voice.
(C) Ask your officer if it is all right for you to purchase a sign stating "No Peddlers."
(D) Let the peddler show you all his merchandise and then have him or her leave.

The correct answer is C. A sign would provide information to peddlers that is informative and appropriate. Answer (A) may or may not solve your problem initially, but it will continue to occur for weeks and months to follow. Answer (B) demonstrates rude and inappropriate behavior by the firefighter toward the peddler. Answer (D) will disturb your concentration for quite some time and it will also not solve your problem.

## Question 2

You are a probationary firefighter out for the night with several other firefighters from your company. While ordering a drink at the bar, you accidentally spill some of it on an inebriated young man standing next to you. The person gets angry and calls you an imbecile. He also requests that you both go outside to settle things. What is your best course of action in this situation?

(A) Call him an idiot but don't go outside.
(B) Apologize for the accident and move back to your fellow firefighters.
(C) Take the guy up on his offer to go outside.
(D) Buy the guy a drink and try making friends with him.

The correct answer is (B). The apology is the appropriate action; also the less time you stay with the alcohol-affected patron, the better for everybody's safety. Calling the patron a derogatory name, as suggested in answer (A), will only make matters worse. Going outside the bar to fight, as suggested in answer (C), may lead to serious criminal charges and affect your job status. Buying the guy a drink, as suggested in answer (D), is not what you should do; buying a person a drink who has already had one too many is an unacceptable action. What if he leaves the bar after spending time with you and gets behind the wheel of a vehicle?

## REVIEW QUESTIONS

*Circle the letter of your choice.*

**1.** A firefighter working next to you inside a fire building picking up hose line at the conclusion of operations stumbles and injures her hand. She is unsure how badly she is hurt. What would be your best action in this situation?

(A) Pull off her glove to assess the extent of her injury.
(B) Tell the firefighter to "shake it off" and continue working.
(C) Instruct the firefighter to keep her hand motionless and leave the building.
(D) Suggest to the firefighter that she put down the hose line and perform less strenuous duties.

**2.** School fire drills are required to be held 12 times a year. Six fire drills are conducted during the first three months of the school year. What is the reason for having half the required fire drills conducted during this time frame?

(A) More teachers are available to act as monitors.
(B) More students are attending classes during this time.
(C) The weather is ideal for fire drills during this period.
(D) Students learn fire drill protocols early and often.

**3.** Marine firefighting involves fireboats directly applying salt water on structural fires along wharfs and piers. What is the main advantage to having fireboats use salt water?

(A) It is readily available in abundance.
(B) It can extinguish more fire than fresh water.
(C) It is less corrosive to the contents of the buildings than fresh water.
(D) It flows more readily out of the fireboat's monitor nozzles.

**4.** While inspecting a mall during fire prevention activities, a firefighter discovers a serious fire violation inside a store. The store owner is a former classmate and friend from high school. What should the firefighter do in this situation?

(A) Inform the owner of the violation but take no action.
(B) Leave the store immediately to avoid an embarrassing confrontation.
(C) Issue a stricter penalty than required so as not to be suspected of showing favoritism.
(D) Inform the owner of the violation and issue the appropriate penalty.

**5.** Arson is a crime that can be defined as the intentional and malicious burning of property. Which of the following is an example of arson?

(A) A fire on a patio accidentally started by a barbeque grill
(B) A fire set at the front door of a house to intimidate a neighbor
(C) A fire started by a candle knocked over by the family pet
(D) A fire in a bedroom closet caused by a child carelessly playing with matches

**6.** A firefighter driving to work in the early morning hours notices a storefront doorway ajar and a burglar alarm system ringing. What action should the firefighter take in this situation?

(A) Don't get involved for fear of being late for work.
(B) Assume the owner is inside opening the store for the day.
(C) Notify the police via the nearest public phone or his personal cell phone.
(D) Continue on to the firehouse and inform her supervisor of the matter.

**7.** A firefighter operating a fire department vehicle is hit from behind while stopped in traffic by a civilian vehicle. The firefighter is not injured in the accident. What is the first action the firefighter should take during this event?

(A) Issue a summons.
(B) Admonish the civilian driver but take no summary action.
(C) Call the police.
(D) Determine if anyone in the civilian vehicle is injured.

**8.** Two visitors from a foreign country are outside your quarters and ask you, a firefighter, if they can take a picture of themselves sliding the firehouse pole. What would it be best for you to do?

(A) Allow only one visitor to slide the pole and only one picture to be taken.
(B) Allow both to slide the pole and take pictures since it would show good public relations.
(C) Explain to the visitors that you don't have the authority to allow them to slide the pole.
(D) Tell them that you don't know what they do in their country, but in the United States civilians are not allowed to slide the pole.

**9.** Firefighters take a "no rush" approach when operating during fires in vacant buildings. Why?

(A) Vacant buildings have no monetary value.
(B) The life hazard is generally minimal.
(C) Utilities are usually already shut down inside the building.
(D) Fires are usually small and easily extinguished.

**10.** During large-scale firefighting operations in Fulton County, fire companies in neighboring Elton County are temporarily relocated to the firehouses in Fulton County. What is the major reasoning behind this strategy?

(A) To provide security for the firehouses in Fulton County
(B) To allow Elton County firefighters to become familiar with the Fulton County structures and hydrant system
(C) To have Elton County fire chiefs critique Fulton County firefighting tactics
(D) To provide fire coverage for Fulton County

**11.** What is the best justification for firefighters to work in pairs during firefighting and emergency operations?

(A) It enhances safety.
(B) It reduces duplication of effort.
(C) It improves communication.
(D) It minimizes mistakes.

**12.** Instructing hardware store managers during fire inspection duties to place oil-based paints and varnishes inside metal cabinets will have what type of effect on fire prevention?

(A) A negative effect since the managers may feel imposed on by such action
(B) A positive effect in reducing the number of fires caused by combustible liquid spills
(C) A positive effect in keeping aisle space unobstructed
(D) No effect because oil-based paints and varnishes do not pose a fire hazard

**13.** What is the primary reason for using water to extinguish most fires?

(A) It flows freely through water mains and fire hoses.
(B) It is a relatively heavy and stable liquid.
(C) It is plentiful and has a high cooling capacity.
(D) It is the traditional extinguishing agent used by firefighters.

**14.** What is the main reason the use of elevators during fire operations is not recommended?

(A) They can malfunction during fire conditions.
(B) They are normally being used by building occupants during the fire.
(C) Fires usually start within the elevator shafts.
(D) Firefighters need the exercise provided by climbing stairs.

**15.** You are a firefighter working in the fire department tool repair shop. Department rules state that a tool cannot be accepted for repair without an official form signed by a company officer. A fellow firefighter approaches you with an axe that has a broken handle. The firefighter does not have the required form but states it was filled out and left back at the firehouse. In this scenario, which of the following would it be best for you to do?

(A) Allow the firefighter to submit the tool but insist on the presentation of the form when he or she returns to pick up the repaired tool.
(B) Call the firefighter's firehouse to see if the signed form is, in fact, there, and if so, allow him or her to submit the tool for repair.
(C) Check the firefighter's official identification, and if legitimate, allow him or her to submit the tool for repair.
(D) Refuse to accept the tool for repair until he or she presents the official signed form.

**16.** Firefighters are expected to obey the orders of their superiors. Which of the following situations presents a possible exception to this rule?

(A) When the orders conflict with what you believe to be right
(B) During unforeseen circumstances that endanger your life
(C) When it does not seem to matter whether you perform the action your way or your superior's way
(D) When your superior has made mistakes in the past

**17.** During a very hot summer afternoon, a firefighter observes a group of children down the block from quarters unlawfully opening a fire hydrant. What should the firefighter immediately do?

(A) Scare the children away from the hydrant.
(B) Help the children to safely open the hydrant fully.
(C) Call all members to get onto the apparatus and chase the children home.
(D) Advise the children to stop because they are violating the law.

**18.** In the fire service, traffic violations other than parking violations are commonly referred to as moving violations. What then are moving violations?

(A) Traffic violations
(B) Parking violations
(C) Traffic violations other than parking violations
(D) Parking violations and moving violations

19. When interacting with the public, a firefighter must be congenial, courteous, and tactful. What is the primary reason firefighters should behave in this manner?

(A) To provide the public with superior, professional service
(B) To avoid being disciplined by their superior for inappropriate behavior
(C) To impress on the public the importance of the job of firefighter
(D) To acquire civilian commendations leading to promotion

20. Which of the following is the best policy to attain compliance with unpopular building regulations under the enforcement jurisdiction of the fire department?

(A) Use of selective enforcement
(B) Use of public relations educational techniques
(C) Use of specially trained squads to force compliance
(D) Forgoing warnings and issuing immediate summonses

21. You are standing in front of your firehouse when a civilian confronts you and asks to see the officer in charge. He states that he has important, confidential fire safety information to discuss only with the officer. Which of the following is the most appropriate action for you to take?

(A) Inform the officer of the person's request.
(B) Refuse the request, since the officer cannot be expected to see everyone who wishes to talk to him or her.
(C) Instruct the person to go into the firehouse unescorted and see the officer inside the kitchen.
(D) Request that the person tell you the information for relay to the officer.

22. You are a firefighter giving a lecture on safety to schoolchildren. A child asks you a question about firefighting. You are unsure of the appropriate answer. What should you do in this scenario?

(A) Explain to the children that you will only answer questions pertaining to the subject matter in the lecture.
(B) Ignore the question.
(C) Give an answer you think is accurate.
(D) Admit you are unsure of the answer but state you will get back to them with the correct information.

23. Following a fire in a residential apartment building, the occupant of the dwelling on the floor directly above the apartment where the fire originated complains to you, a firefighter, about damage caused during firefighting operations. He states that there is no fire damage inside his apartment, yet all the windows are broken and the lock on the entrance door has been removed. What is the best answer to give the occupant?

(A) Explain the reasons why firefighters forced entry into his dwelling and ventilated the apartment.
(B) Tell the occupant you are busy and suggest he take his complaints to the chief officer in charge of the fire.
(C) Inform the occupant that his fire insurance policy should cover damages.
(D) Instruct the occupant to put his complaints in writing and address them to the fire commissioner.

24. A firefighter at a suspicious fire is asked by a private insurance investigator for some information concerning the cause of the fire. In this situation the firefighter should take what action?

(A) Tell the insurance investigator everything he wants to know.
(B) Ask the insurance investigator for identification and credentials prior to supplying information concerning the cause of the fire.
(C) Refer the insurance investigator to the officer in command of the fire.
(D) Volunteer your services to help investigate the cause of the fire.

25. Your superior officer orders you to perform an assignment that you do not particularly like doing. What should you do in this situation?

(A) Inform the officer that you are ill to avoid having to do the task.
(B) Pay no attention to the orders given.
(C) Acknowledge the orders but do not carry out the assignment.
(D) Acknowledge and obey the orders.

26. What is the main reason for prohibiting drinking of alcohol inside the firehouse, whether on or off duty?

(A) Some firefighters cannot "hold their liquor."
(B) It can interfere with work duty performance and safety.
(C) Company parties should be held at locations other than the firehouse.
(D) It can lead to boisterous behavior.

27. What is the primary reason why probationary firefighters are not allowed to watch television while working at the firehouse?

(A) They have too much to do and too much to learn.
(B) They must be taught self-sacrifice and discipline.
(C) They are not looked upon favorably by fellow firefighters.
(D) They often tend to break the remote control channel changers.

**28.** You are at the scene of a fire and are ordered by your company officer to go to the roof of the fire building and perform ventilation duties. On your way up to the roof a chief officer stops you and orders you to assist with a hose line being stretched to the top floor. What should you do in this situation?

(A) Perform the task ordered by the chief officer.
(B) Inform the chief of your assigned duties and await his or her reply.
(C) Acknowledge the chief's orders but continue on to the roof.
(D) Ignore the chief officer but inform your officer of the request.

**29.** While entering a residential building to investigate a reported fire involving food atop a stove, you are met in the staircase by an elderly woman who is seriously burned and in need of medical attention. What is the most appropriate action for you to take?

(A) Direct the woman to walk down the stairs and seek emergency medical personnel at the command post.
(B) Ask the woman to take you to the place where she sustained the burns for documentation purposes.
(C) Radio emergency medical personnel for assistance and begin medical treatment immediately.
(D) Assume that the fire atop the stove has been extinguished.

**30.** During fire drills, some schools block one or more exit doors for the duration of the exercise. What is the probable reason for this practice?

(A) It reduces the time it takes to complete the fire drill.
(B) It makes students aware of secondary means of egress.
(C) It frightens the students into acting irrationally.
(D) It lets school administrators see if they can block the exit doors during normal school hours.

# Answer Key

| | | |
|---|---|---|
| **1.** C | **11.** A | **21.** A |
| **2.** D | **12.** B | **22.** D |
| **3.** A | **13.** C | **23.** A |
| **4.** D | **14.** A | **24.** C |
| **5.** B | **15.** D | **25.** D |
| **6.** C | **16.** B | **26.** B |
| **7.** D | **17.** D | **27.** A |
| **8.** C | **18.** C | **28.** B |
| **9.** B | **19.** A | **29.** C |
| **10.** D | **20.** B | **30.** B |

## ANSWER EXPLANATIONS

**1.** Injury to firefighter equates to stopping work, leaving the building, and getting medical attention. Pulling the glove off the firefighter's injured hand could cause more damage to the hand.

**4.** A serious fire violation could create a life-threatening situation at a later time. Regardless of your relationship with the owner, you are duty bound to act on the violation.

**5.** The key words here are "intentional" and "malicious." Answer (B) meets both criteria.

**6.** Indications of a crime must be acted upon immediately.

**7.** Life is the number-one priority. The first action to take is to ensure no one in the civilian vehicle is injured.

**8.** Sliding the pole can be extremely dangerous, even for firefighters who use it every day. Only a superior officer has the authority to allow civilians to slide the pole.

**15.** Fire department rules are promulgated for a reason and should be followed.

**21.** The civilian has confidential and important fire safety information for your officer's ears only. Inform your officer of the civilian's request. Don't, however, allow the civilian to walk into the firehouse without an escort.

**22.** Don't give out incorrect information to the public. If unsure of the answer, admit it, and make a mental note to return with accurate information.

**23.** Smoke can travel throughout a building far from where the fire started and potentially kill or seriously injure occupants remote from the fire. Search and forcible entry, as well as ventilation, are tactics used by firefighters in an attempt to save lives.

**24.** Fire marshals investigate suspicious fires. Providing any information to non–fire department personnel can harm the fire marshal's investigation. Also, in general, firefighters are not given the authority to talk about fires and emergencies to investigators outside the fire department.

**25.** The fire department is a quasi-military organization. Superior officers give orders and they expect subordinates to carry out those orders.

**28.** Communication is a key facet in firefighting operations. If the chief directs you to perform a different duty from what your officer ordered you to do, ensure you let your officer know about it so he or she can have another firefighter cover your assignment.

CHAPTER 15

# Reading Comprehension

## Your Goals for This Chapter

- Learn tips and techniques for improving your reading comprehension
- Practice with sample reading comprehension questions

The reading comprehension section of a civil service exam is designed to test the cognitive ability of the candidate. Several reading passages, under various subject headings that may or may not deal with firefighting and firematics, are used to determine how well you read, reason, remember, think, and process the information given in the reading passages. For the most part, the reading passages are on fire-related topics without getting too technical and may include firefighting procedures, tools and equipment, management theory, firefighting lore stories, life in the firehouse, and technical skills. However, you do not need any prior experience in the fire service to understand the text passages and answer the questions correctly. Other topics, such as public relations, world issues, current events, and what may be considered obscure text are also sometimes included to ensure that it is your reading comprehension that is being tested, not your knowledge of a particular subject. The content is typically taken from academic journals and manuals and contains a great deal of information in a formal compact style. The test is at the level commensurate with the educational requirements listed on the notice of examination.

At the end of each reading passage, there are questions designed to test your thinking ability and how well you can concentrate under pressure. In general, unless the reading material is very short in length (one or two paragraphs), it is not wise to read the questions prior to reading the passage. When there are many questions pertaining to a reading, you will find it difficult and distracting to try to remember all the information being asked about in the questions.

Save your time to read more of the passage deliberately and to review once you have completed reading the passage.

Skilled readers are "active" readers. They don't just read; rather, they interact with the reading material; this leads to greater comprehension. Good readers use their prior knowledge and experience to process the words and sentences, determine their meaning, understand the passage, and possibly foresee what will be stated next. Tools, such as highlighting, underlining, and making marginal notes help in doing this.

Before I provide some general tips and some specific strategies for improving reading comprehension, I would like to remind you that there is an abundance of self-help literature on the Internet and in bookstores that can provide guidance on enhancing reading comprehension skills. I recommend purchasing fundamental reading comprehension soft cover books. Using these books will exercise your mind in cognitive reasoning as it relates to reading passages and answering questions about the subject matter presented.

## GENERAL TIPS ON IMPROVING READING COMPREHENSION

- **Broaden your background knowledge.** Read newspapers, magazines, journals, and books on diverse subjects. Reading the editorial sections in your local newspaper will help you learn about the major issues in your area.
- **Build a strong vocabulary.** Vocabulary has long been recognized as an extremely important component of reading comprehension. The best way to improve your vocabulary is to use a dictionary regularly. Carry around a pocket dictionary and use it to look up new words when you're reading. Buy an address book that is divided alphabetically to keep track of new words you learn in your reading. Recording a sentence in which a new word is used will help you remember its meaning.
- **Preview the passage by reading the first sentence of each paragraph.** Generally, the main topic of each paragraph is contained in the first sentence. Reading the first sentence of each paragraph will give you an overview of the text and a summary of what you are about to read. Since most of the reading passages on the test will be short, previewing the lead sentences will not use up much time.
- **Familiarize yourself with the paragraph structure.** Paragraphs in the text will normally have a beginning, middle, and end. As stated previously, the first sentence usually provides an overview or framework for the rest of the sentences in the paragraph. The middle sentences elaborate on the subject matter of the first sentence and provide details. The last sentence may summarize the information and transition toward a new topic in the following paragraph.
- **Find the main idea of the text.** Think of finding the main idea of a reading passage as a problem-solving task, and approach it strategically. The main idea will direct your focus toward any prior knowledge and experience of the topic you have, aid in a better understanding of the subject matter, and

help you to anticipate what questions will most likely be asked at the end of the reading. Main idea questions ask the candidate to identify the text's overall theme as opposed to supporting and technical information. In these types of questions, answer choices that emphasize factual information can usually be eliminated, as can answer choices that are too narrow or too broad. The answer choice that contains key words and concepts from the main idea presented in the text is usually the correct selection.

## SPECIFIC TECHNIQUES TO IMPROVE READING COMPREHENSION

Several techniques have been shown to be effective in improving reading comprehension.

- **Think aloud.** Studies have shown that thinking aloud improves reading comprehension on exams. Thinking aloud requires you to stop periodically to recognize the strategies being employed in the text and to become aware of how you are processing and understanding the passage. It allows you to relate orally (and quietly to yourself) to what you are doing cognitively. The thinking-aloud process also gives you the opportunity to ask yourself questions regarding the meaning of the text and to rethink the information. This technique helps you to evaluate your own thought processes, a key component of learning that allows you to adjust your strategy, if need be, for greater success.
- **Highlight text and make marginal notes.** Stop at the end of a paragraph and review the most important points in the passage. Try to distinguish between what the main ideas being presented are and what ideas play a supporting role in the text. Then, highlight, underline, or circle key phrases and words. (Do not emphasize entire sentences and don't mark up the text as you go or you will end up with too many markings on the page.) Your highlights and notes act as "memory pegs," or mental pictures, that will aid you in remembering information and identifying what is important. Highlights and marginal notes will also aid you in reformatting the passage according to your own style of thinking and remembering and will help steer you in the right direction when you answer the questions at the end of the text. Review your special notations as necessary when answering the questions. You will find them to be of great value.
- **Review your notes.** When you finish reading the text, briefly review your highlighting and marginal notes. Skim over the first sentence of each paragraph to help you summarize what you have just read. This review process helps you to coordinate the ideas and information in the text into a major theme. The connection of key points provides a stronger understanding of the reading. Then, you are ready to start reading and answering the questions that follow.
- **Apply critical reasoning skills (mini-reading comprehension).** When encountering short text (one or two paragraphs) followed by just a few questions, working backward—that is, reading the questions before reading the passage—may save time. Read the questions carefully, determining what answers you should be looking for before reading the passage.

Then, you may be able to pick out the answers quickly and accurately. Review each sentence in the paragraph(s) individually, noting whether the information being presented answers the questions you read before. Once you complete the reading, go back to the questions and try to answer them without even looking at the answer choices given. This will enhance your focus in finding the correct answer.

- **Employ the SQR3 method—survey, question, read, recite, review.**
  - **Survey.** Look over the title of the text and any other major section headings that may be in the reading. The heading(s) provide clues about the nature of the material and emphasize key areas. Note the length of the passage and the number of questions. Set a time limit for approximately how long it should take you to read the information and answer all the questions.
  - **Question.** After reading each paragraph, briefly ask yourself the following: "What is the main point of the paragraph?" "What bits of information in the paragraph support the main point?" "How does the paragraph relate to the major theme of the text?" Answer each of your questions using the thinking-aloud strategy discussed earlier.
  - **Read.** Actively read the paragraphs, concentrating on the first sentences of each paragraph at the onset. Become familiar with the paragraph structure and the writing style of the author. Organize the material in your own way to help discover the main idea of the text as well as important supporting information.
  - **Recite.** Use the thinking-aloud strategy to help you focus on the plan you have developed to fully understand the reading material. Reflect on the information and verbalize to yourself the answers to the questions you may have concerning key points and facts.
  - **Review.** After reading the passage, and before moving on to the questions, mark the text using your highlighting and note-making strategies. This will help you to format and reconstruct the information with your own style of remembering.

# PRACTICE READING PASSAGES

The following reading passages provide you with a diversified sample of topics. These reading samples are formatted in just one of the many ways in which the reading passages on the actual examination may test your reading ability and cognitive understanding of the printed word. In each question set, circle the letter of each answer you choose.

## Reading Passage 1

### CARBON MONOXIDE, THE INVISIBLE KILLER

Carbon monoxide (CO) gas accounts for more fire-related deaths than any other toxic product of combustion. This gas is colorless, tasteless, and odorless. It is invisible to the naked eye. It is present at every fire. The more inefficient the burning of the fire, the more carbon monoxide gas is generated.

Black smoke is rich in particulate carbon and carbon monoxide because of incomplete combustion. However, large amounts of this gas may be present in areas where no smoke at all is visible!

The human body is affected by carbon monoxide in a number of ways. Normally, hemoglobin in red blood cells combines with oxygen taken in by the lungs from the atmosphere to form oxyhemoglobin, which is carried to body tissues where the oxygen is released. When carbon monoxide is present, it combines with the hemoglobin 200 times more readily than does oxygen. The resulting carboxyhemoglobin significantly reduces the amount of oxygen transported. Additionally, the oxygen that does manage to leave the lungs and become attached to the hemoglobin is greatly inhibited from leaving the red blood cells to go to the body's tissues. These two mechanisms combine to starve the body's tissues of oxygen, the fuel required for them to function and survive.

Concentrations of carbon monoxide in air above five hundredths of one percent (0.05 percent) can be dangerous. Physiological symptoms demonstrated by victims breathing in higher than normal concentrations of carbon monoxide range from headache, nausea, and dizziness at 0.30 percent to 0.60 percent CO in air to immediate unconsciousness and danger of death in one to three minutes at 1.25 percent CO in air.

Measuring carbon monoxide concentrations in the air is not the ideal way to predict possible physiological symptoms. Measuring the concentration of carboxyhemoglobin in the blood is the true indicator because this level is what determines oxygen starvation in the body and the likelihood of physiological symptoms. A victim's overall physical condition, age, degree of physical activity, and length of exposure all affect the development of symptoms.

## REVIEW QUESTIONS

**1.** Which of the following characteristics is NOT true about carbon monoxide?

(A) It is present at every fire.
(B) It is colorless, tasteless, and odorless.
(C) It is invisible to the naked eye.
(D) It is nontoxic.

**2.** According to the passage, the more inefficient combustion is,

(A) the more carbon monoxide is produced
(B) the less carbon monoxide is produced
(C) the less black smoke will be generated
(D) None of the above

3. Fill in the blank: Carbon monoxide combines with hemoglobin in the blood's red blood cells about _______ times more readily than oxygen.

(A) 20
(B) 100
(C) 200
(D) 400

4. Oxyhemoglobin is formed when oxygen taken in by the lungs from the atmosphere combines with

(A) carbon monoxide
(B) hemoglobin in the red blood cells
(C) the body's tissues
(D) none of the above

5. A process mentioned in the article that starves the body's tissues of oxygen is

(A) oxygen taken into the lungs from the atmosphere
(B) oxygen leaving the lungs and attaching to red blood cells
(C) oxygen inhibited from leaving red blood cells
(D) oxygen utilized by the tissues of the body

6. Carboxyhemoglobin is most accurately defined as

(A) the exchange of oxygen between the lungs and the hemoglobin in the red blood cells
(B) the percent of carbon monoxide in the air
(C) the compound resulting from the combining of carbon monoxide with hemoglobin
(D) the percent of oxygen in the air

7. According to the passage, concentrations of carbon monoxide in the air over what percent are considered dangerous?

(A) 0.05 percent
(B) 0.005 percent
(C) 0.01 percent
(D) 0.0005 percent

8. A physiological symptom not normally experienced by people breathing in concentrations of carbon monoxide in the 0.30 percent to 0.60 percent range is

(A) nausea
(B) dizziness
(C) headache
(D) unconsciousness

**9.** The ideal way to predict physiological symptoms in victims of carbon monoxide is by

(A) measuring the carbon monoxide concentration in air
(B) measuring the concentration of carbon monoxide in the blood
(C) measuring the concentration of red blood cells
(D) measuring the concentration of oxygen in the atmosphere

**10.** According to the passage, the likelihood of physiological symptoms from breathing in above-normal levels of carbon monoxide is affected by all of the following variables EXCEPT

(A) overall physical condition
(B) age
(C) degree of physical activity
(D) gender

# Reading Passage 2

## GENESIS OF FIRE

Greek mythology states that the world was without fire until Prometheus resolved to improve mankind's plight. Prometheus was a lesser Greek god and a descendant of the Sun. He was the creator of man. Prometheus carefully crafted man in the image and shape of the gods. Zeus, the ruling Greek god, was against all ideas Prometheus had regarding helping mankind. Zeus took no interest in the mortal race of men on Earth. He intended for them to live as primitives until the day they would all die off. Zeus claimed that knowledge and divine gifts would only bring misery and despair to the mortals. He insisted that Prometheus not interfere with his wishes. Prometheus did not listen to Zeus. He gave the mortals an abundance of gifts, including numbers, healing drugs, the alphabet, how to tell the seasons by the stars, and all art.

As time passed, Prometheus again took pity on human beings shivering inside their caves on cold winter nights. Zeus decreed that mortals must eat their food raw and uncooked. Prometheus could not control his emotions and decided to compound his crime of disobedience. He ventured to steal fire from the gods. Prometheus climbed Mount Olympus and stole the divine fire from the chariot of Apollo. He carried the fire back down the mountain in the stalk of a fennel plant, which burns slowly and thereby was ideal for this task. Prometheus gave mankind the gift of fire and thus mankind was warm.

The gift of divine fire unleashed an abundance of inventiveness, productivity, and respect for the immortal gods by mortals. Soon, culture and literacy permeated Earth. When Zeus heard of the deception, he was outraged. Zeus swore to punish Prometheus for such an act of defiance. He decreed that Prometheus be chained and shackled to a large rock in the desolate Caucasus mountain range for 30,000 years. Each day, Prometheus would be tormented by Zeus's eagle as it tore at his flesh and tried to eat his liver. Each cold night, the torn flesh would heal so the eagle could once again resume its attack at the beginning of the new dawn.

About 30 years into the punishment, Heracles (known as Hercules in Roman mythology) passing by in the course of his eleventh of Twelve Labors (to find the golden apples of Hesperides), freed Prometheus. Although Prometheus was later invited to return to Olympus, he still had to carry with him the rock that was chained to his body.

As the introducer of fire to mankind, Prometheus is revered as the patron of human civilization.

## REVIEW QUESTIONS

**1.** Prometheus can best be described as

(A) the ruling Greek god
(B) a descendant of the Sun
(C) the creator of fire
(D) a mortal

**2.** An accurate statement concerning Zeus is that he

(A) was a lesser Greek god
(B) gave the mortals an abundance of gifts
(C) took no interest in the mortal race of men
(D) feared all mortals living on Earth

**3.** Which gift was given to mankind by Prometheus prior to fire?

(A) Raw meat
(B) Apples
(C) The alphabet
(D) Warmth

**4.** What decree made by Zeus caused Prometheus to steal fire from the gods?

(A) All mortals should eventually die off.
(B) All mortals should eat their food raw and uncooked.
(C) Prometheus should be shackled to a rock.
(D) All mortals should shiver inside of their caves at night.

**5.** Where did Prometheus locate the gift of fire upon Mount Olympus?

(A) In Apollo's chariot
(B) On the stalk of a fennel plant
(C) Adjacent to a large rock
(D) Under an eagle's nest

**6.** The gift of fire unleashed what attributes among the mortals on Earth?

(A) Inventiveness and productivity
(B) Deception and outrage
(C) Kindness and forgiveness
(D) Peace and harmony

**7.** What punishment did Zeus inflict on Prometheus for his flagrant disobedience?

(A) 30 years of hard labor
(B) To be burned at the stake
(C) To roam the Caucasus mountain range for eternity
(D) To be chained to a large rock for 30,000 years

**8.** Each day during the punishment of Prometheus, he would be tormented by

(A) Hercules
(B) Zeus's eagle
(C) Apollo
(D) Hesperides

**9.** How many years of punishment did Prometheus actually serve in the Caucasus Mountain range?

(A) 30,000 years
(B) 12 years
(C) 3,000 years
(D) 30 years

**10.** As the introducer of fire to mankind, Prometheus is known as

(A) the patron of civilization
(B) the giver of great gifts
(C) the greatest of the immortals
(D) the patron saint of firefighters

# Reading Passage 3

## PHILADELPHIA AND COLONIAL FIRE PROTECTION

As a young boy, Benjamin Franklin lived in Boston. The city of Boston had been greatly affected by fire, experiencing conflagrations in 1653, 1676, and 1711. At the age of six, Franklin witnessed the Great Fire of 1711. Franklin moved to Philadelphia at the age of 18 and was eager to make his new home a safer place against the perils of fire.

In 1682, William Penn founded what is now the city of Philadelphia. He had lived in England and seen firsthand the devastation of the Great London Fire of 1666. Penn was determined to reduce the possibility of fire in his new town. He ensured that buildings were adequately separated by wide streets, which would act as fire breaks and prevent conflagrations. Penn also encouraged the use of noncombustible materials (brick and stone) instead of wood to construct buildings and enacted a fire ordinance requiring chimney cleaning, thereby making chimneys less susceptible to fire and fire spread. In 1718, Philadelphia bought its first fire engine, but it was not placed into service until after the most disastrous fire to rage in the town's history destroyed the Fishbourn wharf area in 1730.

After the blaze, the city ordered fire equipment, including leather buckets, fire hooks, ladders, and engines, from England. Franklin wrote about the dangers of fire and the need for organized fire protection in his newspaper, the *Pennsylvania Gazette*. After another conflagration ravaged the city in 1736, Franklin formed a fire brigade known as the Union Fire Company with 30 volunteers. This was the first such organized volunteer fire company in America. Franklin was designated the fire chief. Quickly, the idea of volunteer fire companies gained popularity, and additional brigades were formed not only in Philadelphia but also throughout the Colonies. Some famous Americans who served as volunteer firefighters include George Washington, Thomas Jefferson, Samuel Adams, John Hancock, Paul Revere, Alexander Hamilton, and John Jay.

## REVIEW QUESTIONS

**1.** Where did William Penn originally learn about the devastation of fire?

(A) London
(B) Boston
(C) Philadelphia
(D) New York

**2.** What event caused the city of Philadelphia to order fire equipment from England?

(A) The Boston conflagration of 1711
(B) The Great London Fire of 1666
(C) The Fishbourn wharf fire of 1730
(D) The conflagration of 1736

**3.** When did Philadelphia place its first fire engine into service?

(A) As a result of the conflagration in 1736
(B) After the Fishbourn wharf fire of 1730
(C) In 1736 when Franklin formed the Union Fire Company
(D) In 1718

**4.** According to the passage, Benjamin Franklin did all of the following EXCEPT

(A) writing about the dangers of fire in the *Pennsylvania Gazette*
(B) enacting a fire ordinance requiring chimney cleaning
(C) forming a fire brigade
(D) being designated a fire chief

**5.** What did William Penn do to reduce the possibility of fire in his new town?

(A) He organized the Union Fire Company.
(B) He encouraged the use of wood in building construction.
(C) He had buildings built close together on narrow streets.
(D) He had buildings constructed with noncombustible materials.

**6.** From the information in the passage, approximately what year did Franklin move from Boston to Philadelphia?

(A) 1723
(B) 1726
(C) 1728
(D) 1729

**7.** Subsequent to the 1730 fire that destroyed the Fishbourn wharf area, what article of colonial firefighting equipment listed below was purchased?

(A) Speaking trumpet
(B) Life net
(C) Leather buckets
(D) Fire hose

**8.** The word "conflagration," as used in the passage, means most nearly

(A) a large, major fire
(B) a natural disaster
(C) a storm warning
(D) a volunteer fire company

# Reading Passage 4

## THE SIGNIFICANCE OF THE MALTESE CROSS

The Maltese cross is a symbol that dates back to the time of the Crusades. It is a badge of honor that was first issued to courageous soldiers known as the Knights of St. John, who lived on Malta, a small island in the Mediterranean Sea, and fought gallantly in their war against the Saracens for possession of the Holy Land. A key weapon used by the Saracens was known as Greek fire, made from fuel oil and a secret mixture of chemicals. The Saracens poured this flammable concoction into jars and pots and threw them, like hand grenades, at their enemy. Barrels were also filled with Greek fire and catapulted across great distances.

During the battles, many knights risked their lives in attempts to save their companions in arms from being killed by the flames produced by these weapons. For their heroic efforts, these knights were recognized by fellow Crusaders and awarded a badge of courage—The Maltese cross.

The Maltese cross represents bravery under the threat of fire. The cross is used throughout the fire service as a symbol of protection and bravery. Firefighters who wear or carry this cross are performing the same acts of valor, saving victims from fire, as the brave Knights of St. John did so many years ago.

## REVIEW QUESTIONS

**1.** Which of the following is another appropriate title for this passage?

(A) Greek Fire and Weapons of the Middle Ages
(B) The Saracens
(C) The Crusaders and the Holy Land
(D) The Knights of St. John

**2.** According to the passage, what did the Saracens use as a weapon in their battles against the Knights of St. John?

(A) Greek fire
(B) Crossbows
(C) Hand grenades
(D) Long spears

**3.** The Maltese cross represents all of the following EXCEPT

(A) self-sacrifice
(B) protection
(C) honor
(D) weaponry

4. The origin of the Knights of St. John can be traced back to

   (A) the Holy Land
   (B) Malta
   (C) Greece
   (D) the Saracen Empire

5. The war between the Saracens and the Crusaders was fought for what reason?

   (A) Control of the Mediterranean Sea
   (B) Use of the Maltese cross as a heroic symbol
   (C) Possession of the Holy Land
   (D) Knowledge of the secret ingredients to making Greek fire

# Reading Passage 5

## FIRE MARKS

Fire marks were first used in London by insurance companies in 1667. The Fire Office, which was the first fire insurance company formed, used a phoenix rising out of flames as its fire mark emblem in order to identify policyholders and their properties. To minimize claims, insurance companies also organized their own fire brigades to extinguish fires quickly and reduce damage and loss. In London, insurance company fire brigades only extinguished fires in structures that had the respective insurance company fire mark. This arrangement went on until the late 1880s. In 1886, however, with the cost of compensation becoming too expensive, the government was asked to take over responsibility for protecting life and property from fire. The Metropolitan Fire Brigade was established.

In America, however, organized firefighting existed outside of insurance companies. Insurance companies organized fire patrols to help in salvage operations at a fire, but not to fight fires themselves. The first American insurance company fire mark consisted of four clasped and crossed leaden hands mounted on a wooden shield. A fire mark was more of a promotional item and was not needed to show a property was insured. Volunteer fire companies would attempt to extinguish fires whether or not a property displayed a fire mark. In fact, most American insurance companies did not issue fire marks. One possible advantage of displaying a fire mark was that it might have deterred an arsonist from intentionally destroying property. The fire mark signaled that the owner would be compensated for damages and that law enforcement would likely attempt to find the arsonist.

The use of fire marks spread throughout the country and peaked from 1850 to 1870. Some of the better known fire marks and their known history include:

> **U.F. Fireman**—This 1860 fire mark of the United Fireman's Insurance Co. of Philadelphia depicts one of the first steam-powered fire engines ever used.

**Star**—A six-pointed star fire mark first issued in 1794. It is one of the rarest of the early fire marks.

**F&A Hose**—An original still remains in Independence Hall.

**Sun**—A fire mark of unknown origin. This fire mark indicated the policy number on a small plaque at the bottom.

## REVIEW QUESTIONS

**1.** In what city were fire marks first used?

(A) Philadelphia
(B) New York
(C) London
(D) None of the above

**2.** Why did insurance companies in London organize their own fire brigades?

(A) To minimize claims
(B) To deter arson
(C) To impress future clients
(D) None of the above

**3.** Name the primary reason why the Metropolitan Fire Brigade was established.

(A) Need for expansion
(B) The cost of compensation was too expensive
(C) Government insisted on it
(D) All of the above

**4.** Fire patrols in America were organized to perform what type of work?

(A) Firefighting
(B) Overhaul
(C) Fire prevention
(D) Salvage

**5.** Describe the first American insurance company's fire mark.

(A) A six-pointed star
(B) A fire hydrant and hose
(C) A sun
(D) Four clasped and crossed leaden hands

**6.** In America, why were fire marks a possible deterrent for arson?

(A) The owner would be compensated for damages.
(B) They were made of noncombustible materials.
(C) They were reflective and could be seen in the dark.
(D) None of the above

7. Select from the choices below a year when fire marks in America were most plentiful.
   (A) 1794
   (B) 1860
   (C) 1886
   (D) 1667

## Reading Passage 6

### WOMEN PIONEERS OF THE UNITED STATES FIRE SERVICE

The first recognized female firefighter of the United States was Molly Williams. She was a slave from New York City who was highly regarded by her fellow firefighters during the early 1800s. She was a member of the Oceanus Engine Company No. 11. Firefighter Williams was remembered for pulling the pumper to fires through heavy snow during the blizzard of 1818. Williams could be easily spotted on the fireground since she often wore a calico dress and checked apron!

Marina Betts followed Williams into firefighting. She was a volunteer firefighter in Pittsburgh in the 1820s. Firefighter Betts was known for her hard work and extraordinary dedication to fighting fires. It was said that firefighter Betts never missed an alarm during her 10 years of service. She is additionally remembered for pouring buckets of water over the heads of male bystanders who refused to assist firefighters at fires. Lillie Hitchcock Coit, an honorary member of the Knickerbocker Engine Company No. 5 in San Francisco in 1863, is also considered to be one of the first female firefighters in America. As a teenager, she helped her company haul the engine to a fire on Telegraph Hill.

At the beginning of the next century, there were women's volunteer fire companies located in Maryland and California. During World War II, women served as firefighters in America replacing firemen who were overseas fighting in the war. In fact, during WWII, two fire departments in Illinois were all-female. Additionally, in 1942 the first all-female forest firefighting crew in California was created. The first known female fire chief in the United States was Ruth E. Capello. In 1973, she became fire chief of the Butte Falls fire department in Oregon.

## REVIEW QUESTIONS

**1.** Why is Molly Williams considered a woman pioneer of the U.S. fire service?

(A) She was a slave.
(B) She worked in a snowstorm.
(C) She was the first recognized female firefighter.
(D) None of the above

**2.** Why was firefighter Williams easy to recognize at fire scenes?

(A) She wore a calico dress and checked apron.
(B) She was hardworking and dedicated to her company.
(C) She was highly regarded by her fellow firefighters.
(D) All of the above

**3.** Select the female pioneer firefighter who is known for never missing an alarm during her years of service.

(A) Molly Williams
(B) Marina Betts
(C) Lillie Hitchcock Coit
(D) Ruth E. Capello

**4.** Lillie Hitchcock Coit was an honorary member of what fire company?

(A) Knickerbocker Engine Company No. 5
(B) Butte Falls
(C) Oceanus Engine Company No. 11
(D) California hot shots

**5.** What was the main reason behind the formation of all-female fire departments in the 1940s?

(A) Male bystanders refused to assist firefighters at fires.
(B) Male citizens were not interested in the job.
(C) Males were serving overseas in the armed forces.
(D) All of the above

**6.** Why is Ruth E. Capello considered a woman pioneer of the U.S. Fire Service?

(A) She worked for the Butte Falls fire department in Oregon.
(B) She was the first female firefighter in Oregon.
(C) She was the first female fire chief.
(D) None of the above

# Reading Passage 7

## KNOW YOUR BUILDINGS

It is important that firefighters know the various types of building construction in their community districts since firefighting strategy and tactics are based upon it. Recognizing construction type is key to a successful operation at fire scenes. Some building types burn faster than other types. Collapse potential as a result of fire is also dependent upon building construction type. Determining the type of building construction incorrectly can be dangerous for firefighters. Many firefighters have been seriously injured as a result. Ordering an aggressive interior hose line attack inside a vacant building with combustible load-bearing walls that is fully involved in fire is just one example of how firefighters can get hurt or worse. This reading passage seeks to summarize the five types of building construction for firefighter candidates.

### Type I—Fire-Resistive Buildings

Load-bearing walls (both exterior and interior), columns, beams, girders, trusses, and arches are made of noncombustible materials that have the maximum fire-resistive rating as prescribed by code. They are commonly composed of steel and/or masonry. The roof and floors are also fire-resistive rated although generally not to the degree of the structural members (load-bearing building components) mentioned in the first sentence of this paragraph. Examples of fire-resistive construction include: high-rise office buildings and shopping centers.

### Type II—Noncombustible Buildings

Load-bearing walls (both exterior and interior), columns, beams, girders, trusses, arches, and floors can have the same fire-resistive rating, although not to the degree of fire-resistive buildings. The roof and floors may or may not have a fire-resistive rating. Commonly these buildings have masonry walls and lightweight steel roofs. Examples of noncombustible construction include warehouses and automobile repair shops.

### Type III—Ordinary Buildings

These types of building are commonly referred to as brick and joist buildings. Masonry or brick fire-resistive load-bearing walls enclose interior wooden structural members (columns, beams, girders, trusses, and joists), which may or may not be required to be fire resistive. The roof and floors also may or may not be required to be fire resistive. Examples of ordinary construction include retail stores and apartment buildings.

### Type IV—Heavy Timber Buildings

These types of building, also known as mill buildings, were constructed mainly for factories. Exterior load-bearing walls are fire resistive and made of masonry or brick. Interior columns, beams, girders, and joists are made of large-dimensional lumber to help resist the ravages of fire for long periods of time before failure. In "hybrid" heavy timber buildings, however, cast-iron columns are used in place of wooden columns. The roof and floors may or may not be required to be fire resistive.

### Type V—Wood Frame Buildings

This is the only type of construction to have combustible load-bearing exterior walls. Interior columns, beams, girders, and joists consist of nominal dimensional lumber. The roof and floors, like the structural members mentioned in the first two sentences of this paragraph, may or may not be required to be fire resistive. The one-family home is an example of this type of construction.

## REVIEW QUESTIONS

**1.** Why is it important for firefighters to know about building construction?

(A) To pass promotional exams
(B) To help differentiate between occupied and vacant buildings
(C) To formulate strategy and tactics
(D) To facilitate getting water on a fire

**2.** What can be the consequences of determining the type of building construction inaccurately?

(A) Embarrassment
(B) Serious injury to firefighters
(C) A successful operation
(D) Less manpower requirements

**3.** In the first paragraph of the reading passage, combustible load-bearing walls are mentioned. What type of building construction is most likely being referred to?

(A) Type V
(B) Type IV
(C) Type II
(D) Type I

**4.** What is the primary goal of the reading passage?

(A) To explain in-depth high-rise building construction
(B) To review ways buildings collapse in fires
(C) To validate the need for firefighters to perform building inspections
(D) To summarize the five types of building construction

**5.** What type of buildings are commonly called Type III—ordinary buildings?

(A) Heavy timber buildings
(B) Mill buildings
(C) Brick and joist buildings
(D) Wood frame buildings

**6.** A one-family home is an example of what type of building construction?

(A) Type V
(B) Type IV
(C) Type III
(D) Type II

**7.** What type of building listed below is most likely to have its roof and floors fire-resistive rated?

(A) Warehouse
(B) Apartment building
(C) High-rise office building
(D) Factory

**8.** What does the term "structural member" most likely mean?

(A) Fire-resistive rated
(B) Collapse potential
(C) An interior wall that does not support a load or weight
(D) A load-bearing building component

**9.** In high-rise office buildings, what construction material commonly makes up the exterior walls?

(A) Wood
(B) Steel
(C) Brick
(D) Large-dimensional lumber

**10.** What makes "hybrid" heavy timber construction unique?

(A) Use of large-dimensional lumber
(B) Use of nominal dimensional lumber
(C) Use of steel for the exterior walls
(D) Use of cast-iron columns

**11.** Heavy timber buildings are also called what other kind of buildings?

(A) Mill
(B) Brick and joist
(C) Ordinary
(D) Wood frame

**12.** Select from the choices below what you would consider to be a combustible material.

(A) Masonry
(B) Steel
(C) Wood
(D) Brick

# Answer Key

### Reading Passage 1: Carbon Monoxide, the Invisible Killer

1. D
2. A
3. C
4. B
5. C
6. C
7. A
8. D
9. B
10. D

### Reading Passage 2: Genesis of Fire

1. B
2. C
3. C
4. B
5. A
6. A
7. D
8. B
9. D
10. A

### Reading Passage 3: Philadelphia and Colonial Fire Protection

1. A
2. C
3. B
4. B
5. D
6. A
7. C
8. A

### Reading Passage 4: The Significance of the Maltese Cross

1. D
2. A
3. D
4. B
5. C

### Reading Passage 5: Fire Marks

1. C
2. A
3. B
4. D
5. D
6. A
7. B

### Reading Passage 6: Women Pioneers of the United States Fire Service

1. C
2. A
3. B
4. A
5. C
6. C

### Reading Passage 7: Know Your Buildings

1. C
2. B
3. A
4. D
5. C
6. A
7. C
8. D
9. B
10. D
11. A
12. C

CHAPTER 16

# Deductive and Inductive Reasoning

**Your Goals for This Chapter**

- Understand the thought process behind deductive reasoning
- Review everyday deductive arguments
- Understand the thought process behind inductive reasoning
- Review everyday inductive arguments
- Test your deductive and inductive reasoning skills with review questions

Deductive and inductive reasoning are two ways of arriving at a conclusion. Employers in the fire service are increasingly interested in finding young employees who exhibit good problem-solving skills and show sound judgment when decisions need to be made both in the workplace as well as on the fireground. The deductive and inductive thought processes are used to formulate opinions based upon incomplete information. They allow for analysis of facts and figures in an attempt to solve problems. The difference between deductive and inductive arguments is that a deductive argument's conclusion is a guarantee from its reasoning, while the inductive argument's conclusion is probable from its reasoning.

A deductive and inductive mindset may be helpful in solving problems that involve spatial reasoning or algebraic and number relationships, or to determine if a given argument is valid and identify errors in a given strategy or proof.

## DEDUCTIVE REASONING

Deductive reasoning occurs when a person works from the more general information to the more specific. This is often called the "top-down" approach because the thinking starts at the top with a very broad range of information and works downward to a specific conclusion. It can begin with a theory concerning a point of interest. From there, a statement or prediction about the relationship between two or more variables is drawn. The prediction is then narrowed down further when observations are gathered in order to test the logic behind the thought process. Deductive reasoning is based on a general premise or assumption, and if the premise is true, then the reasoning will be valid. If the initial general premise is not accurate, however, results may be unexpected. Deductive reasoning, however, is logically valid, and it is the fundamental way in which mathematical facts are proven to be true. Researchers also utilize deductive reasoning to develop hypotheses (educated guesses) on how things work in the world we live in.

Deductive reasoning requires the problem solver to use at least two facts or rules in order to draw a valid conclusion based on given information. Deduction rules have two important components. They can be either universal or particular. Universal deduction rules apply to all groups to which they refer. Particular rules apply to only some groups or entities, not all to which they refer. An example of a universal statement about all groups of a given set would be: All horses have four legs. An example of a particular statement that applies to only some of a given group is: Some horses are thoroughbreds. You must differentiate between a universal rule and a particular rule, or you may make mistakes in your conclusions about the information you have read. If the rule is particular, the conclusion using deductive reasoning must always include the caveat: "maybe" or "perhaps."

An example of deductive reasoning can be seen in this set of statements: Every day, I (a firefighter) leave home on my motorcycle to go to the firehouse at 0815 hours. Every day, the drive to the firehouse takes 20 minutes, allowing me to be on time and ready to line up for roll call beginning at 0900 hours. Therefore, if I leave for work from home every day at 0815 hours, I will be on time and ready to line for roll call.

This example of deductive reasoning makes logical sense, yet it cannot be relied upon to be perfectly accurate at all times. What if the motorcycle gets a flat tire on the way to work requiring the firefighter to get the motorcycle towed to a repair shop? The firefighter in this scenario will be late for roll call. This is why a hypothesis based upon deductive reasoning should not be completely relied upon. There is always the possibility for the initial premise to be wrong.

## EXAMPLES OF DEDUCTIVE ARGUMENTS

- In mathematics, if A = B and B = C, then A = C.
- All peaches are fruits, all fruits grow on trees; therefore, all peaches grow on trees.
- Humans are mortal, and I am a human, therefore I am mortal.

- All whales are mammals, all mammals have lungs; therefore all whales have lungs.
- All squares are rectangles, and all rectangles have four sides, so all squares have four sides.
- If John does not go to work and at work there is a party, then John will miss the party.
- All numbers ending in 0 or 5 are divisible by 5. The number 40 ends with a 0, so it is divisible by 5.
- Saturn is a planet, and all planets orbit a sun, therefore Saturn orbits a sun.
- All students who have at least 128 credits will earn a bachelor's degree.
- It is dangerous to drive when streets are wet. It is raining now, so it is dangerous to drive now.
- Lizards are reptiles and reptiles are cold-blooded; therefore, lizards are cold-blooded.
- Acute angles are less than 90 degrees and this angle is 60 degrees, so this angle is acute.
- All noble gases are stable and argon is a noble gas, so argon is stable.

# INDUCTIVE REASONING

Inductive reasoning works the opposite way from deductive reasoning. The thought process moves from specific observations to broader generalities and theories. This is often known as the "bottom-up" approach. A person starts with specific observations and then begins to detect patterns and regularities. Inductive reasoning then formulates some tentative predictions in order to ultimately derive some general conclusions or theories. It will allow the thinker to make an educated guess based upon multiple observations.

Inductive reasoning is a discovery process where the observation of specific cases leads a person to assume strongly (though without absolute logical certainty) that some general principle is true. Although inductive reasoning can lead to false conclusions, it is a good tool for making predictions that may be verified using deductive reasoning.

An example of inductive reasoning can be seen in this set of statements: Today, I (a firefighter) left home on my motorcycle to go to the firehouse at 0815 hours and was on time to line up for roll call at 0900 hours. Therefore, every day that I leave my house to go to work at 0815 hours, I will arrive at the firehouse on time to line up for roll call.

Specific observations often do not correlate to valid general principles, and therefore inductive reasoning is not logically valid. In the example above, firefighters often work weekends and holidays. If the day in question is a Sunday or holiday, the time frame to go from home to work may be misleading. Less traffic could allow the firefighter to get to work sooner than, let's say, on a Monday at the beginning of the workweek. It is illogical to assume a general premise because one specific observation seems to suggest it.

# EXAMPLES OF INDUCTIVE ARGUMENTS

- Joseph leaves for day camp at 0630 hours and is on time. Joseph assumes, then, that he will always be on time if he leaves at 0630 hours.
- Larry is a fire captain. All the fire captains I have known are nice. Therefore, Larry will be nice.
- All observed soccer players are short, so all soccer players are short.
- All firefighters at the fire academy are right-handed, so all firefighters are right-handed.
- All observed cats have fluffy fur, so all cats must have fluffy fur.
- All observed girls like to play with dolls. All girls, therefore, enjoy playing with dolls.
- The water in the ocean has always been approximately 70 degrees in August. It is August. The water will, therefore, be around 70 degrees.
- All observed firefighters are under 40 years old. Tom is a firefighter. Tom is under 40 years old.
- Barry is a football player. All observed football players are taller than 6 feet. Barry is assumed to be more than 6 feet tall.
- All observed wrestlers have a muscular build. Sam wrestles. It is assumed that Sam is muscular.
- All observed houses on Front Street are in poor structural condition. Charles lives on Front Street. The house that Charles lives in is in danger of collapse.

## REVIEW QUESTIONS

1. Use inductive reasoning to find the probable next term in the pattern:
   5, 15, 20, 30, 35, 45, 50, 60, . . .

   (A) 70
   (B) 65
   (C) 75
   (D) 80

2. Using inductive reasoning, what is the probable next term in the pattern below?
   3, 6, 9, 12, 15, . . .

   (A) 16
   (B) 17
   (C) 18
   (D) 19

For questions 3 through 8, indicate whether the statement is an example of deductive or inductive reasoning.

3. Since today is Wednesday, tomorrow will be Thursday.______

4. It has rained every Easter Sunday for six years; therefore, it will rain on Easter Sunday this year.__________

5. A woman observes a dozen roses, all of which are red. She concludes that all roses are red.__________

6. Dan received a B grade on his first three science exams. He concludes he will earn a B on his next science exam.__________

7. If $3x = 12$, then $x = 4$.__________

8. All math teachers are over 5 feet, 8 inches tall. Professor Davis is a math teacher. Therefore, Mr. Davis is over 5 feet, 8 inches tall.__________

Answer questions 9 and 10 using True or False.

9. Conclusions made on inductive reasoning will always be true.__________

10. Deductive reasoning requires the problem solver to use only one fact or rule in order to draw a valid conclusion based on given information. __________

Use inductive reasoning to determine the probable next number in each list for questions 11 through 13.

**11.** Probable next number in sequence: 3, 7, 11, 15, 19, 23, 27, . . .

(A) 28
(B) 29
(C) 30
(D) 31

**12.** Probable next number in sequence: 1, 1, 2, 3, 5, 8, 13, 21, . . .

(A) 28
(B) 30
(C) 34
(D) 38

**13.** Probable next number in sequence: 1, 2, 4, 8, 16, 32, . . .

(A) 64
(B) 58
(C) 56
(D) 54

**14.** Which of the following would make the best major premise for a deductive argument?

(A) No one knows if a comet will strike the Earth.
(B) There are no comets.
(C) Those who believe comets will strike the Earth have overexcited imaginations.
(D) Scientists have proven comets will not strike the Earth.

**15.** Change the following invalid conclusion to make it valid. In Harry's home state you must be 16 years old to get a learner's permit for driving an automobile. Harry's sixteenth birthday is today. Therefore, Harry can now purchase an automobile.

(A) Harry can now drive an automobile.
(B) Harry can now get a driver's license.
(C) Harry can now get a learner's permit.
(D) Harry can now get a chauffeur's license.

**16.** What is the four-digit number in which the first digit is one-third the second, the third is the sum of the first and second, and the last is three times the second?

(A) 1,349
(B) 2,689
(C) 1,358
(D) 3,478

**17.** To every question there is an answer. From this statement, which of the following is not possible?

(A) There is an answer that does not address any question.
(B) If there is a question then it has an answer.
(C) Wayne answered me although I did not ask a question.
(D) The teacher discussed questions that have no answer.

**18.** From the following two statements, what conclusion can be derived?
Only fish oil contains omega 3.
Only foods that contain omega 3 help with brain development.

(A) The only food that helps with brain development is fish oil.
(B) Only what contains omega 3 is fish oil.
(C) There are fish oils that help with brain development.
(D) All fish oils help with brain development.

**19.** How do researchers use deductive reasoning?

(A) To validate their existence
(B) To develop hypotheses
(C) To create new ideas
(D) To control experiments

**20.** Use inductive reasoning to make an educated guess in drawing the next figure in the pattern going from left to right.

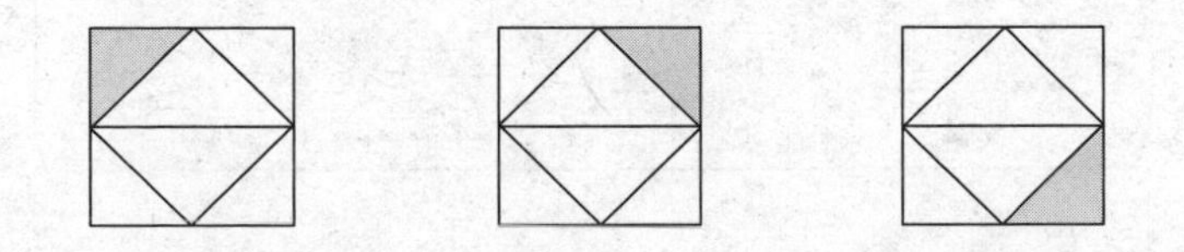

**21.** Draw the next set of dots on the right-hand side using inductive reasoning.

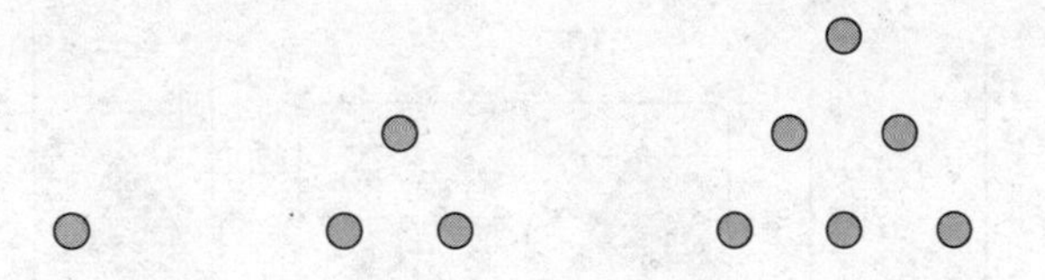

For questions 22 through 25, indicate whether the statement is an example of deductive or inductive reasoning.

**22.** I have never met a poodle with a nasty temperament. I bet there aren't any in existence._________

**23.** Since some apples are red, and all apples are fruit, some fruit is red.__________

**24.** Obama will make a great president. After all, he was a great state senator._________

**25.** It was the hot dog that made my stomach upset. What else could it be? I was not ill prior to eating it.__________

A combination of deductive and inductive reasoning is required to solve questions 26 through 30. Use your knowledge of logical thinking to determine the proper sequence of figures. Input the correct figure at the question mark (?) designation.

**26.**

**27.**

**28.**

**29.**

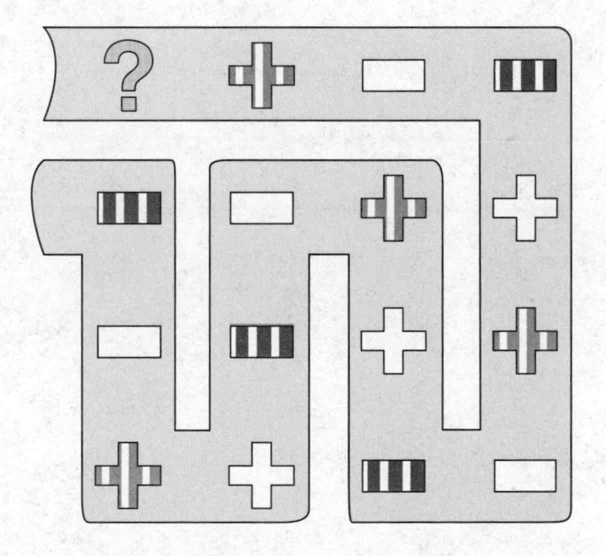

30.

# Answer Key

1. B. It appears the numbers are sequentially increasing by 5 and 10.
2. C. It appears the numbers are increasing sequentially by 3.
3. Deductive
4. Inductive
5. Inductive
6. Inductive
7. Deductive
8. Deductive
9. False
10. False
11. D. It appears each number in the list is obtained by adding 4 to the previous number.
12. C. Beginning with the third number in the list, each number is obtained by adding the two previous numbers in the list.
13. A. It appears here that in order to obtain each number after the first, we must double the previous number.
14. C. It is stated as a generalization.
15. C
16. A
17. D
18. A. Combine the two statements provided in reverse: helps with brain development—contains omega 3—fish oil.
19. B
20. 

21. 

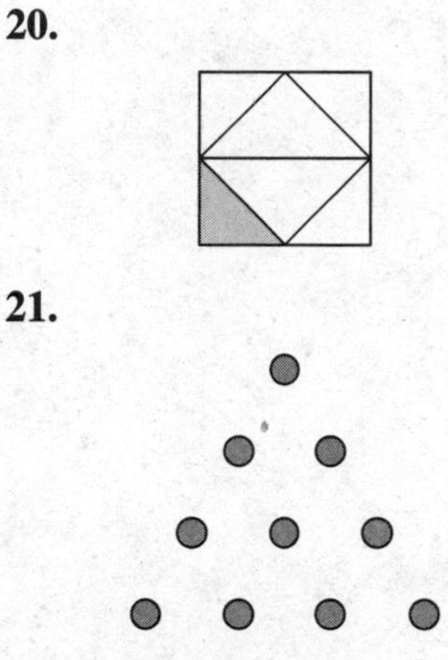

22. Inductive
23. Deductive
24. Inductive
25. Inductive
26. 

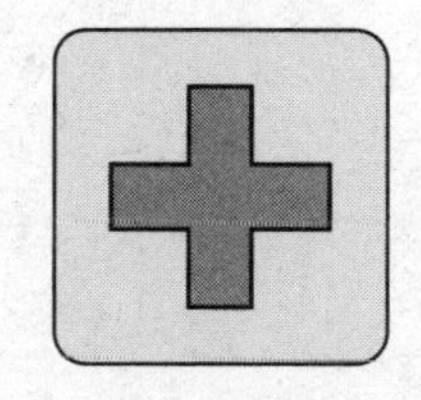

27. 

28. 

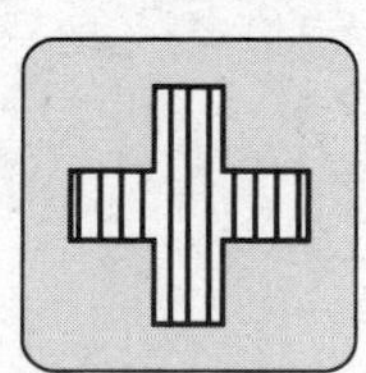

**29.**

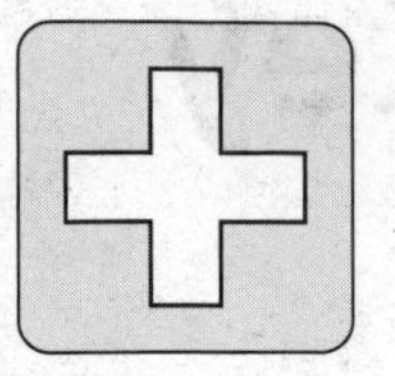

**30.**

# ANSWER EXPLANATIONS

**3.** Deductive reasoning is based on a general premise or assumption, and if the premise is true, then the reasoning will be valid. If the initial general premise is not accurate, however, results may be unexpected. This is often called the "top-down" approach because the thinking starts at the top with a very broad range of information and works downward to a specific conclusion.

**4.** Inductive reasoning works the opposite way from deductive reasoning. The thought process moves from specific observations to broader generalities and theories. This is often known as the "bottom-up" approach.

**9.** Inductive reasoning can lead to false conclusions. Specific observations often do not correlate to valid general principles, and therefore inductive reasoning is not logically valid.

**10.** Deductive reasoning requires the problem solver to use at least two facts or rules in order to draw a valid conclusion based on given information.

**19.** Hypotheses are suppositions or proposed explanations made on the basis of limited evidence as a starting point for further investigation.

**20.** Darkened corner moves clockwise as you look at the given figures from left to right.

**21.** Dots increase at the base (1-2-3) as you review the given figures from left to right. Dot rows decrease from the base by one as you move upwards from the base.

**26.** The symbols run in pairs as they are drawn inside the darkened pathway.

**27.** Follow the symbol pattern found in the third vertical row as you view the drawing from left to right.

**28.** The symbols alternate from arrow to cross throughout the drawing.

**29.** The symbols (crosses and rectangles) are drawn in pairs with one of the paired symbols being white while the other paired symbol is partially darkened in a vertical configuration.

**30.** The symbols (stars, crosses, and triangles) are drawn in pairs with one of the paired symbols being white while the other paired symbol is black.

CHAPTER 17

# Problem Sensitivity, Information Ordering, and Written Expression

## PROBLEM SENSITIVITY

Problem sensitivity is the ability to recognize when something is wrong or is likely to go wrong. It does not necessarily involve solving a problem—only recognizing there is a problem. Questions dealing with problem sensitivity challenge the candidate to identify the whole problem, as well as the parts that make up the whole. An example of this ability might be a fire officer determining that additional police presence is needed for traffic control at an automobile fire on an express highway. A common question consists of particulars of a fire provided by witnesses or victims. The information, however, may be contrary to that supplied by other witnesses and victims.

A second type of problem sensitivity question often starts with the presentation of rules, regulations, procedures, or recommended practices. This information is followed by an incident description in which the given rules should be applied. Based on the situation described in the question, you will be tasked with identifying a problem or the most serious of several problems. Identifying key information in the narrative provided will assist in answering these questions correctly.

## INFORMATION ORDERING

Information ordering is the ability to arrange ideas or actions in a certain order or pattern according to a specific regulation or set of rules. When correctly answering these types of questions you should evaluate how actions should be performed in a particular order. For example, in a training exercise involving stretching an attack hose line, the candidate is initially given a set of instructions concerning what should be done in sequential order (first, second, third, and so on). By accurately following the presented standard procedures, the hose line will be successfully placed into proper position and ready for advancement onto the fire. This ability is also employed when determining what information must be considered before making a decision or taking a particular action. For example, at a structural fire, if positioning a fire apparatus on the street where the fire is located will block out other essential vehicles, consideration should be given to alternative placement before taking any action.

## WRITTEN EXPRESSION

Written expression is the ability to use English words or sentences in writing so that it is easily understandable to others. It includes vocabulary, spelling, understanding of differences among words, and knowledge of grammar and the way words are ordered. Written expression questions require that you recognize the best way to communicate a particular thought or idea to another individual. Alternative language chosen should accurately represent the meaning of the original idea in a clear and concise manner.

Another type of written expression question requires you to arrange your thoughts in a logical order so that others will understand you. Such questions commonly start with a list of statements to be made by a fire officer or firefighter in writing. The statements may be descriptions of various actions or events that occurred in sequential order at a fire scene, for example. They may not be presented, however, in correct order.

Response alternatives will be presented in the question with several possible ways to arrange the statements. When working with this type of written expression question, examine the content of each statement. Determine whether a statement can stand alone; if so, seek out statements that must come before or after. Ask yourself the questions: what happened first and what happened next? Do not, however, try to determine the correct order of all of the statements before viewing the alternatives provided since there may be several logical ways to arrange the statements with only one way presented as the correct answer.

## PROBLEM SENSITIVITY REVIEW QUESTIONS

1. Fire Marshal Jones interviewed four witnesses to an arson fire. Each of the witnesses viewed a suspicious-looking person fleeing a burning vacant building in the early evening hours. The suspect was described as follows:

   Witness 1—white male, 6 feet tall and approximately 200 pounds, wearing a white, long-sleeve shirt and black pants

Witness 2—white or Hispanic male, 5 feet 10 inches tall, 190 pounds, wearing a white tank top, gold bracelet on his right arm and dark pants
Witness 3—Hispanic male, 6 feet 1 inch tall, 200 pounds wearing a light-colored, winter jacket and grey pants
Witness 4—white male, 5 feet 11 inches tall, 185 pounds wearing a tan long-sleeve shirt and brown pants

**Given the eyewitness information, the fire marshal should know that there is a major problem with the description provided by witness:**

(A) 1
(B) 2
(C) 3
(D) 4

**2.** Standard operating procedures (SOPs) for both the fire department and police department during bomb threats within buildings are as follows:

1 – Police are in charge of overall operations and decision making
2 – No radio transmissions are to be made at the scene
3 – Only the ranking police chief may address the media
4 – Evacuation can be performed by building management under orders from the police chief
5 – Chance of a secondary device is always possible

**Based upon the SOPs listed, what action could have the most serious life-and-death consequences?**

(A) On arrival, the fire lieutenant orders members to stretch a hose line into the building where an explosion occurred
(B) The police captain speaks to the press concerning the incident
(C) The fire chief receives radio transmissions from the operations center
(D) The police chief orders evacuation of the damaged building

**3.** Use the information given here to answer the question accurately. First responders with police power should follow department guidelines presented next when dealing with individuals suspected of driving while under the influence of alcohol (DUI).

1. A separate citation should be issued for any traffic offense that originally brought the driver to the attention of the first responder.
2. If the driver submits to a blood alcohol test and scores above the legal limit, the first responder should order the driver to surrender his or her license and issue a citation for DUI.
3. If the driver scores below the legal limit on the blood alcohol test and passes field sobriety test(s), the license should not be confiscated.
4. If the driver scores below the legal limit on the blood alcohol test, the driver can still be charged with DUI if the first responder can justify the charge through the use of field sobriety testing. In this situation, the license should be confiscated.

## Scenario

A first responder with police power views a vehicle swerving in and out of traffic on a busy roadway. The car is stopped and the driver is asked by the first responder to submit to a blood alcohol test for DUI. The driver takes the test and scores below the legal limit. Subsequent to the blood alcohol test two field sobriety tests are given. The driver fails both field tests. Based upon the results of the field tests, the first responder charges the driver with DUI and orders the driver to surrender his or her license.

**What first responder action was according to department guidelines at this incident?**

(A) Driver was issued just one citation
(B) Driver's license shouldn't be confiscated
(C) Driver's license should be confiscated
(D) If the driver passes the blood alcohol test, he or she cannot be charged with DUI

**4.** According to fire department rules and regulations, firefighters are not authorized to speak to the media regarding firefighter death and serious injury in the line of duty without the permission of the officer in charge.

## Scenario

Firefighter Jackson sustains a life-threatening injury at a three-alarm structural fire located in the Bronx. Upon return to the firehouse, Firefighter Norton receives a call from the brother of Firefighter Jackson, asking to speak to him over the phone. Firefighter Norton informs Jackson's brother that he is in the hospital with a significant head injury he received while operating at a major fire.

**Firefighter Norton's action was:**

(A) Improper, this type of notification requires the permission of the officer in charge
(B) Improper, this type of notification should only be made face to face
(C) Proper, because Jackson's brother asked him for information
(D) Proper, because next of kin should be notified of serious injury without delay

**5.** Firefighters use metal portable ladders for window access, window egress, ventilation, and search and rescue.

## Scenario

At a fire scene, a woman is observed yelling for help out a second-floor window. A heavy rain is falling and tree branches are covering the upper pane of the window. In addition, live electrical wires are nearby and might interfere with the raising of the ladder.

**In this scenario, the greatest drawback to the use of metal ladders can be found in which choice?**

(A) The top pane of the window is closed
(B) A wet metal ladder is difficult to climb
(C) Tree branches are in the vicinity of the window to be entered
(D) Live overhead electrical wires are close to the tip of the ladder

**6.** A fire company is inspecting a gas station located in their district. The following conditions are observed:

(A) Motorists are allowed to fill up their vehicles without assistance from employees
(B) An electric heater is being used inside the repair garage
(C) There are incorrect types of portable fire extinguishers in the garage
(D) A mechanic is repairing a half-filled gas tank while smoking a cigar

**Of the four conditions, which one is the most dangerous from a fire hazard point of view?**

**7.** The owner of a building where several apartments suffered fire damage complains to you (a firefighter) that many apartment door locks were broken for no apparent reason. The most appropriate action you should take in this case is:

(A) Tell the owner to stop complaining and be thankful that nobody was killed
(B) Explain the reason why apartment doors were forced
(C) Inform the owner to tell his story to the police department
(D) Tell the owner to put all his concerns in a letter to the fire commissioner and mail it to fire department headquarters

**8.** During building inspection in late January, fire protection inspectors visit an unoccupied swim club that is closed for business during the winter season. They observed the following violations to the fire code:

(A) Defective boiler
(B) Locked fire exits
(C) Accumulation of combustible materials and flammable liquids
(D) Obstructed windows

**What violation listed should the fire protection inspectors tell the owner to correct immediately?**

**9.** A fire lieutenant has been observing the behavior of a firefighter assigned to her company. The lieutenant notices that the firefighter is normally cheerful when at work but on occasion, the firefighter demonstrates mood swings and restlessness. The lieutenant assumes that the firefighter is having problems at home. The following week, the firefighter was acting particularly agitated. Upon questioning the firefighter if everything was alright at home, the firefighter snaps at the lieutenant, stating that she should mind her own business. Several hours later during the tour of duty the firefighter apologizes to the lieutenant and says that everything is fine and that he doesn't have any problems.

**Based upon the information, if the firefighter was having a problem it would most likely be:**

(A) Children related
(B) Drug related
(C) Spouse related
(D) Financial

**10.** At a kitchen fire (food on the stove) inside a dwelling unit, a firefighter notices the female occupant with old bruises on her arms as well as a fresh black eye. The female occupant provides information to the firefighter concerning how the fire started. The male occupant, however, does not say a word and leaves the apartment to smoke a cigarette. The firefighter asks the female occupant how she acquired the marks on her arms and face. The woman nervously replies she was injured while trying to extinguish the fire.

**Considering all of the information provided, if the female occupant was having a problem it would most likely be:**

(A) Work-related stress
(B) Nervousness
(C) Domestic violence
(D) Children related

**11.** Fire apparatus must be inspected on both a daily and weekly basis. The two following lists demonstrate what type of checks are needed to be performed and how often?

Daily Inspections

Battery (maintain a charge)

Check engine oil level

Check gas gauge

Check chassis

Weekly Inspections

Ensure nuts and bolts are tight around doors

Maintain the brake air lines and connections to the wheels

Check tire air pressure

## Scenario

Firefighter Milton is assigned to inspect the fire engine, which has not been run for 72 hours. Milton performs a variety of inspections, but his work is cut short when a medical emergency occurs in front of quarters and he must respond.

**The following list reveals the inspections that were not yet performed. Which one is the most critical?**

(A) Did not check the tire pressure
(B) Failed to maintain the brake air lines to the wheels
(C) Failed to check engine oil level
(D) Failed to check the battery for charge

**12.** Fire Marshal Wilson interviewed four witnesses to an automobile fire that started as a result of a collision with a tree. The incident was described by the witnesses in this fashion:

Witness 1—Subsequent to the accident, the driver fled the vehicle. Fire started in the rear, adjacent to the gas tank.
Witness 2—After the collision with the tree, smoke issued from under the hood and suddenly flames appeared from the engine compartment. The driver fled.
Witness 3—The gas tank area was on fire right after the collision. The driver was lucky to escape unharmed.
Witness 4—The fire started near the gas tank in the rear of the vehicle. This was the area where the car made contact with the tree.

**Given the information, the fire marshal should be aware that there is a problem with the description of the accident given by:**

(A) Witness 1
(B) Witness 2
(C) Witness 3
(D) Witness 4

**13.** Emergency medical technicians (EMTs) have strict protocols of what actions to take when they reach the location of a suspicious death prior to the police being on scene. Information gathering should start at once and brought to the attention of police officers upon their arrival. The most important actions are as follows:

1 – Obtain the name of the deceased, if known
2 – Confirm the address/location of the suspected crime scene
3 – Note the apparent cause of death and approximate time of death
4 – Gather names and contact numbers of witnesses
5 – Ensure no potential evidence at the scene is taken
6 – Note the position of the victim's body

## Scenario

Two EMTs arrive at the scene of a deceased person lying in the front lobby of a building. On examination, the death seems suspicious. The EMTs acquire the name of the victim and record the apparent cause of death. The address is confirmed, and the names of witnesses are ascertained along with their contact information. When the police arrive, the valuable information is transferred over to them.

**The one point of information not obtained by the EMTs is:**

(A) Name of the deceased person
(B) Building address
(C) Witness names
(D) Approximate time of death

**14.** Review the protocol actions listed in Question 13 prior to answering this question.

## Scenario

The EMTs, while driving back to their station house, observe a possible homicide located on an unlit roadway. A phone call is made to the local police station, and they provide the desk sergeant with the name and home address of the victim. They also state that the victim apparently died of a gunshot wound to the chest approximately two hours ago. Prior to hanging up, they provide their callback number and urge the police officer to phone them if more information is required. They remain on the scene and await the arrival of the police.

**According to this passage, the EMTs failed to provide what important piece of information to the police?**

(A) Name and home address of the victim
(B) Apparent cause of death
(C) Approximate time of death
(D) Location of the suspected crime scene

**15.** A two-passenger plane crashed into a river near a small airport. Fire trucks from the local town as well as the airport were dispatched. Foam was applied on leaking fuel to inhibit vaporization and prevent ignition. Search of the plane wreckage produced two bodies. Information from the airport was not definitive as to what caused the crash, but there was evidence that the pilot was taking medication for migraine headaches and blackouts.

**Based on the passage, there is a good chance that the cause of the crash was due to:**

(A) Pilot's recent health problems
(B) Mechanical failure
(C) Inclement weather
(D) Leaking fuel

# INFORMATION ORDERING REVIEW QUESTIONS

**1.** Firefighters responding to a medical emergency where the patient has a severe burn injury should perform the following department first aid procedures in the order given:

1 – Check the victim's airway
2 – Prevent contamination of the wound
3 – Assess for shock
4 – Treat for shock, if necessary
5 – Cover the wound with a moist sterile dressing
6 – Transport the victim to nearest hospital

## Scenario

Firefighter Smith is administering first aid to a burn victim. She is following proper department procedure and determines that the patient does not need to be treated for shock.

**Which one of the following actions should Firefighter Smith take next?**

(A) Check the victim's airway
(B) Prevent contamination of the wound
(C) Cover the wound with a moist sterile dressing
(D) Transport the victim to the nearest hospital

**2.** Protective clothing is the first line of defense for firefighters at a structural fire incident. This gear must be donned prior to getting on the fire apparatus in sequential order as shown here:

1 – Pants
2 – Boots
3 – Coat
4 – Gloves
5 – Helmet

**You are a firefighter inside your firehouse preparing to put on your protective gear prior to mounting the fire apparatus. Following department procedures, what piece of clothing should you put on after donning your coat?**

(A) Helmet
(B) Gloves
(C) Pants
(D) Boots

**3.** Pressurized plain water portable fire extinguishers must be refilled and recharged after each use. Firefighters must perform the following actions in the order given:

1 – Release residual air pressure by inverting the extinguisher and squeezing the lever
2 – Remove the top portion (head assembly) of the extinguisher
3 – Fill the extinguisher with 2½ gallons of water (up to the "fill" mark)
4 – Replace head assembly
5 – Use air hose to pressurize extinguisher to 100 psi (check gauge reading)
6 – If gauge reads 100 psi, label extinguisher with date of recharging

**You are a firefighter recharging a portable fire extinguisher after use at a rubbish fire in your district. What should you do next if, after pressurizing the extinguisher, the gauge reads 90 psi?**

(A) Label the extinguisher with the date of recharging
(B) Add more water
(C) Invert the extinguisher
(D) Use the air hose to pressurize the extinguisher to 100 psi

**4.** Pressurized foam portable fire extinguishers also must be refilled and recharged after each use. Firefighters must perform the following actions in the order given:

1 – Release residual air pressure by inverting the extinguisher and squeezing the lever
2 – Remove the top portion (head assembly) of the extinguisher
3 – Fill the extinguisher with 2 gallons of water
4 – Add to the extinguisher 10 ounces of foam concentrate
5 – Slowly add water to reach 2½ gallons (up to the "fill" mark)
6 – Replace head assembly
7 – Use air hose to pressurize extinguisher to 100 psi (check gauge reading)
8 – Invert extinguisher several times to ensure proper mixture of foam and water
9 – Label extinguisher with date of recharging

**Select an accurate statement concerning refilling and recharging both plain water and foam portable fire extinguishers.**

(A) Actions for refilling and recharging are identical
(B) Initial action for plain water and foam extinguishers is different
(C) Foam portable extinguishers hold more water than plain water types
(D) Both extinguishers are pressurized to 100 psi

**5.** Using the guidelines for refilling a portable foam fire extinguisher found in Question 4, what next action should be taken after removing the head assembly?

(A) Fill the extinguisher with 2½ gallons of water
(B) Fill the extinguisher with water up to the "fill" mark
(C) Fill the extinguisher with 2 gallons of water
(D) Add 10 ounces of foam concentrate

**6.** When removing trapped victims from burning buildings, firefighters must follow department procedures in sequential order as noted:

1 – Use interior stairs, if feasible, as your first priority
2 – Lateral movement to a safe area should be your second choice for removal
3 – Fire escapes are a valid tertiary selection
4 – Apparatus and portable ladders positioned to the roof and windows can also be used for rescue

5 – As a last resort, due to the perilous nature of the procedure, the life-saving rope (LSR) should be used from the roof to pluck a victim from a window located below.

# Scenario

A firefighter in a ladder company is attempting to use the interior stairs of a building to remove an injured occupant from a second-floor landing. The stairs, however, are blocked by fire.

**In this situation, based upon department procedures listed earlier, what action should the firefighter take to complete the rescue attempt?**

(A) Cover the occupant's head with your coat and proceed through the fire
(B) Radio to your officer to initiate an LSR rescue operation
(C) Move the occupant laterally to a room not involved in fire
(D) Move vertically with the occupant to the third floor in search for fire escapes

**7.** When reviewing trapped victim removal procedures listed in Question 6, why do you believe the use of the LSR to be the last resort?

(A) Not all buildings have fire escapes
(B) Radio communication is not always easy to understand
(C) Portable ladders are difficult to position
(D) Extremely dangerous for both rescue firefighter and victim

**8.** Firefighters communicate with each other using handie-talkie portable radios. The terms MAYDAY and URGENT are used when critical information (necessary to protect life and serious injury) needs to transmitted.

MAYDAY is used only in the following situations:

1 – Imminent collapse of a building is feared
2 – Structural collapse of a building has occurred
3 – Unconscious firefighter
4 – Missing member
5 – Firefighter trapped
6 – Firefighter lost

URGENT is used only in the following situations:

1 – Firefighter suffers injury that is not immediately life threatening but requires medical attention
2 – Interior attack at a structural fire is discontinued and an exterior fire operation is being instituted
3 – Structural problem indicating the danger of collapse
4 – Fire entering a neighboring structure
5 – Loss of water endangering operating firefighters

**Which of the following situations would dictate a MAYDAY radio transmission?**

(A) Firefighter falls down a flight of stairs and loses consciousness
(B) Crack in exterior wall indicating the danger of collapse
(C) Fire extends from the building of origin to another building
(D) Loss of water at a structural fire

**9.** Reviewing the information provided in Question 8, an URGENT radio message is most appropriate for what type of condition?

(A) Aggressive interior attack is being discontinued and a more cautious exterior fire operation is being established
(B) Structural roof collapse is observed by a firefighter inside a tower ladder basket
(C) Water is leaking out the front brick wall of the fire building and firefighters are afraid the wall may collapse
(D) Company officer becomes aware during search and rescue operations that one of his members is missing

**10.** Firefighters commonly get their water to fight fires from fire hydrants. The following list provides department guidelines that must be performed in sequential order for engine apparatus hydrant positioning, supplying water for the hose line, and stretching the hose line to the fire.

1 – Engine stops next to fire hydrant located in front of the fire building
2 – Firefighters remove enough hose line from the engine hose bed to reach the fire
3 – Attach nozzle to the lead end of the hose line
4 – Advance the hose line to the fire
5 – Use hydrant hose to connect engine to hydrant
6 – Attach the other end of the hose line that was stretched to the fire to the proper engine water outlet

## Scenario

At a fire in a commercial occupancy (clothing store) engine company firefighters arriving first on the scene position their apparatus adjacent to a nearby fire hydrant. They then remove enough hose line from the engine hose bed to reach the fire.

**What action should be performed next?**

(A) Attach the other end of the hose line to the apparatus
(B) Attach the nozzle to the lead end of the hose line
(C) Advance the hose line to the fire
(D) Use hydrant hose to connect engine to the hydrant

**11.** Fire marshals interview witnesses at the scene of a suspicious fire to collect valuable information when attempting to solve a case. The way questions are asked can influence how a witnesses answers. Fire marshals commonly take the following steps during the interview process:

1 – Ask objective questions in chronological order
2 – Identify themselves and state the purpose of the interview
3 – Document the interview in their notebook
4 – Establish a rapport with the witness
5 – Show appreciation to the witness for cooperating

**The most logical order of the steps listed is:**

(A) 4, 1, 2, 5, 3
(B) 2, 1, 3, 4, 5
(C) 4, 2, 1, 3, 5
(D) 2, 4, 1, 5, 3

**12.** As it pertains to Question 11, if a tape recording is made during the conversation with the witness, which of the steps is not necessary?

(A) 4
(B) 3
(C) 1
(D) 2

**13.** It is important that firefighters triage injured victims at a fire. Their first priority should be providing medical care for heart failure, deep wound bleeding, and shock (in that order); second priority injuries include negligible burns and broken bones. Minor cuts, abrasions, and sprains/strains have the lowest priority.

**Use the information found in the paragraph to select the type of injury that should be treated last.**

(A) Shock
(B) Deep bleeding wound
(C) Sprained ankle
(D) Negligible burn

**14.** Determine the priority order in which you would treat injured victims at a fire incident based on the information located in Question 13.

(A) C, B, A, D
(B) B, D, A, C
(C) B, A, D, C
(D) A, B, D, C

**15.** A volunteer fire department has established "out of the box" fireplace fire procedures that are enumerated here. They are listed in the order in which they must be performed.

1 – If the fire has not communicated outside to the room, place a canvas cover in front of the fireplace to protect the flooring
2 – Use a portable water fire extinguisher to douse the flames
3 – Examine adjoining walls for fire spread
4 – Ready the fire extinguisher for further use
5 – If fire is inside the walls, open them up using axes
6 – Use fire extinguisher to put out any fire inside the walls

## Scenario

In a private home, a fireplace fire has spread outside its containment to room contents. Firefighters arrive on scene and are ordered by their officer to follow department procedures very strictly. Upon entering the room where the fireplace is located, they observe a couch on fire. The firefighters begin operating according to their guidelines.

**Which action listed is in agreement with department procedures?**

(A) Place a canvas cover in front of the fireplace
(B) Tear the walls open in search of fire upon entering the room
(C) Use fire extinguisher to put out fire in the walls
(D) Use fire extinguisher to douse the flames

# WRITTEN EXPRESSION REVIEW QUESTIONS

For **Questions 1–8**, fill in the blank with the correct answer provided.

**1.** Are you able to _______ CDs on your computer?

(A) melt
(B) toast
(C) cook
(D) burn

**2.** The steak is _____________. Can I have the recipe?

(A) tender
(B) gentle
(C) mild
(D) soft

**3.** I'm sorry, but you are not supposed __________ photos at this fire academy.

(A) to be taken
(B) to take
(C) taking
(D) to have taken

**4.** I'll do my best to _________ there on time.

(A) have got
(B) be getting
(C) get
(D) getting

**5.** Police have been sent in to try to restore __________ in the area.

(A) regulation
(B) organization
(C) harmony
(D) order

**6.** Did you have any problems __________ our firehouse?

(A) finding
(B) to find
(C) for finding
(D) find

7. The training instructor asked if ________ to bring our school manuals to class.

(A) all we had remembered
(B) we had all remembered
(C) had we all remembered
(D) had all we remembered

8. The discovery of a large number of matchboxes ______ that the fire could have been intentionally set.

(A) indicate
(B) indicates
(C) designates
(D) decided

Select the correctly spelled word in **Questions 9–12**.

9. (A) Foreign
(B) Forein
(C) Forine
(D) Fareign

10. (A) Pessinger
(B) Passinger
(C) Passenger
(D) Passengar

11. (A) Forecast
(B) Forrcast
(C) Forcast
(D) Forcaste

12. (A) Rigarous
(B) Riggerous
(C) Rigorous
(D) Regerous

In **Questions 13–15**, four words are given. Select the word that is spelled incorrectly.

13. (A) Commend
(B) Appraise
(C) Metainance
(D) Evaluate

14. (A) Inflamable
(B) Combustible
(C) Incendiary
(D) Incandescent

15. (A) Economics
(B) Chemisty
(C) Communications
(D) Geometry

In **Questions 16–21**, the four **bold type** words found in (A), (B), (C), and (D) make up a sentence. Words in bold type may be either inappropriate in the context of the sentence or wrongly spelled. Choose the bold type word that is either unsuitable or misspelled as your answer.

16. (A) At a recent fire, members were quiet
(B) **amazed** at the
(C) **turn** of
(D) **events**

17. (A) The officer's decision was based on
(B) **adequate** and
(C) **acurate**
(D) **information**

18. (A) She was polite
(B) but **ferm**
(C) in her **dealings**
(D) with the **public**

19. (A) The faces of the
(B) **twin** firefighters were
(C) so **identical** that the
(D) chief officer could not **differentiate** between them

20. (A) To solve a problem
(B) **requires**
(C) **intelligent** and
(D) **determination**

21. (A) The respiratory tract
(B) is **vulnerable** to injury
(C) and gases **asociated** with fires
(D) must not be **inhaled**

**22.** A firefighter is working at a fire inside a private dwelling that is two stories high. The residential structure also has a cellar floor. The fire originated within the cellar boiler room located to the rear of the building. Fire spread out of the boiler room up the interior stairs and into the first-floor kitchen area.

**While writing the information concerning the fire in her notebook, the best way to describe the incident can be found in which choice?**

(A) Fire started in combustibles located in the cellar and spread to first floor
(B) Fire was caused by a defective boiler and spread to the first floor
(C) Fire started in the kitchen and spread downward into the cellar
(D) Fire started inside the boiler room and spread to the first floor

**23.** A report about a gas leak located inside a building and subsequent explosion that includes the following statements would best be formulated using what order listed?

1 – I placed a splint on the victim's broken arm
2 – Force of the blast knocked down an exterior wall of the building
3 – During medical treatment of the victim, more units were called for assistance
4 – I saw a pedestrian covered in rubble
5 – While procuring the evening meal with my fire company I heard an explosion

(A) 5, 2, 4, 1, 3
(B) 5, 2, 1, 4, 3
(C) 5, 2, 4, 3, 1
(D) 2, 5, 4, 1, 3

**24.** In a given fire department there are more squad companies than rescue companies. To conclude that the number of firefighters assigned to squad companies exceeds the number of firefighters assigned to rescue companies is to make a basic assumption. The most correct statement validating the basic assumption referred to in this question is:

(A) There are four firefighters working on squad companies
(B) A squad company has fewer firefighters than a rescue company
(C) Squad apparatus cost less money than rescue apparatus
(D) Same number of firefighters are assigned to each type company

**25.** There are two types of rope fibers used by a Midwestern fire department: sisal and Manila. A firefighter describing the characteristics of these two fibers in a report used the following statements:

1 – Manila rope fiber is the stronger of the two
2 – Sisal fiber therefore is used more often for making smaller-diameter rope
3 – Manila fiber comes from the Philippines

4 – Sisal fiber comes from Mexico
5 – Rope is an essential piece of equipment in the fire service
6 – Both Manila and sisal have great strength with a minimum of stretch

**Select the proper order the statements should follow in the report.**

(A) 5, 6, 1, 2, 3, 4
(B) 1, 2, 3, 4, 6, 5
(C) 5, 3, 2, 4, 1, 6
(D) 6, 5, 4, 3, 1, 2

In **Questions 26–35** commonly used words, **in bold**, are placed in sentences to enhance your vocabulary as well as improve your writing and conversation skills. Select the choice that most closely denotes its meaning.

**26.** Due to unforeseen circumstances, the fire at 24 Henry Street turned into a **fiasco**.

(A) Failure
(B) Success
(C) Triumph
(D) Victory

**27.** The firefighter will one day **rue** his decision to transfer to another city agency.

(A) Appreciate
(B) Regret
(C) Recall
(D) Forget

**28.** **Copious** amounts of water were used on the fire.

(A) Small
(B) Miniscule
(C) Abundant
(D) Insignificant

**29.** The officer was **fastidious** in training the firefighters of his unit on how to use the safety rope.

(A) Careless
(B) Frivolous
(C) Uninformed
(D) Careful

**30.** The chief instructed the captain of the incoming company to carry out his orders **forthwith**.

(A) Hesitantly
(B) Ambitiously
(C) Immediately
(D) Eagerly

**31.** The firefighter was in a **quandary** over his upcoming promotion.

(A) Dilemma
(B) Panacea
(C) Harbinger
(D) Homage

**32.** All was lost in the **conflagration**.

(A) Tsunami
(B) Earthquake
(C) Hurricane
(D) Fire

**33.** The young cadet was told not to **malign** his instructors.

(A) Follow
(B) Copy
(C) Trust
(D) Slander

**34.** During holidays, the fire department would **augment** their work force.

(A) Increase
(B) Reduce
(C) Redeploy
(D) Arrange

**35.** Smoke in the room did not **dissipate**.

(A) Darken
(B) Disperse
(C) Propagate
(D) Grow

# PROBLEM SENSITIVITY REVIEW QUESTIONS ANSWER KEY AND EXPLANATIONS

**1.** B Witness 2 is the only eyewitness to describe a shirt or jacket that is not long-sleeved.

**2.** A (police decision, potential injury and death of firefighters)

**3.** C
A See guideline 1
B See guideline 3
D See guideline 4

**4.** A

**5.** D

**6.** D
A Not a hazard
B Not a hazard if heater is far enough away from combustibles
C Violation

**7.** B

**8.** C
A No hazard
B No immediate hazard, swim club is closed
D No immediate hazard, swim club is closed

**9.** B Firefighter's erratic behavior are clues he may be using drugs

**10.** C Female occupant does not provide a credible reason for her injuries

**11.** D Battery should be checked every day; fire engine has not been run for 72 hours; if battery is low/dead, fire engine will not be able to respond to a call
A Weekly inspection
B Weekly inspection
C Not as serious a mistake as D

**12.** B There is agreement among Witnesses 1, 3, and 4 regarding the area of the car (gas tank) where the fire started. Witness 2 is the only person denoting the engine compartment as the place of fire origin.

**13.** D

**14.** D

**15.** A

# INFORMATION ORDERING REVIEW QUESTIONS ANSWER KEY AND EXPLANATIONS

**1.** C

**2.** B

**3.** D

**4.** D
A Different
B The same
C Less water

**5.** C

**6.** C
A Not feasible
B Last resort
D Incorrect

**7.** D

**8.** A
B URGENT message
C URGENT message
D URGENT message

**9.** A
B MAYDAY message
C MAYDAY message
D MAYDAY message

**10.** B

**11.** D

**12.** B

**13.** C

**14.** C

**15.** D
Listed procedures are in the order which they must be performed.
A Canvas is only employed if fire has not communicated outside to the room
B Use of portable fire extinguisher precedes opening up walls; also the walls must first be examined prior to any action taken
C Use of portable fire extinguisher should be for dousing visible flames, not extinguishing fire inside the walls at this time

# WRITTEN EXPRESSION REVIEW QUESTIONS ANSWER KEY AND EXPLANATIONS

**1.** D

**2.** A

**3.** B

**4.** C

**5.** D

**6.** A

**7.** B

**8.** B The subject (discovery) is singular; therefore, the verb also needs to be singular.

**9.** A

**10.** C

**11.** A

**12.** C

**13.** C Maintenance

**14.** A Inflammable

**15.** B Chemistry

**16.** A quite

**17.** C accurate

**18.** B firm

**19.** C similar

**20.** C intelligence

**21.** C associated

**22.** D
A No causes are given
B No causes are given
C Fire started in the cellar

**23.** A Statement 5 introduces the incident topic (gas leak and explosion). Statement 2 describes the results of wall collapse leading to civilian injury. Statement 4 deals with the discovery of a victim, and Statement 1 explains the type of treatment administered. Statement 3 denotes action taken after medical treatment of the victim.

**24.** D Validates the assumption, as Choices A and B are both insufficient information to validate the assumption. Choice C does not deal with the number of firefighters.

**25.** A Statement 5 can stand alone. Statement 6 mentions both types of ropes and therefore must come before all other statements.

**26.** A Fiasco can also be a debacle, disaster, or failure.

**27.** B As a verb, rue can also mean to regret, be remorseful, or feel sorry about.

**28.** C Copious means abundant, plentiful, or profuse.

**29.** D Fastidious in this context means careful. However, it can also mean fussy or finicky.

**30.** C Forthwith can mean immediately, without delay, or at once.

**31.** A Quandary is a dilemma, predicament, or a sticky situation.

**32.** D A large in magnitude fire, blaze, inferno, brush fire, or forest fire can all define a conflagration.

**33.** D To malign someone is to criticize, slander, or smear someone in a hurtful way.

**34.** A Augment means to supplement, add to, enlarge, expand, or enhance.

**35.** B To dissipate, something has dispersed, scattered, or dispelled.

CHAPTER 18

# Fire History Chronology

Major fires and natural or manmade disasters resulting in the development and improvement of firefighting departments, equipment, procedures, and standards are summarized below.

| | |
|---|---|
| **300 BC** | A fire service, *Familia Publica*, is established in ancient Rome. It later becomes the Corps of Vigiles, appointed by the Roman emperor. |
| **200 BC** | Ctesibius and Heron, both of Alexandria, invent and improve a portable piston pump used in organized firefighting. |
| **64** | Fire rages in Rome for eight days under Emperor Nero's reign. |
| **871–899** | *Curfew* ordinances, instituted by Alfred the Great, required that fires used for heat and cooking be put out during the night. |
| **1189** | London's first lord mayor issues an ordinance banning thatched roofs. |
| **1212** | The first great London fire. |
| **1631** | The Boston conflagration leads to the first fire ordinance in America being adopted. It prohibited thatched roofs and wooden chimneys. |
| **1648** | Governor Stuyvesant (New Amsterdam) appoints four fire wardens. |
| **1666** | The Great Fire of London burns for five days, destroying more than 13,000 homes. |
| **1667** | Nicholas Barbon establishes the first fire insurance company (Fire Office) in London. |
| **1679** | Fire in Boston leads to the formation of the first paid fire department in North America. |
| **1680** | London fire insurance companies establish fire brigades. |
| **1718** | Mutual fire societies (forerunner of fire insurance companies in the United States) formed in Boston. |

| | |
|---|---|
| **1729** | A Constantinople (Turkey) conflagration kills 7,000 people. The city has experienced more major fires through history than any other in the world. |
| **1736** | The Union Fire Company is founded by Benjamin Franklin. |
| **1743** | Thomas Lote (New York) builds America's first successful pumping engine. |
| **1752** | Philadelphia Contributionship, the first fire insurance company in America, is started. |
| **1803** | Wooden hydrants are installed in Philadelphia. |
| **1809** | The first fireboat in New York is put into service. The era of the American fireboat is born. |
| **1812** | A Moscow fire lasts five days, ravaging nine-tenths of the city (the French-Russian War). |
| **1829** | John Braithwaite and John Ericsson (England) build the first fire engine (the *Novelty*) using steam to pump water. |
| **1835** | The Great New York Fire, in which 530 buildings burn, the largest fire in an English-speaking country since the Great Fire of London in 1666. |
| **1841** | Paul Hodge builds the first steam fire engine in the United States. |
| **1845** | The second great New York fire. |
| **1851** | "Black Thursday" forest fires in Australia damage 50,000 square miles. |
| **1853** | The first successful public demonstration of a steam fire engine ("Uncle John Ross") in Cincinnati; leads to the first salaried fire department in the United States. |
| **1861** | New York City firefighters organize the first Fire Zouaves regiments (the Civil War). |
| **1870** | San Francisco firefighter Daniel Hayes develops the first successful aerial ladder apparatus. |
| **1871** | A Paris conflagration (Commune Revolution); the Great Chicago Fire kills 300 and leaves 100,000 homeless; forest fires (firestorm) destroy the town of Peshtigo, Wisconsin, killing more than 1,100 people. |
| **1896** | The National Fire Protection Association is created. |
| **1904** | The Great Fire of Baltimore leads to the development of National Standard hose threads. |
| **1906** | The San Francisco earthquake and resulting fire kills 3,000 and leaves 300,000 people homeless. |
| **1911** | The first National Fire Prevention Day; Weeks Act initiates federal-state cooperation in forest fire protection. |
| **1925** | The first National Fire Prevention Week. |
| **1939–1945** | Incendiary bombs (World War II) ravage European and Japanese cities. |
| **1945** | The atomic bombing and subsequent conflagrations of Hiroshima and Nagasaki. |
| **1986** | The Chernobyl nuclear reactor accident in the Ukraine. The main casualties are firefighters, whose exposure to radiation causes 28 deaths in the first four months and 19 subsequently. |

**1988** More than 25,000 wildland firefighters operate at 50 forest fires throughout the year in Yellowstone National Park.

**1991** Oil well fires (Kuwait) following the Persian Gulf War; a major wildland-urban interface firestorm in Oakland, California, kills 25 and does an estimated $1.5 billion in damage.

**1992** Los Angeles riots and arson fires cause more than $1 billion in property damage.

**2001** The terrorist attack using hijacked commercial airliners to strike the New York City World Trade Center Twin Towers and the Pentagon. The resulting fires and eventual collapse of the World Trade Center buildings and nearby buildings kill nearly 3,000 people, including 343 firefighters, 23 New York Police Department officers, and 37 Port Authority police officers. In Arlington, the death toll is 125, with major sections of the U.S. Department of Defense building destroyed.

**2004** Indian Ocean earthquake and tsunami kills approximately 230,000 people.

**2005** Hurricane Katrina devastates the Gulf of Mexico, killing more than 1,800 people and causing more than $80 billion in damage.

**2007** California wildfires rage for over two weeks, causing more than $2 billion in property damage.

**2008** Earthquake 7.9 in magnitude kills nearly 70,000 people in central China.

**2009** Bushfires in Victoria, Australia, kill more than 170 people and burn down thousands of homes.

**2010** Russian wildfires span 450 square miles across the western part of the country.

**2011** Japan earthquake causes tsunami leaving more than 15,000 dead and 3,200 missing.

**2012** Wildland fires char more than 7 million acres in the United States. Garment factory fire in Pakistan kills 300 behind locked doors.

**2013** Brazil nightclub fire kills 233. Nineteen "hotshot" crew firefighters killed battling Arizona wildfire.

**2014** Forest fires in Chile kill 16 and destroy more than 2,000 homes.

**2015** Indonesian forest fires result in 100,300 premature deaths.

**2016** Oakland warehouse fire (Ghost Ship) kills 36.

**2017** London's Grenfell Tower fire kills more than 80 residents. Hurricane Harvey drops more than 50 inches of rain in one week on Houston, Texas, and surrounding counties. Estimated recovery cost is placed at $160 billion. Hurricane Maria (Category 5) kills more than 100 people in Puerto Rico and Dominica causing greater than $100 billion in property damage; Thomas Wildfire in California burns approximately 300,000 acres and destroys greater than 1,000 homes; Residential building fire in the Bronx (NYC) kills 13 people.

**2018** Kemerovo (Russia) shopping mall fire kills 64 people, 41 of them children.

## PART IV

# PRACTICE EXAMINATIONS

# PRACTICE EXAM 1

Answer **questions 1 through 10** based on the reading passage below. Use a worksheet to draw the scenario to make the location and firefighter actions clearer.

On a cold morning in April, firefighters arrive at a structural fire inside a two-story building with the address, 113 State Street. The cross street intersecting State Street where the fire building is located is Caton Street. Addresses on Caton Street range from 304 to 322. The fire is located on the first floor at the front of the building. Flames can be seen coming out of two windows of a clothing store. Facing the fire building from State Street, a pizza store is situated to the left of the clothing store and a laundry to the right of the clothing store. All three stores have the same address. A three-story residential building is located to the left of the pizza store. The address for the residential building is 121 State Street. A fire escape is installed on the front of the three-story building. Windows on the third floor of this building are boarded up with plywood. A young child can be seen crying for help from a second floor window of the residential building. There are two fire hydrants located in the vicinity of the fire building and an Engine company is utilizing one of them and stretching four lengths of hose to fight the fire. There are five firefighters assigned to the Engine company. Two firefighters from a Ladder company, also on the scene, are placing a portable ladder to the window where the child is located for a rescue attempt. The other three members of this Ladder company are forcing entry into the store of fire origin to provide access for the Engine company with the hose line. A Battalion Chief observing the situation upon arrival transmits a message to the communications dispatcher that "all hands" are at work and an additional Engine company and Ladder company should be dispatched to the fire.

**1.** How many firefighters are at the scene of the fire?

(A) 12
(B) 11
(C) 8
(D) 5

**2.** What action are Engine company firefighters taking at this fire?

(A) Stretching hose
(B) Positioning a portable ladder to a window
(C) Forcing entry to the store of fire origin
(D) Laddering the roof of the residential building

**3.** What is the address of the building with the fire escape?

(A) 113 State Street
(B) 304 Caton Street
(C) 322 Caton Street
(D) 121 State Street

**4.** Where is the fire located?

(A) Laundry
(B) Residential building
(C) Clothing store
(D) Pizza store

**5.** What is the address of the pizza store?

(A) 113 State Street
(B) 121 State Street
(C) 304 Caton Street
(D) The address is not mentioned in the passage.

**6.** What building has boarded-up windows?

(A) The two-story building
(B) The buildings on Caton Street
(C) The residential building
(D) None of the above

**7.** How many fire hydrants are in the vicinity of the fire building?

(A) One
(B) Two
(C) Three
(D) Four

**8.** How many firefighters are involved in the rescue attempt?

(A) 10
(B) 5
(C) 4
(D) 2

**9.** What fire service member gave the status of the fire to the communications dispatcher?

(A) Engine company officer
(B) Ladder company officer
(C) Battalion Chief
(D) None of the above

**10.** What aspect of the fire scenario should take top priority?

(A) Extinguishing the fire in the clothing store
(B) Cutting a ventilation hole above the stores
(C) Rescuing the child at the window
(D) Checking for fire extension in the buildings on Caton Street

**GO ON TO THE NEXT PAGE**

Answer **questions 11 through 15** based on the reading passage below.

Housewatch duty is a task all firefighters must perform during their tours of work. It consists of a three-hour span when a firefighter monitors the department radio, telephone, and other communications equipment. When on housewatch, you are also responsible to greet all visitors to quarters and determine the nature of their business. While on housewatch firefighters are required to wear their uniform cap and display a nameplate. When a call comes in for a fire or emergency, the firefighter on housewatch must acknowledge the receipt of the alarm and notify all members of the need to respond. This task is the most important task of the housewatch duty. A buzzer or ring system is used to communicate what firefighter personnel are being summoned to respond. For example, one ring may indicate that all Engine company firefighters are to respond, while two rings could mean Ladder company firefighter must go. Three rings may mean both units are being called out. Another important duty of the housewatch firefighter is to maintain a journal or log of all activities carried out during the time frame of duty. Alarm responses, maintenance of apparatus by chauffeurs, conduction of training drills by company officers, and similar actions are just a few of the entries required. Housewatch duty helps keep the firefighters constantly vigilant and on guard in the protection of the community they serve.

**11.** What are firefighters required to wear while on housewatch duty?

(A) Gloves
(B) Long sleeve shirt
(C) Tie
(D) Cap

**12.** Select a duty, mentioned in the reading passage, required to be performed by the housewatch firefighter.

(A) Check their personal e-mail
(B) Greet all visitors to quarters
(C) Maintenance of apparatus
(D) Conduct training drills

**13.** From the choices below, which one is NOT a company journal entry?

(A) Raising overhead doors to sweep out apparatus floor
(B) Maintenance of apparatus by chauffeurs
(C) Apparatus response
(D) Training being conducted by company officer

**14.** What is the most important task performed by the firefighter on housewatch duty?

(A) Answering the telephone
(B) Greeting visitors
(C) Listening to firefighting activity on the department radio
(D) Receipt of alarms and notification of members to respond

**GO ON TO THE NEXT PAGE**

**15.** In the alarm buzzer/ring system mentioned in the reading passage, one ring would indicate what type of alarm response?

(A) Ladder company only
(B) Both Ladder company and Engine company
(C) Engine company only
(D) None of the above

Personal protection equipment (PPE) is essential in keeping firefighters safe during fire and emergency incidents. PPE includes pants and boots, handie-talkie radio, hood, coat, gloves, helmet, and positive pressure self-contained breathing apparatus (SCBA). Firefighters must also don PPE in the order listed above. With this information answer **questions 16 through 18**.

**16.** What PPE item must be donned immediately prior to putting on your hood?

(A) SCBA
(B) Handie-talkie radio
(C) Boots
(D) Pants

**17.** What PPE item must be put on subsequent to donning a SCBA?

(A) Coat
(B) Gloves
(C) Helmet
(D) None of the above

**18.** Select the INACCURATE sequence regarding PPE donning requirements from the choices below.

(A) Hood, coat, gloves, and helmet
(B) Pants and boots, handie-talkie radio, and hood
(C) Coat, helmet, gloves, and SCBA
(D) Handie-talkie radio, hood, coat, and gloves

**GO ON TO THE NEXT PAGE**

Review the apartment floor layout for one minute, then cover the drawing and answer **questions 19 through 24**.

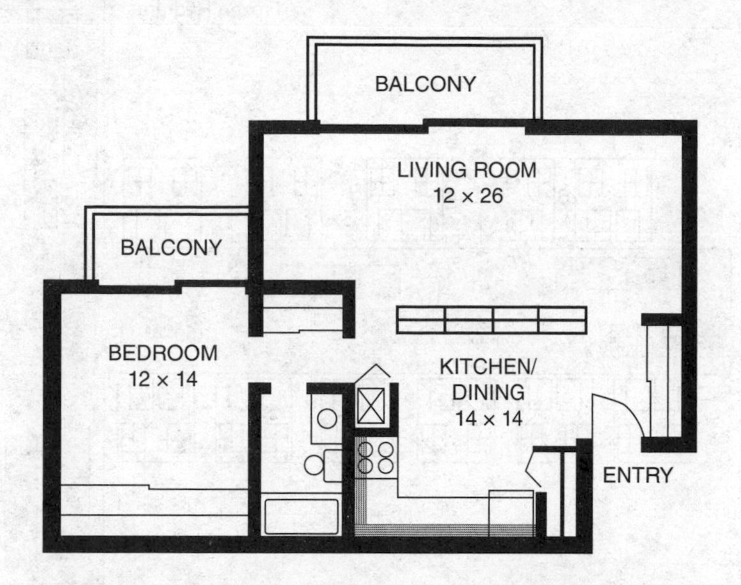


**19.** What room inside the apartment is furthest from the entry hall?

(A) Living room
(B) Bedroom
(C) Kitchen/dining room
(D) Bathroom

**20.** What room listed below doesn't have direct access to a balcony?

(A) Bathroom
(B) Bedroom
(C) Living room
(D) All of the above

**21.** What room is adjoining the larger balcony?

(A) Bathroom
(B) Kitchen/dining room
(C) Entry hall
(D) Living room

**22.** Select the accurate dimensions listed for the kitchen/dining room.

(A) 12 × 14
(B) 14 × 14
(C) 12 × 26
(D) 12 × 16

**23.** How many square feet is the bedroom?

(A) 168
(B) 196
(C) 312
(D) 192

**24.** What room is closest to the entry door of the apartment?

(A) Kitchen/dining room
(B) Bathroom
(C) Bedroom
(D) Living room

**GO ON TO THE NEXT PAGE**

Use the drawing below to answer **questions 25 through 30**.

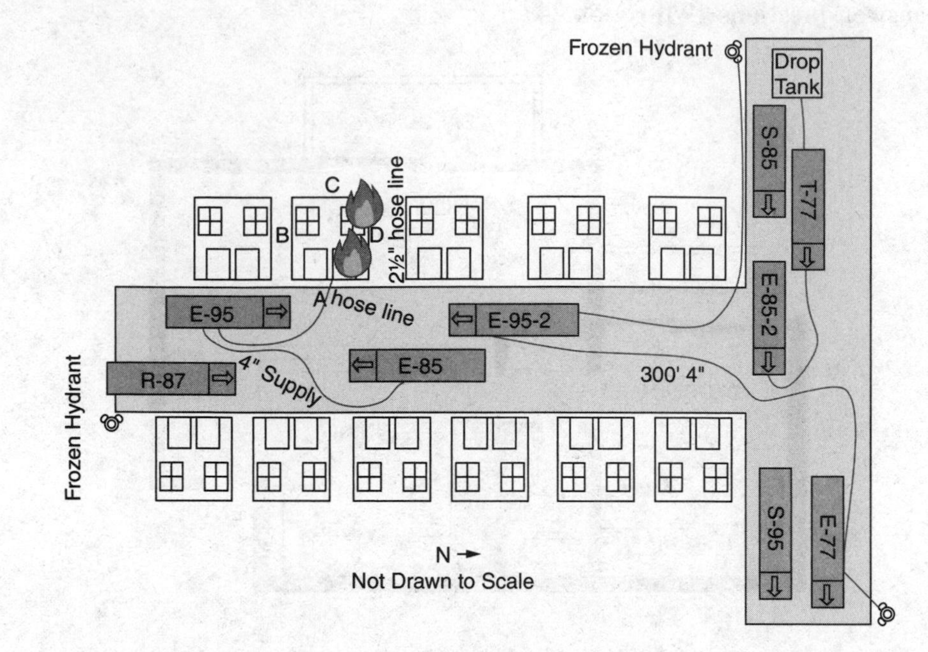


**25.** The fire building has how many floors of fire?

(A) Four
(B) Three
(C) Two
(D) One

**26.** Choose the Engine (E) apparatus that has hose lines stretched into the fire building.

(A) E-95-2
(B) E-85
(C) E-85-2
(D) E-95

**27.** What apparatus is connected to a frozen fire hydrant?

(A) E-95-2
(B) R-87
(C) E-77
(D) T-77

**28.** How many feet of hose is E-77 using to supply water to E-95-2?

(A) 100 feet
(B) 200 feet
(C) 300 feet
(D) 400 feet

**29.** What Engine apparatus has a 4-inch water supply hose line directly into E-95?

(A) E-95-2
(B) E-85
(C) E-85-2
(D) E-77

**GO ON TO THE NEXT PAGE**

**30.** What apparatus is utilizing the northernmost water supply source?

(A) R-87
(B) E-95-2
(C) T-77
(D) E-77

Use the private dwelling floor layout below to answer **questions 31 through 36**.

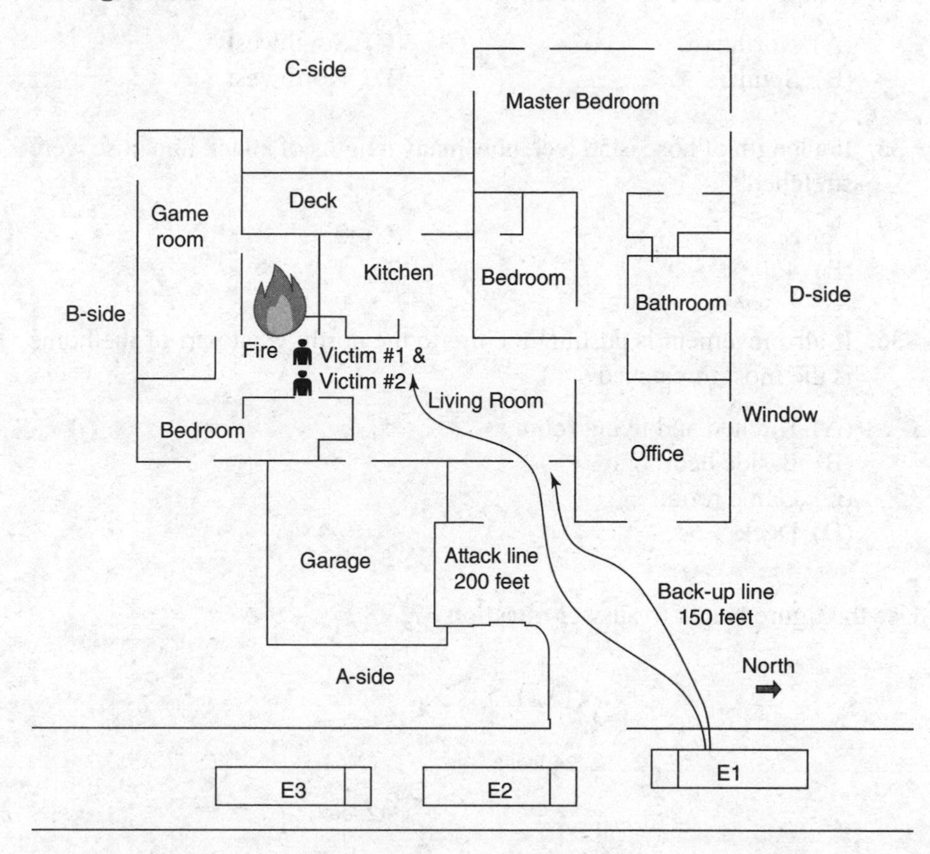


**31.** On what side of the building is the fire located?

(A) The A side
(B) The D side
(C) The B/C side
(D) The C/D side

**32.** Where are the two victims located?

(A) In the garage
(B) In the game room
(C) In the living room
(D) In the bathroom

**GO ON TO THE NEXT PAGE**

33. What action did E1 take at this fire?

(A) Stretched two attack hose lines
(B) Stretched an attack hose line only
(C) Stretched a backup hose line only
(D) Stretched an attack line and a backup line

34. Where were the two victims located in relation to the origin to the fire?

(A) Northeast
(B) South
(C) Southwest
(D) Northwest

35. If a length of hose is 50 feet, how many lengths of attack line hose were stretched?

(A) 8
(B) 4
(C) 2
(D) 1

36. If air movement is pushing the fire to the north, what part of the home is the most threatened?

(A) Kitchen and living room
(B) B-side bedroom
(C) Game room
(D) Deck

Use the figure below to answer **question 37.**

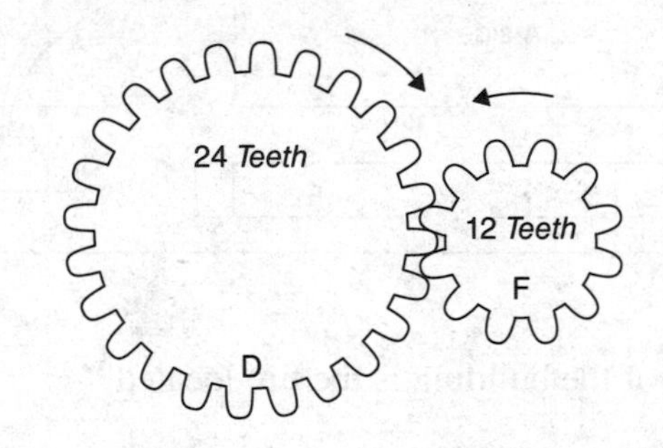


37. Gear D has 24 teeth and Gear F has 12 teeth. If Gear D makes 8 complete turns, how many turns will Gear F make?

(A) 8
(B) 12
(C) 14
(D) 16

38. A fire building being examined by investigating fire marshals needs to be roped off to prevent unauthorized entry. If the area is 90 feet long and 30 feet wide, how many yards of rope will be required?

(A) 60
(B) 70
(C) 80
(D) 90

**GO ON TO THE NEXT PAGE**

**39.** What is the perimeter (in feet) of a rectangle that has sides equal to 6 inches, 12 inches, 6 inches, and 12 inches?

(A) 1
(B) 2
(C) 3
(D) 4

**40.** How many lengths of 100-foot hose can you make if you have 6 lengths of 25-foot hose and 7 lengths of 50-foot hose?

(A) 5
(B) 6
(C) 7
(D) 8

**41.** If you had on your hose bed 6 lengths of 100-foot hose, 10 lengths of 50-foot hose, and 12 lengths of 25-foot hose, how many feet of hose would you have?

(A) 1,000
(B) 1,100
(C) 1,200
(D) 1,400

**42.** What is the average of the following three numbers: 10, 15, and 50?

(A) 20
(B) 25
(C) 30
(D) 35

**43.** What is the average of the following six numbers: 20, 45, 47, 53, 72, and 75?

(A) 50
(B) 52
(C) 54
(D) 60

**44.** Plastics are categorized as what type of materials?

(A) Class A
(B) Class B
(C) Class D
(D) Class K

**45.** A correct example of a Class D material can be found in what choice below?

(A) Gasoline
(B) Aluminum
(C) Rubber
(D) Plastics

**46.** What type of fire is one involving energized computers?

(A) Class A
(B) Class B
(C) Class C
(D) Class D

**47.** What type of extinguishing agent uses smothering to exclude oxygen from the burning material?

(A) Water
(B) Foam
(C) Dry chemical
(D) None of the above

**GO ON TO THE NEXT PAGE**

**48.** A 120,000 gallon tank is full of water. The discharge system allows it to be emptied at the rate of 200 gallons per minute. How long (in hours) would it take to completely empty the tank?

(A) 4 hours
(B) 6 hours
(C) 8 hours
(D) 10 hours

**49.** What is the area of a triangle having a 3-inch base and a height of 4 inches? Use the formula:

$A = B \times H \div 2$

where A is the area of the triangle
B is the base of the triangle
H is the height of the triangle

(A) 4 square inches
(B) 6 square inches
(C) 8 square inches
(D) 12 square inches

**50.** In the gears shown below, which of the following statements accurately describes their rotation?

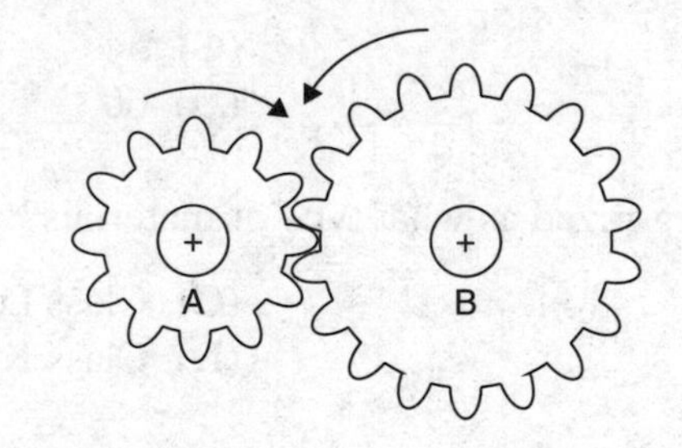


(A) As gear B turns counterclockwise, gear A will turn in a clockwise direction at a faster speed
(B) As gear A turns counterclockwise, gear B will turn clockwise at a slower speed
(C) As gear B turns counterclockwise, gear A will turn in a clockwise direction at a slower speed
(D) None of the above

**GO ON TO THE NEXT PAGE**

**51.** For the pulley system shown below, what is the reason for the crossed belt configuration?

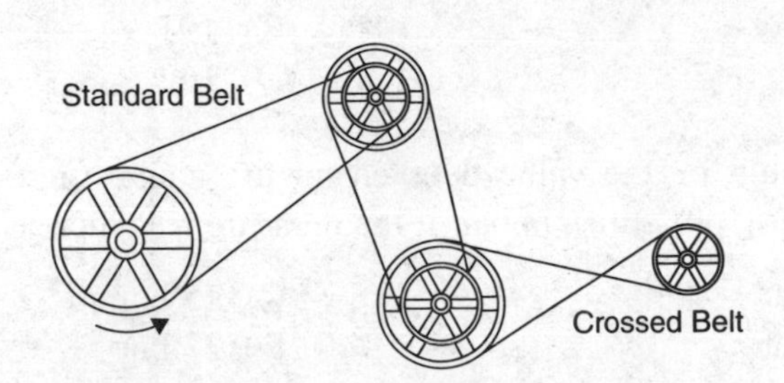


(A) It makes the small pulley turn slower.
(B) It makes the small pulley turn faster.
(C) It changes the direction of the small pulley.
(D) It reduces the force on the small pulley.

Provide answers to **questions 52 through 55** based upon the information in this reading passage and the air pressure gauge shown below the passage.

Vehicle tire gauges are zero-referenced to atmospheric pressure, which means they measure the pressure above atmospheric pressure (approximately 14.7 pounds per square inch or 1 bar). This is gauge pressure and is often referred to as bar (g). Absolute pressures, however, are zero-referenced to a complete vacuum (0 pounds per square inch). In the United States, where pressures are still often expressed in pounds per square inch (symbol *psi*), gauge pressures are referred to as *psig*, and absolute pressures are referred to as *psia*.

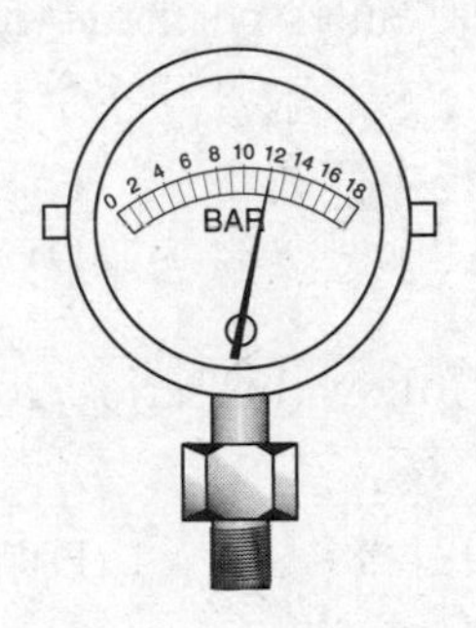


**GAUGE READS 12 BAR**

**52.** What is the psig value denoted on the gauge (zero-referenced to atmospheric pressure)?

(A) 176
(B) 195
(C) 210
(D) 243

**GO ON TO THE NEXT PAGE**

**53.** What is the psia value denoted on the gauge (zero-referenced to a complete vacuum)?

(A) 221
(B) 215
(C) 191
(D) 162

**54.** What would be the value denoted on the gauge (zero-referenced to atmospheric pressure), in bar, if the pressure was increased by 44 psi?

(A) 20 bar
(B) 18 bar
(C) 15 bar
(D) 13 bar

**55.** What would be the value denoted on the gauge (zero-referenced to a complete vacuum), in bar, if the pressure was increased by 88 psi?

(A) 18 bar
(B) 24 bar
(C) 27 bar
(D) 30 bar

**56.** How many items listed below are EXACT duplicates of one another?

| | |
|---|---|
| 88322 | 88322 |
| 987fxNGS32 | 987FxNGS32 |
| HL876245 | HI876245 |
| Povcx3091 | povcx3091 |

(A) One
(B) Two
(C) Three
(D) Four

**57.** A car is traveling at 60 miles per hour (mph). How far will it go in 20 minutes?

(A) 15 miles
(B) 20 miles
(C) 25 miles
(D) 30 miles

**58.** Which number should follow this series: 20/20, 15/20, 8/16, _____?

(A) 10/10
(B) 2/8
(C) 1/3
(D) 6/10

**59.** How many sides does the 3-D image shown below have?

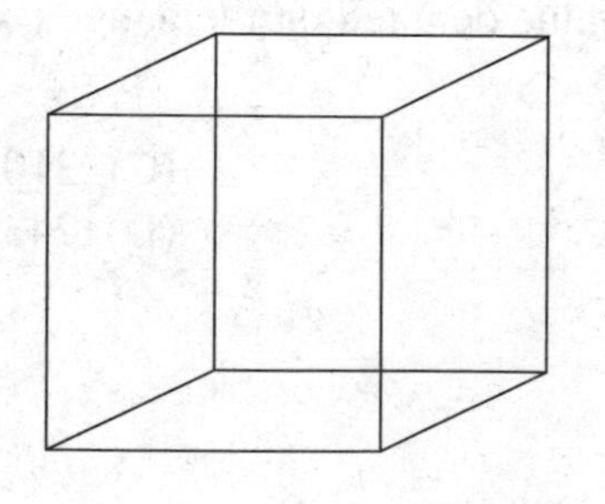

(A) Two
(B) Three
(C) Four
(D) Six

**GO ON TO THE NEXT PAGE**

Answer **questions 60 through 64** based upon the figure below.

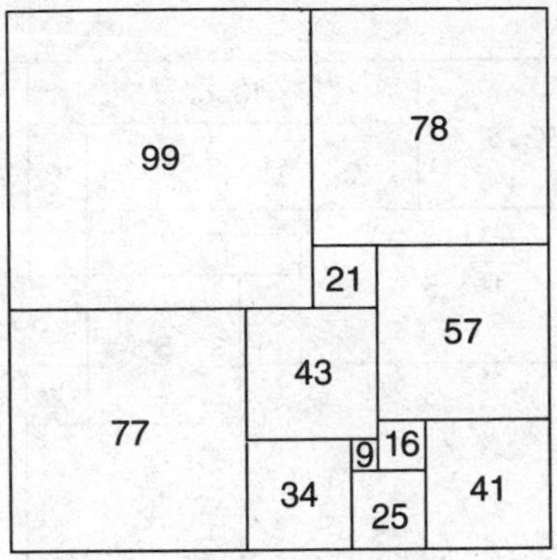

**60.** How many boxes does square 43 directly touch on?

(A) Four
(B) Six
(C) Seven
(D) Eight

**61.** Which two boxes have the second and third largest dimensions?

(A) 9 and 78
(B) 9 and 77
(C) 77 and 78
(D) 78 and 57

**62.** Which box has the second smallest dimensions?

(A) 9
(B) 25
(C) 21
(D) 16

**63.** Which three boxes when added together give a sum total of more than 100?

(A) 43, 9, and 34
(B) 34, 43, and 21
(C) 16, 43, and 41
(D) 9, 16, and 78

**64.** Which box listed below touches on the least number of boxes?

(A) 77
(B) 9
(C) 25
(D) 43

GO ON TO THE NEXT PAGE

Use the figure below to answer **questions 65 through 69**. Solve all multiplication equations where required.

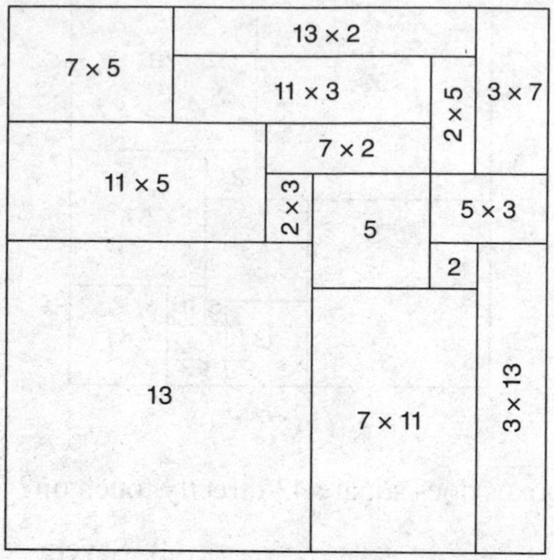

7 × 5
13 × 2
11 × 3
2 × 5
3 × 7
11 × 5
7 × 2
2 × 3
5
5 × 3
2
13
7 × 11
3 × 13

**65.** How many boxes have a number or product between 1 and 12?

(A) Four
(B) Three
(C) Two
(D) One

**66.** How many boxes have a number or product above 36?

(A) Four
(B) Three
(C) Two
(D) One

**67.** What is the sum of the numbers in the single-digit boxes?

(A) 13
(B) 5
(C) 20
(D) 22

**68.** How many boxes have a product that is divisible by 7?

(A) Four
(B) Three
(C) Two
(D) One

**69.** What box has the largest product number?

(A) $11 \times 5$
(B) $7 \times 11$
(C) $3 \times 13$
(D) $7 \times 5$

GO ON TO THE NEXT PAGE

Review the home layout below for one minute, then answer **questions 70 through 77** without referring back to it.

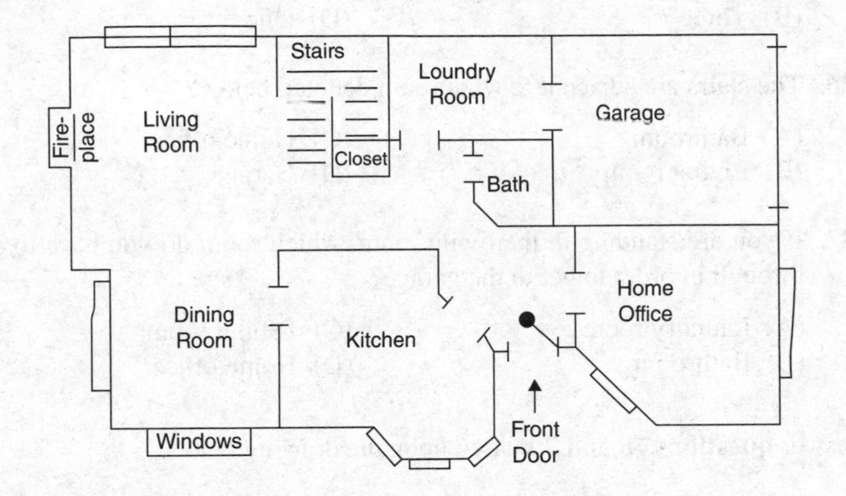


**70.** Which room has a fireplace?

(A) Dining room
(B) Living room
(C) Kitchen
(D) Home office

**71.** Which room listed below does NOT have a window?

(A) Bathroom
(B) Home office
(C) Kitchen
(D) Dining room

**72.** What room listed is closest to the front door?

(A) Living room
(B) Dining room
(C) Kitchen
(D) Laundry room

**73.** Where is the home office situated as you enter the front door of the home?

(A) To the left of the person entering
(B) To the right of the person entering
(C) Straight ahead
(D) Straight ahead and then to the left

**74.** How many entranceways lead into the laundry room?

(A) Four
(B) Three
(C) Two
(D) One

**GO ON TO THE NEXT PAGE**

**75.** The kitchen has how many windows?

(A) Four
(B) Three
(C) Two
(D) One

**76.** The stairs are adjacent to what room denoted below?

(A) Bathroom
(B) Living room
(C) Home office
(D) Garage

**77.** If you are standing in the living room, which room do you have to go through in order to get to the garage?

(A) Laundry room
(B) Bathroom
(C) Dining room
(D) Home office

Answer **questions 78 and 79** based upon the drawing below.

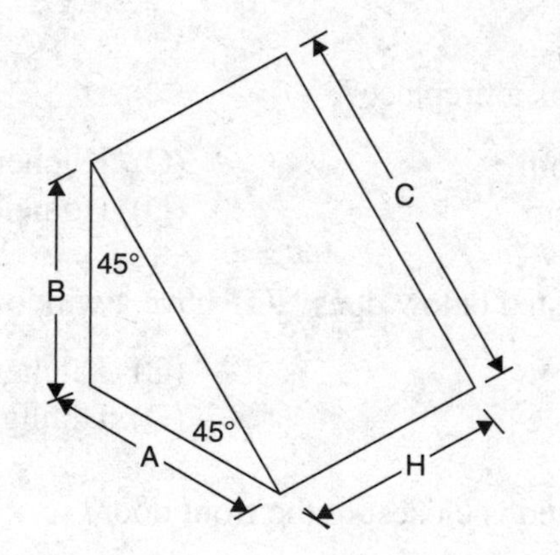


**78.** What is the angle formed at the joint connection between lines B and A?

(A) 45 degrees
(B) 75 degrees
(C) 90 degrees
(D) 180 degrees

**79.** How many sides does the above figure have?

(A) Eight
(B) Six
(C) Five
(D) Three

GO ON TO THE NEXT PAGE

Use the figure below to answer **questions 80 and 81**.

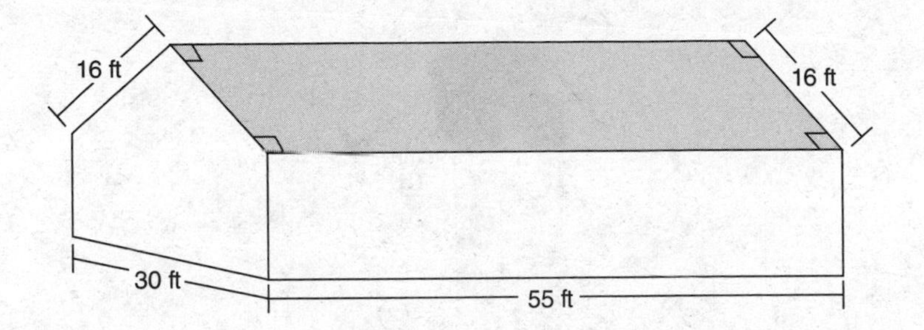


**80.** What is the square footage of the house?

(A) 1,500
(B) 1,550
(C) 1,600
(D) 1,650

**81.** Facing the house and starting from the right-side corner, how many 25-foot lengths of hose line would be needed to bring the line around to the left to the back side of the building?

(A) Six
(B) Five
(C) Four
(D) Three

**82.** What is the volume (in cubic feet) of the rectangular prism having the dimensions listed in the drawing

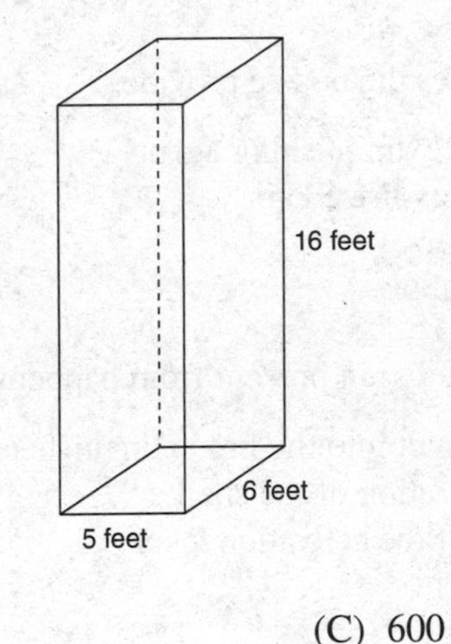


(A) 480
(B) 520
(C) 600
(D) 660

**GO ON TO THE NEXT PAGE**

Use the figure below to answer **questions 83 through 86**.

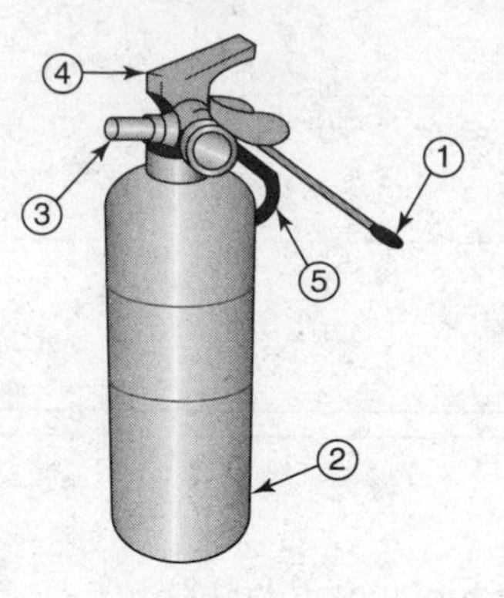


Components:

1. Safety pin
2. Shell
3. Nozzle
4. Top handle
5. Bottom activation lever

**83.** What component of the portable fire extinguisher holds the extinguishing agent?

(A) Bottom lever
(B) Top handle
(C) Shell
(D) Nozzle

**84.** What function does the nozzle provide?

(A) Discharge of extinguishing agent
(B) Storage of pressurized air
(C) Safety for user
(D) None of the above

**85.** What does the safety pin prevent from happening?

(A) Overfill of agent into the fire extinguisher
(B) Overpressurization of the shell
(C) Depression of the activation lever
(D) Theft

**86.** Where on a portable fire extinguisher would you find valuable information concerning operational guidance and applicable use?

(A) Underneath the activation lever
(B) On the shell
(C) Atop the handle
(D) Engraved on the safety pin

**GO ON TO THE NEXT PAGE**

Review the figure below and read the following passage concerning the operation of fire sprinkler systems to answer **questions 87 through 92.**

A fire sprinkler system consists of piping, sprinkler heads, and valves (to open and close the water supply at the source). Sprinkler heads are fitted into the piping at intervals along ceilings and sometimes walls. Sprinkler heads with fusible links or other type actuator devices operate independently of each other. The figure above depicts a sprinkler head with a heat-sensitive bulb filled with a red liquid. The bulb shatters when the sprinkler head is exposed to a certain temperature. The bulb is connected to a seal that prevents water from discharging when the sprinkler head is nonoperational. During a fire, however, high heat will cause the pressure of the liquid in the bulb to rise. At a predetermined temperature, the bulb will break, releasing the seal and allowing water out of the piping. The water strikes the deflector, which breaks the solid stream into small droplets that fall over a wide area to control and extinguish fire.

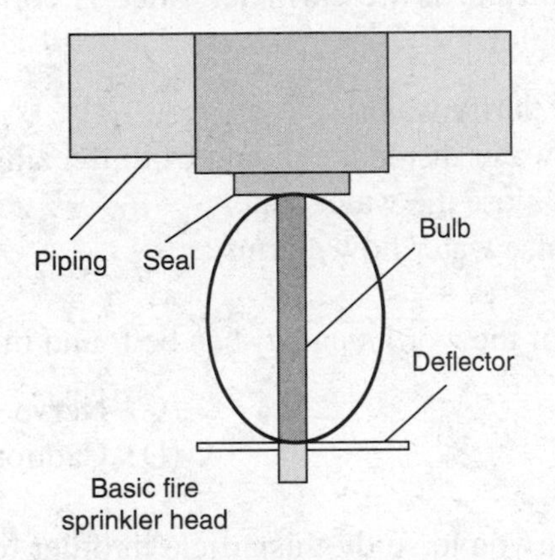

Basic fire sprinkler head

**87.** Why is there red liquid in the bulb?

(A) To match the curtains in the room where installed
(B) To expand under high temperatures and break the glass
(C) To add more water onto the fire
(D) For better visibility during maintenance intervals

**88.** What is the function of the sprinkler head deflector?

(A) Keep the water in the piping from freezing in cold weather
(B) Shield the bulb from heat
(C) Break up the water stream into droplets
(D) Prevent water from discharging out from the seal

**89.** What are the three major fire sprinkler system components mentioned in the reading passage?

(A) Bulbs, seals, and piping
(B) Piping deflectors and red liquid
(C) Valves, sprinkler heads, and ceilings
(D) Piping, sprinkler heads, and valves

**GO ON TO THE NEXT PAGE**

**90.** What prevents water from discharging out of the piping when the sprinkler head is in the nonoperational mode?

(A) The red liquid
(B) The glass bulb
(C) Both the red liquid and the glass bulb
(D) The seal

**91.** Fire sprinkler piping is installed along what areas of a building?

(A) Ceilings only
(B) Walls only
(C) Both ceiling and walls
(D) None of the above

**92.** The functionality of a fire sprinkler valve is correctly stated in what choice listed below?

(A) Filter and purify water
(B) Keep the water inside the piping from freezing
(C) Open and close the water supply
(D) Prevent false water flow alarms

**93.** The meaning of the word *temerity* can be found in what choice below?

(A) Love
(B) Hate
(C) Nerve
(D) Caution

**94.** It will behoove you to study this article in order to pass the upcoming exam. What is the meaning of the word *behoove* in the prior sentence?

(A) Hurt
(B) Challenge
(C) Inspire
(D) Benefit

**95.** Firefighters searched the contiguous buildings on the block in search of victims. What is the meaning of the word *contiguous* in the prior sentence?

(A) Adjacent
(B) Identical
(C) Colorful
(D) Large

**96.** Your salary is contingent on the quality of work you perform. What is the meaning of the word *contingent* in the prior sentence?

(A) Dependent
(B) Examined
(C) Compared
(D) Contrary to

**GO ON TO THE NEXT PAGE**

Firefighters must be able to read charts and reports to understand work schedules as well as training manuals. Read the Fire Department Report below in order to answer **questions 97 through 100**.

| **Fire Department Report** | | | |
|---|---|---|---|
| **Company** | **Total Hours Performing Building Inspections** | **Total Number of Buildings Inspected** | **Number of Inspections Needing Follow-up** |
| E-18 | 11 | 13 | 3 |
| L-2 | 16 | 20 | 5 |
| L-19 | 8 | 12 | 2 |
| E-33 | 9 | 15 | 4 |

**97.** What was the average number of hours the fire department spent performing building inspections at these four companies?

(A) 7
(B) 8
(C) 11
(D) 13

**98.** What was the average number of buildings inspected by the four companies of the fire department?

(A) 14
(B) 15
(C) 16
(D) 17

**99.** What company performed the most building inspections per hour?

(A) E-18
(B) L-2
(C) L-19
(D) E-33

**100.** What company must do the most follow-up inspection work?

(A) E-33
(B) L-19
(C) L-2
(D) E-18

**STOP. THIS IS THE END OF PRACTICE EXAM 1**

# Answer Key

1. B
2. A
3. D
4. C
5. A
6. C
7. B
8. D
9. C
10. C
11. D
12. B
13. A
14. D
15. C
16. B
17. D
18. C
19. B
20. A
21. D
22. B
23. A
24. A
25. C
26. D
27. A
28. C
29. B
30. D
31. C
32. C
33. D
34. A
35. B
36. A
37. D
38. C
39. C
40. A
41. D
42. B
43. B
44. A
45. B
46. C
47. B
48. D
49. B
50. A
51. C
52. A
53. C
54. C
55. A
56. A
57. B
58. B
59. D
60. C
61. C
62. D
63. D
64. A
65. A
66. B
67. C
68. A
69. B
70. B
71. A
72. C
73. B
74. C
75. B
76. B
77. A
78. C
79. C
80. D
81. C
82. A
83. C
84. A
85. C
86. B
87. B
88. C
89. D
90. D
91. C
92. C
93. C
94. D
95. A
96. A
97. C
98. B
99. D
100. C

# ANSWER EXPLANATIONS

Following you will find explanations to specific questions that you might find helpful.

**37.** D Gear F (12T) rotates two times for every one rotation for Gear D (24T).
Small gear completes 24/12 = 2 rotations for every one rotation of the large gear. This is called a gear reduction, or gear ratio of 1:2.
Gear D = 8 then Gear F = 16.

**38.** C Convert to yards: 90 ft (L) + 30 ft (W) = 30 yds + 10 yds = 40 yds
40 yds + 40 yds = 80 yds

**39.** C Convert to feet: 12 in (L) = 1ft; 6 in (W) = .5 ft;
$P = 2L + 2W = 2(1) + 2(.5) = 2 + 1 = 3$

**40.** A $6 \times 25$ ft = 150 ft; $7 \times 50$ ft = 350 ft
$500 \div 100 = 5$

**41.** D 100 ft $\times$ 6 = 600 ft; 50 ft $\times$ 10 = 500 ft; 25 ft $\times$ 12 = 300 ft
600 ft + 500 ft + 300 ft = 1400 ft

**42.** B $10 + 15 + 50 = 75$
$75 \div 3 = 25$

**43.** B $20 + 45 + 47 + 53 + 72 + 75 = 312$
$312 \div 6 = 52$

**48.** D 200 gals $\cdot$ 60 min = 12,000 gals per hour
120,000 $\div$ 12,000 gals per hour = 10 hours

**49.** B $A = B \times H \div 2$
A = 3 in $\times$ 4 in $\div$ 2 = 6 in$^2$

**52.** A 1 bar = 14.7 psi
14.7 psig $\cdot$ 12 = 176 psig

**53.** C 176 psig + 14.7 psi = 191 psia

**54.** C 44 psi = 3 bar
12 bar + 3 bar = 15 bar

**55.** A 88 psi = 6 bar
12 bar + 6 bar = 18 bar

**56.** A One; 88322

**57.** B 60 miles $\div$ 3 = 20 miles

**58.** B 20/20 = 1; 15/20 = ¾; 8/16 = ½ ; 2/8 = ¼

**59.** D A cube is a three-dimensional solid object bounded by six sides

**60.** C Square 43 directly touches on 77, 99, 21, 57, 9, 16, and 34

**64.** A Box 77 only touches on 99, 43, and 34
Box 9 touches on 43, 34, 25, and 16
Box 25 touches on 41, 16, 9, and 34
Box 43 touches on 77, 99, 21, 57, 9, 16, and 34

**65.** A Four: 2, 5, 2 × 5, and 2 × 3

**66.** B Three: 7 × 11, 11 × 5, and 3 × 13

**67.** C 13 + 5 + 2 = 20

**68.** A Four: 7 × 5, 7 × 2, 3 × 7, and 7 × 11

**78.** C 45° + 45° + x = 180°

$90° + x = 180°; x = 90°$

**79.** C Five: Two faces are triangles and three faces are rectangles

**80.** D 55 ft · 30 ft = 1,650 $ft^2$

**81.** C 55 ft + 30 ft = 85 ft; 25 ft hose · 4 = 100 ft

**82.** A Volume = (width)(length)(height) = 5 ft · 6 ft · 16 ft = 480 $ft^3$

**93.** C Audacity, nerve, effrontery, impudence, impertinence, cheek, gall, boldness

**94.** D Be incumbent on, be obligatory for, be required of, be expected of, be appropriate for

**95.** A Adjacent, neighboring, adjoining

**96.** A Conditional on, subject to, determined by, hinging on, resting on

**97.** C 11 hours + 16 hours + 8 hours + 9 hours = 44 hours ÷ 4 = 11 hours

**98.** B 13 inspections + 20 inspections + 12 inspections + 15 inspections = 60 ÷ 4 = 15

**99.** D E-18 = 13/11 = 1.18; L-2 = 20/16 = 1.25; L-19 = 12/8 = 1.5; E-33 = 15/9 = 1.66

# PRACTICE EXAM 2

The floor layout drawing below depicts two apartments side by side. The apartment on the right has a fire. Review the drawing for three minutes, and then cover it when answering **questions 1 through 10**.

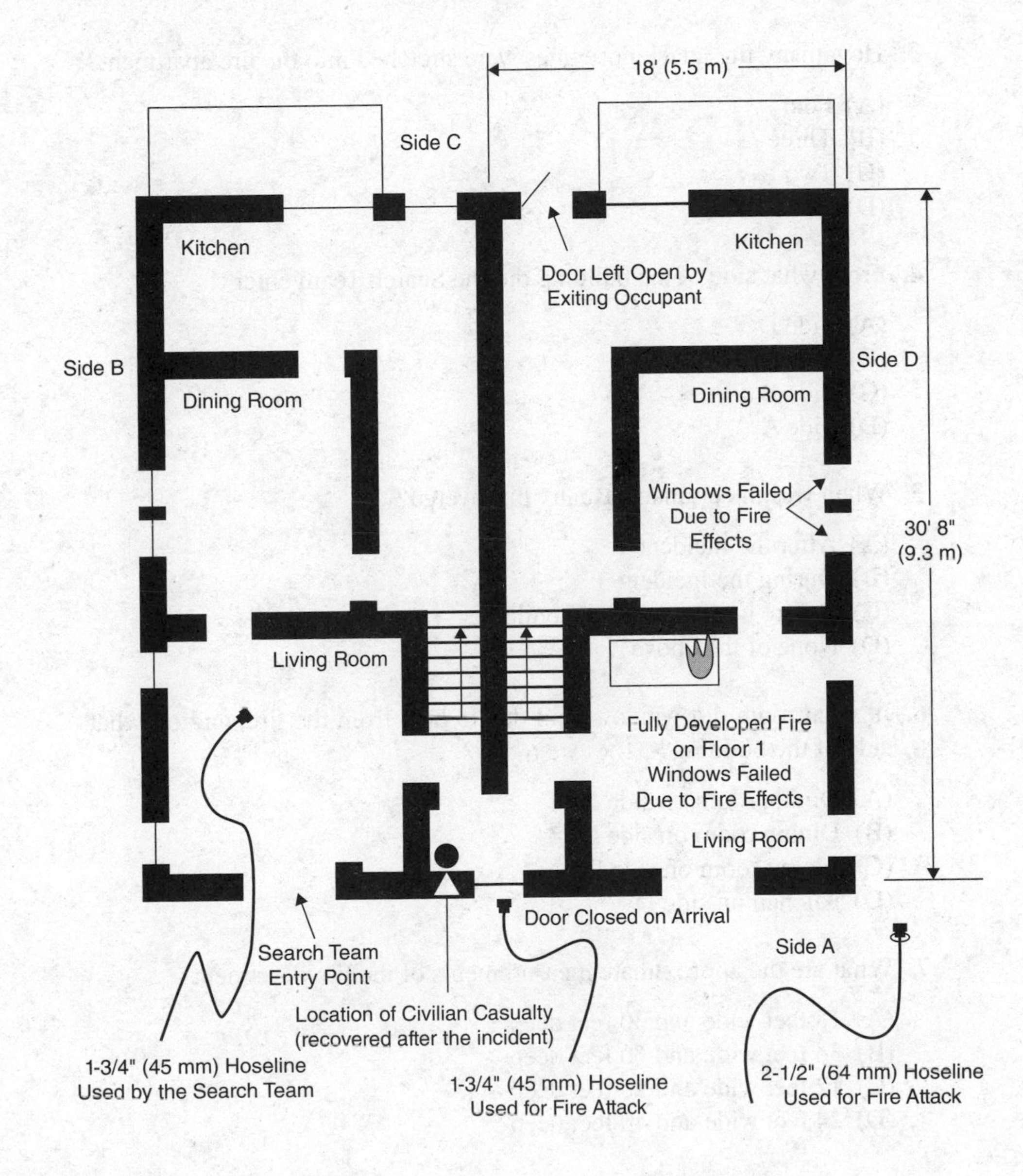

1. What apartment has the fire, and in what room is it located?

(A) Apartment on the right in the kitchen
(B) Apartment on the right in the living room
(C) Apartment on the left in the living room
(D) Apartment on the left in the kitchen

2. On what side of the building was the door left open by the exiting occupant?

(A) Side D
(B) Side C
(C) Side B
(D) Side A

3. How many fire attack hose lines were stretched into the fire apartment?

(A) Four
(B) Three
(C) Two
(D) One

4. From what side of the building did the Search Team enter?

(A) Side D
(B) Side C
(C) Side B
(D) Side A

5. When was the civilian casualty discovered?

(A) After the incident
(B) During the incident
(C) Before units entered the building
(D) None of the above

6. In what room did windows fail due to heat from the fire, and on what side of the building?

(A) Dining room on side B
(B) Dining room on side D
(C) Living room on side D
(D) Kitchen on side D

7. What are the approximate measurements of the fire apartment?

(A) 18 feet wide and 30 feet deep
(B) 36 feet wide and 60 feet deep
(C) 12 feet wide and 20 feet deep
(D) 24 feet wide and 40 feet deep

**GO ON TO THE NEXT PAGE**

8. What apartment did the Search Team enter, and from what side of the building?

(A) Apartment on the right and from side A
(B) Apartment on the right and from side C
(C) Apartment on the left and from side A
(D) Apartment on the left and from side B

9. What size hose line(s) was used for attack on the fire?

(A) 1¾-inch hose line only
(B) 2½-inch hose line only
(C) Both 1¾-inch and 2½-inch hose line
(D) None of the above

10. What size hose line(s) was used by the Search Team?

(A) 1¾-inch hose line only
(B) 2½-inch hose line only
(C) Both 1¾-inch and 2½-inch hose line
(D) Search Team did not use a hose line

Read the informational passage below to answer **questions 11 through 16**.

In 2012, U.S. fire departments responded to an estimated 1,375,000 fires according to a National Fire Protection Association (NFPA) survey. The survey further stated that these fires resulted in 2,855 civilian fire fatalities, 16,500 civilian fire injuries, and an estimated $12,427,000,000 in direct property loss. Additional information from the NFPA report included: home fires caused 2,380, or 83 percent, of the civilian fire deaths; fires accounted for 4 percent of the 31,854,000 total calls; 7 percent of the calls were false alarms; and 68 percent of the calls were for aid such as EMS.

11. What type calls were most prevalent?

(A) Calls to report a house fire
(B) Calls for medical assistance
(C) Calls for water leaks
(D) Calls to report a hazardous material spill

12. What was the sum total of civilian fire deaths and injuries?

(A) More than 19,000
(B) Less than 19,000
(C) Less than 15,000
(D) More than 23,000

**GO ON TO THE NEXT PAGE**

**13.** What type of incident caused the most civilian fire deaths?

(A) Home fires
(B) Commercial occupancy fires
(C) Church fires
(D) School fires

**14.** Fires accounted for how many of the total calls?

(A) Less than 1,200,000
(B) More than 1,200,000
(C) Less than 1,000,000
(D) More than 2,000,000

**15.** False alarms accounted for how many of the total calls?

(A) More than 2,200,000
(B) Less than 2,200,000
(C) More than 3,000,000
(D) Less than 2,000,000

**16.** If the indirect loss from fires was 10 times more than direct property loss, what would be the estimated monetary figure?

(A) Less than 80 billion dollars
(B) Less than 100 billion dollars
(C) More than 100 billion dollars
(D) More than 180 billion dollars

Review the pie chart to answer **questions 17 through 19**.

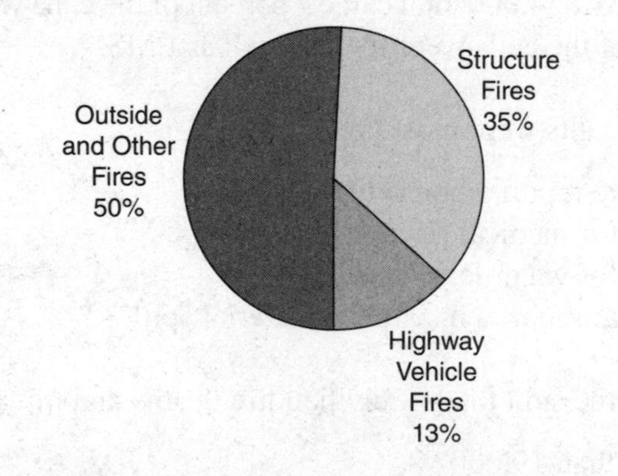


**FIRES IN THE UNITED STATES DURING 2012**

**17.** Approximately what percentage of fires were nonstructural?

(A) 50 percent
(B) 13 percent
(C) 63 percent
(D) 35 percent

**GO ON TO THE NEXT PAGE**

**18.** Outside and other type fires could include what choice listed below?

(A) Highway vehicle fires
(B) Structure fires
(C) Brush fires
(D) None of the above

**19.** What type of fire would NOT be included in structural fire statistics?

(A) Warehouse fire
(B) House trailer fire
(C) Storage shed
(D) Train car

Use the road map below to answer **questions 20 through 23**. Assume the top of the map is north.

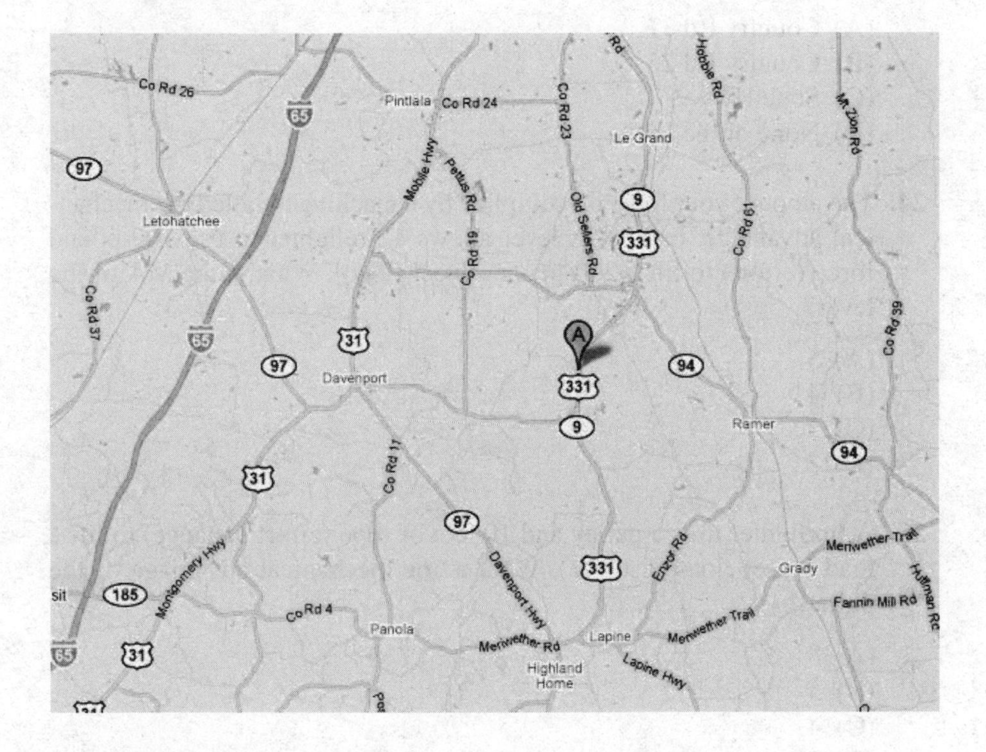


**20.** From Davenport, what is the best direction (shortest distance) to travel to get to Route 331?

(A) East
(B) North
(C) South
(D) West

**GO ON TO THE NEXT PAGE**

**21.** What road does not lead to Davenport?

(A) U.S. Hwy 31 (Montgomery Hwy)
(B) State Hwy 97 (Davenport Hwy)
(C) Interstate Hwy 65
(D) U.S. Hwy 31 (Mobile Hwy)

**22.** From your location (denoted by the letter A), what is the best direction to go to get to State Hwy 94 on your way to Ramer?

(A) South on U.S. Hwy 331
(B) North on U.S. Hwy 331
(C) West on State Hwy 9
(D) None of the above

**23.** If traveling north on Interstate Hwy 65 (at the intersection of State Hwy 185), what road is best to take to get to Letohatchee?

(A) Country Rd 37
(B) Country Rd 26
(C) State Hwy 97
(D) None of the above

**24.** The amount your force is multiplied by a machine is called the mechanical advantage, or MA. A lever allows a firefighter to use 100-pound force (effort) to lift a 200-pound object (load). What is the MA of the lever?

(A) 5
(B) 4
(C) 3
(D) 2

**25.** A firefighter uses a pulley and 10 feet of rope (effort distance) to lift a load 2 feet (load distance). What is the mechanical advantage of the pulley?

(A) 8
(B) 5
(C) 4
(D) 2

**26.** In a class 1 lever, the fulcrum is between the load and the effort. The lever has an MA when the fulcrum is closer to the load than to the effort. If the MA is 4, what force (effort) is required to lift a 200-pound object (load)?

(A) 30 pounds
(B) 50 pounds
(C) 75 pounds
(D) 100 pounds

**GO ON TO THE NEXT PAGE**

**27.** In a class 2 lever, the load is between the effort and the fulcrum. The effort arm is as long as the whole lever, but the load arm is shorter. Class 2 levers always have a mechanical advantage. If the effort distance of a wheelbarrow is 4 feet, and the load distance to the center of the load is 2 feet, what is the MA?

(A) 2
(B) 3
(C) 4
(D) 6

**28.** In a class 3 lever, the effort is between the load and the fulcrum. A tong is a good example. The length of the effort arm and the load arm are calculated from the fulcrum, as with the class 2 lever. What is the MA of the class 3 lever if the effort is 4 pounds and the load is 2 pounds?

(A) 3
(B) 2
(C) 0
(D) 0.5

Use the figure below to answer **question 29**.

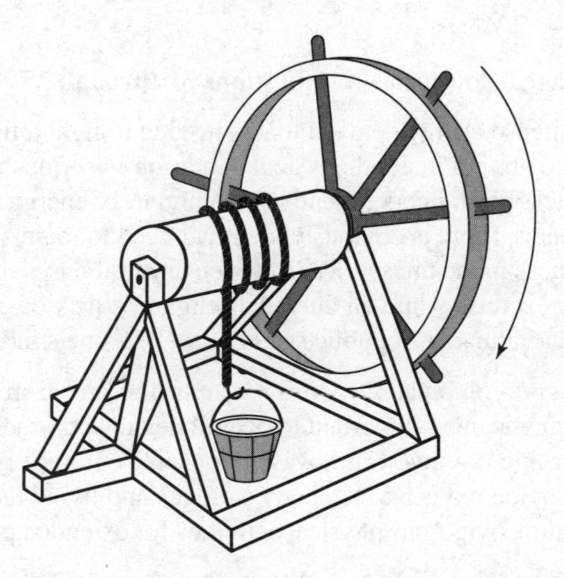

**29.** The MA of a wheel and axle is the ratio of the radius of the wheel to the radius of the axle. If the radius of the wheel is 24 inches and the radius of the axle is 4 inches, what is the MA of the assembly?

(A) 6
(B) 4
(C) 3
(D) 2

**GO ON TO THE NEXT PAGE**

Use the figure below to answer **question 30**.

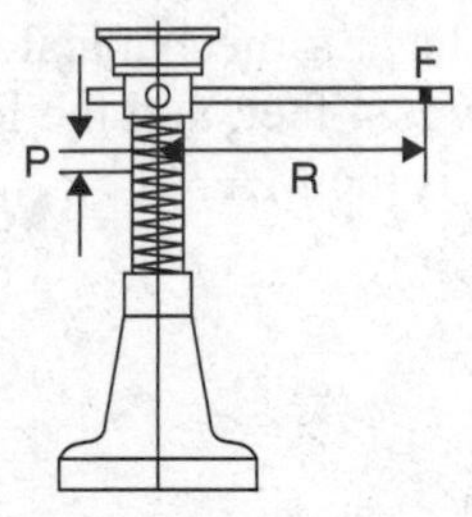


**30.** The jackscrew shown above has a handle radius (R) of 18 inches. The distance it advances in one complete turn or pitch (P) is 1/8 inch. The weight to be lifted is 100,000 pounds. What is the force (F) required to be applied at the end of the handle? Use the formula:

$$F = \frac{W \times P}{2\pi R}$$

(A) 100 lb
(B) 110 lb
(C) 140 lb
(D) 165 lb

Read the passage below to answer **questions 31 through 35**.

The sudden, intense energy demand that is needed to fight a fire is what puts firefighters who are not in good physical condition in serious danger of having a heart attack. Firefighters expend large amounts of energy during a major fire or emergency. There is obviously a critical need to maintain a high level of physical fitness in the fire service. At sedentary firehouses firefighters may get little or no exercise while on duty. A firefighter's lack of physical fitness can be viewed as a matter of public safety as well as one's individual health.

Optimal fitness is a combination of lifestyle, nutrition, and exercise. Regarding the latter, nothing is more important to overall health and fitness than cardiovascular or aerobic training. Cardiovascular exercise improves the ability of the lungs to provide oxygen to the body's organs and living tissue. This type of exercise requires vigorous physical activities for extended periods of time.

The maximal heart rate declines with age. The equation 220 minus your age equals your maximum heart rate is generally accepted as the basis for establishing the danger zone for people who are exercising or working too hard. Most people can only sustain a high percentage (90 percent) of their heart rate for a short duration. To increase your cardiovascular fitness, you must undertake a regular regime of sustained aerobic exercise. The most effective exercises for producing an improvement in cardiovascular fitness are those that are performed continuously while using large muscle groups. Activities that meet these criteria include jogging, fast walking, cycling, stair climbing,

**GO ON TO THE NEXT PAGE**

rope skipping, aerobics, and swimming. If performed the proper way, a good dose of exercise will make the firefighter's life safer and healthier.

**31.** Why is cardiovascular training so important to a firefighter's health and fitness?

(A) It is relatively inexpensive to do.
(B) It improves the lungs' ability to provide oxygen to the body's organs.
(C) It eliminates the threat of heart attack.
(D) All of the above

**32.** What will the maximum heart rate be for a 50-year-old firefighter?

(A) 150
(B) 160
(C) 170
(D) 185

**33.** Select the exercise activity below that does NOT meet the criteria for producing an improvement in cardiovascular fitness.

(A) Jogging
(B) Cycling
(C) Swimming
(D) Horseshoe throwing

**34.** Why are firefighters at high risk for having a heart attack?

(A) They don't eat healthy foods in the firehouse.
(B) They often skip meals due to being at fires during lunchtime.
(C) They use up large amounts of energy in short periods of time.
(D) They get little rest during their tours of duty.

**35.** Which of the following is not one of the three components that combine to produce optimal fitness?

(A) Sudden energy demand
(B) Lifestyle
(C) Nutrition
(D) Exercise

**36.** $1.65 - 0.4 =$

(A) 1.2
(B) 1.25
(C) 1.69
(D) 1.21

**37.** $3/7 \times 7/9 =$

(A) 1/3
(B) 2/3
(C) 10/63
(D) 1/6

**38.** $12/x = 60/15$, solve for $x$

(A) 3
(B) 4
(C) 8
(D) 9

**GO ON TO THE NEXT PAGE**

**39.** What percent of 80 is 4?

(A) 10
(B) 8
(C) 6
(D) 5

**40.** What does 55% look like as a decimal?

(A) 55.0
(B) 5.50
(C) 0.55
(D) 5550

**41.** 2,438 subtracted from 9,112 =

(A) 6,674
(B) 7,326
(C) 6,734
(D) 7,474

**42.** 9 is 20% of what number?

(A) 81
(B) 90
(C) 54
(D) 45

**43.** $8^2 + 7 =$

(A) 11
(B) 23
(C) 71
(D) 65

**44.** $3.2 \div 1.6 =$

(A) .2
(B) 2.2
(C) 2
(D) 22

**45.** What number below can replace both question marks in the following equation?

5/? = ?/20

(A) 5
(B) 10
(C) 12
(D) 20

**GO ON TO THE NEXT PAGE**

**46.** A firefighter is training with three ropes. One rope is 4 yards, 2 feet, and 8 inches; the second rope is 2 yards, 1 foot, and 9 inches; while the third rope is 3 yards, 2 feet, and 2 inches. What is the sum total of the three ropes in yards, feet and inches?

(A) 12 yards, 2 feet, and 7 inches
(B) 11 yards, 1 foot, and 3 inches
(C) 11 yards and 7 inches
(D) 12 yards and 4 inches

**47.** Three firefighters are respectively 6'3", 5'11", and 6'4" tall. What is the average height of the three firefighters in feet and inches?

(A) 6'3"
(B) 6'2"
(C) 6'1"
(D) 6'0"

**48.** At a firehouse, 4 firefighters weigh the following amounts: 175, 214, 166, and 205 pounds. What is the average weight of the firefighters?

(A) 190
(B) 194
(C) 195
(D) 198

**49.** Firefighter Jones bought a pair of pliers for $16.25. Six months later he sold it for a 25% loss. What price did Firefighter Jones get for the used pliers?

(A) $13.03
(B) $12.86
(C) $12.19
(D) $11.42

**50.** A fire engine apparatus is pumping water at 90 psi. An increase of 20% is ordered by the company officer. What will be the adjusted pressure after the increase?

(A) 120 psi
(B) 117 psi
(C) 111 psi
(D) 108 psi

**51.** A high school has 250 students enrolled. During a recent fire at the school, 12 students were absent. Forty students were evacuated by firefighters via the stairs; another 40 were taken from windows via ladders. The remaining students were sheltered in place inside the auditorium. How many students were sheltered in place?

(A) 180
(B) 173
(C) 170
(D) 158

**GO ON TO THE NEXT PAGE**

**52.** Two fire engines are pumping water at a fire. One fire engine is pumping water at 350 gallons per minute (gpm), while the other fire engine is pumping water at 400 gpm. If the engines are pumping water for 12 minutes, how much more water did the engine pumping at 400 gpm deliver on the fire than the other engine in operation?

(A) 500 gallons
(B) 550 gallons
(C) 600 gallons
(D) 650 gallons

**53.** In the fire scenario in question 52, how many gallons of water were delivered onto the fire by both fire engines in their 12 minutes of operation?

(A) 4,800
(B) 9,000
(C) 9,800
(D) 4,200

**54.** The required volume of water to supply sprinkler heads at a fire is based upon the formula Q = A ÷ 4; where Q = water in gpm and A = area in square feet. What is the estimated volume of water (Q) required to supply sprinkler heads in an area equal to 400 square feet?

(A) 80
(B) 100
(C) 110
(D) 120

**55.** In the scenario in question 54, if five sprinkler heads activated, how many gpm flowed out of each sprinkler head?

(A) 20
(B) 25
(C) 15
(D) 12

**56.** At a fire in a private dwelling, Engine 16 stretched 4 lengths of 1¾-inch hose. Engine 20 stretched 5 lengths of 2½-inch hose off the hose bed of Engine 16. Additionally, Engine 30 stretched 5 lengths of 1¾-inch hose to the adjoining building. How many lengths of 2½-inch hose were stretched at this fire?

(A) 14
(B) 10
(C) 9
(D) 5

**57.** A fire company went to 70 fires during 2010, 88 fires during 2011, and 122 fires in 2012. Of the 70 fires during 2010, 42 were auto fires. In 2011, the fire company had 35 auto fires and 18 brush fires. The fire company in 2012 had 13 house fires and 26 auto fires. What was the total number of fires during this three-year span?

(A) 280
(B) 414
(C) 388
(D) 290

**58.** In the scenario in question 57, how many auto fires did the fire company respond to during 2011 and 2012?

(A) 103
(B) 77
(C) 61
(D) 68

**GO ON TO THE NEXT PAGE**

Use the information from the rope chart below to answer **questions 59 through 63**.

| **Rope Diameter in Inches** | **Max Lift Load** | **Breaking Point** |
|---|---|---|
| 1 inch | 1,000 lbs | 3,000 lbs |
| ¾ inch | 830 lbs | 2,790 lbs |
| ½ inch | 710 lbs | 2,130 lbs |
| 3/8 inch | 560 lbs | 1,680 lbs |
| ¼ inch | 415 lbs | 1,245 lbs |

**59.** True or False, as the diameter of the rope increases, the maximum lift load decreases.

(A) True
(B) False

**60.** What would be the maximum lift load if you doubled the diameter of a ¼-inch rope?

(A) 1,000 lbs
(B) 830 lbs
(C) 710 lbs
(D) 560 lbs

**61.** For 3/8-inch diameter rope, what is the lbs difference between the maximum lift load and the breaking point?

(A) 2,130 lbs
(B) 1,120 lbs
(C) 830 lbs
(D) 710 lbs

**62.** What is the safety factor (breaking point compared to maximum lift load) for all the ropes listed?

(A) 5:1
(B) 4:1
(C) 3:1
(D) 2:1

**63.** If you need to lift a weight of 850 lbs safely, what size diameter rope should be utilized?

(A) 1 inch
(B) ¾ inch
(C) ½ inch
(D) Any of the above listed ropes should be used.

**64.** The formula for ideal water flow rate is: GPM = L (length) × W (width) × H (height) ÷ 100. For a fire inside a rectangular room with the dimensions 16 feet long, 14 feet wide, and a ceiling height of 10 feet, what would be the GPM required to extinguish the fire?

(A) 18.8
(B) 22.4
(C) 25.7
(D) 30.9

**GO ON TO THE NEXT PAGE**

Use the information below to answer **questions 65 and 66.**

| Drug Facts | Ibuprofen (200 mg) | Aspirin (325 mg) |
|---|---|---|
| Purpose | pain/fever reliever | pain/fever reliever |
| Uses | headache, muscle pain | headache, muscle pain |
| Warning | may cause severe allergic reaction | may cause liver damage |
| Ask doctor if you have | ulcers | asthma |

**65.** Which drug may cause rashes, blisters, or skin reddening?

(A) Both drugs
(B) Neither drug
(C) Ibuprofen
(D) Aspirin

**66.** For what drug should you seek a doctor's opinion if you have problems with asthma?

(A) Both drugs
(B) Neither drug
(C) Ibuprofen
(D) Aspirin

Read the nutrition facts information below for hard pretzels, black raisins, potato chips, and cashews to answer **questions 67 through 78.**

| Food | Hard Pretzels | Black Raisins | Potato Chips | Cashews |
|---|---|---|---|---|
| Serving size | 1 oz (28 g) | ¼ cup (40 g) | 1 oz | 1 oz |
| Calories | 108 | 120 | 150 | 160 |
| Fat (total) | 0.75 g (1%)* | 0 g (0%)* | 10 g (16%)* | 13 g (20%)* |
| Cholesterol | 0%* | 0%* | 0%* | 0%* |
| Carbs (total) | 22 g (8%)* | 32 g (11%)* | 15 g (5%)* | 9 g (3%)* |
| Protein | 3 g | 1 g | 2 g | 4 g |
| Vitamin C | 0%* | 2%* | 10%* | 0%* |
| Calcium | 1%* | 2%* | 0%* | 2%* |

*Percent daily value (allowance) based upon a 2,000 calorie diet

**67.** How many calories is the percent daily value (allowance) predicated on?

(A) 1,000
(B) 2,000
(C) 4,000
(D) 4,500

**GO ON TO THE NEXT PAGE**

**68.** Which food has the fewest calories per serving?

(A) Hard pretzels
(B) Black raisins
(C) Potato chips
(D) Cashews

**69.** How many carbohydrates (carbs) are needed to attain the percent daily value (allowance)?

(A) 410
(B) 380
(C) 330
(D) 300

**70.** How much total fat is required to attain the percent daily value (allowance)?

(A) 80
(B) 75
(C) 65
(D) 50

**71.** How many servings of potato chips would you have to eat to reach the percent daily value (allowance) for vitamin C?

(A) 1
(B) 5
(C) 8
(D) 10

**72.** How many calories would be consumed if a firefighter eats 2 servings of hard pretzels and 1 serving of potato chips?

(A) 366
(B) 333
(C) 288
(D) 275

**73.** True or False: A firefighter eating 1 serving of potato chips and 1 serving of cashews would exceed 1/3 of the percent daily value for fat (total).

(A) True
(B) False

**74.** One cup of black raisins is equivalent to how many grams?

(A) 40
(B) 80
(C) 120
(D) 160

**75.** What food listed below provides the highest amount of calcium per serving?

(A) Potato chips
(B) Hard pretzels
(C) Both black raisins and cashews
(D) Both potato chips and hard pretzels

**GO ON TO THE NEXT PAGE**

**76.** A firefighter eats 2 servings of hard pretzels and 2 servings of cashews. How many total grams of fat were consumed?

(A) 26.7
(B) 27.5
(C) 14.5
(D) 14.7

**77.** How many ounces of cashews are needed to provide 12 grams of protein?

(A) 1
(B) 2
(C) 3
(D) 4

**78.** What combination of foods listed below will give you more than 50% of the percent daily value (allowance) for fat.

(A) 2 servings of potato chips, 1 serving of black raisins, and 1 serving of cashews
(B) 2 servings of hard pretzels, 1 serving of black raisins, and 1 serving of cashews
(C) 3 serving of hard pretzels, 3 servings of black raisins, and 2 servings of potato chips
(D) 1 serving of potato chips, 3 servings of black raisins, and 1 serving of cashews

Choose the best definition for the words typed in bold for **questions 79 through 87**.

**79. Clad**

(A) Wounded
(B) Covered
(C) Negative
(D) Bent

**80. Detrimental**

(A) Basic
(B) Stingy
(C) Sad
(D) Harmful

**81. Insidious**

(A) Treacherous
(B) Porous
(C) Unusual
(D) Painful

**GO ON TO THE NEXT PAGE**

**82. Cognizant**

(A) Needy
(B) Reasonable
(C) Common
(D) Aware

**83. Valid**

(A) Honorable
(B) True
(C) Peaceful
(D) Tender

**84. Cadre**

(A) Group
(B) Box
(C) Uniform
(D) Protocol

**85. Adroit**

(A) Hopeful
(B) Meaningful
(C) Skillful
(D) Lonely

**86. Adamant**

(A) Brave
(B) Calm
(C) Knowing
(D) Inflexible

**87. Edifice**

(A) Mountain
(B) Structure
(C) Division
(D) Yacht

**GO ON TO THE NEXT PAGE**

For **questions 88 through 100**, choose the answer that best reflects your own personal opinion. Suggested answers located in the answer key are based upon the opinion of the author.

**88.** I dislike strenuous physical activity.

(A) Agree strongly
(B) Agree up to a point
(C) Disagree
(D) Disagree strongly

**89.** I like to rest on my couch and watch TV during weekends.

(A) Agree strongly
(B) Agree up to a point
(C) Disagree
(D) Disagree strongly

**90.** I enjoy tough challenges and am motivated to take them on.

(A) Agree strongly
(B) Agree up to a point
(C) Disagree
(D) Disagree strongly

**91.** I enjoy working with others in a team concept.

(A) Agree strongly
(B) Agree up to a point
(C) Disagree
(D) Disagree strongly

**92.** When working in a group, I tend to let others make decisions regarding appropriate actions.

(A) Agree strongly
(B) Agree up to a point
(C) Disagree
(D) Disagree strongly

**93.** I get along well with others at home and in the work environment.

(A) Agree strongly
(B) Agree up to a point
(C) Disagree
(D) Disagree strongly

**GO ON TO THE NEXT PAGE**

**94.** I seriously dislike working nights since I do my best work during the day.

(A) Agree strongly
(B) Agree up to a point
(C) Disagree
(D) Disagree strongly

**95.** I am a self-motivated individual.

(A) Agree strongly
(B) Agree up to a point
(C) Disagree
(D) Disagree strongly

**96.** I am empathetic to others' pain and suffering.

(A) Agree strongly
(B) Agree up to a point
(C) Disagree
(D) Disagree strongly

**97.** I get angry when my work is criticized by my boss.

(A) Agree strongly
(B) Agree up to a point
(C) Disagree
(D) Disagree strongly

**98.** I get flustered when confronted with a chaotic situation.

(A) Agree strongly
(B) Agree up to a point
(C) Disagree
(D) Strongly disagree

**99.** The sight of blood makes me feel ill.

(A) Agree strongly
(B) Agree up to a point
(C) Disagree
(D) Strongly disagree

**100.** I'm afraid of heights and like when my feet are firmly planted on the ground.

(A) Agree strongly
(B) Agree up to a point
(C) Disagree
(D) Strongly disagree

**STOP. THIS IS THE END OF PRACTICE EXAM 2**

# Answer Key

1. B
2. B
3. C
4. D
5. A
6. B
7. A
8. C
9. C
10. A
11. B
12. A
13. A
14. B
15. A
16. C
17. C
18. C
19. D
20. A
21. C
22. B
23. C
24. D
25. B
26. B
27. A
28. D
29. A
30. B
31. B
32. C
33. D
34. C
35. A
36. B
37. A
38. A
39. D
40. C
41. A
42. D
43. C
44. C
45. B
46. C
47. B
48. A
49. C
50. D
51. D
52. C
53. B
54. B
55. A
56. D
57. A
58. C
59. B
60. C
61. B
62. C
63. A
64. B
65. C
66. D
67. B
68. A
69. D
70. C
71. D
72. A
73. A
74. D
75. C
76. B
77. C
78. A
79. B
80. D
81. A
82. D
83. B
84. A
85. C
86. D
87. B
88. D
89. D
90. A
91. A
92. D
93. A
94. D
95. A
96. A
97. D
98. D
99. D
100. D

# ANSWER EXPLANATIONS

Following you will find explanations to specific questions that you might find helpful.

**17.** C Structural fires = 35%;
Outside fires and other = 50% + Highway vehicle fires = 13% = 63%

**24.** D MA = load/effort = 200 lbs/100 lbs = 2

**25.** B Pulley
MA = effort (force) distance/resistance (load) distance = 10 ft/ 2 ft = 5

**26.** B Class 1 Lever
MA = 4;
4 = 200 lbs (load) = $4x$ = 200 lbs = $x$ = 50 lbs
$x$ (effort)

**27.** A Class 2 Lever
MA = effort (force) distance/resistance (force) distance = 4 ft/2 ft = 2

**28.** D Class 3 Lever
MA = resistance force (load)/effort (force) = 2 lbs/4 lbs = 0.5

**29.** A Wheel and axle
MA = wheel radius/axle radius = 24 in/2 in = 6

**30.** B
F = W × P = 100,000 lbs × 0.125 in = 12,500 lbs = 110 lbs
$2\pi R = (2)(3.14)\ 18 = 113$

**36.** B 1.65
−0.4
1.25

**37.** A 3/7 × 7/9 – 21/63 = 3/9 = 1/3

**38.** A 12/$x$ = 60/15; cross-multiply
60$x$ =180; $x$ = 180/60 = 3

**39.** D Percent = Part ÷ Whole · 100
Percent = 4 ÷ 80 · 100 = 0.05 · 100 = 5

**40.** C 55 · 1/100 = 55/100 = 0.55

**41.** A 9,112
−2,438
6,674

**42.** D Whole = Part ÷ % · 100
Whole = (9) ÷ (20) · 100 = 0.45 · 100 = 45

**43.** C $8^2 + 7 = 64 + 7 = 71$

**44.** C To divide one decimal into another, move the decimal in the divisor to the right until it is a whole number and then move the decimal to the right the same number of places in the dividend.

3.2 ÷ 1.6 = 32 ÷ 16 = 2

**45.** B 5/? = ?/20
5/10 =1/2
10/20 =1/2

**46.** C 4 yards + 2 yards + 3 yards = 9 yards
2 ft + 1 ft + 2 ft = 5 ft = 1 yard + 2 ft
8 in + 9 in + 2 in = 19 in = 1 ft + 7 in
2 ft + 1 ft = 1 yard
Total: 11 yards and 7 in

**47.** B 75 in + 71 in + 76 in = 222 in
222 in ÷ 3 = 74 in = 6'2"

**48.** A 175 lbs + 214 lbs + 166 lbs + 205 lbs = 760 lbs ÷ 4 = 190 lbs

**49.** C What is 16.25 decreased by 25%?
Part = % · Whole ÷ 100
Part = (25)($16.25) ÷ 100
Part = $406.25 ÷ 100 = $4.06; $16.25 – $4.06 = $12.19

**50.** D What is 90 psi increased by 20%?
Part = % · Whole ÷ 100
Part = (20)(90 psi) ÷ 100
Part = 1,800 psi ÷ 100 = 18 psi; 90 psi + 18 psi = 108 psi

**51.** D 250 students – 12 = 238
40 evacuated via stairs + 40 taken from windows = 80
238 – 80 = 158

**52.** C 350 gpm · 12 min = 4,200 gallons; 400 gpm · 12 min = 4,800 gallons
4,800 gallons – 4,200 gallons = 600 gallons

**53.** B 4,800 gallons + 4,200 gallons = 9,000 gallons

**54.** B Q = A ÷ 4; Q = water in gpm; A= area in $ft^2$
Q = 400 $ft^2$ ÷ 4 = 100 gallons

**55.** A 100 gallons ÷ 5 = 20 gallons

**57.** A 2010 = 70 fires; 2011 = 88 fires; 2012 = 122 fires = 280 total fires

**58.** C 2011 = 35 auto fires; 2012 = 26 auto fires = 61 auto fires

**59.** B As diameter of rope increases, the maximum lift load increases

**60.** C ¼" · 2 = ½" = 710 lbs maximum lift load

**61.** B 3/8" = 560 lbs maximum lift load and 1,680 lbs breaking point
1,680 lbs – 560 lbs = 1,120 lbs

**62.** C 3,000 lbs:1,000 lbs = 3:1; 2,790 lbs:830 lbs = 3:1; 2,130 lbs:710 lbs = 3:1
1,680 lbs:560 lbs = 3:1; 1,245 lbs:415 lbs = 3:1

**63.** A 1" diameter rope = 1,000 lbs maximum lift load

**64.** B GPM = (L)(W)(H) ÷ 100
GPM = (16 ft)(14 ft)(10 ft) = 2240 $ft^3$ ÷ 100 = 22.4 gpm

**69.** D potato chips = 15 g (5%)
15 g · 20 = 300 g; 5% · 20 = 100%

**70.** C Cashews = 13 g (20%)
13 g · 5 = 65 g; 20% · 5 = 100%

**71.** D 1 oz = 1 serving size = 10%;
1 oz · 10 = 10 oz; 10% · 10 = 100%

**72.** A Hard pretzels = 108 calories per serving · 2 = 216 calories
Potato chips = 150 calories per serving
216 calories + 150 calories = 366 calories

**73.** A 1 oz potato chips = 16% fat (total) + 1 oz cashews = 20% fat (total) = 36% fat (total)

**75.** C Black raisins and cashews are both 2% calcium

**76.** B Hard pretzels (2 servings) + cashews (2 servings)
0.75 g · 2 = 1.5 g; 13 g · 2 = 26 g
1.5 g + 26 g = 27.5 g

**77.** C 1 oz cashews = 4 g protein; 3 oz cashews = 12 g protein

**78.** A Potato chips (2 servings) = 16 · 2 = 32%; black raisins = 0%; cashews = 20% = 52%

**79.** B Dressed, clothed, attired, garbed

**80.** D Damaging, injurious, hurtful, inimical, deleterious, destructive

**81.** A Stealthy, subtle, surreptitious, cunning, crafty, sly, wily, shifty

**82.** D Conscious, mindful, knowledge of, alertness

**83.** B Well founded, sound, reasonable, rational, logical, justifiable, defensible, viable

**84.** A Corps, body, team, nucleus, core

**85.** C Adept, dexterous, deft, able, capable, expert

**86.** D Unshakable, immovable, resolute, unwavering, uncompromising,

**87.** B Building, property, pile, complex

# PRACTICE EXAM 3

*For each question, circle the letter of your answer choice.*

**Directions:** Study the fireground situation below for 3 minutes. At the end of that allotted time, cover the picture. Then answer **questions 1 to 16**.

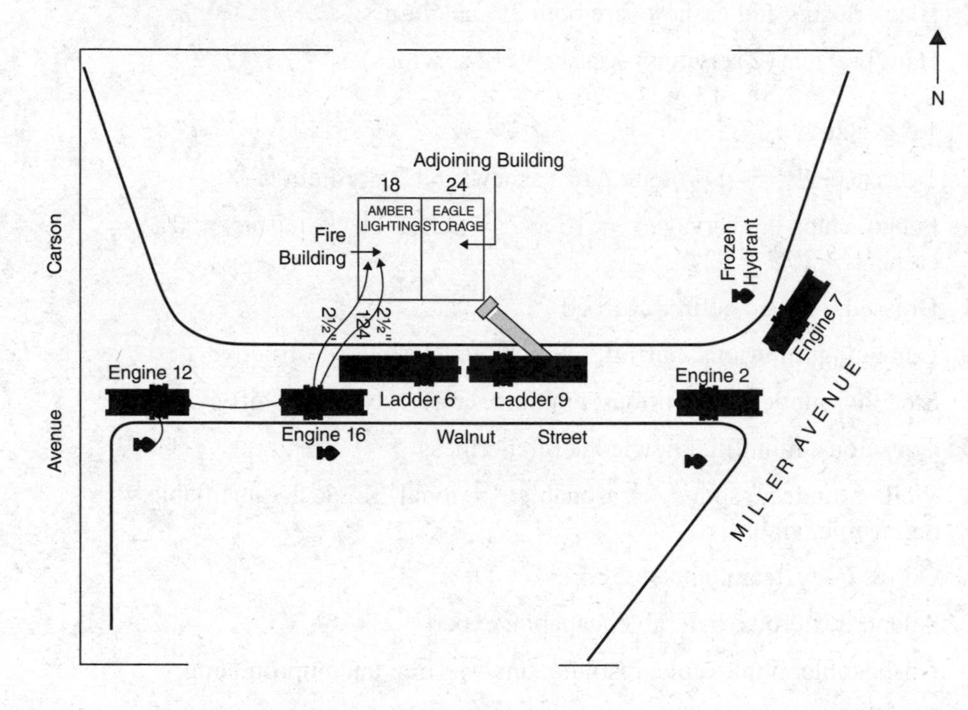


1. What is the name of the building on fire?

   (A) Eagle Storage
   (B) Amber Lighting
   (C) Carson Electric
   (D) Miller Plumbing

2. What is the name of the building adjoining the fire building?

   (A) Carson Electric
   (B) Amber Lighting
   (C) Miller Plumbing
   (D) Eagle Storage

**3.** What is the designation of the apparatus raising a ladder?

(A) Engine 7
(B) Ladder 9
(C) Ladder 6
(D) Engine 16

**4.** What apparatus below is situated directly in front of the fire building?

(A) Ladder 9
(B) Ladder 6
(C) Engine 12
(D) Engine 2

**5.** From what apparatus are hose lines stretched into the building on fire?

(A) Engine 16
(B) Engine 12
(C) Engine 2
(D) Engine 7

**6.** What size diameter hose is being used to extinguish the fire?

(A) 1 inch
(B) 1½ inch
(C) 2 inch
(D) 2½ inch

**7.** What is the address of the building on fire?

(A) 24 Walnut Street
(B) 18 Carson Avenue
(C) 18 Walnut Street
(D) 24 Miller Avenue

**8.** How many fire hydrants are there in the picture?

(A) 1
(B) 2
(C) 3
(D) 4

**9.** How many fire hydrants are frozen?

(A) 1
(B) 2
(C) 3
(D) 4

**GO ON TO THE NEXT PAGE**

**10.** What is the designation of the apparatus connected to a hydrant?

(A) Engine 7
(B) Engine 2
(C) Engine 16
(D) Engine 12

**11.** What activity listed below is Engine 12 performing?

(A) Raising a ladder to the building adjoining the fire building
(B) Supplying water to Engine 16
(C) Supplying water to the fire department connection at the fire building
(D) Supplying water to Ladder 6

**12.** What apparatus is adjacent to a frozen hydrant?

(A) Engine 7
(B) Engine 16
(C) Ladder 9
(D) Engine 12

**13.** How many fire hydrants are located on the south side of Walnut Street?

(A) 4
(B) 3
(C) 2
(D) 1

**14.** How many fire apparatus are there in the picture?

(A) 9
(B) 8
(C) 7
(D) 6

**15.** What is the ratio of Engines to Ladders found in the picture?

(A) 3:1
(B) 1:3
(C) 2:1
(D) 1:2

**16.** What is the best way for Engine 2 to obtain water?

(A) Connect to the hydrant at the SE corner of Walnut St. and Miller Ave.
(B) Connect to the hydrant at the NE corner of Walnut St. and Miller Ave.
(C) Receive water from Ladder 9
(D) Receive water from Engine 12

**GO ON TO THE NEXT PAGE**

**17.** What is 40 increased by 40%? (Use Part = % · Whole ÷ 100.)

(A) 84
(B) 72
(C) 68
(D) 56

**18.** What is 70 decreased by 20%? (Use Part = % · Whole ÷ 100.)

(A) 53
(B) 56
(C) 58
(D) 59

**19.** The fraction 3/5 equals what percent?

(A) 60%
(B) 54%
(C) 48%
(D) 46%

**20.** 28% equals what fraction (in lowest terms)?

(A) 1/3
(B) 2/5
(C) 7/25
(D) 3/10

**21.** What is the missing number in the arithmetic sequence below?

4, 9, __, 19, 24, 29, 34, 39 . . .

(A) 17
(B) 15
(C) 14
(D) 11

**22.** A Boy Scout troop is composed of 18 boys and 3 men. What is the ratio of boys to men?

(A) 3:1
(B) 4:1
(C) 5:1
(D) 6:1

**GO ON TO THE NEXT PAGE**

**23.** A box of candy contains 60 mints, of which 12 are filled with chocolate. What is the probability of selecting a mint with chocolate at random from the box on your initial try? (Use: Probability = event/possible outcomes.)

(A) 1/6
(B) 1/5
(C) 5/1
(D) 6/1

**24.** A dresser drawer contains 7 different colored shirts, 2 pants (jeans and dress), and 5 dissimilar patterned ties. How many different shirt/pant/tie combinations can be worn from the clothes in the drawer?

(A) 70
(B) 60
(C) 55
(D) 40

**25.** $9^3 =$

(A) 27
(B) 81
(C) 243
(D) 729

**26.** $\sqrt{100} =$

(A) 5
(B) 10
(C) 20
(D) 50

**27.** $\sqrt{49} - \sqrt{16} =$

(A) 65
(B) 33
(C) 8
(D) 3

**28.** A painter is paid $12 an hour. If in a week she worked 4 eight-hour days and 1 half-day (four hours), how much money did the painter earn for the week?

(A) $384
(B) $418
(C) $432
(D) $476

**GO ON TO THE NEXT PAGE**

**29.** A part-time bartender earns \$71.98 in tips on Friday night and \$63.21 on Saturday night. How much more cash did the bartender earn on Friday night than on Saturday night?

(A) \$8.77
(B) \$8.97
(C) \$135.19
(D) \$137.85

**30.** A sharpshooter taking target practice is able to squeeze off 8 shots in a 15-second time frame. If the shooter continues for two minutes, how many shots will be fired?

(A) 30
(B) 32
(C) 54
(D) 64

**31.** The price of a pizza is \$12.00. If three people share the cost of the pie equally, how much money would each person have to pay?

(A) \$4.50
(B) \$4.00
(C) \$3.50
(D) \$3.00

**32.** 35°C equals what temperature on the Fahrenheit scale? (Use: $F = 9/5C + 32$.)

(A) 120.6°F
(B) 102°F
(C) 95°F
(D) 91°F

**33.** 68°F equals what temperature on the Celsius scale? (Use $C = 5/9(F - 32)$.)

(A) 5.4
(B) 8
(C) 18
(D) 20

**GO ON TO THE NEXT PAGE**

**Directions:** Read the passage below to answer **questions 34 to 43**.

## Confined Spaces

Rescue operations carried out by firefighters inside locations that are below ground and restricted from natural ventilation are known as confined space incidents. Examples of confined spaces are water tunnels, utility trenches, ship holds, and storage tanks. The safety of the rescue workers is the most important factor in these situations. Recognizing the inherent hazards of the confined space prior to entry is paramount to a safe and successful operation.

Atmospheric condition hazards that must be addressed by rescue workers include oxygen deficiency, combustible and flammable gas vapors, and toxic gases. Physical hazards may also be present inside the confined space. Limited access and egress, unstable structural conditions, liquids in depth, and open gas lines or live electric lines are just a few of the physical hazards encountered.

No firefighter should enter a confined space location alone. The wearing of all personal protective equipment (PPE) and the use of positive pressure breathing apparatus (SCBA) is mandatory for firefighters working inside confined spaces. Long air supply hoses may be utilized and attached to the SCBA to allow firefighters to work for longer periods of time inside the space. Each firefighter entering the confined space must be attached to a lifeline (rope) and constantly monitored by a safety officer. The safety officer must keep a log of all members inside the confined space, how long each member has been operating, and how much oxygen each rescuer has remaining to ensure a safe return. A standby team must be assembled by the safety officer equal in number to the rescuers working inside the confined space. This reserve team must have the tools and equipment necessary to safely enter, medically treat, transport, and remove any and all rescue workers inside the confined space should the need arise.

A communications system must be established between the rescue workers and the safety officer. Portable radios may not be the method of choice if rescue workers are inside a combustible or flammable atmosphere for fear of an explosion as a result of radio transmissions. One way of communicating is via the **OATH** method. Each letter represents a single tug from the lifeline. One pull (**O**) on the rope by both exterior and interior members in reply stands for the verification that the members inside the confined space are **OK** and not in any danger. Two tugs (**A**) from each group represent the **advancement** of air supply line. Conversely, three pulls (**T**) from each end of the rope calls for the **taking up** of air supply line. Four tugs (**H**) is a distress signal and a call for **help** from the members inside the confined space with confirmation (four tugs) from members outside the confined space that help is on the way.

The establishment of a command post and staging area is also critical at confined space operations. A command post is required for the incident commander (ranking chief officer) to coordinate manpower and equipment resources. From the command post, the incident commander will also meet with other agency

**GO ON TO THE NEXT PAGE**

leaders who will be needed for assistance that is essential to a successful operation. The staging area should also be formed in an adequate space far enough away from the incident as not to interfere but close enough for resources to arrive at the command post in a timely fashion. A staging area chief officer should be designated in charge to ensure resources are recorded and properly managed both to and from the incident. Confined space operations are inherently dangerous to both trapped victims and firefighters. No firefighter should enter a confined space without these safety procedures being implemented. It is the responsibility of all supervisors to ensure that these rules are adhered to by all firefighters under their command. The special precautions listed in this article will not guarantee a successful operation but will enhance the safety of all firefighters.

**34.** Which of the following is NOT a true statement about the OATH method used to communicate with firefighters inside a confined space?

(A) One pull at each end verifies that firefighters inside the space are OK.
(B) Two pulls at each end represents the stretching of a hose line.
(C) Three pulls at each end calls for taking up the air supply line.
(D) Four pulls is a distress signal.

**35.** Long air supply hoses attached to SCBA allow firefighters inside the confined space to

(A) contact the safety team
(B) contact the safety officer
(C) work longer
(D) keep a log of all members inside the confined space

**36.** Which of the following atmospheric hazards is NOT mentioned in the reading?

(A) Oxygen deficiency
(B) Oxygen-enriched atmosphere
(C) Toxic gases
(D) Flammable gas vapors

**37.** Which of the following is NOT considered a confined space?

(A) A ship's hold
(B) A utility trench
(C) A water tunnel
(D) A high-rise office building

**GO ON TO THE NEXT PAGE**

38. According to the passage, a command post is important to the incident commander of a confined space operation for all but which of the following reasons?

(A) It is used to record resources.
(B) It is a place to meet with other agency leaders.
(C) It helps to utilize manpower.
(D) It helps to utilize equipment.

39. When entering a confined space, firefighters should wear and use certain protective equipment and devices to ensure their safety. Which of the following is NOT an item of protective gear?

(A) Self-contained positive pressure breathing apparatus
(B) PPE
(C) Lifeline
(D) Nozzle

40. Recognizing the inherent hazards of a confined space is important to ensure

(A) a safe and successful operation
(B) advancement of hose lines
(C) camaraderie among the firefighters working inside the confined space
(D) None of the above

41. A staging area chief officer at a confined space incident should be designated in charge to perform which of the following duties?

(A) Ensure resources arrive promptly at the command post.
(B) Monitor the oxygen of each rescuer inside the confined space.
(C) Establish the command post close to the staging area.
(D) Inspect the lifeline prior to it being used inside the confined space.

42. Which of the following is NOT an example of a physical hazard inside a confined space?

(A) Toxic gases
(B) Combustible gases
(C) Personal protective equipment
(D) Liquids in depth

43. Which of the following is NOT a responsibility of the safety officer?

(A) Monitor firefighters inside the confined space
(B) Keep a log
(C) Carry and use rescue tools
(D) Assemble a standby rescue team

**GO ON TO THE NEXT PAGE**

**Directions:** The following figures show the Rule of Nines used to estimate the area of the body burned in a fire. Use information from the figures to answer **questions 44 to 50**.

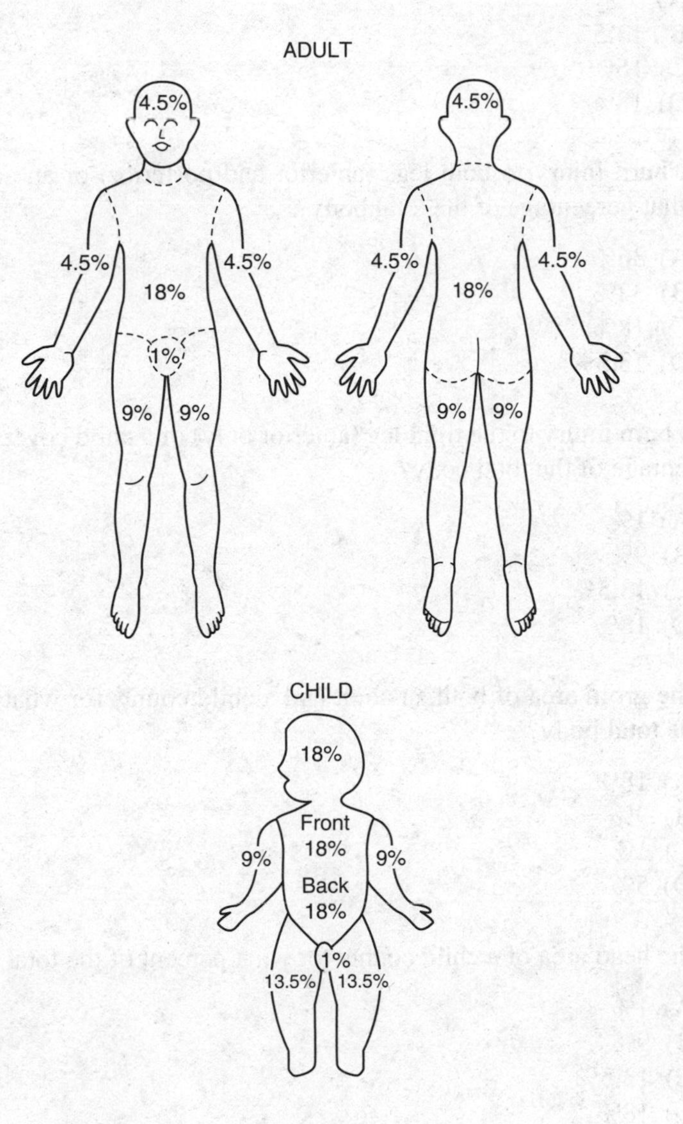


**44.** Which of the following parts of the adult body encompasses the most area?

(A) Right arm (anterior and posterior)
(B) Chest
(C) Left leg (anterior only)
(D) Groin

**45.** A burn injury to the head covers what percent of an adult's total body?

(A) 18%
(B) 9%
(C) 4.5%
(D) 1%

**GO ON TO THE NEXT PAGE**

**46.** A burn injury to both arms (anterior and posterior) of an adult covers what percent of the total body?

(A) 9%
(B) 13.5%
(C) 15%
(D) 18%

**47.** A burn injury to both legs (anterior and posterior) of an adult covers what percentage of the total body?

(A) 36%
(B) 32%
(C) 18%
(D) 13.5%

**48.** A burn injury to the right leg (anterior only) of a child covers what percentage of the total body?

(A) 1%
(B) 9%
(C) 13.5%
(D) 18%

**49.** The groin area of both an adult and a child counts for what percent of the total body?

(A) 18%
(B) 9%
(C) 1%
(D) 5%

**50.** The head area of a child counts for what percent of the total body?

(A) 1%
(B) 9%
(C) 13.5%
(D) 18%

**GO ON TO THE NEXT PAGE**

**Directions:** Use the figure below to answer **question 51**.

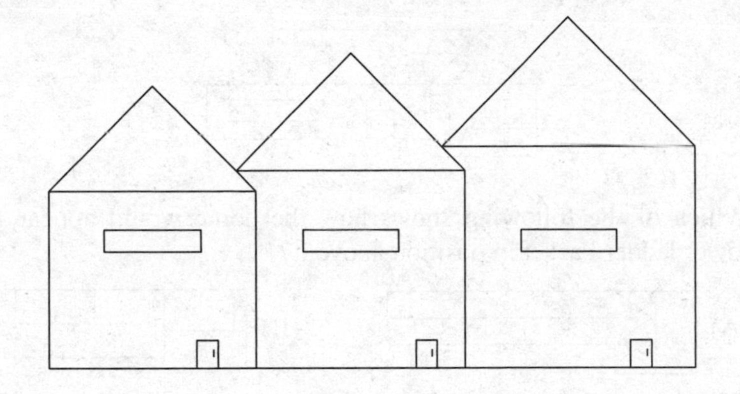

**51.** Which of the following shows how the buildings would appear from above?

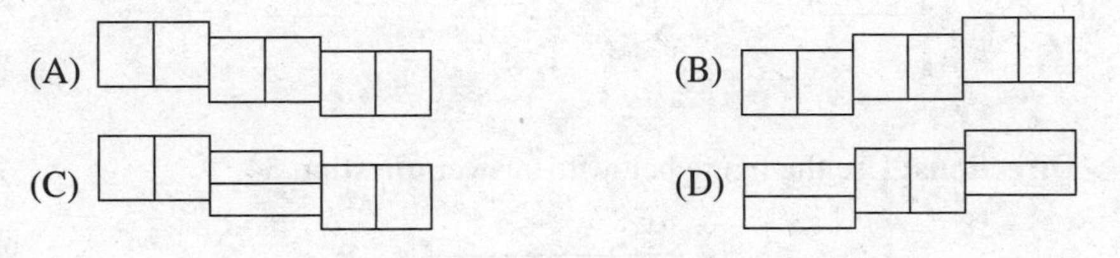


**Directions:** Use the figure below to answer **question 52**.

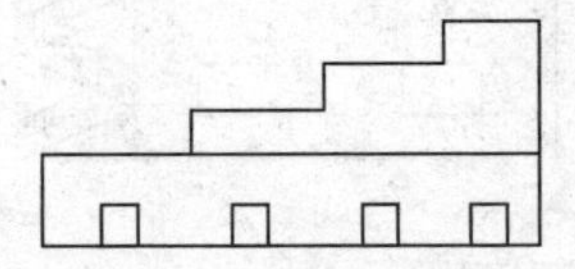

**52.** Which of the following shows how the building would appear from the rear?

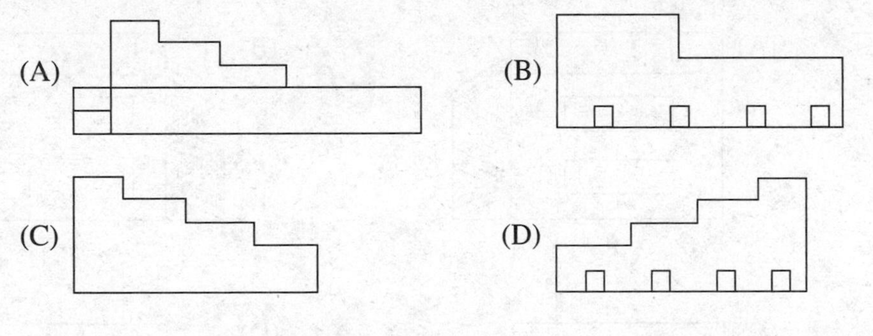


**GO ON TO THE NEXT PAGE**

**Directions:** Use the figure below to answer **question 53**.

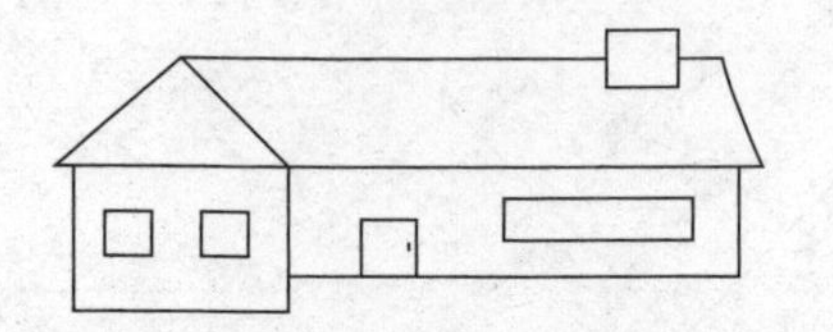

**53.** Which of the following shows how the home would appear from a tower ladder basket in position above it?

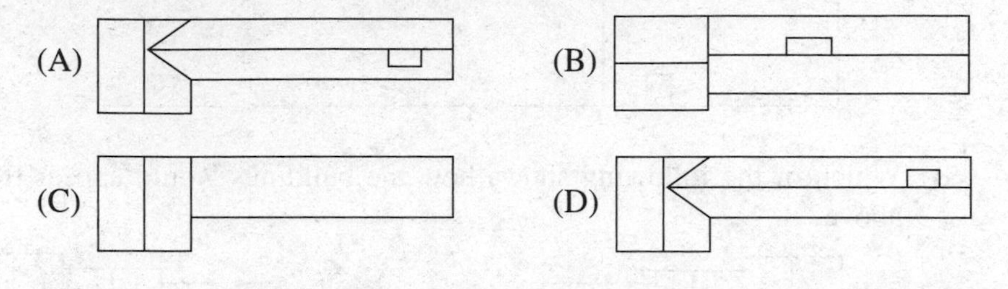


**Directions:** Use the figure below to answer **question 54**.

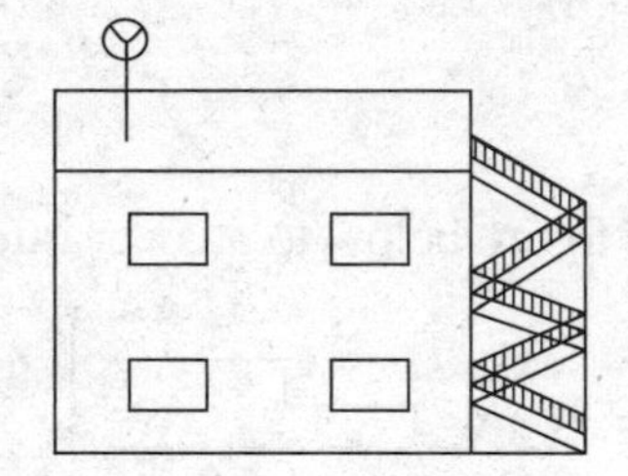

**54.** Which of the following shows how the structure would appear from the rear?

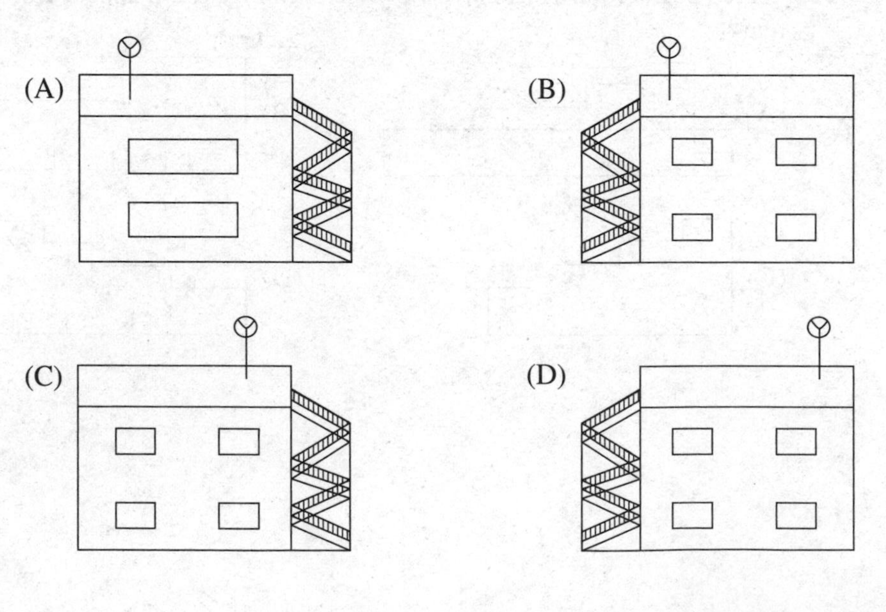


GO ON TO THE NEXT PAGE

**Directions:** Use the figure below to answer **question 55**.

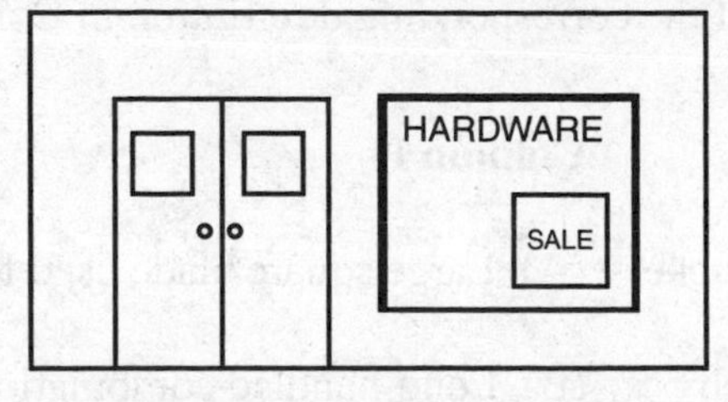


**55.** Which of the following shows how the front of the hardware store would appear from inside the store?

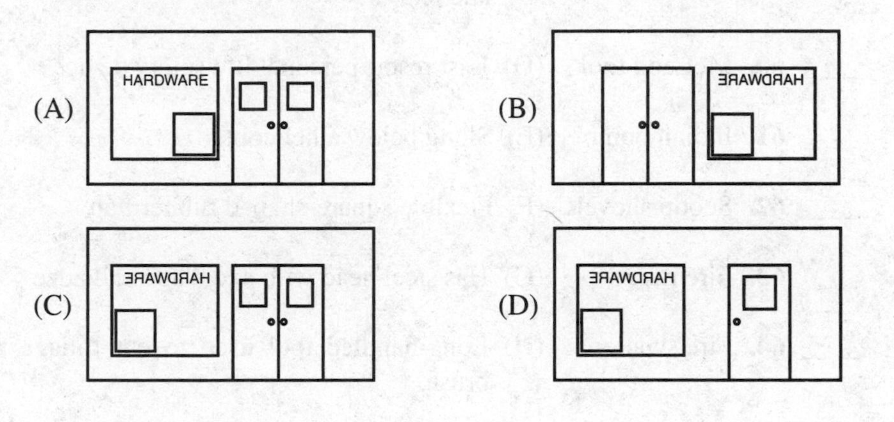


**Directions:** Use the figure below to answer **question 56**.

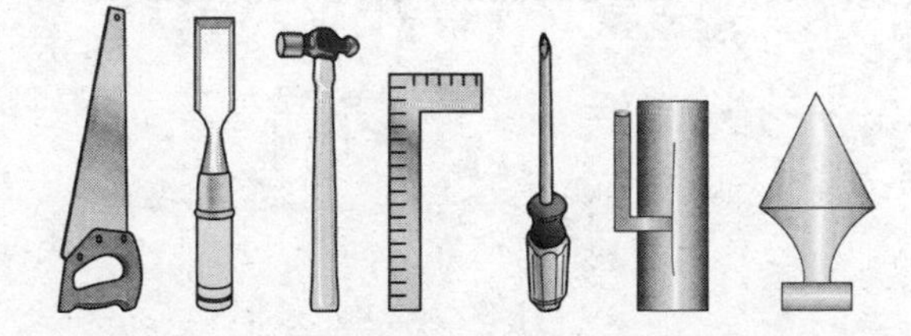

**56.** The figure shows a set of tools lying on a tabletop in front of you. Which of the following shows how the tools would look from behind the table?

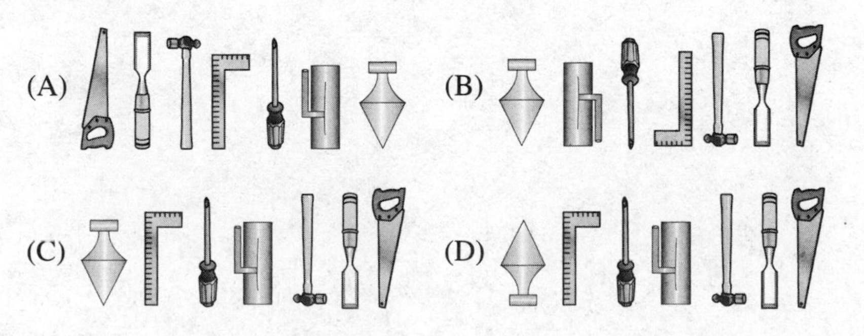


**GO ON TO THE NEXT PAGE**

**Directions:** For each item of wildland firefighting equipment listed in Column A, write the letter of the corresponding description in Column B in the space provided.

| Column A | Column B |
|---|---|
| ____ **57.** Bambi bucket | (A) Large square blade used for lifting fire debris |
| ____ **58.** Drip torch | (B) Long-handled combination rake and hoe |
| ____ **59.** Pulaski tool | (C) Designed for direct attack on fires in grass and leaves |
| ____ **60.** McLeod tool | (D) Last resort personal life safety device |
| ____ **61.** Indian pump | (E) Slung below a helicopter |
| ____ **62.** Scoop shovel | (F) Flexible square-shaped rubber flap |
| ____ **63.** Fire rake | (G) Has steel head with axe blade and adze |
| ____ **64.** Fire swatter | (H) Long-handled tool used to cut foliage and brush |
| ____ **65.** Fire broom | (I) Used to ignite foliage and brush |
| ____ **66.** Fire shelter | (J) Backpack-mounted water tank |

**GO ON TO THE NEXT PAGE**

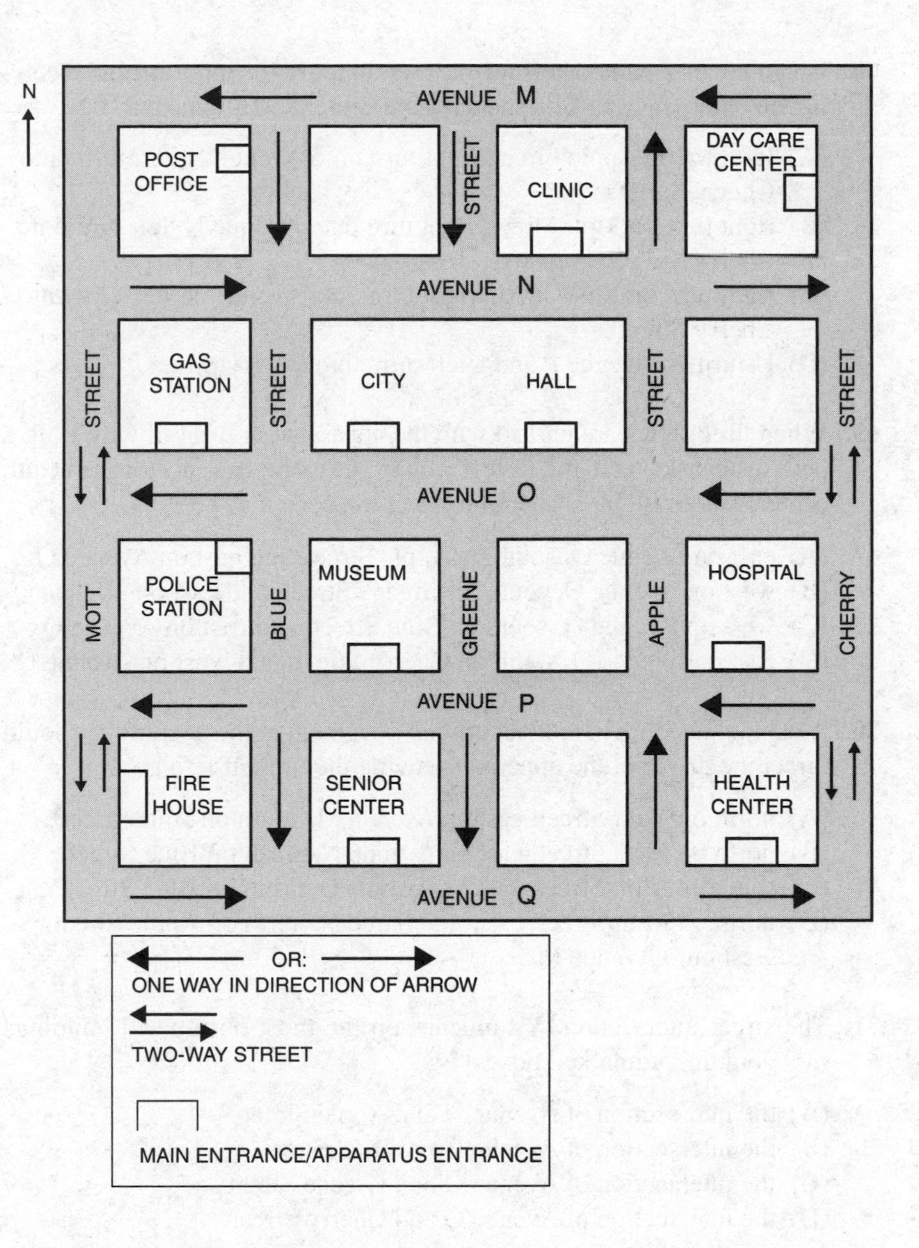


**Directions:** Study the road map shown above to answer **questions 67 to 71**. All responses by the fire apparatus must be performed in a legal manner as designated by road map directional signs.

**67.** The quickest way for the fire apparatus to arrive from quarters to a sprinkler water flow alarm at the hospital is to make a

(A) left turn out of quarters, right turn onto Avenue Q, left turn onto Apple Street
(B) right turn out of quarters, right turn onto Avenue N, right turn onto Cherry Street
(C) left turn out of quarters, left turn onto Avenue Q, left turn onto Apple Street
(D) right turn out of quarters, right turn onto Avenue P

**GO ON TO THE NEXT PAGE**

**68.** An ambulance parked in front of the entrance to the museum and receiving an emergency call for the day care center should make a

(A) left turn on Apple Street, right turn onto Avenue N, left turn onto Cherry Street
(B) right turn on Mott Street, right turn onto Avenue N, left turn onto Cherry Street
(C) right turn on Blue Street, right turn onto Avenue N, left turn onto Cherry Street
(D) U-turn on Avenue P and a left turn onto Cherry Street

**69.** When firefighters are parked with the apparatus in front of City Hall, a pedestrian asks them the fastest way to get to the entrance of the health center. The firefighters should instruct the person to go

(A) east on Avenue O, south on Apple Street, and east on Avenue Q
(B) west on Avenue O, south on Greene Street, and east on Avenue Q
(C) west on Avenue O, south on Blue Street, and east on Avenue Q
(D) east on Avenue O, south on Cherry Street, and west on Avenue Q

**70.** A captain needing to mail an official document at the post office should direct the driver of the apparatus leaving the firehouse to go

(A) north on Mott Street, east on Avenue M, south on Blue Street
(B) north on Mott Street, east on Avenue N, north on Blue Street
(C) south on Mott Street, east on Avenue Q, north on Blue Street
(D) north on Mott Street, east on Avenue N, north on Apple Street, west onto Avenue M

**71.** The street intersection that touches on the most designated buildings (not counting unmarked boxes) is

(A) the intersection of Avenue N and Apple Street
(B) the intersection of Avenue P and Blue Street
(C) the intersection of Avenue O and Greene Street
(D) the intersection of Avenue Q and Cherry Street

**GO ON TO THE NEXT PAGE**

**72.** A pry bar used to lift out a nail as shown in the figure is acting as what class lever system?

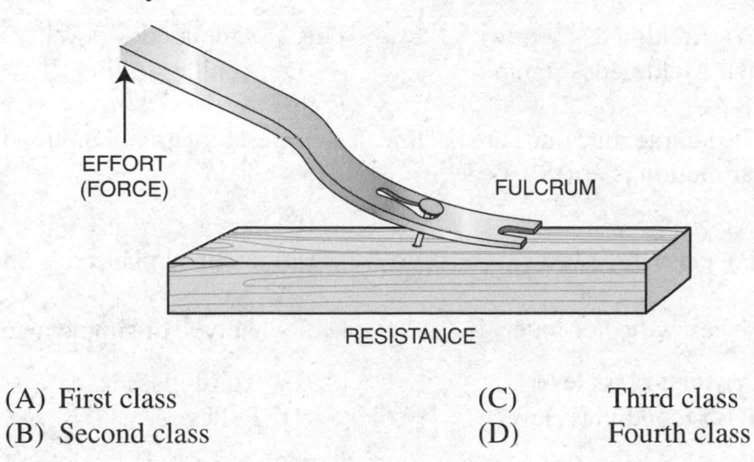


(A) First class
(B) Second class
(C) Third class
(D) Fourth class

**73.** It takes 100 N of force to raise a sail on a yacht. The load is also 100 N. What is the mechanical advantage?

(A) 0
(B) 1
(C) 2
(D) 4

**74.** Effort force is applied to gear A, which is interlocked with gear B. If gear A has 48 teeth, the resulting effect on gear B with 16 teeth is

(A) a speed advantage of 2
(B) a force advantage of 2
(C) a speed advantage of 3
(D) a force advantage of 3

**75.** The block and tackle pulley system shown has a mechanical advantage of

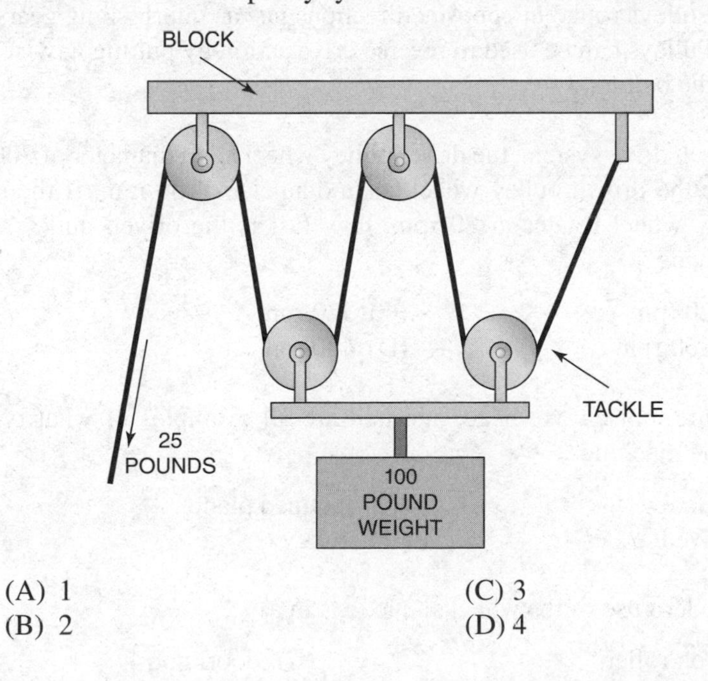


(A) 1
(B) 2
(C) 3
(D) 4

**GO ON TO THE NEXT PAGE**

**76.** A simple machine invented to draw out water from a ship's hold is called

(A) Archimedes' screw
(B) Archimedes' ramp
(C) Archimedes' lever
(D) Achilles' pulley

**77.** The simple machine listed below that converts rotational motion to linear motion is a(n)

(A) wedge
(B) class three lever
(C) screw
(D) inclined plane

**78.** A seesaw (teeter-totter) is an example of what type of simple machine?

(A) First-class lever
(B) Second-class lever
(C) Third-class lever
(D) Pulley

**79.** A warehouse worker rolls a wooden barrel weighing 80 pounds up an inclined plane. If the distance along the plane is 8 feet and the height of the far end of the plane is 2 feet, how much effort (force) was exerted by the worker?

(A) 80 pounds
(B) 60 pounds
(C) 40 pounds
(D) 20 pounds

**80.** Which of the following statements about pulley drive systems is INCORRECT?

(A) Pulleys work in a similar way to gears but are not directly joined.
(B) Pulleys can be used at a distance from each other.
(C) Pulleys rotate in opposite directions just as interlocking gears do.
(D) Pulleys can be used to reverse drive action by putting a twist into the belt.

**81.** In a belt drive system, the driver pulley wheel has a diameter of 200 mm while the driven pulley wheel has a diameter of 50 mm. If the driver pulley wheel rotates at 80 rpm, how fast is the driven pulley wheel revolving?

(A) 20 rpm
(B) 160 rpm
(C) 320 rpm
(D) 400 rpm

**82.** A knife, chisel, axe head, and nail are all examples of what type of simple machine?

(A) Screw
(B) Wedge
(C) Inclined plane
(D) Pulley

**83.** A modern use of the wheel and axle is in a

(A) log roller
(B) tongs
(C) doorknob
(D) log splitter

**GO ON TO THE NEXT PAGE**

**Directions:** Study the apparatus accident situation below for 3 minutes. At the end of the time, cover the picture. Then answer **questions 84 to 91**.

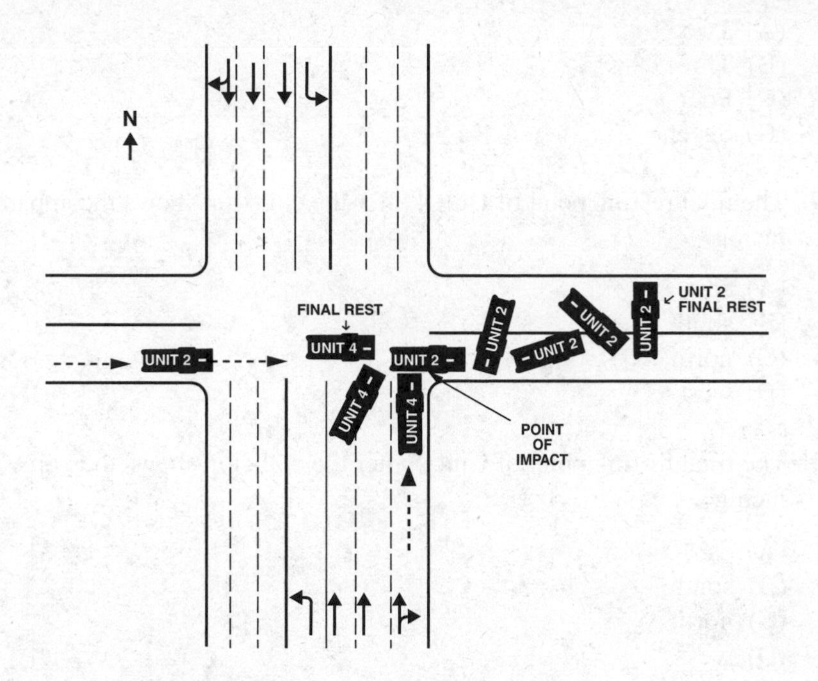


**84.** What type of accident has occurred?

(A) Rear-end collision
(B) Right-angle collision
(C) Front-end collision
(D) Sideswipe collision

**85.** In what direction was Unit 4 traveling prior to the collision?

(A) East
(B) South
(C) North
(D) West

**86.** In what direction was Unit 2 traveling prior to the collision?

(A) East
(B) South
(C) North
(D) West

**87.** How many northbound traffic lanes, including turning lanes, are south of the intersection?

(A) Two
(B) Three
(C) Four
(D) Five

**GO ON TO THE NEXT PAGE**

**88.** How many northbound traffic lanes, including turning lanes, are north of the intersection?

(A) Two
(B) Three
(C) Four
(D) Seven

**89.** The final resting point of Unit 4 after the collision shows the apparatus facing

(A) east
(B) south
(C) north
(D) west

**90.** The final resting point of Unit 2 after the collision shows the apparatus facing

(A) east
(B) south
(C) north
(D) west

**91.** Which of the following accurately describes this apparatus accident?

(A) Unit 4 was struck by Unit 2 as Unit 4 entered the intersection.
(B) Unit 2 was struck on the driver side of the apparatus.
(C) Unit 2 struck Unit 4 in the front (cab) section of the apparatus.
(D) Unit 4 was in the right-hand lane.

**92.** A flammable gas generated at most structural fires is

(A) chlorine
(B) fluorine
(C) carbon monoxide
(D) helium

**93.** A boiler failure is an example of what type of explosion?

(A) Nuclear
(B) Endothermic
(C) Backdraft
(D) Mechanical

**GO ON TO THE NEXT PAGE**

**94.** A component of the fire triangle listed below is

(A) fuel
(B) uninhibited chemical chain reaction
(C) foam
(D) water

**95.** What characteristic of most solid materials causes them to be more sus ceptible to ignition?

(A) High moisture content
(B) High surface area to mass ratio
(C) Low resistance to gravity
(D) Irregular shape

**96.** The ratio of the weight of a liquid to the weight of an equal volume of water is called the liquid's

(A) flash point
(B) specific gravity
(C) ignition temperature
(D) boiling point

**97.** A flammable liquid has a flash point of

(A) less than 100°F
(B) 100°F or greater
(C) 100°F to 200°F
(D) greater than 200°F

**98.** Liquids with low flash points tend to

(A) evaporate slowly
(B) flow evenly
(C) vaporize readily
(D) not burn

**99.** The property used to determine whether vapors/gases will hug the ground or rise up into the atmosphere is

(A) reactivity
(B) combustibility
(C) solubility
(D) vapor density

**100.** The lower and upper flammable limits of vapors/gases in air are used to determine

(A) if the vapor/gases will dissolve in water
(B) if the vapor/gases are lighter/heavier than air
(C) if the vapor/gases will ignite
(D) if the vapor/gases are toxic

**STOP. THIS IS THE END OF PRACTICE EXAM 3**

# Answer Key

1. B
2. D
3. B
4. B
5. A
6. D
7. C
8. D
9. A
10. D
11. B
12. A
13. B
14. D
15. C
16. A
17. D
18. B
19. A
20. C
21. C
22. D
23. B
24. A
25. D
26. B
27. D
28. C
29. A
30. D
31. B
32. C
33. D
34. B
35. C
36. B
37. D
38. A
39. D
40. A
41. A
42. C
43. C
44. B
45. B
46. D
47. A
48. C
49. C
50. D
51. B
52. C
53. A
54. D
55. C
56. B
57. E
58. I
59. G
60. B
61. J
62. A
63. H
64. F
65. C
66. D
67. C
68. B
69. A
70. D
71. B
72. B
73. B
74. C
75. D
76. A
77. C
78. A
79. D
80. C
81. C
82. B
83. C
84. B
85. C
86. A
87. C
88. B
89. A
90. C
91. D
92. C
93. D
94. A
95. B
96. B
97. A
98. C
99. D
100. C

# ANSWER EXPLANATIONS

Following you will find explanations to specific questions that you might find helpful.

**17.** D Part = % · Whole ÷ 100
Part = (40)(40) ÷ 100 = 1,600 ÷ 100 = 16
40 + 16 = 56

**18.** B Part = % · Whole ÷ 100
Part = (20)(70) ÷ 100 = 1,400 ÷ 100 = 14
70 – 14 = 56

**19.** A 3 ÷ 5 = 0.6 · 100 = 60%

**20.** C 28% = 28/100 = 7/25

**21.** C numbers increase by 5
4, 9, 14, 19, 24, 29, 34, 39

**22.** D 18:3 = 6:1

**23.** B Probability = event/possible outcomes = 12/60 = 1/5

**24.** A 7 shirts · 2 pants · 5 ties = 70 combinations

**25.** D $9^3 = 9 \cdot 9 \cdot 9 = 729$

**26.** B $\sqrt{100} = 10$

**27.** D $\sqrt{49} - \sqrt{16} = 7 - 4 = 3$

**28.** C (4)(8 hours) = 32 hours + 4 hours = 36 hours · $12 = $432

**29.** A 71.98 – $63.21 = $8.77

**30.** D 8 shots in 15 seconds;
8 shots · 4 = 32 shots per minute
32 shots · 2 minutes = 64 shots

**31.** B $12.00 ÷ 3 = $4.00

**32.** C F = 9/5(C) + 32
F = 9/5(35) + 32 = 63 + 32 = 95°

**33.** D C = 5/9(F – 32)
C = 5/9(68 – 32) = 5/9(36) = 20°

**44.** B Adult: chest = 18%; right arm (anterior and posterior) = 9%; left leg (anterior only) = 9%; groin = 1%

**45.** B Adult: 4.5% + 4.5% = 9%

**46.** D Adult: both arms (anterior and posterior) = 9% + 9% = 18%

**47.** A Adult: both legs (anterior and posterior) = 18% + 18% = 36%

**48.** C Child: right leg (anterior and posterior) = 13.5%

**49.** C Adult and child: groin = 1%

**50.** D Child: head = 18%

**51.** B Peaked roofs (front to back) and staggered from left to right (left building in front)

**52.** C Flip the drawing and tiered superstructure is at the left with no extension beyond the highest tier

**53.** A Chimney located at the right side of the building in front of the peak

**54.** D Flip the drawing and the exterior staircase is to the left and the roof antenna is to the right

**55.** C Flip the drawing and double doors are to the right and "Hardware" is spelled backwards

**56.** B Turn the drawing upside down in front of you

**72.** B Resistance (load) is between the effort (force) and the fulcrum

**73.** B MA = resistance force (load)/effort (force) = 100 N/100 N = 1

**74.** C When computing gear ratio, always compare the larger gear rotating once to the smaller gear, regardless of whether it is a driver or driven gear. The gear ratio of a gear train, also known as its speed ratio, is the ratio of the angular velocity of the input gear to the angular velocity of the output gear. The gear ratio can be calculated directly from the numbers of teeth on the gears in the gear train.
Driver gear (48 teeth):Driven gear (16 teeth)
For every one rotation of the driver gear, the driven gear rotates three times

**75.** D Visually count the number of ropes supporting the load

**78.** A A first-class lever has the fulcrum between the effort (force) and the resistance force (load).

**79.** D Effort × Effort distance = Resistance force × Resistance distance
Effort force × Length of inclined plane = Load × Height
Effort force × 8 ft = 80 lbs × 2 ft

$$\text{Effort force} = \frac{80 \text{ lbs} \times 2 \text{ ft}}{8\text{ft}} = \frac{80 \text{ lbs}}{4} = 20 \text{ lbs}$$

**80.** C Pulleys rotate in the same direction (unlike gears, which do the opposite).

**81.** C

Speed (Driver wheel) · Diameter (wheel) = Speed (Driven wheel) · Diameter (wheel)

$$80 \text{ rpm} \cdot 200 \text{ mm} = x \cdot 50 \text{ mm}$$

$$\frac{80 \text{ rpm} \cdot 200 \text{ mm}}{50 \text{ mm}} = x$$

$$80 \text{ rpm} \cdot 4 = x$$

$$320 \text{ rpm} = x$$

# PRACTICE EXAM 4

*For each question, circle the letter of your answer choice.*

**Directions:** Read the following reading passage to answer **questions 1 through 10**.

## Saving Lives

Many of the calls to the fire department regarding life-threatening situations do not involve fires. They may relate to medical emergencies where a victim is having a heart attack or choking. Firefighters will use their first aid, CPR, and CFR-D training in conjunction with resuscitator and defibrillator equipment to stop bleeding, splint broken bones, and restore lung and heart functions.

Firefighters are also called to motor vehicle accidents where people are injured from the impact of the collision or trapped inside the vehicle unable to extricate themselves from the wreckage. Manual forcible entry tools like the axe and halligan in conjunction with reciprocating saws and the Jaws of Life are used by firefighters to enter the vehicle and free victims from the metal around them. Public transportation accidents involving airplanes and trains resulting in multiple casualties are dealt with by firefighters using an abundance of manpower that may involve mutual aid assistance from surrounding jurisdictions to supplement operational needs.

Utility emergencies are yet other situations to which fire departments respond. Occupants may be overcome by the gases of incomplete combustion, such as carbon monoxide, emitted from their heating systems and stoves. Firefighters will have to don self-contained breathing apparatus (SCBA) to enter the building in order to gain access to the faulty heating appliance to shut it down and ventilate. Stuck elevators cause people inside to panic and fear for their safety and timely removal. The fire service may use special keys to open stuck hoistway doors or lower portable ladders from the floor above the stuck elevator car to reach and remove trapped occupants.

Natural disasters, including tornados, earthquakes, storms, floods, and hurricanes, endanger the lives of thousands and require fire service commitment for long periods of time.

Large municipalities may form preestablished incident management teams (IMT) composed of firefighters of all ranks to help manage and coordinate the rescue effort. The fire service also conducts lifesaving efforts on our nation's rivers, lakes, and seas utilizing both large vessels and swift-water rafts. These watercraft are used to rescue people from boats in distress as well as to save victims from drowning.

The list of non-fire-related situations where firefighters are needed to save lives is long. Firefighters must be ready to perform lifesaving acts of skill and bravery that do not involve actually extinguishing fires on a daily basis. The age of firefighters just going and operating at fires is a thing of the past. Firefighters are truly diversified first responders of the modern age.

**1.** Another appropriate title for this article would most likely be

(A) Multifaceted Heroes
(B) Fires and Firefighters
(C) Fire Extinguishment
(D) Medical Emergencies

**2.** An example of a medical emergency is

(A) a victim unable to exit a vehicle
(B) an occupied stalled elevator
(C) a choking victim
(D) a boat in distress

**3.** The Jaws of Life would most likely be used by firefighters in which of the following emergency situations?

(A) Motor vehicle accidents
(B) A drowning incident
(C) A stuck elevator
(D) Splinting broken bones

**4.** A deadly gaseous by-product of incomplete combustion is

(A) oxygen
(B) water vapor
(C) nitrogen
(D) carbon monoxide

**5.** A tool utilized by firefighters to free trapped occupants inside a stalled elevator is a

(A) reciprocating saw
(B) portable ladder
(C) defibrillator
(D) raft

**6.** Which of the following best describes the meaning of mutual aid?

(A) First aid training to stop a victim's bleeding
(B) Specific firefighting procedures at public transportation accidents
(C) Outside assistance to supplement operational needs
(D) None of the above

**GO ON TO THE NEXT PAGE**

**7.** A Halligan is an example of what type of tool used in the fire service?

(A) Jaws of Life
(B) First aid tool
(C) Reciprocating tool
(D) Forcible entry tool

**8.** Preestablished incident management teams may be formed by large municipalities during events that require a firefighting commitment for an extensive period of time. The IMT is specifically utilized to perform what function?

(A) Extinguish a fire
(B) Coordinate the rescue effort
(C) Investigate the cause of a public transportation accident
(D) Mitigate weather conditions

**9.** According to the passage, SCBA is most commonly used in what type of life-threatening emergency?

(A) A utility emergency
(B) A medical emergency
(C) A drowning incident
(D) An elevator emergency

**10.** Which of the following would not be considered a natural disaster?

(A) Tornado
(B) Earthquake
(C) Flood
(D) Heart attack

**Directions:** Use the information and table given below to answer **questions 11 through 18.**

When trying to lose weight, one must strive to expend energy in greater amounts than caloric intake. Use the table below showing the correlation between food (calories consumed) and the exercise required to burn off the calories to answer questions 11 through 18.

| Food | Calories | Exercise | Equivalent Time |
|---|---|---|---|
| Cake (1 piece) | 250 | rope skipping | 30 min |
| Apples (1 cup) | 65 | walking | 15 min |
| Cola (12 oz) | 160 | rowing | 15 min |
| Lamb chops (2 oz) | 210 | tennis | 30 min |
| Butter (1/4 cup) | 400 | bicycling | 60 min |
| Italian bread (slice) | 100 | aerobic class | 15 min |
| Flounder (3 oz) | 120 | yoga | 30 min |
| Lentils (1/2 cup) | 100 | swimming | 10 min |
| Ice cream (1 cup) | 400 | jogging | 60 min |

**GO ON TO THE NEXT PAGE**

11. Which exercise burns the most calories during a 30-minute workout?

(A) Swimming
(B) Rowing
(C) Yoga
(D) Jogging

12. Which of the following exercises burns the least calories during a 30-minute workout?

(A) Aerobic class
(B) Bicycling
(C) Walking
(D) Tennis

13. What food listed has the highest caloric content?

(A) Cola (6 oz)
(B) Lamb chops (2 oz)
(C) Italian bread (2 slices)
(D) Ice cream (1/4 cup)

14. Which type of workout would you have to do to burn off the equivalent caloric intake contained in three ounces of flounder and a half cup of lentils?

(A) Aerobic class (15 min) and yoga (30 min)
(B) Bicycling (30 min) and swimming (10 min)
(C) Jogging (30 min) and tennis (15 min)
(D) Rope skipping (15 min) and rowing (15 min)

15. Which exercise will burn off 600 calories in 1 hour?

(A) Bicycling
(B) Tennis
(C) Rope skipping
(D) Swimming

16. Which food grouping below provides the lowest caloric intake?

(A) Lamb chops (2 oz) and apples (1/2 cup)
(B) Flounder (3 oz) and cake (1 piece)
(C) Lentils (1/2 cup) and Italian bread (2 slices)
(D) Cola (12 oz) and ice cream (1/2 cup)

GO ON TO THE NEXT PAGE

17. A triathlete preparing for an upcoming event swims for 20 minutes, jogs for 30 minutes, and bicycles for an additional 1 hour. How many calories did the athlete burn during the workout?

(A) 1,200
(B) 1,000
(C) 800
(D) 700

18. How many minutes would a yoga practitioner have to exercise to burn off 360 calories?

(A) 60
(B) 75
(C) 90
(D) 120

19. Generally, which fuel below would contain the most heat energy per pound?

(A) Hydrogen gas
(B) Paper
(C) Wood
(D) Coal

20. Which is NOT considered a thermal heat unit?

(A) BTU
(B) Calorie
(C) Fahrenheit degree
(D) Joule

21. How many whole number increments are there on the Centigrade (Celsius) scale between the freezing point of water and its boiling point?

(A) 32
(B) 100
(C) 180
(D) 212

22. The Fahrenheit degree is equal to the

(A) Celsius degree
(B) BTU
(C) Kelvin degree
(D) Rankine degree

**GO ON TO THE NEXT PAGE**

**23.** Absolute zero is the hypothetical point at which

(A) water flows freely
(B) all materials explode
(C) all molecular movement ceases
(D) liquids change to gas

**24.** The amount and speed of heat transfer through an object is dependent on

(A) thermal conductivity
(B) shininess
(C) shape of the object
(D) color

**25.** Which of the following is an accurate statement concerning heat transfer via radiant waves?

(A) The darker/duller the object, the less heat it will absorb.
(B) The lighter/shinier the object, the more heat it will absorb.
(C) Radiant waves can pass through air, water, and glass.
(D) Radiant waves are limited to a distance of about 20 feet or less.

**26.** The fire process is divided into three phases. The initial phase is known as the

(A) decay phase
(B) cold phase
(C) free-burning phase
(D) growth phase

**27.** The graph that aids in the visualization of the heat energy and temperature attained during the three phases of fire is called the

(A) flame spread index
(B) standard time/temperature curve
(C) anatomy of fire
(D) fire phase scale

**28.** How many classes of fire are firefighters challenged to extinguish daily?

(A) 2
(B) 3
(C) 4
(D) 5

**GO ON TO THE NEXT PAGE**

**29.** Rubber products are listed under what classification of fire?

(A) Class A
(B) Class B
(C) Class D
(D) Class K

**30.** Which of the following is NOT considered to be a Class D material?

(A) Zirconium
(B) Uranium
(C) Sodium
(D) Iron

**31.** An ABC multipurpose dry chemical portable fire extinguisher is designed to put out fires in all of the choices listed below except

(A) vegetable cooking oils
(B) energized electrical equipment
(C) textiles
(D) diesel oil

**32.** When responding to fire alarms, officers are permitted to talk to chauffeurs driving the apparatus only under exceptional circumstances. What is most likely the reason for this?

(A) It allows the chauffeur to concentrate on driving the apparatus.
(B) It permits the officer to critique the chauffeur's driving ability.
(C) It maintains the rank structure between officer and firefighter.
(D) It enables the firefighters in the crew cab to hear urgent radio transmissions.

**33.** Fire prevention inspections ideally should be carried out at irregular hours and intervals. The best justification for this practice is that

(A) it permits firefighters to inconvenience business owners
(B) it enables firefighters to issue more violations and summonses
(C) it allows firefighters to beef up their inspection statistics
(D) it lets firefighters observe occupancies in their normal condition

**34.** A firefighter operating a hose line inside a large mazelike, one-story warehouse discovers the hose line has been cut and is no longer delivering water on the fire. Large amounts of smoke fill the area, and it is now time to leave the building. The best way for the firefighter to find the way to the outside of the building is

(A) find a wall and follow it to a side exit
(B) find a window and climb through it
(C) follow the hose line back to the entrance door
(D) move forward in the direction of the hose stream

GO ON TO THE NEXT PAGE

**35.** A firefighter on the way to work is confronted by a civilian complaining of blocked aisles in a department store a few blocks away. Of the following, the most appropriate action the firefighter should take in this situation is

(A) go with the civilian to the store to investigate the complaint
(B) assure the civilian that the violation is not serious and continue on your way
(C) obtain the address of the store and tell the civilian you will relay the nature of the complaint to your supervisor
(D) tell the civilian to return to the store and take pictures of the violation to substantiate the complaint and deliver the photographs to the firehouse

**36.** You are a firefighter with a new idea on how to force entry through entrance doors in commercial occupancies that you think will aid fellow firefighters. Your new method, however, is not sanctioned by the fire department where you work. Your best action to take concerning your new procedure is

(A) dismiss the idea since your department probably will not accept it
(B) train others in your new method on an unofficial basis
(C) use your new procedure at fire operations to allow others to witness its effectiveness
(D) talk to your supervisor concerning your new idea

**37.** Firefighters should answer the official department telephone on the first ring. What is the main reason for this?

(A) It demonstrates professionalism.
(B) It shows firefighters are quick with their hands.
(C) It gives the appearance that firefighters are waiting to be called to a fire.
(D) It will impress the caller, especially if it is a superior officer.

**38.** A firefighter on the way to work is standing at a bus stop when suddenly a pedestrian is seen choking on a hot dog. The civilian is having difficulty breathing and is unable to talk. In this instance the firefighter's action should be to

(A) run to his or her firehouse and seek assistance
(B) administer first aid if necessary to help restore normal breathing
(C) call for an ambulance via cell phone and board the arriving bus
(D) tell the choking victim to continue walking to the closest emergency room approximately four blocks away

**GO ON TO THE NEXT PAGE**

**39.** At the start of day tour, your superior officer gives you, a firefighter, an order that you do not fully understand. The assignment is not too complex and you don't want to make it seem that you were inattentive when the order was given. In this scenario you should most likely

(A) perform the task to the best of your ability
(B) ask a senior firefighter to help explain the order given
(C) ask the officer to please repeat the order
(D) wait till lunch to ask the officer to repeat the order

**40.** Standpipe system hose outlets are strategically placed inside large-area buildings allowing firefighters to connect to them and conduct hose line operations. The primary reason why such outlets are advantageous in fighting fires is

(A) different diameter hose lines can be connected to them
(B) they have male threads
(C) they may or may not have pressure reducer devices installed within their orifice
(D) they allow firefighters to stretch less hose line to reach the fire

**41.** What is the main reason why water is not recommended as an initial extinguishing agent for fires inside U.S. mailboxes?

(A) It generally is not successful in putting out the fire.
(B) The mail usually reacts with water causing a violent chemical reaction.
(C) Water is not used on fires located within confined spaces.
(D) The water may cause unburned mail to be undeliverable.

**42.** You are a probationary firefighter responding to your first fire. While on the apparatus, you turn on your self-contained breathing apparatus (SCBA) only to find that it is defective and will not allow air into the face piece. As you arrive on the scene of the fire, your initial action should be to

(A) tell the officer that your SCBA is defective and seek a replacement
(B) team up with a fellow firefighter in case you need to use a SCBA
(C) avoid the officer so as to not miss a chance at fighting the fire
(D) stay outside the building and notify your officer at the completion of his or her duties

**GO ON TO THE NEXT PAGE**

**43.** You are a newly assigned firefighter. You answer the department phone and the caller asks you for the address and phone number of a firefighter also assigned to the firehouse but who is not presently working. The caller states that it is important and that she is a relative of the firefighter she wishes to contact. In this situation, your best action is most likely

(A) to give the caller the information she seeks
(B) to tell the caller you cannot give out any personal information over the phone and instruct her to try a telephone operator
(C) to refer the caller to your superior
(D) to take down the caller's telephone number and give it to the firefighter she wishes to contact upon his or her return to work

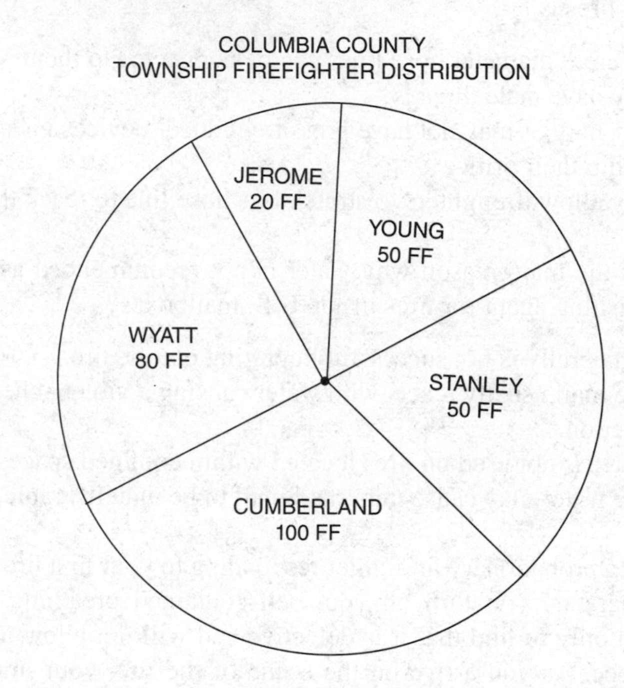


**Directions:** The pie chart above shows the distribution of firefighters in Columbia County. Use this information to answer **questions 44 through 49**.

**44.** Which two fire departments combined have more than 50% of the firefighters in the entire county?

(A) Wyatt and Stanley
(B) Jerome and Young
(C) Cumberland and Wyatt
(D) Stanley and Cumberland

**GO ON TO THE NEXT PAGE**

**45.** Which two fire departments have the same number of firefighters as Cumberland Township?

(A) Jerome and Wyatt
(B) Stanley and Jerome
(C) Wyatt and Stanley
(D) Young and Wyatt

**46.** Compared to the county, approximately what percentage of firefighters does the Stanley Township fire department have?

(A) 13.3%
(B) 16.6%
(C) 21.2%
(D) 26.1%

**47.** Together Jerome and Wyatt Townships have approximately what percentage of firefighters in the county?

(A) 20%
(B) 26.6%
(C) 33.3%
(D) 38.7%

**48.** Jerome Township, with the lowest number of firefighters, has approximately what percentage of firefighters in the county?

(A) 6.6%
(B) 13.3%
(C) 16.6%
(D) 20%

**49.** The total number of firefighters from Wyatt, Young, and Cumberland Townships is

(A) 190
(B) 200
(C) 210
(D) 230

**50.** How much 3% foam concentrate solution should be added to 100 gallons of 6% foam concentrate solution to obtain a 4% foam concentrate solution?

(A) 50 gallons
(B) 100 gallons
(C) 150 gallons
(D) 200 gallons

**51.** The sum of two consecutive integers is 23. If $x$ is the first integer and $x + 1$ is the second integer, what are the numbers?

(A) 11 and 12
(B) 9 and 14
(C) 7 and 16
(D) 5 and 18

**52.** The sum of two numbers is 38. If one number is six more than the other number, what is the smaller number of the two?

(A) 12
(B) 16
(C) 18
(D) 22

**GO ON TO THE NEXT PAGE**

**53.** What is the average of the following set of numbers: 8, 40, 61, 14, 7.

(A) 20
(B) 22
(C) 26
(D) 29

**54.** The total sum of a set of numbers is 84. The average of the numbers is 21. How many numbers are there in the set?

(A) 10
(B) 8
(C) 6
(D) 4

**55.** A student gets a 75, 82, and 90 on three 100-point exams in English class during the semester. What grade will be needed on the final 100-point exam to obtain an 85 average?

(A) 86
(B) 88
(C) 90
(D) 93

**56.** Paul is three times older than his grandson Bill. Bill is 10 years older than his sister, Sharon. The sum of Paul's, Bill's, and Sharon's ages is 140. What is Paul's age?

(A) 90
(B) 88
(C) 80
(D) 77

**57.** In question 56 above, what is Sharon's age?

(A) 30
(B) 21
(C) 24
(D) 20

**58.** Larry can complete apparatus quarterly maintenance by himself in 2 hours. Keith can perform this task in 3 hours. If they work together, how long will it take?

(A) 1 hour
(B) 1 hour and 12 minutes
(C) 1 hour and 20 minutes
(D) 1 hour and 30 minutes

**59.** What does the $W$ represent in the formula: $W = r \cdot t$?

(A) Weight
(B) Whole
(C) Work
(D) Wedge

**60.** David can paint a large shed in 6 hours. George can complete the task in 3 hours. If they work together, how long will it take?

(A) 5 hours
(B) 4 hours
(C) 3 hours
(D) 2 hours

**GO ON TO THE NEXT PAGE**

**61.** Two cars pass each other on a country road moving in opposite directions. Car A is traveling north at 40 mph, and Car B is traveling south at 30 mph. When will the two cars be 140 miles apart?

(A) 1 hour
(B) 2 hours
(C) 3 hours
(D) 4 hours

**62.** An athlete can run at 5 mph when there is no wind. With the wind at her back, she can run 21 miles in 3 hours. What is the wind speed?

(A) 1 mph
(B) 2 mph
(C) 3 mph
(D) 4 mph

**63.** A car travels past a road sign at 40 mph. Fifteen minutes later, a second vehicle, traveling in the same direction, passes the same road sign at 55 mph. How long will it take for the second vehicle to overtake the first?

(A) 20 minutes
(B) 30 minutes
(C) 40 minutes
(D) 50 minutes

**64.** Two tour buses pass each other going in opposite directions. One bus is going at 50 mph while the other bus is traveling at 40 mph. In how much time will the buses be 180 miles apart?

(A) 2 hours
(B) 2 hours and 30 minutes
(C) 3 hours
(D) 3 hours and 20 minutes

**65.** Carl has $20 in quarters and dimes and a total of 104 coins. How many quarters does he have?

(A) 32
(B) 40
(C) 58
(D) 64

**66.** In question 65 above, how many dimes does Carl have?

(A) 37
(B) 40
(C) 42
(D) 45

**67.** Morris earns two times more money per hour than Jones. Jones, however, makes $5 more per hour than Porter. Together they earn $80 per hour. How much money does Porter earn per hour?

(A) $51.25
(B) $47.50
(C) $16.25
(D) $12.75

**GO ON TO THE NEXT PAGE**

**68.** What is the probability that a king will be selected at random from a deck of 52 playing cards?

(Use formula: P = event/possible outcomes)

(A) 1:52
(B) 1:26
(C) 1:13
(D) 1:4

**69.** An iron column weighs 600 pounds and is 16 feet long. What is the weight of a similar column that is 10 feet long?

(A) 515 pounds
(B) 476 pounds
(C) 375 pounds
(D) 355 pounds

**Directions:** Use the standard time/temperature curve to answer **questions 70 through 78**.

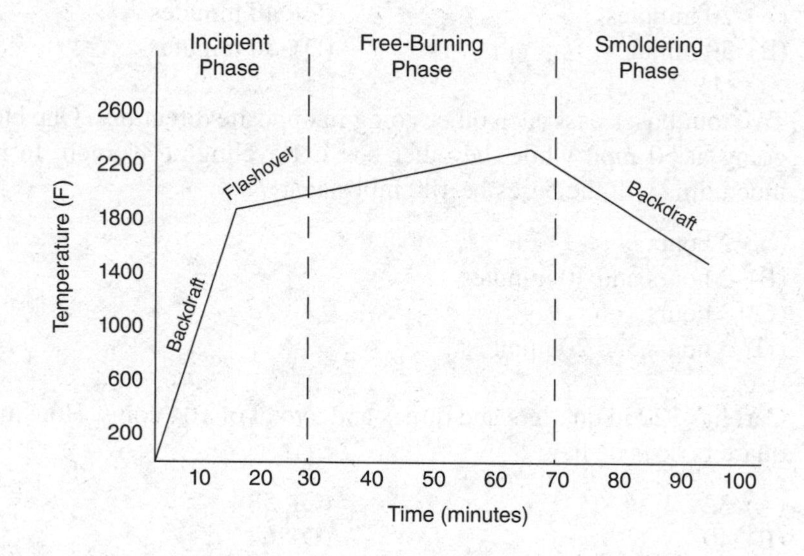


**70.** During which phase does temperature rise at the most rapid rate?

(A) Incipient
(B) Free burning
(C) Smoldering
(D) Temperature rate is equal in all phases

**71.** During which phase(s) are the highest temperatures attained?

(A) Incipient
(B) Free burning
(C) Smoldering
(D) Both incipient and smoldering

**GO ON TO THE NEXT PAGE**

**72.** What is the approximate highest temperature reached during the three phases of fire?

(A) 1,400°F
(B) 1,800°F
(C) 2,200°F
(D) 2,600°F

**73.** How long does a fire take to reach the smoldering phase?

(A) 50 minutes
(B) 70 minutes
(C) 90 minutes
(D) 120 minutes

**74.** During which phase does temperature remain fairly constant?

(A) Incipient
(B) Free burning
(C) Smoldering
(D) The temperature remains fairly constant during all phases

**75.** What is the approximate temperature 15 minutes into a fire?

(A) 1,000°F
(B) 1,400°F
(C) 1,800°F
(D) 2,600°F

**76.** During which phase(s) does flashover occur?

(A) Incipient only
(B) Incipient and smoldering
(C) Free burning only
(D) Free burning and smoldering

**77.** During which phase(s) of a fire does backdraft occur?

(A) Incipient only
(B) Free burning and smoldering
(C) Smoldering only
(D) Incipient and smoldering

**78.** The most significant aspect of the standard time/temperature curve is

(A) slow buildup of temperature during the early phases of fire
(B) rapid buildup of temperature during the smoldering phase
(C) erratic changes in temperature during the free-burning phase
(D) rapid buildup of temperature during the incipient phase

**GO ON TO THE NEXT PAGE**

**79.** Driver gear A has 60 teeth, while driven gear B has 12 teeth. What is the gear ratio of gear A to gear B?

(A) 1:5
(B) 1:4
(C) 5:1
(D) 4:1

**80.** When using simple machines, the trading of effort (force) for distance is called

(A) mechanical advantage
(B) momentum
(C) torque
(D) power

**81.** In what class lever system does the load move in the direction opposite to the effort (force)?

(A) Third class
(B) Second class
(C) First class
(D) None of the above

**82.** Driver gear A has 30 teeth and revolves at 70 rpm. What is the rpm for the driven gear B that has 10 teeth?

(A) 70
(B) 120
(C) 140
(D) 210

**83.** A 600-pound weight is placed on the end of a plank 2 feet from a fulcrum. To obtain equilibrium, how much effort (force) would a firefighter have to exert on the opposite end of the plank that is 10 feet from the fulcrum?

(A) 100 pounds
(B) 120 pounds
(C) 150 pounds
(D) 200 pounds

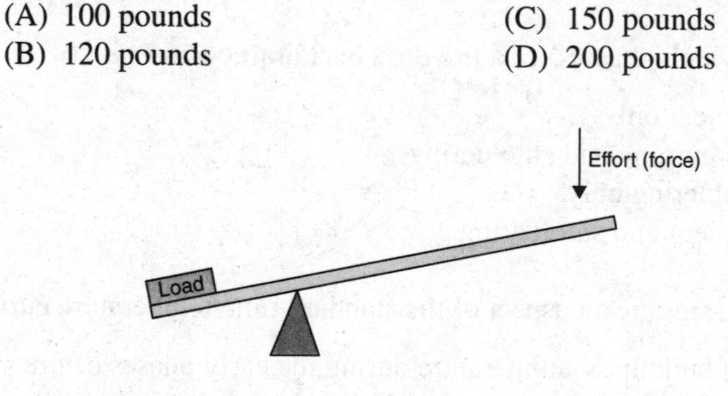


**84.** A lever system as shown above has a mechanical advantage of 4. If an effort (force) is applied at a distance 1 foot from the fulcrum, what is the resistance force (load) distance?

(A) 4 feet
(B) 1 foot
(C) 4 inches
(D) 3 inches

GO ON TO THE NEXT PAGE

**85.** The transfer of energy through motion is called

(A) mass
(B) weight
(C) work
(D) length

**86.** A hose reel has a handle attached to a 1-inch radius axle. The turning circumference of the crank is 12 inches. How much force is required to lift three lengths of hose that together weigh 100 pounds?

(A) 43 pounds
(B) 52 pounds
(C) 68 pounds
(D) 74 pounds

**87.** A pulley system is used to lift a 1,500-pound load. The system applies an effort (force) of 500 pounds. The mechanical advantage of the pulley system is

(A) 3
(B) 2
(C) 1
(D) 0

**88.** If the resistance force (load) of an object is 2,400 pounds, what would be the effort (force) required by a wedge with a mechanical advantage of 6 to lift it?

(A) 1,400 pounds
(B) 1,200 pounds
(C) 800 pounds
(D) 400 pounds

**89.** The circumference of a screw multiplied by the number of times it is turned is equal to its effort (force). A screw with a circumference of 7 mm is turned six times, causing it to move through a piece of gypsum board a distance of 3 mm. What is the mechanical advantage of the screw?

(A) 14
(B) 7
(C) 6
(D) 3

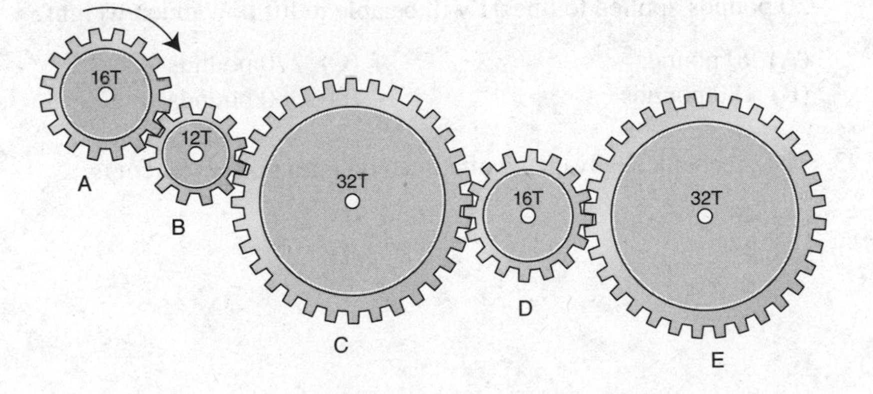


**GO ON TO THE NEXT PAGE**

**90.** In the gear drive system shown in the figure above, how does gear E compare to driver gear A?

(A) It rotates faster in the opposite direction.
(B) It rotates faster in the same direction.
(C) It rotates more slowly in the opposite direction.
(D) It rotates more slowly in the same direction.

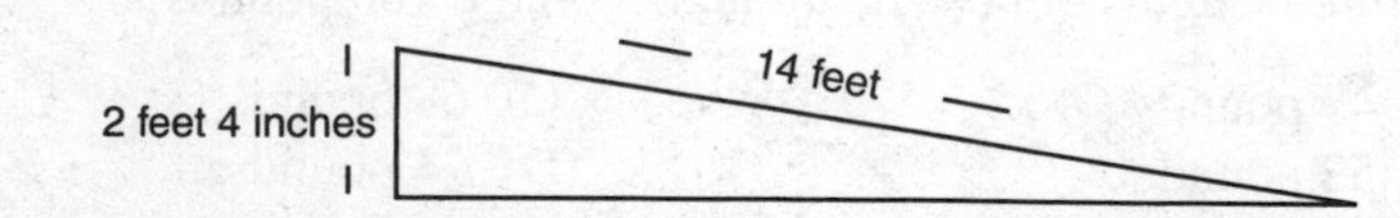


**91.** The inclined plane in the figure above has a mechanical advantage of

(A) 8 to 1
(B) 7 to 1
(C) 6 to 1
(D) 5 to 1

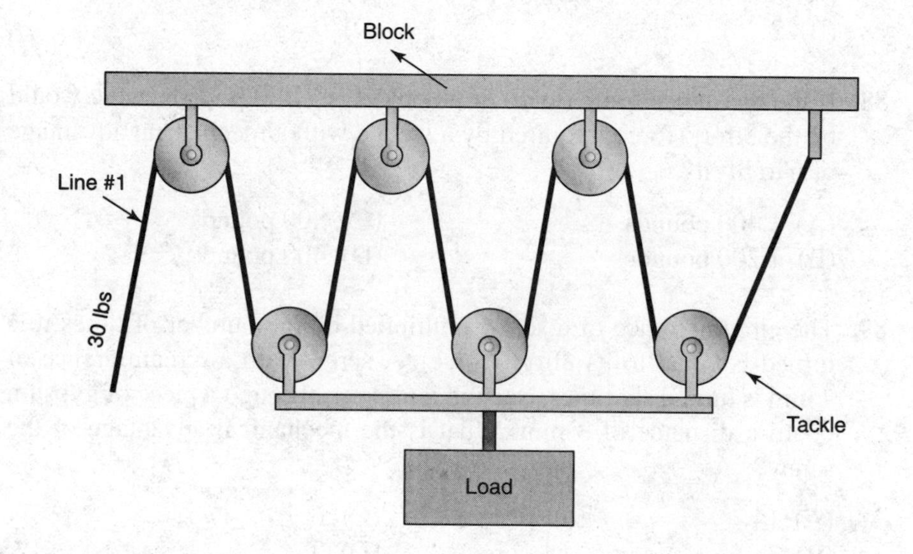


**92.** In the block and tackle pulley system in the figure above, a pull equal to 30 pounds applied to line #1 will be able to lift how much weight?

(A) 90 pounds
(B) 180 pounds
(C) 270 pounds
(D) 360 pounds

**93.** The mechanical advantage of a fixed (single) pulley system is

(A) 4
(B) 3
(C) 2
(D) 1

GO ON TO THE NEXT PAGE

**94.** In a fixed pulley system, if the effort (force) is applied in a downward direction, the load being lifted will travel

(A) in the same direction
(B) in the opposite direction
(C) in either the same or opposite direction
(D) perpendicular to the effort (force) being applied

**95.** A wheelbarrow holds a load weighing 40 pounds. The load is 18 inches from the wheels. How much effort (force) is required to lift the load if the handles of the wheelbarrow are 36 inches from the load?

(A) 40 pounds
(B) 30 pounds
(C) 20 pounds
(D) 10 pounds

**96.** In question 95 above, what is the mechanical advantage of the wheelbarrow?

(A) 4 to 1
(B) 3 to 1
(C) 2 to 1
(D) 1 to 1

**97.** An example of a third-class lever is a

(A) broom
(B) bottle opener
(C) bellows
(D) nutcracker

**98.** A pulley system lifts 160 pounds with an effort (force) of 40 pounds. The pulley rope where the effort (force) is being applied moves 4 inches. How many inches does the load move?

(A) 1 inch
(B) 4 inches
(C) 8 inches
(D) 16 inches

**99.** A pulley system has a mechanical advantage of 4:1. What is the efficiency of the system if the rope end where effort (force) is being applied moves 10 inches while the load is lifted 2 inches?

(A) 50%
(B) 65%
(C) 75%
(D) 80%

**100.** How much force is required to roll a 250-N barrel up an inclined plane that is 5 meters in length to a platform 2 meters in height?

(A) 125 N
(B) 100 N
(C) 80 N
(D) 75 N

**STOP. THIS IS THE END OF PRACTICE EXAM 4**

# Answer Key

1. A
2. C
3. A
4. D
5. B
6. C
7. D
8. B
9. A
10. D
11. B
12. C
13. B
14. A
15. D
16. A
17. C
18. C
19. A
20. C
21. B
22. D
23. C
24. A
25. C
26. D
27. B
28. D
29. A
30. D
31. A
32. A
33. D
34. C
35. C
36. D
37. A
38. B
39. C
40. D
41. D
42. A
43. C
44. C
45. A
46. B
47. C
48. A
49. D
50. D
51. A
52. B
53. C
54. D
55. D
56. A
57. D
58. B
59. C
60. D
61. B
62. B
63. C
64. A
65. D
66. B
67. D
68. C
69. C
70. A
71. B
72. C
73. B
74. B
75. C
76. A
77. D
78. D
79. A
80. A
81. C
82. D
83. B
84. D
85. C
86. B
87. A
88. D
89. A
90. D
91. C
92. B
93. D
94. B
95. C
96. C
97. A
98. A
99. D
100. B

# ANSWER EXPLANATIONS

Following you will find explanations to specific questions that you might find helpful.

**11.** B Rowing = 160 calories for 15 minutes × (2) = 320 calories
Swimming = 100 calories for 10 minutes × (3) = 300 calories
Yoga = 120 calories for 30 minutes
Jogging = 600 calories for 60 minutes ÷ 2 = 300 calories

**12.** C Walking = 65 calories for 15 minutes × 2 = 130 calories
Aerobics class = 100 calories for 15 minutes × 2 = 200 calories
Bicycling = 400 calories for 60 minutes ÷ 2 = 200 calories
Tennis = 210 calories for 30 minutes

**14.** A 3 oz flounder = 120 calories; ½ cup of lentils = 100 calories = 220 calories
Aerobics class = 100 calories for 15 minutes + yoga = 120 calories for 30 minutes = 220 calories

**15.** D Swimming = 100 calories for 10 minutes × 6 = 600 calories

**16.** A Lamb chops (2 oz) = 210 calories + [apples (1 cup) = 65 calories ÷ 2] = 32.5 calories = 242.5
Flounder (3 oz) = 120 calories + cake (1 piece) = 250 calories = 370
Lentils (1/2 cup) = 100 calories + Italian bread (2 slices) = 2 × 100 calories = 200 calories = 300
Cola (12 oz) = 160 calories + ice cream (1 cup) = 400 calories ÷ 2 (1/2 cup) = 200 calories = 360

**17.** C Swimming (20 min) = 100 calories (10 min) × 2 = 200 calories
Jogging (30 min) = jogging for 60 min = 400 calories ÷ 2 = 200 calories
Bicycling (60 min) = 400 calories
200 + 200 + 400 = 800 calories

**18.** C Yoga (30 min) = 120 calories
30 min × 3 = 90 min = 360 calories

**19.** A Fahrenheit (F) degrees is a temperature unit

**25.** C
A The darker/duller the object, the more heat it will absorb.
B The lighter/shinier the object, the less heat it will absorb.
D Large amounts of radiated heat can travel large distances (50 to 100 feet).

**28.** D Class A, Class B, Class C, Class D, and Class K

**30.** D Iron is not a combustible metal. It will heat up when contacted by fire, but will not ignite.

**31.** A Vegetable cooking oil is a Class K material.

**44.** C Cumberland Township 100 FF + Wyatt Township 80 FF = 180 FF
Wyatt Township 80 FF + Stanley Township 50 FF = 130 FF
Jerome Township 20 FF and Young Township 50 FF = 70 FF
Stanley Township 50 FF + Cumberland Township 100 FF = 150 FF

**45.** A Cumberland Township = 100 FF
Jerome Township 20 FF + Wyatt Township 80 FF = 100 FF

**46.** B What percent of 300 (total number Columbia County FF) is 50 (number FF in Stanley Township)?
$\% = 50 \div 300 \cdot 100 = 16.6\%$

**47.** C Jerome Township = 20 FF + Wyatt Township = 80 FF = 100 FF
What percent of 300 is 100?
$\% = 100 \div 300 \cdot 100 = 33.3\%$

**48.** A Jerome Township = 20 FF
What percent of 300 is 20?
$\% = 20 \div 300 \cdot 100 = 6.6\%$

**49.** D Wyatt Township = 80 FF + Young Township = 50 FF + Cumberland Township = 100 FF = 230 FF

**50.** D 100 gallons of 6% foam concentrate solution
Add 100 gallons of 3% foam concentrate solution = 5% foam concentrate solution
Add 100 gallons more of 3% foam concentrate solution = 4% foam concentrate solution

**51.** A Sum of two consecutive integers is 23
$x + (x + 1) = 23$
$2x + 1 = 23$
$2x = 22$
$x = 11; x + 1 = 12$
Check your answer: $11 + (11 + 1) = 11 + 12 = 23$

**52.** B Sum of two numbers is 38
$x + (x + 6) = 38$
$2x + 6 = 38$
$2x = 32$
$x = 16; x + 6 = 22$
Check your answer: $16 + (16 + 6) = 16 + 22 = 38$

**53.** C $8 + 40 + 61 + 14 + 7 = 130 \div 5 = 26$

**54.** D Total sum of numbers = 84; average of the numbers = 21
What number divided into 84 equals 21?
$84 \div 4 = 21$
4

**55.** D $75 + 82 + 90 = 247/300 \cdot 100 = 82.3\%$
$247 + 86 = 333/400 \cdot 100 = 83.2\%$
$247 + 90 = 337/400 \cdot 100 = 84.2\%$
$247 + 93 = 340/400 \cdot 100 = 85\%$

**56.** A Sum of all ages = 140 years
Paul = $3x$
Bill = $x$
Sharon = $x - 10$
$3x + x + (x - 10) = 140$
$4x + (x - 10) = 140$
$5x - 10 = 140$
$5x = 150$
$x = 30$; Paul = 90 yrs

**57.** D Sharon = x – 10
Sharon = 30 – 10 = 20 yrs

**58.** B Larry = 2 hours; Keith = 3 hours
The variable $t$ stands for how long both Larry and Keith take to do the job together. Then they can do: $1/t$ per hour
½ + 1/3 = 5/6
$1/t = 5/6$; flip the equation:
$t/1 = 6/5 = 1.2$ hours; change .2 hours to minutes; change decimal to fraction and multiply by 60
$2/10 \cdot 60/1 = 12$ minutes
1 hour and 12 minutes

**60.** D Let $x$ = time to mow lawn together
David does 1/6 the work in one hour
George does 1/3 the work in one hour
$1/x = 1/6 + 1/3$; find a common denominator ($18x$) by multiplying the denominators
$18 = 6x + 3x$
$18 = 9x$
$18/9 = x$
$2 = x$
It will take David and George 2 hours working together

**61.** B $d = r \cdot t$
40 mph + 30 mph = 70 mph
Let $t$ = the number of hours the cars will travel after passing each other
140 miles = 70 mph · $t$
$$\frac{140 \text{ miles}}{70 \text{ mph}} = 2 \text{ hours}$$

**62.** B $d = r \cdot t$
$r$ = wind speed
$(r + 5)$ = speed running with wind at his back
21 miles = $(r + 5)$ (3 hours) = $3r + 15$
$6 = 3r$
$2 = r$

**63.** C $d = r \cdot t$

Car A = (40 mph) $\frac{(15 \text{ minutes})}{(60 \text{ minutes})}$

Car A = (40)¼ = 10 miles traveled after Car B reaches the same road sign

Car B is gaining at a rate of 55 mph – 40 mph = 15 mph

10 miles = 15(t): divide both sides of equation by 15

$10/15 = t$

$2/3 = t$ = 40 minutes

**64.** A $d = r \cdot t$

Bus A = 50 mph; Bus B = 40 mph = 90 mph

180 miles = 90 mph $\cdot\, t$

180 miles/90 mph = $t$

2 hours = $t$

**65.** D $20; 104 coins

$x = 0.10$ (number of dimes in a dollar amount)

$104 - x$ = number of quarters; $0.25(104 - x)$ – dollar amount

$0.10x + 0.25(104 - x) = \$20$

$0.10 + 26 - 0.25x = \$20$

$-0.15x = -6$

$-x = -40$

$x = 40$ (dimes)

$104 - 40 = 64$ (quarters)

**66.** B See Question 65 explanation.

**67.** D Let $x$ = Porter; $x + 5$ = Jones; $2(x + 5)$ = Morris

$x + (x + 5) + 2(x + 5) = \$80$

$x + x + 5 + 2x + 10 = \$80$

$4x + 15 = \$80$

$4x/4 = 65/4 = x = \$16.25$ (Porter)

$x + 5 = 16.25 + 5 = \$21.25$ (Jones)

$2\,(x + 5) = 2\,(16.25) + 10 = 32.50 + 10 = \$42.50$ (Morris)

Check your answer: 16.25 + 21.25 + 42.50 = $80

**68.** C P = event/possible outcomes

4/52 = 1/13 = 1: 13

**69.** C 600 lbs ÷ 16 ft = 37.5 lbs per ft

10 ft · 37.5 lbs = 375 lbs

**79.** A When computing gear ratio, always compare the larger gear rotating once to the smaller gear, regardless of whether it is a driver or driven gear.

Gear A rotates one time while Gear B rotates five times

$$\frac{\text{Distance moved by effort (force)}}{\text{Distance moved by resistive force (load)}} = \frac{1}{5} = 1{:}5$$

**82.** D Gear A rotates one time while Gear B rotates three times

Gear A = 70 rpm

Gear B = 70 rpm · 3 = 210 rpm

**83.** B $\text{MA} = \dfrac{\text{effort (force) distance}}{\text{resistance force (load) distance}} = \dfrac{10\text{ ft}}{2\text{ ft}} = 5$

$5 = \dfrac{600\text{ lbs}}{\text{effort}} = \dfrac{600\text{ lbs}}{x}$

$5x = 600$ lbs

$x = 120$ lbs

**84.** D MA = 4

$4 = \dfrac{\text{effort (force) distance}}{\text{resistance force (load) distance}} = \dfrac{12\text{ in}}{x}$

$4 = 12\text{ in}/x$

$4x = 12$ in

$x = 3$ in

Effort (force) · Crank wheel circumference = Resistive force (load) · Circumference of axle

| | | | | |
|---|---|---|---|---|
| $x$ | 12 in | = | 100 lbs | $2\pi r$ |
| $x$ | 12 in | = | 100 lbs | 6.28 in |
| | $12x$ | = | 628 lbs | |

$x = 628\text{ lbs}/12 = 52$ lbs

**87.** A $\text{MA} = \dfrac{\text{resistive force (load)}}{\text{effort (force)}} = \dfrac{1{,}500\text{ lbs}}{500\text{ lbs}} = 3$

**88.** D MA = 6

2,400 lbs/6 = 400 lbs

**89.** A 7 mm · 6 = 42 mm

$\text{MA} = \dfrac{\text{effort (force) distance}}{\text{resistance force (load) distance}} = \dfrac{42\text{ mm}}{3\text{ mm}} = 14$

**91.** C $\text{MA} = \dfrac{\text{effort (force) distance}}{\text{resistance force (load) distance}} = \dfrac{14\text{ ft}}{2\text{ft }4\text{ in}} = \dfrac{168\text{ in}}{28\text{ in}} = 6$

**92.** B Visually counting the number of ropes supporting the load = 6 = MA

30 lbs · 6 = 180 lbs

**95.** C effort (force) · effort distance = load · resistance distance

| | | | | |
|---|---|---|---|---|
| $x$ | · | 36 in | = | 40 lbs · 18 in |
| $x$ | · | 2 in | = | 40 lbs · 1 in |
| $2x$ | | | = | 40 lbs |

$x = 20$ lbs

**96.** C If 20 lbs of effort (force) is required to move a load of 40 lbs, the MA = 2

**98.** A $\text{MA} = \dfrac{\text{resistance force (load)}}{\text{effort (force)}} = \dfrac{160 \text{ lbs}}{40 \text{ lbs}} = 4$

MA = 4. The trade-off is that the effort (force) distance is greater than the resistance force (load) distance.
If the pulley rope where the effort (force) is being applied moves 4 inches, the load moves 1 inch.

**99.** D MA = 4

$\text{Velocity ratio (VR)} = \dfrac{\text{effort (force) distance}}{\text{load distance}} = \dfrac{10 \text{ in}}{2 \text{ in}} = 5$

$\text{MA} \div \text{VR} = 4 \div 5 = 0.8 \cdot 100 = 80\%$

**100.** B

effort (force) · effort (distance) = resistance (force) · resistance (distance)

$$x \cdot 5 \text{ m} = 250 \text{ N} \cdot 2 \text{ m}$$
$$x = 250 \text{ N} \cdot 2 \text{ m}/5 \text{ m}$$
$$x = 250 \text{ N} \cdot 0.4$$
$$x = 100 \text{ N}$$

# PRACTICE EXAM 5

*For each question, circle the letter of your answer choice.*

**Directions:** Read the following reading passage to answer **questions 1 through 15**.

## The Firefighter's Role in Determining the Cause of a Fire

Usually a combination of factors causes a fire—the type of fuel ignited, the form and source of the heat of ignition, and the human action or omission of action (if a person is involved).

Firefighters responding to and operating at the scene of a fire can play a significant role in assisting the fire investigator (fire marshal) in his or her investigation of the cause and origin of the fire. Fire investigators are rarely present on the scene while firefighters are battling the fire. They cannot observe who is at the scene watching the firefighters upon their arrival. They are seldom present to talk with occupants and witnesses while the fire is still burning and when people are more apt to discuss particulars of the incident that could point to the cause of the fire. Fire investigators are also not there to limit the amount of overhauling (search and complete extinguishment of hidden fire) that can destroy important evidence.

Through observation, trained firefighters can gather important information that, when communicated to the investigators, can aid greatly in determining valid facts about the fire and how it started. On receipt of the alarm, firefighters should record the time of day and weather conditions. Time of day could indicate whether occupants of residential buildings are at home sleeping or awake, fully clothed or in pajamas. If the outside temperature is high, the heating unit may not be in operation. Conversely, if the temperature is low, the windows should be in a closed position. Firefighters arriving at the scene of a fire should try to observe any conditions contrary to normal expectations.

Firefighters arriving on the scene of the fire within five minutes of the initial alarm to find a house fully involved in fire should be suspicious as to why the fire spread so fast and be on the lookout for signs of arson. Fires found in more than one location in and around the building are definite indicators of an incendiary (arson) fire.

Occupants leaving the scene of a fire should also be noted. Firefighters should try to remember the make, model, and color of vehicles fleeing the area upon their arrival. They should also try to make a mental picture of how people at the incident look and dress. People observing the action from a distance may be enjoying themselves and smiling. This information should be relayed to investigators. Time of arrival and extent of fire are other key bits of information that will prove useful to the investigator.

Firefighters should also observe the color of smoke and flames emanating from doors and windows. These observations can help determine what type of fuel is burning and thus help the investigators in their search for the cause and origin of the fire. This information may also assist the fire investigators in determining whether the type of fuel burning should or should not be in the structure under ordinary circumstances.

Firefighters should also report to investigators any signs of forcible entry prior to their arrival. Are there discarded burglary tools or empty containers with residual flammable liquid in or around the perimeter of the building? Are doors and windows locked or unlocked? Are doors and windows covered over to delay the discovery of fire inside the building?

During actual firefighting operations, firefighters should continue to observe conditions that could lead to the determination of the cause and origin of the fire. Unusual odors (gasoline, paint thinner, kerosene, etc.) could indicate that accelerants were used to start and enhance the fire. Incendiary devices (items used by arsonists to start fires) found inside the building are extremely important clues for the fire investigator to prove that the fire is incendiary (arson). If at all possible, these devices should not be disturbed by firefighters during their extinguishment duties. Information concerning the absence of furniture, appliances, clothing, and personal belongings should also be communicated to fire investigators for their review and evaluation.

At the conclusion of fire operations, all information gathered by firefighters should be reported to the chief officer-in-command of the fire incident. The chief officer-in-command will record all information and relay it to the fire investigators assigned to the case. The chief officer should also perform on a limited basis preliminary interviews with owners, occupants, and witnesses to obtain supplementary information needed by the fire investigators upon their arrival. The chief officer should seek base-level information, including positive identification, contact numbers, and a brief description of events and actions prior to and during the fire. At no time should the chief officer make any attempt to cross-examine a potential arson suspect.

Firefighters trained in observational skills upon receipt of an alarm, en route to the fire, on arrival at the fire scene, and during actual firefighting operations are tremendous assets to the fire investigator. The information gathered and relayed to the chief officer-in-command will play an important role in the ultimate findings and conclusions of the fire investigator.

**GO ON TO THE NEXT PAGE**

1. Which of the following is NOT considered a factor in the cause of a fire?

(A) Fuel ignited
(B) Form and source of the heat of ignition
(C) Human action or omission
(D) Firefighters operating at the scene of fires

2. Based on the passage, an accurate statement concerning fire investigators is that they are

(A) rarely present on the scene while firefighters are battling the blaze
(B) able to observe onlookers watching firefighters at work
(C) able to talk to witnesses while the fire is still burning
(D) able to limit overhauling by firefighters

3. Occupants are more apt to discuss particulars concerning the fire with investigators during what time frame?

(A) Two days after the fire
(B) While the fire is still burning
(C) The day after the fire
(D) None of the above

4. Upon the receipt of the alarm at the firehouse, firefighters should record what pertinent information?

(A) Make, model, and color of vehicles leaving the scene
(B) Physical description of onlookers
(C) Time of day and weather conditions
(D) Position of windows, open or closed

5. According to the passage, if outside temperatures are high, it can be assumed that

(A) windows will be in the closed position
(B) doors may be found unlocked
(C) the heating unit may not be in operation
(D) fire spread will be unusually fast

6. The color of smoke and flame coming from the fire building can indicate to the fire investigator

(A) the type of fuel burning
(B) the weather conditions
(C) the appliance involved in the fire
(D) how much firefighter overhauling will be required

GO ON TO THE NEXT PAGE

**7.** Firefighters arriving at the fire scene should make a mental note of all of the following bits of information for relay to investigators EXCEPT

(A) occupants leaving the scene of the fire
(B) smoke and flame colors coming from doors and windows
(C) signs of forcible entry prior to their arrival
(D) positive identification of the owner of the building

**8.** According to the article, unusual odors encountered inside the fire building could indicate to firefighters that

(A) accelerants were used to start and enhance the fire
(B) the fire is accidental
(C) the fire was set by juvenile delinquents
(D) burglary tools will be found around the perimeter of the building

**9.** The term *incendiary device* means

(A) the cause and origin of the fire
(B) windows covered over to delay the discovery of fire
(C) an item used by arsonists to start a fire
(D) absence of personal belongings

**10.** At the conclusion of fire operations, all information gathered by fire-fighters should be relayed to

(A) fire investigators directly
(B) the chief officer-in-command of the fire
(C) the owner, occupants, and witnesses during interview sessions
(D) the arson suspect

**11.** Forcible entry, as used in this reading passage, most likely means

(A) unlawful access into a premises
(B) unnecessary entry by firefighters into a building on fire
(C) damage done to the roof of buildings by falling trees
(D) None of the above

**12.** Another name for fire investigator is

(A) arsonist
(B) firefighter
(C) fire marshal
(D) chief officer

**13.** All but which of the following are listed in the passage as possible accelerants?

(A) Gasoline
(B) Ammonia
(C) Kerosene
(D) Paint thinner

**GO ON TO THE NEXT PAGE**

**14.** Which of the following is the best definition for the term *arson*?

(A) People observing firefighters working at a fire
(B) A house fully involved in fire
(C) A deliberately set fire
(D) An empty container found outside the fire building

**15.** Another appropriate title for this passage is

(A) Built to Burn
(B) Arson and the Use of Accelerants and Incendiary Devices
(C) Juvenile Delinquents
(D) Facilitating the Work of the Fire Investigator

**16.** A fire truck having an extension ladder that is raised and lowered using the power of the truck is called a (an)

(A) engine
(B) pumper
(C) aerial ladder
(D) water tender

**17.** Ball valves, gate valves, wyes, and water thiefs are commonly termed

(A) alarm systems
(B) appliances
(C) hose lines
(D) portable fire extinguishers

**18.** An internal water container found within an engine apparatus is known in the fire service as a

(A) Bambi bucket
(B) air tanker
(C) booster tank
(D) hard suction bed

**19.** A backfire is

(A) a fire located to the rear of firefighters using an attack hose
(B) a fire intentionally set by wildland firefighters
(C) a fire that is threatening to enter a fire shelter
(D) a smoke explosion

**20.** Material that will burn and ignite is referred to as

(A) viscous
(B) inert
(C) miscible
(D) combustible

**21.** In general, lieutenants and captains in the fire service are known as

(A) fire officers
(B) chief officers
(C) fire investigators
(D) staff officers

**GO ON TO THE NEXT PAGE**

**22.** Discussions among first responders that are conducted by professionals after demanding operations and that are intended to help those responders cope with their emotions are called

(A) rest and recuperation
(B) demobilization
(C) reconnaissance
(D) critical incident stress debriefing

**23.** The use of suction hose and the engine apparatus pump to lift water from below the level of the pump is called

(A) dilution
(B) drafting
(C) recycling
(D) positive pressure pumping

**24.** A water source for firefighters consisting of one or more valves and outlets usually connected to a municipal water supply is called a

(A) fire extinguisher
(B) fire hydrant
(C) fire stream
(D) portable reservoir

**25.** A hose line stream having a wide pattern with small droplets of water is called a __________ stream.

(A) fog
(B) foam
(C) solid
(D) straight

**26.** Supply hose used in the fire service

(A) has a large diameter
(B) is used as an attack hose line to extinguish fire
(C) provides a small amount of water to the fire apparatus
(D) is used as a booster hose line

**27.** An advanced life support emergency care provider is called a

(A) proby
(B) chauffeur
(C) paramedic
(D) company officer

**28.** A reduction in the amount of water flowing through hose or piping as a result of inner lining resistance is called

(A) emulsification
(B) inertia
(C) friction loss
(D) None of the above

**29.** An extinguishing agent created by introducing air into a mixture of water and concentrate is called

(A) fog
(B) foam
(C) spray
(D) a clean agent

**GO ON TO THE NEXT PAGE**

**Directions:** Study the building layout shown in the drawing for three minutes. It depicts the top floor of a four-story rectangular motor inn complex. The room configuration is identical on all floors. The room numbering system begins with 101 on the first floor, 201 on the second floor, and so forth. All stairways serve all floors and can be entered from both the exterior parking lot areas and from the interior hallway. The interior courtyard can be entered directly from all interior perimeter rooms (101 to 159) on the first floor and universally via the first floor hallway. The fire is in apartment number 441 as shown. At the end of your allotted time, cover the drawing and you will be asked to answer **questions 30 through 44**.

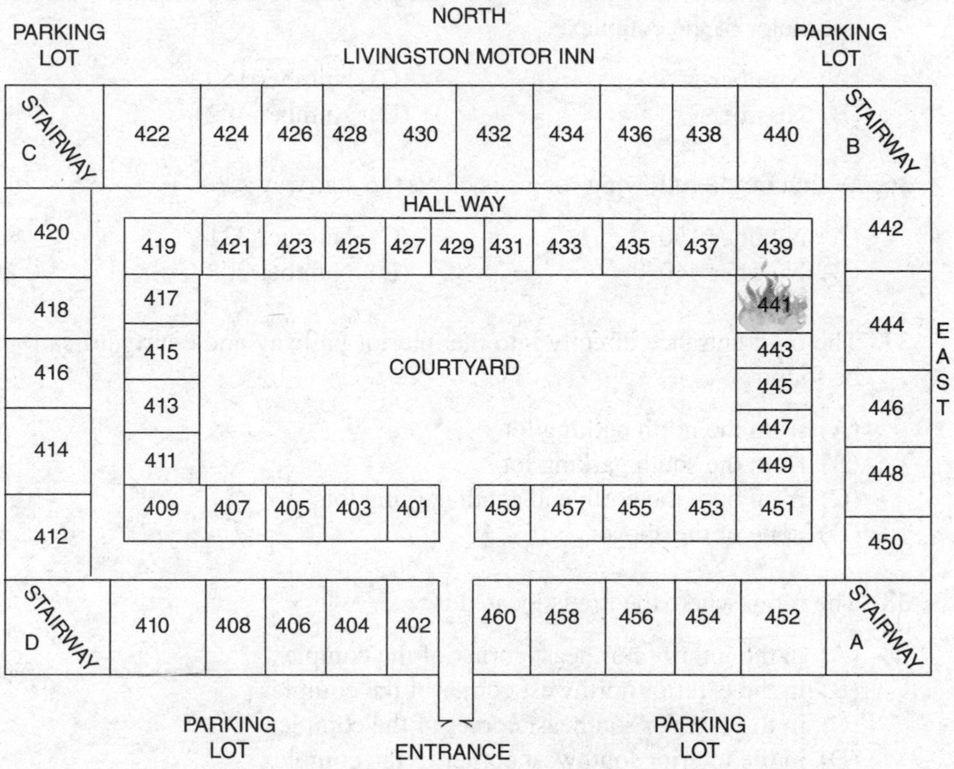

NORTH
LIVINGSTON MOTOR INN
PARKING LOT
PARKING LOT
STAIRWAY C
422 424 426 428 430 432 434 436 438 440
STAIRWAY B
HALL WAY
420
419 421 423 425 427 429 431 433 435 437 439
442
418
417
441
444
416
415
COURTYARD
443
445
E A S T
446
414
413
447
411
449
448
409 407 405 403 401 459 457 455 453 451
412
450
STAIRWAY D
410 408 406 404 402 460 458 456 454 452
STAIRWAY A
PARKING LOT
ENTRANCE
PARKING LOT

**30.** What is the name of the motor inn?

(A) Courtyard
(B) Livingston
(C) Lincoln
(D) Lawrence

**31.** How many stairways are there in the complex?

(A) 6
(B) 5
(C) 4
(D) 2

**32.** How many rooms open directly into the interior courtyard?

(A) 26
(B) 28
(C) 30
(D) 32

**GO ON TO THE NEXT PAGE**

**33.** How many rooms are there on each floor?

(A) 52
(B) 56
(C) 58
(D) 60

**34.** Where is room number 219 located?

(A) In the southeast corner of the complex
(B) In the northwest corner of the complex
(C) In the eastern section of the complex
(D) In the southern section of the complex

**35.** Which of the following rooms would you expect to find in the outer perimeter of the complex?

(A) Number 255
(B) Number 313
(C) Number 159
(D) Number 402

**36.** Which of the following rooms is closest to stairway A?

(A) Number 150
(B) Number 440
(C) Number 321
(D) Number 208

**37.** The only entrance directly into the interior hallway and courtyard can be found

(A) from the north parking lot
(B) from the south parking lot
(C) from both the north and south parking lots
(D) none of the above

**38.** The room where the fire is located is

(A) in the interior northeast corner of the complex
(B) in the exterior northwest corner of the complex
(C) in the exterior southeast corner of the complex
(D) in the interior southwest corner of the complex

**39.** Which staircase should firefighters select to reach the seat of the fire upon arriving at the north parking lot?

(A) Stairway B
(B) Stairway D
(C) Stairway A
(D) Stairway C

**40.** The fire in room number 441 is threatening to

(A) enter into stairway C
(B) enter into room number 440
(C) enter the courtyard shaft
(D) burn itself out

**GO ON TO THE NEXT PAGE**

**41.** Should the fire enter the interior hallway, which of the following rooms is most likely to be threatened?

(A) Room 414
(B) Room 458
(C) Room 444
(D) Room 440

**42.** What room is directly above the fire room?

(A) Number 341
(B) Number 343
(C) Number 444
(D) None of the above

**43.** The fastest way for firefighters in the south parking lot to stretch an attack hose line to extinguish the fire is to

(A) enter the building through the south entrance and into the courtyard
(B) enter the building through the south entrance and into the first-floor hallway
(C) enter stairway D to the fourth floor and into the hallway
(D) enter stairway A to the fourth floor and into the hallway

**44.** Firefighters carrying a charged hose line coming out of stairway A on the fourth floor should travel in what direction to reach the fire room?

(A) North
(B) South
(C) East
(D) West

**Directions:** Choose the best answer to each of the following questions.

**45.** Firefighters should not give unofficial talks at community meeting halls primarily because

(A) they are not professional speakers
(B) they may anger people in attendance
(C) this limits the possibility of citizens getting misinformation
(D) the fire department must have total control of its membership

**46.** During an inspection, you, a firefighter, are asked by the owner of the establishment a fire prevention question. You are unsure of the answer. In this situation you should

(A) tell the owner you are only on the premises to perform an inspection
(B) tell the owner to look the answer up in the Fire Code
(C) ignore the question and continue on with the inspection
(D) admit you are unsure but will return with an answer at a later time

**GO ON TO THE NEXT PAGE**

**47.** Which of the following procedures could help to modernize a fire department?

(A) Lower the entry level age limit.
(B) Extend the time used for training.
(C) Replace antiquated equipment.
(D) Restructure the fire districts.

**48.** While watching a movie at the local theater, you, an off-duty firefighter, are approached by a civilian who states that her child has just found an official fire department badge under a seat. In this situation you should take what action?

(A) Instruct the civilian to leave it where it was found.
(B) Tell the civilian you are off duty and cannot help her.
(C) Allow the child to keep the badge as a souvenir.
(D) Accept the badge and return it to headquarters.

**49.** Firefighters are constantly in training during their entire careers in the fire service. The main reason for this is to

(A) keep firefighters busy during long tours of duty
(B) enhance firefighting skills
(C) improve customer relations
(D) justify budget allocations for training

**50.** An advantage of a competitive, entry-level civil service firefighter examination is that it

(A) provides a fair opportunity for all who meet the test criteria
(B) eliminates candidates who cannot afford the application fee
(C) allows test examiners to be subjective in selecting candidates
(D) causes rivalries and hard feelings among candidates

**51.** Saving lives and protecting property are primary objectives of firefighters. From this statement you can assume that

(A) protecting property is just as important as saving lives
(B) protecting property is not very important
(C) both primary objectives can be obtained by extinguishing fires
(D) fighting fires is not a primary objective of the fire service

**52.** Strategy is a plan of action devised by the chief officer at the scene of a fire or emergency to successfully deal with an incident. Tactics are actions employed by units on the scene of the incident to carry out the plan. From these statements, you can assume that

(A) strategy and tactics are never used together at a single incident
(B) tactics must be carried out by a chief officer
(C) a plan utilizes tactics to implement action
(D) strategy is action performed by units on the scene of the incident

**GO ON TO THE NEXT PAGE**

**53.** You are a firefighter performing cleanup duty in the firehouse when a civilian comes up to you and states that she has locked her keys inside her car with her infant son inside and the motor running. The vehicle is located two blocks from the firehouse. Which of the following is an appropriate action to take?

(A) Instruct the civilian to walk to the locksmith store located down the block.
(B) Tell the civilian to wait until after you complete cleaning the firehouse.
(C) Admonish the civilian for being absentminded.
(D) Notify an officer and all members in preparation for response.

**54.** Fire hydrants are frequently damaged by cars during parking maneuvers. To adequately safeguard fire hydrants for fire department use, which of the following procedures should be employed?

(A) Station firefighters at hydrants to warn motorists.
(B) Install protective bumpers.
(C) Relocate hydrants inside buildings.
(D) Encourage the police to arrest drivers who damage hydrants.

**55.** Many fires that lead to the loss of civilian lives in residential buildings occur at night. A likely reason for this situation is that

(A) occupants are sleeping and initially unaware of the danger
(B) occupants are less aware of their surroundings during the nighttime
(C) occupants tend to keep their doors and windows locked during the nighttime
(D) fires burn with greater intensity during the nighttime

**56.** All probationary firefighters are assigned to a mentor during their initial year in the fire department. A major reason for this procedure is

(A) to keep probationary firefighters out of trouble
(B) to intimidate the probationary firefighter
(C) to provide guidance and instruction
(D) to weed out probationary firefighters who should be terminated

**57.** The siren on the fire apparatus is used by firefighters to alert motorists and obtain the right-of-way when responding to fires and emergencies. Sirens are designed to enhance safety and improve response times. Based on these statements, you can assume that

(A) use of sirens ensures faster response times
(B) sirens should always be used when responding
(C) sirens can aid in the avoidance of apparatus accidents
(D) when sirens are used, no apparatus accidents will occur

**GO ON TO THE NEXT PAGE**

**Directions:** Study the road map shown, then answer **questions 58 through 64**. All responses by the fire apparatus as well as civilian vehicle travel must be performed in a legal manner as designated by road map directional signs.

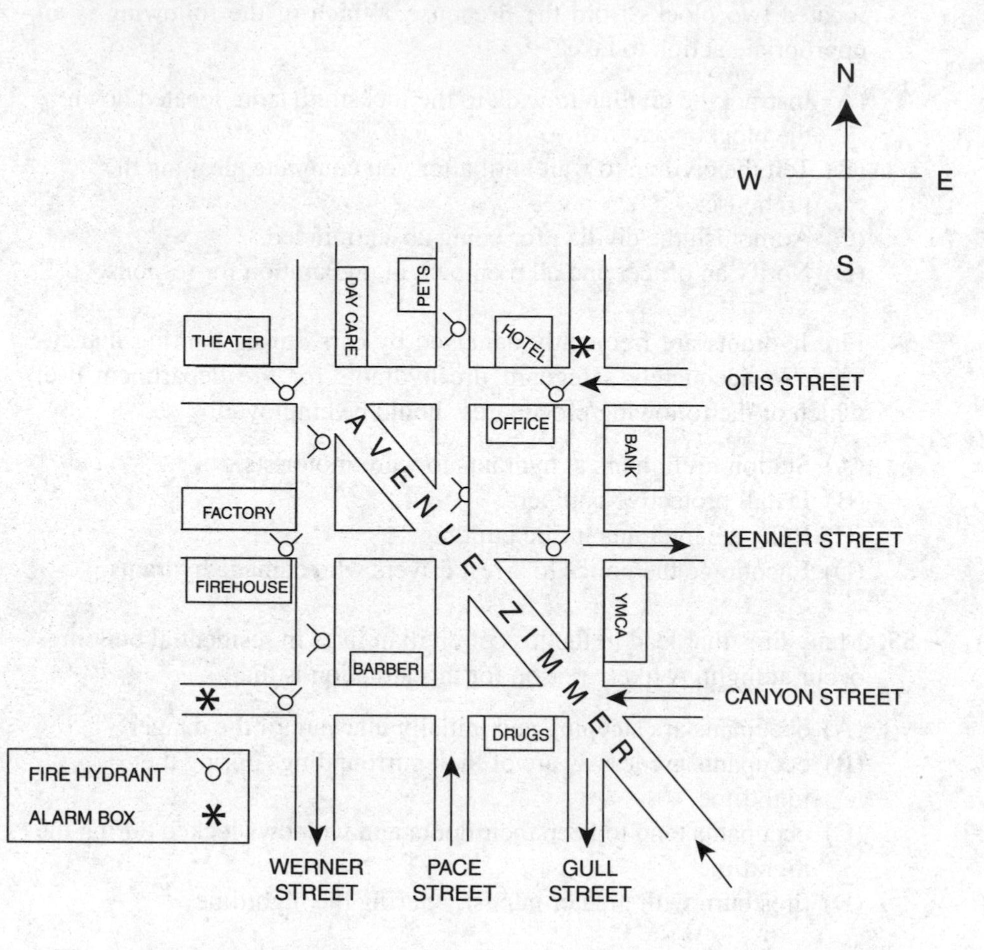


**58.** A fire apparatus traveling on Zimmer Avenue at the intersection of Canyon Street should travel in what direction to perform an inspection at the day care center?

(A) Continue straight to Kenner, turn left, and then turn right on Werner
(B) Turn left on Canyon and left on Werner
(C) Turn left on Canyon and left on Pace
(D) Continue straight ahead

**59.** The proper route to get from the firehouse to the pet store for an alarm activation is

(A) east on Kenner and north on Pace
(B) east on Kenner, north on Werner, east on Otis, and north on Pace
(C) east on Kenner, northwest on Zimmer, east on Otis, and north on Pace
(D) none of the above

**GO ON TO THE NEXT PAGE**

60. The fastest route for a fire apparatus traveling from quarters to a fire located in the theater is

(A) east on Kenner, north on Werner, and west on Otis
(B) east on Kenner, northwest on Zimmer, and west on Otis
(C) east on Kenner, north on Pace, and west on Otis
(D) east on Kenner, north on Pace, and east on Otis

61. An ambulance traveling northwest on Zimmer between Canyon and Kenner receives a radio call from the dispatcher for a medical emergency at the barbershop. The ambulance driver should follow what route to reach the barbershop?

(A) Go north on Pace, west on Otis, south on Werner, and east on Canyon
(B) Go south on Werner and west on Canyon
(C) Go east on Kenner, south on Gull, and west on Canyon
(D) Go south on Werner, east on Kenner, south on Gull, and west on Canyon

62. A passerby in a vehicle traveling south on Gull Street, between Kenner and Canyon Streets, observes smoke coming from the roof of the YMCA. What action should the person take to transmit an alarm to the fire department?

(A) make a left turn at the first intersection
(B) make a right turn at the first intersection
(C) make a U-turn onto Zimmer
(D) continue straight ahead through the first intersection

63. Which of the following buildings is the farthest from a fire hydrant?

(A) Drugstore
(B) Office building
(C) Bank
(D) Factory

64. Which fire hydrant should be utilized by firefighters for a fire in the hotel?

(A) Corner of Canyon and Werner
(B) Corner of Otis and Gull
(C) Corner of Kenner and Werner
(D) Corner of Kenner and Gull

**GO ON TO THE NEXT PAGE**

**Directions:** Now take an additional three minutes to study the road map. Then cover it and answer **questions 65 through 72**.

**65.** What is the northernmost east-west running street on the road map?

(A) Kenner
(B) Canyon
(C) Otis
(D) Gull

**66.** The YMCA is located between which two streets?

(A) Otis and Kenner
(B) Kenner and Canyon
(C) Werner and Pace
(D) Pace and Gull

**67.** The firehouse can be found on which street?

(A) Kenner
(B) Pace
(C) Otis
(D) Gull

**68.** Which of the following buildings is located in the easternmost portion of the road map?

(A) The theater
(B) The factory
(C) The barber shop
(D) The bank

**69.** What is the name of the centrally located avenue on the map?

(A) Pace
(B) Gull
(C) Zimmer
(D) Werner

**70.** Which of the following streets has a fire alarm box?

(A) Werner
(B) Kenner
(C) Pace
(D) Gull

**71.** In what direction does traffic move on Pace Street?

(A) North
(B) South
(C) East
(D) West

**GO ON TO THE NEXT PAGE**

**72.** In what direction does traffic move on Canyon Street?

(A) North
(B) South
(C) East
(D) West

**Directions:** Choose the best answer to each of the following questions.

**73.** Choose the letter that correctly fills the blank.
WVU, TSR, QPO, _ML

(A) K
(B) J
(C) N
(D) I

**74.** Choose the letter and number that correctly fills the blank in the sequence of letters and numbers below.

D4, H8, L12, ___

(A) O14
(B) P16
(C) Q16
(D) R14

**75.** Ladders that are 12 feet long can be climbed to a maximum height of 9 feet. Ladders 20 feet long can be climbed to a maximum height of 15 feet. Ladders 24 feet long can be climbed to a maximum height of 18 feet. Based on this information, what can you infer to be the maximum climbing height for a ladder that is 32 feet long?

(A) 30 feet
(B) 28 feet
(C) 26 feet
(D) 24 feet

**76.** One firefighter can carry a 12-foot ladder, but it takes two firefighters to carry a 24-foot ladder. Three firefighters are needed to carry a 36-foot ladder. Based on this information, which of the following statements can you infer to be true?

(A) A 50-foot ladder will require at least four firefighters to carry it.
(B) The longer the ladder, the fewer firefighters are required to carry it.
(C) There is no correlation between ladder length and weight.
(D) As the length of a ladder doubles, the number of firefighters required to carry it quadruples.

**GO ON TO THE NEXT PAGE**

**77.** A 4-to-1-safety margin must be maintained when using a personal rope in the fire service. A 5/8-inch diameter rope can hold a maximum load of 1,200 pounds. What should be the load restriction for the rope during usage in order to comply with the safety margin?

(A) 4,800 pounds
(B) 1,200 pounds
(C) 400 pounds
(D) 300 pounds

**78.** Rockville City fire department has more firefighters than Ludlum County. Jackson Township has fewer firefighters than Ludlum. Rockville has fewer firefighters than Jackson. If the first two statements are true, the third statement is

(A) true
(B) false
(C) you cannot tell if it is true or false
(D) true only sometimes

**79.** Kimbell fire department is directly south of the Abbey fire department. Barry fire department is northeast of the Abbey fire department. The Barry fire department is west of the Kimbell fire department. If the first two statements are true, the third statement is

(A) true
(B) false
(C) you cannot tell if it is true or false
(D) true only for an observer at the Kimbell fire department

**80.** Heller hose is made of more cotton than Lutz hose but has less cotton than Dylan hose. Lutz hose has more cotton than Kline hose but has less cotton than Utterman hose. Of the five kinds of hose, Kline has the least amount of cotton. If the first two statements are true, the third statement is

(A) true
(B) false
(C) you cannot tell if it is true or false
(D) true only in regard to Lutz hose

**GO ON TO THE NEXT PAGE**

**Directions:** Study the floor plan diagram shown for three minutes. At the end of that allotted time, cover the diagram. Then answer **questions 81 through 94**.

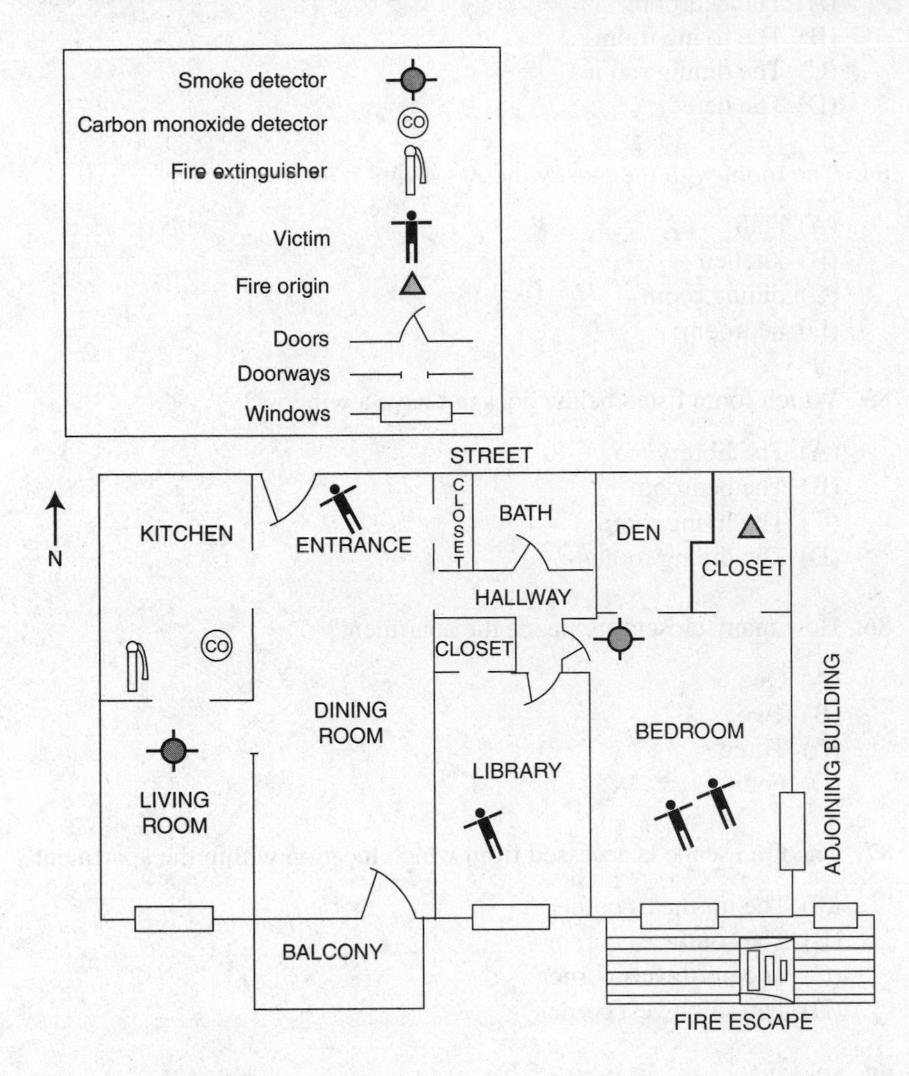


**81.** How many total detectors (smoke/carbon monoxide) are there inside the apartment?

(A) One
(B) Two
(C) Three
(D) Four

**82.** In which room is the fire extinguisher located?

(A) The kitchen
(B) The living room
(C) The dining room
(D) The library

**GO ON TO THE NEXT PAGE**

**83.** Which room leads directly onto the balcony?

(A) The bedroom
(B) The living room
(C) The dining room
(D) The den

**84.** The room with the most windows is the

(A) bath
(B) kitchen
(C) dining room
(D) bedroom

**85.** Which room listed below does not have a window?

(A) The library
(B) The bedroom
(C) The living room
(D) The dining room

**86.** How many closets are inside the apartment?

(A) One
(B) Two
(C) Three
(D) Four

**87.** The fire escape is accessed from which location within the apartment?

(A) The northeast corner
(B) The southeast corner
(C) The northwest corner
(D) The southwest corner

**88.** Visualizing the floor plan with north at the top of the page, the bath, den, bedroom closet, and bedroom have an outline similar to

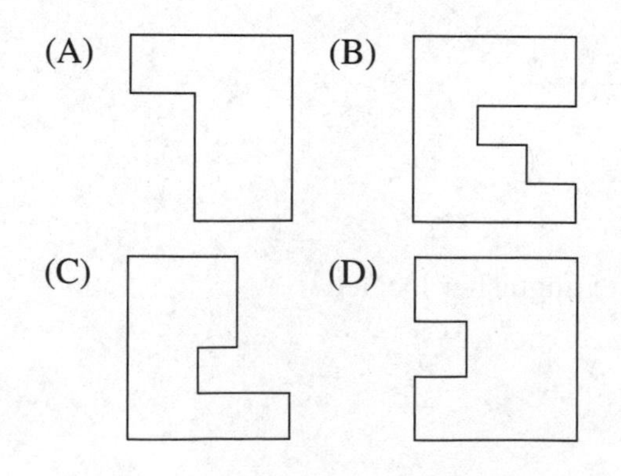


**GO ON TO THE NEXT PAGE**

**89.** In which area of the apartment did the fire start?

(A) The kitchen
(B) The bedroom closet
(C) The bath
(D) The hallway

**90.** How many victims are there inside the apartment?

(A) One
(B) Two
(C) Three
(D) Four

**91.** If the fire inside the apartment spread to the hallway, blocking firefighters from entering the library, how could they still gain access to this area?

(A) Via the living room window
(B) Jumping off the balcony through the closed library window
(C) Breaking through the dining room wall
(D) None of the above

**92.** Firefighters trying to enter the apartment had difficulty opening the door to a fully open position. The reason for this was most likely

(A) defective hinges
(B) the low ceiling
(C) a victim behind the door
(D) the force of air currents generated from the fire

**93.** A fire marshal investigating this fire for probable cause would most likely assume that the fire started due to

(A) food left on the stove
(B) overloaded wiring
(C) smoking in bed
(D) children playing with matches

**94.** The closest access point into the apartment for immediate rescue of the most victims is the

(A) bedroom fire escape window
(B) balcony door
(C) entrance door
(D) window facing the adjoining building

**GO ON TO THE NEXT PAGE**

**Directions:** Choose the best answer to each of the following questions.

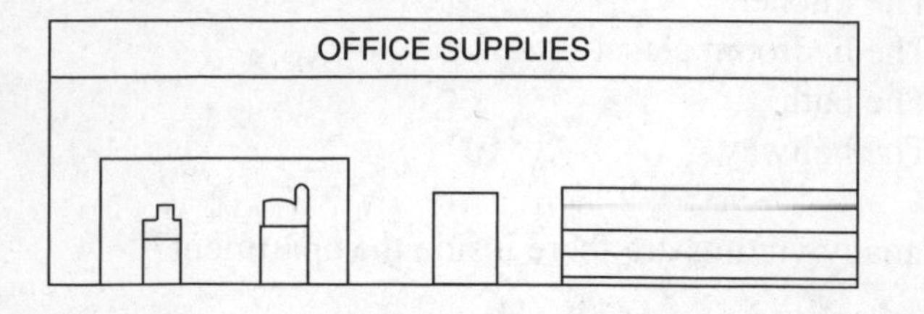


**95.** The figure above shows an office supply store as seen from the outside. If you were inside the store facing the front wall, what would it look like?

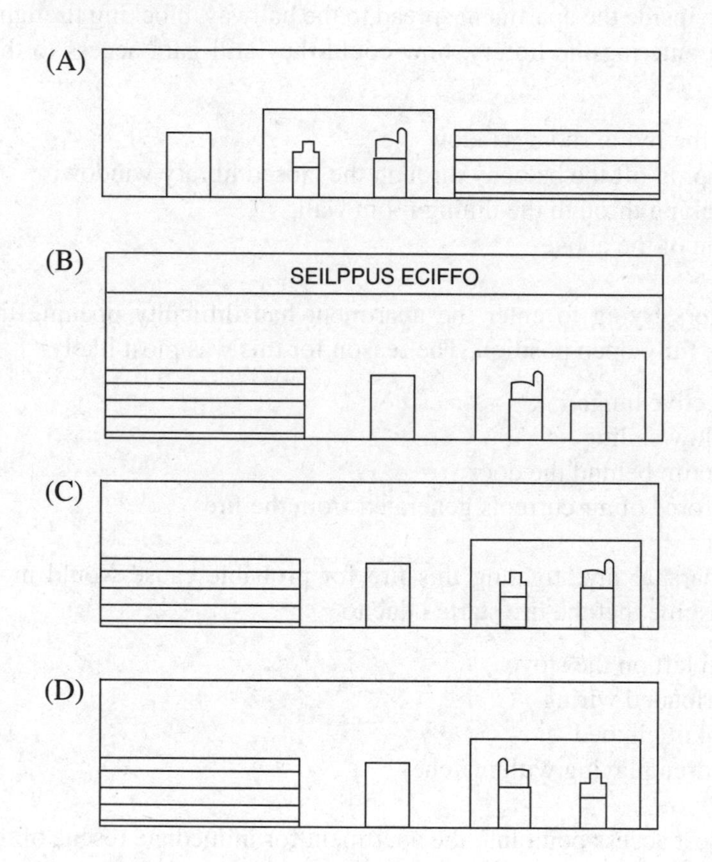


**GO ON TO THE NEXT PAGE**

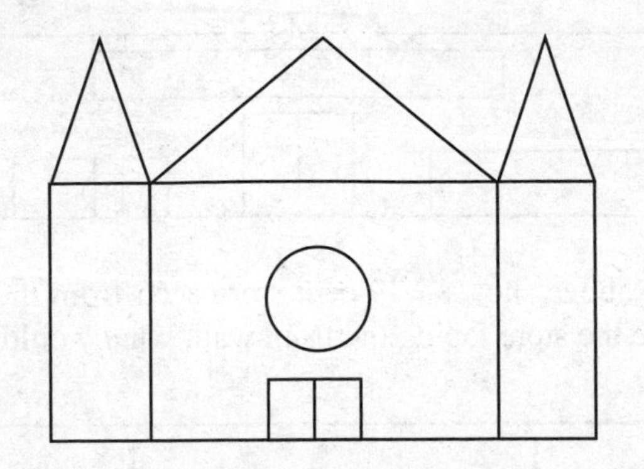

**96.** The figure above shows a church seen from ground level. What would the church look like from an aerial view?

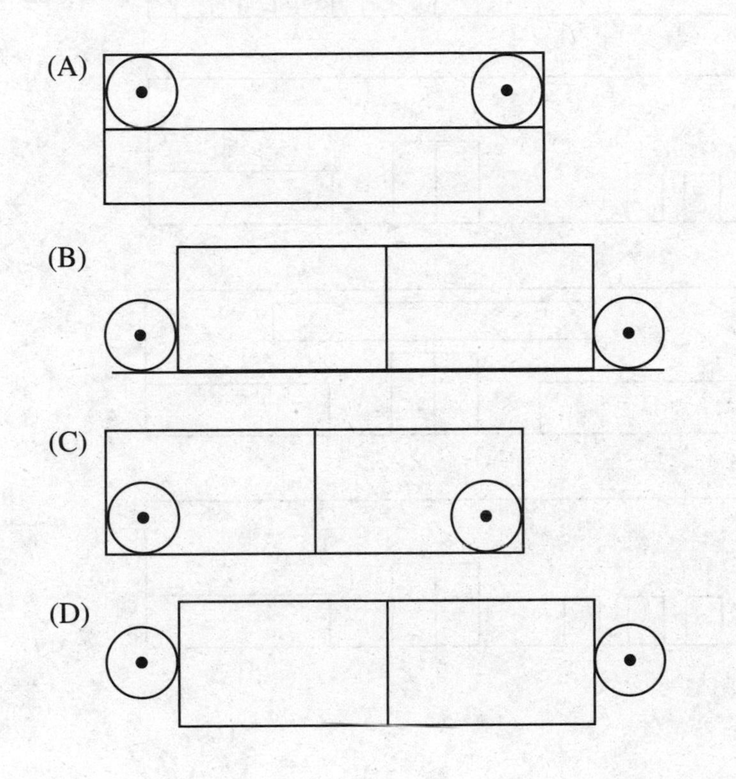


**GO ON TO THE NEXT PAGE**

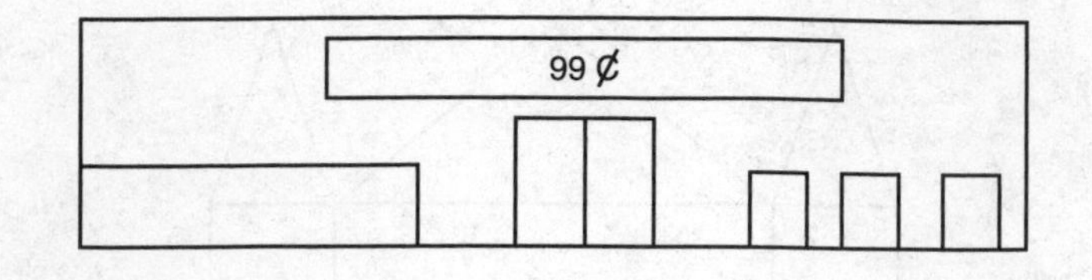


**97.** The figure above shows a 99-cent store seen from the outside. If you were inside the store facing the front wall, what would it look like?

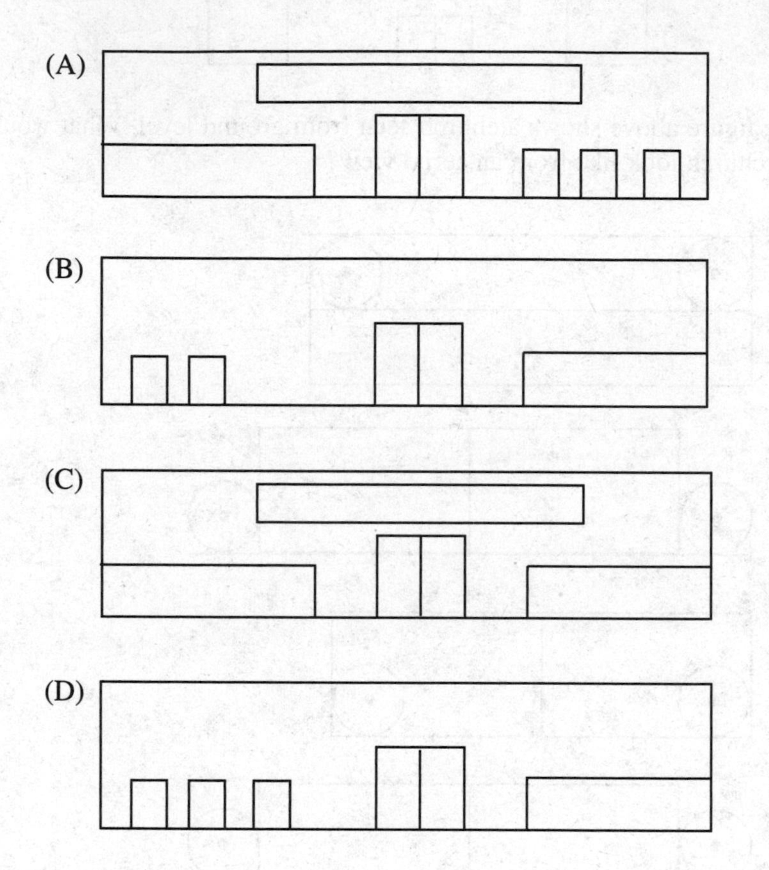


**GO ON TO THE NEXT PAGE**

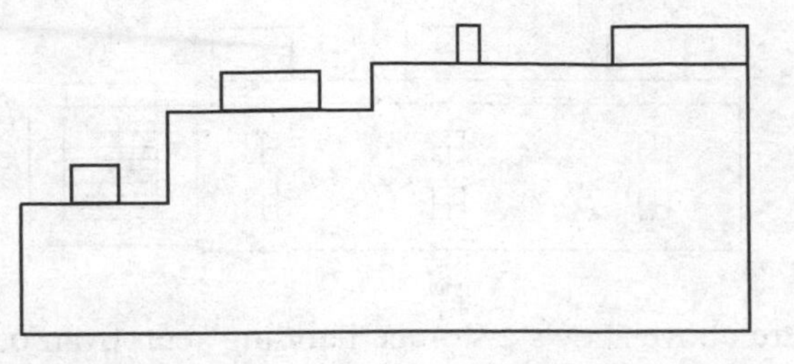

**98.** The figure above shows a factory seen from outside the building. What would the factory look like from an aerial view?

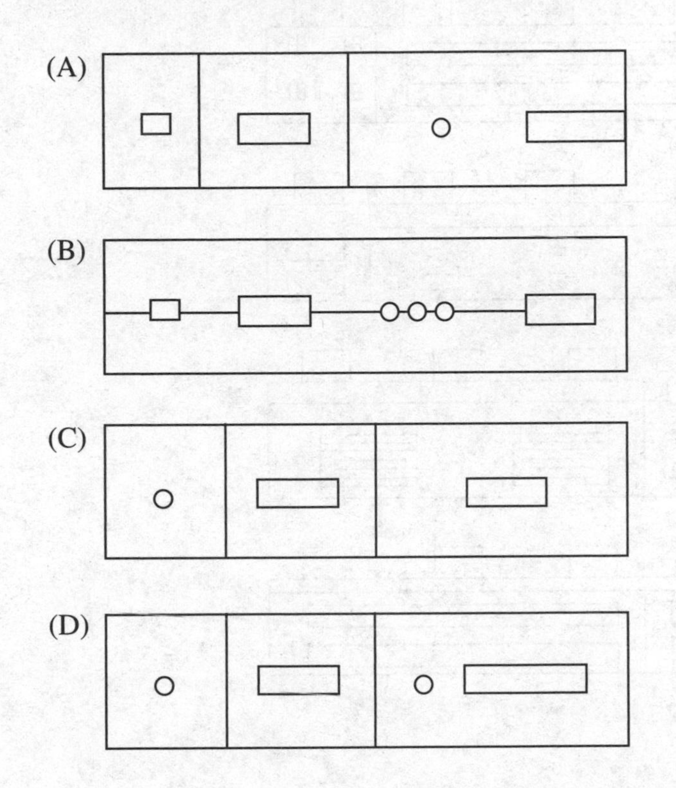

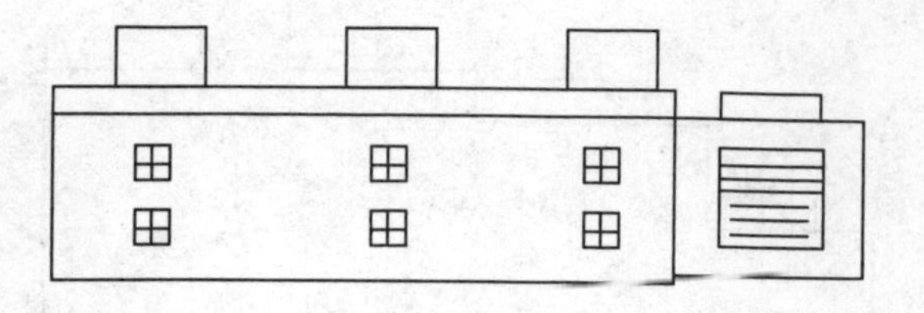

**99.** The figure above shows a storage building seen from outside the front of the building. What would the building look like from the rear?

(A)

(B)

(C)

(D)

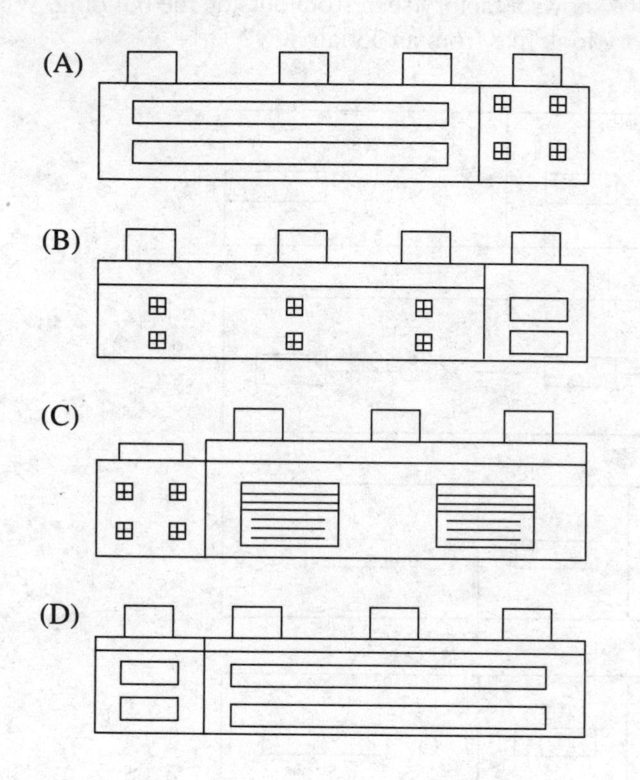

**GO ON TO THE NEXT PAGE**

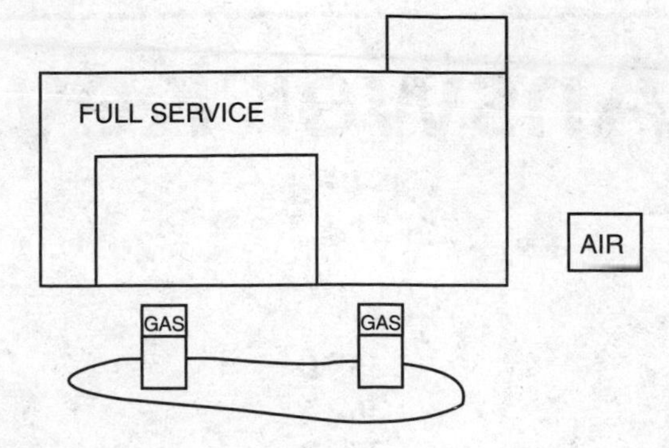


**100.** The figure above shows a gas station seen from outside the front of the station. What would the gas station look like from an aerial view?

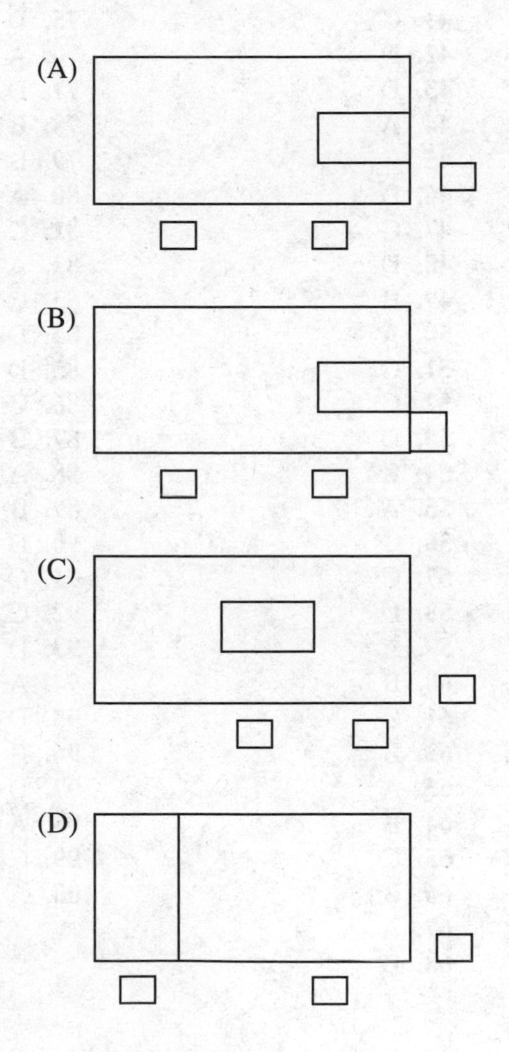


**STOP. THIS IS THE END OF PRACTICE EXAM 5**

# Answer Key

**1.** D
**2.** A
**3.** B
**4.** C
**5.** C
**6.** A
**7.** D
**8.** A
**9.** C
**10.** B
**11.** A
**12.** C
**13.** B
**14.** C
**15.** D
**16.** C
**17.** B
**18.** C
**19.** B
**20.** D
**21.** A
**22.** D
**23.** B
**24.** B
**25.** A
**26.** A
**27.** C
**28.** C
**29.** B
**30.** B
**31.** C
**32.** C
**33.** D
**34.** B
**35.** D
**36.** A
**37.** B
**38.** A
**39.** A
**40.** C
**41.** C
**42.** D
**43.** D
**44.** A
**45.** C
**46.** D
**47.** C
**48.** D
**49.** B
**50.** A
**51.** C
**52.** C
**53.** D
**54.** B
**55.** A
**56.** C
**57.** C
**58.** D
**59.** A
**60.** B
**61.** C
**62.** B
**63.** A
**64.** B
**65.** C
**66.** B
**67.** A
**68.** D
**69.** C
**70.** D
**71.** A
**72.** D
**73.** C
**74.** B
**75.** D
**76.** A
**77.** D
**78.** B
**79.** B
**80.** A
**81.** C
**82.** A
**83.** C
**84.** D
**85.** D
**86.** C
**87.** B
**88.** A
**89.** B
**90.** D
**91.** C
**92.** C
**93.** D
**94.** A
**95.** D
**96.** B
**97.** D
**98.** A
**99.** C
**100.** A

## ANSWER EXPLANATIONS

Following you will find explanations to specific questions that you might find helpful.

**73.** C Letters run backwards in each three-letter sequence.

**74.** B Letters run in a sequence of four, and numbers run in a sequence of four

**75.** D All ladders listed can be climbed only ¾ of their height for safety reasons.

**76.** A One firefighter is required for every 12 feet of ladder.

**77.** D 4:1 safety factor
Maximum load for 5/8-inch-diameter rope is 1,200 lbs
1,200 lbs ÷ 4 = 300 lbs

**78.** B Rockville City has more firefighters than Ludlum County.
Jackson Township has fewer firefighters than Ludlum County.
Therefore, Rockville City has more firefighters than Jackson Township.

**79.** B Kimbell fire department is south of Abbey fire department.
Barry fire department is northeast of Abbey fire department.

**80.** A Heller hose has more cotton than Lutz hose (and therefore Klein hose).
Dylan hose has more cotton than Heller and Lutz (and therefore Klein).
Lutz has more cotton than Klein hose.
Utterman hose has more cotton than Lutz (and therefore Klein).

**95.** D Flip the drawing over from right to left.

**96.** B The peak is in the center of the drawing, and the turrets are located at the front exterior of the main structure.

**97.** D Flip the drawing over from right to left.

**99.** C Flip the drawing over from right to left; the stack on the end (left) is smaller than the three stacks atop the main building.

**100.** A Gas dispensers are located evenly in front of the station; the superstructure is located on the right side of the station; the air dispenser is detached from the station and located to the right.

# PRACTICE EXAM 6

Answer **Questions 1–12** by reviewing the following reading passage.

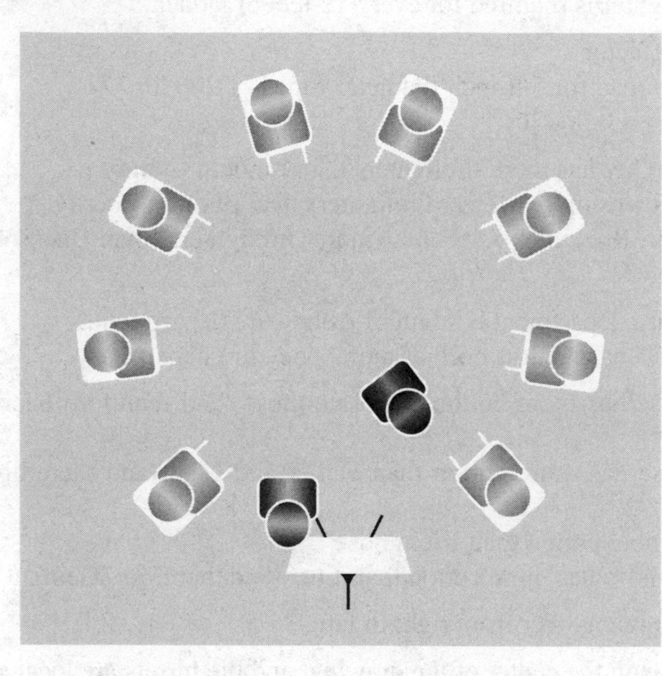

## THE AFTER ACTION REVIEW

The After Action Review (AAR) is a learning tool for developing better firefighters. It is a structured assessment discussion for analyzing operations. The AAR is used to determine what occurred at an incident and why it occurred. Additionally, an AAR can generate information concerning how a fire department can do things better.

The AAR is a review or debriefing process that includes the participants and those responsible for the operation. It may be formal or informal. Both follow the same format and involve the exchange of ideas and observations.

## Formal AAR

Formal AARs are generally scheduled events that are more structured and require planning. Formal AARs were originally developed by the U.S. Army.

They are used today by all U.S. military services, as well as by many other nonmilitary organizations. AARs are additionally being employed by businesses and corporations as a management instrument to build a culture of accountability. The Army, which is a large hierarchical and bureaucratic organization, uses the AAR process to facilitate training.

## Informal AAR

Informal AARs do not need to be scheduled and require a lot less planning. They can be conducted anywhere on the fireground and at any time (upon return to quarters subsequent to the incident) to provide real-time learning lessons. Informal AARs can be held after each identifiable event or major activity. In this way, the AAR becomes a live learning experience. An AAR facilitator, evaluator, or controller provides a task overview. This person leads a dialogue pertaining to the events and activities that focuses on the objectives. An AAR is ideally conducted after a major alarm or complex activity where numerous challenges were encountered. It will allow chief officers as well as the rank and file of the fire department to attain a greater understanding and insight.

## Properly Conducted AAR

A properly conducted AAR should have a positive influence on all participants. Personal attacks should not be allowed. It will enhance the communication process and stimulate feedback, thereby improving the motivation and education of members. It can additionally prevent future confusion relating to organizational goals and priorities. Everyone gets involved, from the lowest-ranking soldiers to commanders. The AAR does not judge success or failure. It does not point fingers or seek to blame individuals for goals and objectives not accomplished. The AAR focuses on strategy and tactics. It encourages all members of an organization to share and learn in order to continually improve.

## U.S. Forest Service

In the U.S. Forest Service, the AAR is employed as a post-shift team discussion at forest fires. It incorporates and integrates both technical information and personnel issues. The AAR summarizes the actions of the tour into a learning cycle. It provides a forum for group conversation in an attempt to analyze crew performance. Strategies are developed for reducing common tactical mistakes. The AAR stimulates communication and facilitates conflict resolution between team members. Group norms are established, emphasized, and reinforced.

**GO ON TO THE NEXT PAGE**

## Topics for Discussion

The following list provides a sample of discussion subjects that may be used during the AAR. Topics will, of course, be commensurate with the type of incident being reviewed as well as factors confronted:

- Adaptation to changing conditions
- Coordination of effort
- Communications
- Environmental impacts on operations
- Equipment performance
- Fatigue and stress
- Lessons learned and reinforced
- Organizational issues
- Perception of events
- Planning
- Roles and responsibilities
- Safety concerns
- Standard operating procedure adherence
- Techniques used

**1.** What is an After Action Review (AAR)?

(A) A meeting to judge success or failure
(B) Encourages all members of an organization to remain silent
(C) Exclusively used by fire departments
(D) A learning tool for developing better firefighters

**2.** Select the correct answer concerning formal AARs.

(A) Formal AARs are generally scheduled events
(B) Formal AARs are less structured than informal AARs
(C) U.S. military services do not use formal AARs
(D) Formal AARs require less planning than informal AARs

**3.** What organization originally developed formal AARs?

(A) U.S. Forest Service
(B) U.S. Army
(C) Fire Service
(D) FBI

**4.** How are AARs being used by businesses and corporations?

(A) Weed out employees demonstrating poor work performance
(B) Enhance discipline in the workplace
(C) Management instrument to build a culture of accountability
(D) Improve inclusion and diversity

GO ON TO THE NEXT P

**5.** Choose an accurate statement regarding fire service informal AARs.

(A) Need to scheduled
(B) Cannot be conducted on the fireground
(C) Can be given upon return to the firehouse at any time
(D) Are not held after a major fire activity

**6.** When is an AAR ideally conducted in the fire service?

(A) After a major alarm or complex activity
(B) Subsequent to company drill
(C) Where limited challenges were presented
(D) During daylight hours

**7.** Which of the following is not the person who provides the task overview during an AAR?

(A) Facilitator
(B) Conductor
(C) Controller
(D) Evaluator

**8.** Pick the accurate statement pertaining to a properly conducted AAR.

(A) Should have a negative influence on all participants
(B) Personal attacks are allowed
(C) Limited involvement of members
(D) Stimulate feedback

**9.** What does an AAR focus on?

(A) Goals and objectives not accomplished
(B) Judging success and failure
(C) Strategy and tactics
(D) Pointing fingers

**10.** How is the AAR employed in the U.S. Forest Service?

(A) Pre-shift team discussion
(B) Post-shift team discussion
(C) Quarterly team discussion
(D) Semiannual team discussion

**11.** Choose the incorrect answer pertaining to a U.S. Forest Service AAR.

(A) Stimulates communication
(B) Strategies are developed for reducing common tactical mistakes
(C) Summarizes the actions of the tour into a learning cycle
(D) Facilitates conflict between team members

**GO ON TO THE NEXT PAGE**

**12.** Which discussion subject mentioned is not listed in the reading passage?

(A) Medical strategy and tactics
(B) Coordination of effort
(C) Environmental impacts on operations
(D) Planning

**13.** What are trained private-sector personnel at chemical plants and airports known as?

(A) Wildland firefighters
(B) Hellfighters
(C) Industrial fire brigades
(D) Hotshots

**14.** Minimal requirements or qualifications to become a firefighter in most municipalities include all but which of the following?

(A) Citizenship
(B) Speak and understand English
(C) Proof of identity (birth certificate)
(D) College education

**15.** An incorrect tip on preparing a résumé can be found in what choice?

(A) Use white or off-white paper only
(B) Use both sides of the paper
(C) Don't use italics
(D) Use no more than two font sizes

**16.** Which selection is not general information that is normally found on a Notice of Examination for firefighter?

(A) How to file for the examination
(B) Cancellation information
(C) Exam questions from the upcoming test
(D) Job functions

**17.** Video-based testing is currently being used in some areas of the United States to replace:

(A) Oral interviews
(B) Psychological test
(C) Physical ability tests
(D) Polygraph test

**GO ON TO THE NEXT PAGE**

**18.** All but which of the following is assessed using video-based testing of firefighter candidates?

(A) Ability to take orders
(B) Actions during routine situations
(C) Ingenuity
(D) Teamwork

**19.** Regarding oral interviews, what type of behavior is inappropriate for a firefighter candidate to display?

(A) Take time to listen to questions
(B) Admit what you do not know
(C) Vary the pitch and tone of your voice
(D) Use slang or jargon

**20.** What does the MMPI-2 test assess?

(A) Mental status of candidates
(B) Candidate's veracity
(C) Personality traits
(D) Physical capabilities of candidates

**21.** Favorable traits and characteristics for firefighter candidates (deemed by the author) include all but which of the following?

(A) Decisive
(B) Integrity
(C) Obstinate
(D) Responsible

**22.** The Candidate Physical Ability Test (CPAT) consists of eight sequential events. Choose the accurate partial order of events from the answers provided.

(A) Ladder raise and extension, Forcible entry, Search, Rescue, Ceiling breach/pull
(B) Stair climb, Equipment carry, Forcible entry, Search, Rescue
(C) Hose drag, Equipment carry, Ladder raise and extension, Search, Rescue
(D) Ceiling breach/pull, Stair climb, Forcible entry, Rescue, Search

**23.** During what CPAT event is a sledgehammer utilized?

(A) Search
(B) Rescue
(C) Equipment carry
(D) Forcible entry

**GO ON TO THE NEXT PAGE**

**24.** Choose the correct point of information about aerobic exercise.

(A) Moderate-intensity exercise sustained over 30 minutes or more
(B) High-intensity, short-duration exercise
(C) Heart rate should be between 75% and 100% of maximum target heart rate
(D) Generally performed two to three times per week

**25.** An example of anaerobic training can be found in what choice?

(A) Jogging
(B) Boxing
(C) Swimming
(D) Martial arts

**26.** Adding 6.2 and 3.8 equals what number?

(A) 9.9
(B) 10
(C) 2.4
(D) 12

**27.** What is the product of multiplying 7.3 by 4.5?

(A) 40.85
(B) 408.51
(C) 32.85
(D) 41.15

**28.** Convert 0.668 to a percent.

(A) 6.68%
(B) 8.6%
(C) 66.8%
(D) 668%

**29.** Convert 22.6% to a decimal.

(A) 02.26
(B) 226.0
(C) 022.6
(D) 0.226

**30.** Using the formula: Part = % · Whole ÷ 100, what is 20% of 30?

(A) 6
(B) 8
(C) 10
(D) 12

**GO ON TO THE NEXT PAGE**

**31.** What percent of 40 is 5? Use the formula: % = Part ÷ Whole · 100.

(A) 125%
(B) 12.5%
(C) 1.25%
(D) 122%

**32.** 3 is 30% of what number? Use the formula: Whole = Part ÷ % · 100.

(A) 15
(B) 13
(C) 10
(D) 9

**33.**

A die is thrown against a wall four times. Each time the number that comes up is 2. If the die is thrown again for the fifth time, what is the probability of the number coming up a 2 again?

(A) 1 out of 3
(B) 1 out of 6
(C) 1 out of 2
(D) 1 out of 4

**34.** Given the temperature on the Fahrenheit scale is 50 degrees, what is the corresponding temperature in degrees on the Celsius scale? Use the formula:

C = 5/9(F – 32)

(A) 20
(B) 15
(C) 10
(D) 5

**GO ON TO THE NEXT PAGE**

**35.** An EMT takes the temperature of a firefighter suffering from heat stroke. The thermometer shows the temperature on the Celsius scale to be 38 degrees. What is the corresponding temperature in degrees on the Fahrenheit scale? Use the formula:

$F = 9/5(C) + 32$

(A) 95.0
(B) 96.8
(C) 98.6
(D) 100.4

**36.** Find the average of the following numbers: 45, 60, 80, 90, and 135.

(A) 75
(B) 80
(C) 81
(D) 82

**37.** Two buses alongside each other are traveling in opposite directions on a desert highway. Bus A is moving at 50 mph. Bus B is going 70 mph. When will the two buses be 240 miles apart?

(A) 1 hour
(B) 2 hours
(C) 3 hours
(D) 4 hours

**38.** In a probationary firefighter class of 60, the ratio of successful promotions to failures is 10:2. How many "probes" failed out of the class?

(A) 12
(B) 10
(C) 8
(D) 6

**39.**

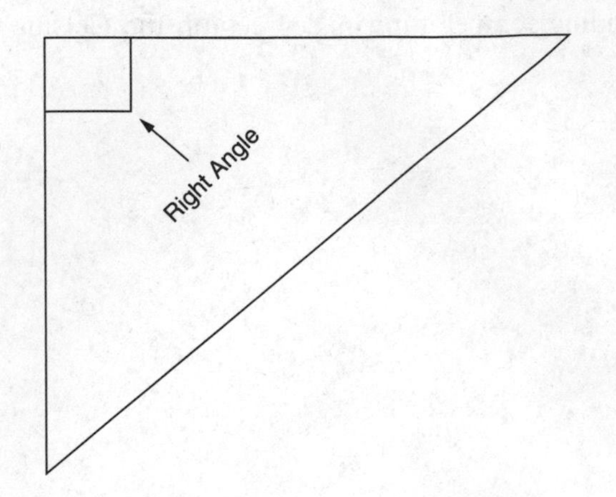


**GO ON TO THE NEXT PAGE**

A right angle measures exactly how many degrees?

(A) 75
(B) 80
(C) 90
(D) 180

**40.**

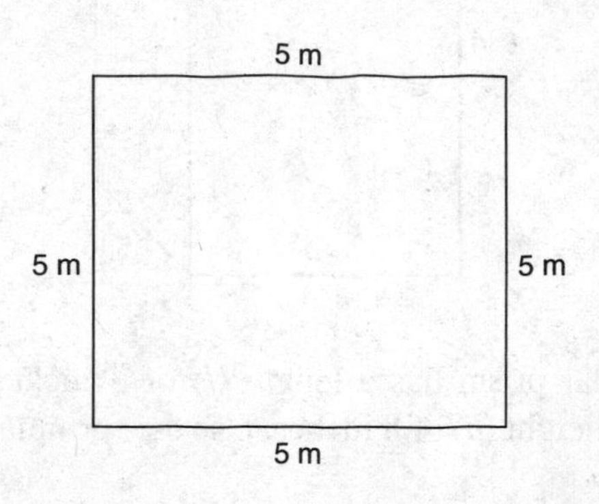


What is the area of the square shown? Use the formula: $A = s^2$.

(A) 5 $m^2$
(B) 20 $m^2$
(C) 25 $m^2$
(D) 50 $m^2$

**41.**

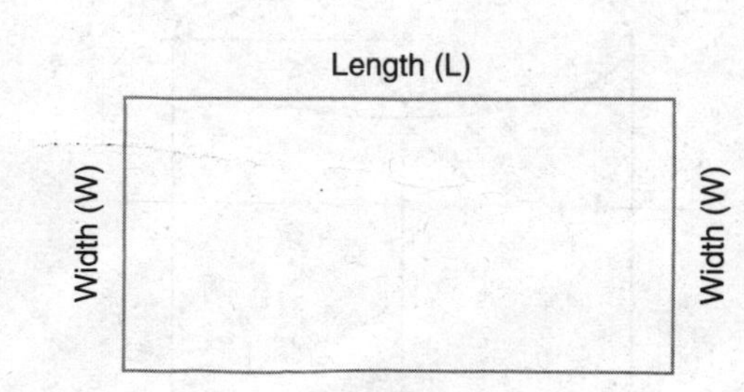


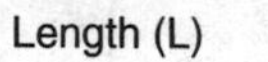


What is the perimeter of the rectangle drawn if the length (L) is 11 inches and the width (W) is 7 inches? Use the formula: $P = 2L + 2W$.

(A) 36 inches
(B) 30 inches
(C) 28 inches
(D) 25 inches

GO ON TO THE NEXT PAGE

**42.**

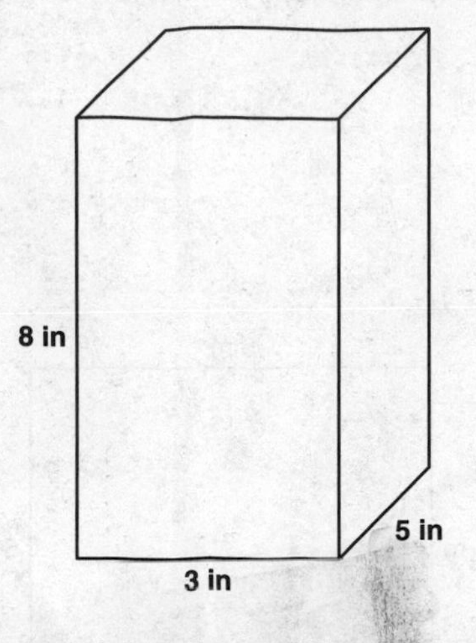


The rectangular prism has a length ($l$) of 3 inches, a width ($w$) of 5 inches, and a height ($h$) of 8 inches. Use the formula: $V = l \cdot w \cdot h$ to find the volume.

(A) 150 $in^3$
(B) 142 $in^3$
(C) 120 $in^3$
(D) 100 $in^3$

**43.**

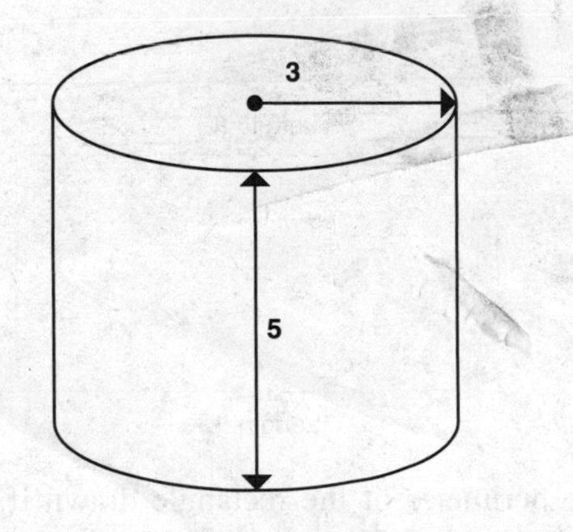


The cylinder has a radius ($r$) of 3 feet and a height of 5 feet. To find the volume of the cylinder, use the formula: $V = \pi \cdot r^2 \cdot h$

(A) 153.6 $ft^3$
(B) 141.3 $ft^3$
(C) 140.7 $ft^3$
(D) 134.2 $ft^3$

**GO ON TO THE NEXT PAGE**

**44.**

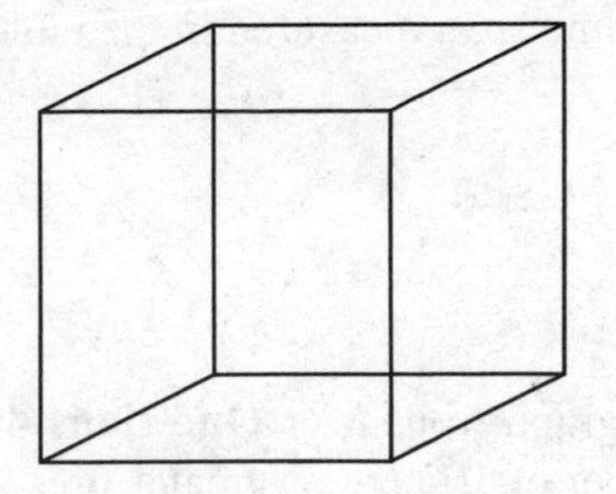

The length of one side of the cube is 4 mm. To determine the surface area of the cube, use the formula: SA = $6 \cdot s^2$

(A) 96 mm$^2$
(B) 103 mm$^2$
(C) 86 mm$^2$
(D) 78 mm$^2$

Use the following table to answer **Questions 45–48**.

**Benjamin County Semiannual (January–June) Firefighter Injury Statistics**

| Fire District | # of Firefighters | Injury Count | Hospitalization |
|---|---|---|---|
| A | 70 | 35 | 17 |
| B | 60 | 20 | 7 |
| C | 60 | 25 | 6 |
| D | 75 | 15 | 3 |

**45.** What district in Benjamin County had the highest number of firefighter injuries?

(A) Fire District D
(B) Fire District C
(C) Fire District B
(D) Fire District A

**46.** Comparing firefighting injuries to hospitalization injuries, what district had the highest percentage of hospitalization firefighter injuries?

(A) Fire District D
(B) Fire District C
(C) Fire District B
(D) Fire District A

**47.** What fire district saw 20% of its injured firefighters go to the hospital?

(A) Fire District D
(B) Fire District C
(C) Fire District B
(D) Fire District A

**GO ON TO THE NEXT PAGE**

**48.** What was the total number of firefighter injuries in all four fire districts for the 6-month time span measured in the table?

(A) 105
(B) 95
(C) 80
(D) 75

Review the line graph to answer **Questions 49–53**. The vertical axis denotes the number of 10-acre or greater fires in a national forest. The horizontal axis plots the months of the year.

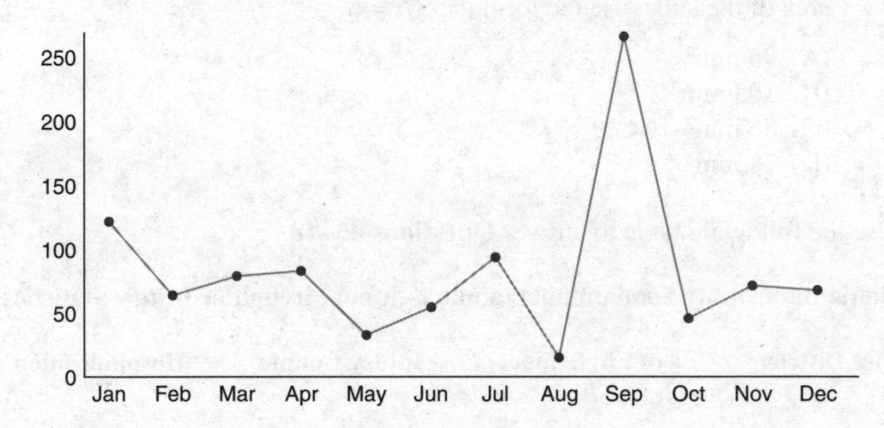


**49.** What two-month period produced the most fires at the national forest?

(A) January/February
(B) March/April
(C) May/June
(D) July/August

**50.** What was the busiest month for fire at the national forest?

(A) July
(B) August
(C) September
(D) October

**51.** July/August had approximately the same number of fires as what two other sequential months?

(A) January/February
(B) May/June
(C) September/October
(D) November/December

**GO ON TO THE NEXT PAGE**

**52.** Approximately how many fires occurred in the national forest from the beginning of the year through July?

(A) More than 500
(B) Fewer than 500
(C) More than 700
(D) Fewer than 400

**53.** Which two consecutive months listed had the greatest disparity regarding number of fires?

(A) February/March
(B) April/May
(C) August/September
(D) September/October

**54.** What type of heat energy is developed when solid objects rub together causing friction?

(A) Chemical
(B) Electrical
(C) Solar
(D) Mechanical

**55.** Which textile material is not considered to be a natural fiber?

(A) Cotton
(B) Rayon
(C) Wool
(D) Hemp

**56.** Nitrogen gas is characterized as what type of gas?

(A) Flammable
(B) Inert
(C) Oxidizer
(D) Reactive

**57.** Common materials that burn store a standard amount of heat energy per pound (latent heat of combustion). What is wood's approximate quantity of stored heat energy per pound?

(A) 7,000 BTU/lb
(B) 12,000 BTU/lb
(C) 16,000 BTU/lb
(D) 23,000 BTU/lb

**GO ON TO THE NEXT PAGE**

**58.** In what two phases of fire can a backdraft situation occur?

(A) Incipient (growth) and free-burning (fully developed)
(B) Free-burning (fully developed) and smoldering (decay)
(C) Incipient (growth) and smoldering (decay)
(D) Smoldering (decay) and flashover

**59.** The wet chemical extinguishing agent was originally developed for what kind of application?

(A) Replacement for halon gas
(B) Extinguish fires in combustible metals
(C) Extinguish Class C fires
(D) Use on unsaturated fat vegetable cooking oils

**60.** The pliers drawn here are an example of what kind of lever?

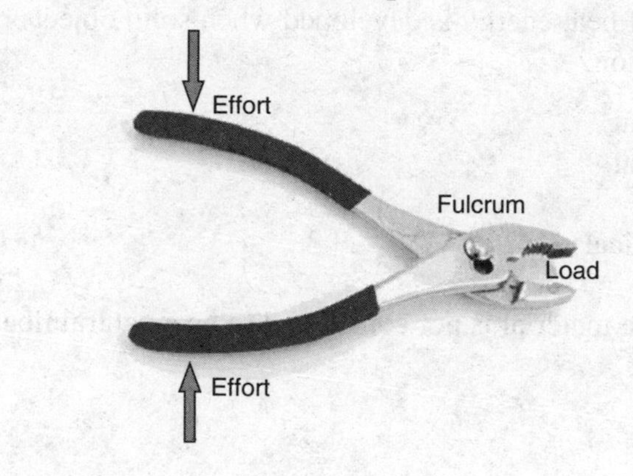


(A) First class
(B) Second class
(C) Third class
(D) None of the above

**61.** In the drawing, the load is 90 lbs and the effort required to lift the load is 30 lbs. What is the mechanical advantage (MA) of this first-class lever simple machine?

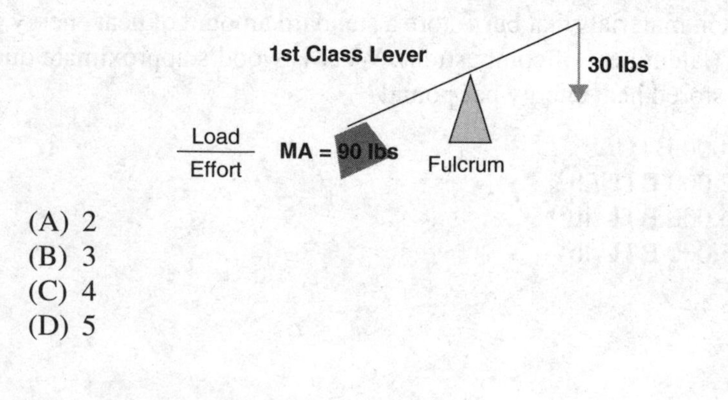


(A) 2
(B) 3
(C) 4
(D) 5

GO ON TO THE NEXT PAGE

**62.** What is the mechanical advantage (MA) provided by the wheel/axle assembly drawn here when the wheel has a radius of 8 inches and the radius of the axle is 2 inches? Use the formula: MA = wheel radius/axle radius.

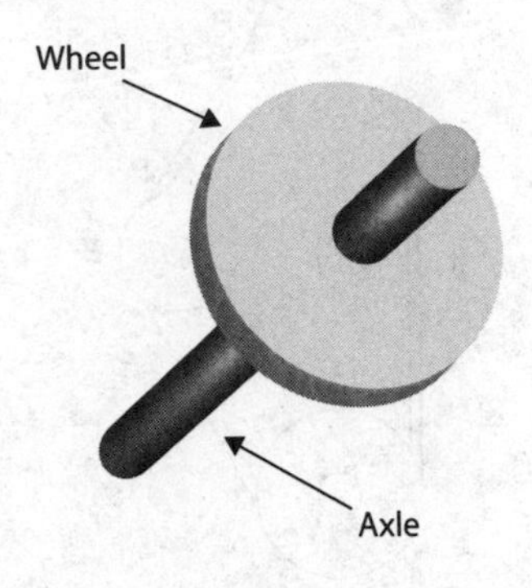


(A) 5
(B) 4
(C) 3
(D) 2

**63.** Reviewing the three pulley systems drawn from left to right, the pulley system on the left has one nonsupporting rope, as does the pulley system in the middle. The pulley system on the right has no nonsupporting ropes. Starting from the left, what are the mechanical advantages provided by each pulley system shown?

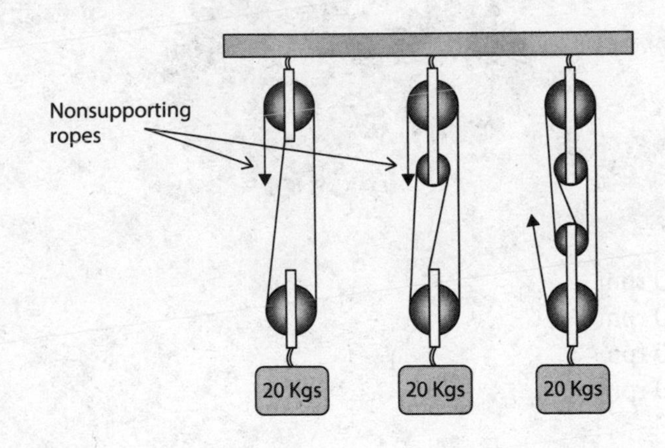


(A) 3, 4, 5
(B) 2, 3, 4
(C) 2, 3, 5
(D) 3, 4, 6

**GO ON TO THE NEXT PAGE**

**64.** Viewing the drawing, if wheel A rotates counterclockwise, which way will wheel B rotate?

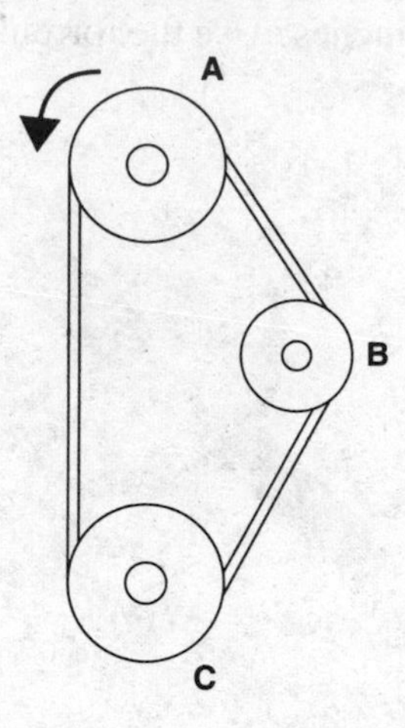


(A) Clockwise
(B) Counterclockwise
(C) Clockwise or counterclockwise
(D) None of the above

**65.** Review the gear drawing. Gear 1 (drive gear) has 20 teeth and revolves at 60 rpm. What is the rpm for Gear 2 (driven gear), which has 40 teeth?

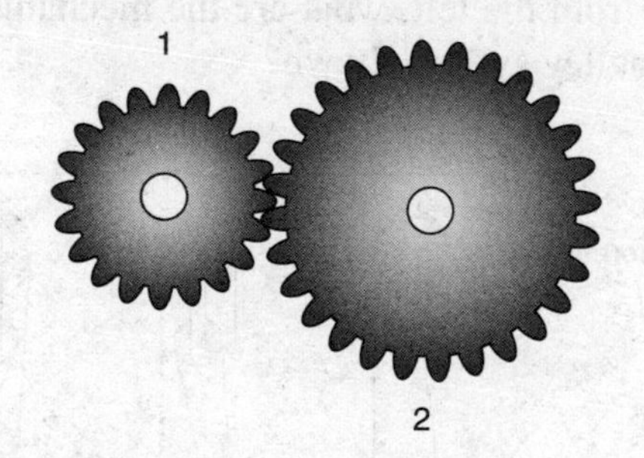


(A) 60 rpm
(B) 50 rpm
(C) 40 rpm
(D) 30 rpm

**66.** What function does a cat's paw tool perform?

(A) Measuring device
(B) Access to obstructed bolts and nuts
(C) Nail puller
(D) Metal cutter

**GO ON TO THE NEXT PAGE**

**67.** An attack hose provides approximately how many gallons per minute (gpm) to the engine company utilizing it for interior fire suppression operations?

(A) 100 to 150
(B) 150 to 250
(C) 250 to 325
(D) Greater than 325

**68.** Which of the following is not considered a fitting device used by engine firefighters in conjunction with hose line couplings to solve hose connection problems?

(A) Gate valve
(B) Double female connection
(C) Reducer
(D) Increaser

**69.** An inaccurate statement regarding the combination ladder can be noted in which choice?

(A) Ranges in length from 24 feet to 35 feet
(B) Can be used as a single (straight) ladder
(C) Can be used as an extension ladder
(D) Can be used as an A-frame ladder

**70.** The rescue knot (bowline on a bight) tied onto a lifesaving rope is particularly used for what purpose?

(A) Tying two ropes of unequal diameter together
(B) Tying two ropes of equal diameter together
(C) Securing hose that has been hoisted along the exterior of a building
(D) Securing the body during lowering and lifting operations

**71.** A Latin square is arranged in such a way so that no row or column contains the same figure/number twice and every row and column contains the same figure/number. Use deductive reasoning to place the correct numbers in the blank squares.

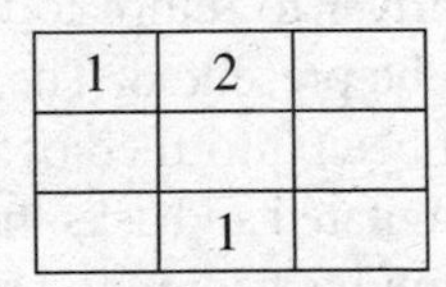


(A) Top right: 3; middle boxes from left to right: 3, 2, 1; bottom left: 2; bottom right: 3
(B) Top right: 3; middle boxes from left to right: 2, 1, 3; bottom left: 2; bottom right: 3
(C) Top right: 3; middle boxes from left to right: 2, 3, 1; bottom left: 3; bottom right: 2
(D) Top right: 1; middle boxes from left to right: 2, 3, 1; bottom left: 2; bottom right: 1

**GO ON TO THE NEXT PAGE**

**72.** Using deductive reasoning, which conclusion can be derived from the combination of the two statements listed?

Only fish oil contains omega-3.
Only foods that contain omega-3 help with brain development.

(A) All fish oils help with brain development
(B) Only what contains omega-3 is fish oil
(C) All that helps with brain development is fish oil
(D) There are fish oils that help with brain development

**73.** Using inductive reasoning, replace the three smiley faces on the bottom row of the drawing with the appropriate figures starting from left to right.

(A) Circle, triangle, square
(B) Cross, circle, triangle
(C) Circle, triangle, cross
(D) Circle, square, triangle

**74.** A police officer observes a vehicle speeding and pulls the motorist over for a traffic stop. The driver of the speeding vehicle steps out of the car with license and registration before the officer can exit his squad car. The officer orders the driver to return to his vehicle and that proof of insurance also needs to be presented. The driver does not comply and explains that he has just been informed of a home emergency. The man appears to be sober but agitated and asks the officer if they can settle the issue immediately without his having to present more information. The officer once again orders the driver to return to his car. This time the driver returns to his vehicle.

**GO ON TO THE NEXT PAGE**

**Based upon this scenario, what do you believe to be the most likely problem the motorist has?**

(A) He is under the influence of illegal drugs
(B) He has something in his car he doesn't want the officer to see
(C) He wants to return home and deal with his family emergency
(D) He does not know the proper actions to take during a traffic stop

**75.** A fire officer is completing an unusual occurrence report on a hit-and-run accident involving his apparatus and a civilian vehicle (late-model Ford). The report will include the following sentences (not listed in correct order).

1. The late-model Ford struck the apparatus in the rear, backed up, and then continued on its way down the road.
2. The apparatus was pulling out from a parked position at the curb when struck.
3. The license plate number of the vehicle was recorded.
4. Two firefighters sitting inside the apparatus were treated by EMTs for bruises.
5. The Ford was traveling approximately 20 miles per hour over the speed limit just prior to the accident.

**The most logical order for these sentences to appear in the report is:**

(A) 2, 3, 5, 4, 1
(B) 2, 5, 1, 3, 4
(C) 5, 3, 1, 4, 2
(D) 3, 2, 1, 5, 4

**76.** Use the information provided to answer the question that follows.

**Firefighters responding to a medical emergency should follow the steps listed in the order given:**

1. Render aid commensurate with training
2. Request an ambulance, when required
3. Notify Borough Dispatch if the victim is wearing a Medic-Alert emblem
4. Wait to direct the ambulance to the scene
5. If the ambulance does not arrive in 20 minutes, make a second call

**GO ON TO THE NEXT PAGE**

## Scenario

Firefighters out on building inspection observe an elderly gentleman fall and hit his head while attempting to cross the street. Medical treatment is administered, and an ambulance is requested to respond. The injured victim is wearing a Medic-Alert emblem.

**According to the steps listed, what is the next action that should be taken?**

(A) Make a second call should the ambulance not arrive in 20 minutes
(B) Send a firefighter to the street corner where the ambulance will be arriving
(C) Turn on apparatus warning lights to enhance visibility
(D) Notify Borough Dispatch of Medic-Alert emblem

**77.** When supplying water through a fire department connection (FDC) to a standpipe fire protection system for a fire on the 40th floor of a high-rise office building, for every 10 feet of height (or every story), a downward pressure of approximately 5 psi is exerted. Engine pump pressure must be increased accordingly. How much pressure escalation is needed to make up for this pressure loss?

(A) 100 psi
(B) 150 psi
(C) 200 psi
(D) 250 psi

**78.** The hand method (2½-inch hose) can be used for calculating friction loss in a 2½-inch-diameter attack hose line within what flow range?

(A) 50 gpm to 300 gpm
(B) 100 gpm to 500 gpm
(C) 200 gpm to 600 gpm
(D) 300 gpm to 650 gpm

**GO ON TO THE NEXT PAGE**

**79.**

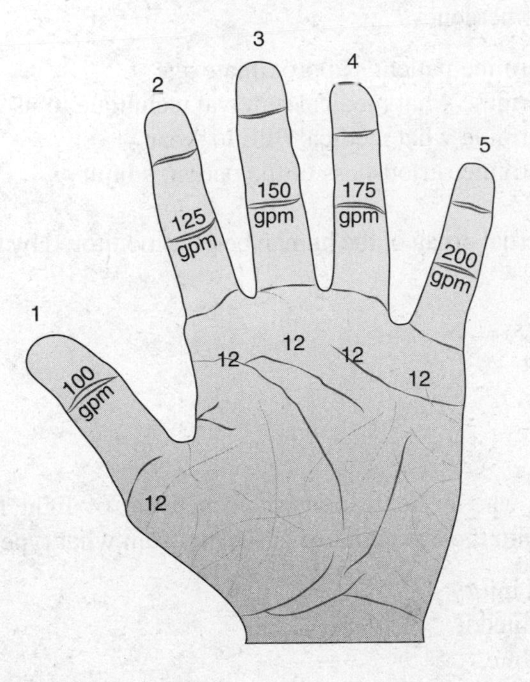


**The hand method (1¾-inch hose)**

Using the hand method (1¾-inch hose) shown, determine the friction loss in a 300-foot hose stretch of 1¾-inch-diameter hose flowing at 200 gpm.

(A) 180 psi
(B) 150 psi
(C) 120 psi
(D) 100 psi

**80.** Extra hazard occupancies have large amounts of highly combustible materials. Review the choices and select the occupancy that is considered to have an extra hazard classification.

(A) Laundry
(B) Hospital
(C) Automobile showroom
(D) Aircraft hangar

**81.** What is the primary reason for fire departments adopting random drug/alcohol testing?

(A) Ensure firefighters come to work clean and sober
(B) Reduce auto accidents when responding to alarms
(C) Demonstrate to the public a strong disciplinary attitude
(D) Expand the role of the medical office

**GO ON TO THE NEXT PAGE**

**82.** How do first responders use the CUPS criteria when arriving upon a medical emergency?

(A) Determine patient's approximate age
(B) Determine what medical removal technique to use
(C) Determine what medical PPE to wear
(D) Determine seriousness of the patient's injury

**83.** What internal organ of the human body is monitored by taking a victim's pulse?

(A) Lungs
(B) Heart
(C) Liver
(D) Kidney

**84.** A victim who is nonresponsive to a beam of light from a penlight directed into the eye could be suffering from what type injury listed?

(A) Head injury
(B) Nosebleed
(C) Frostbite
(D) Heat exhaustion

**85.** When should first responders administer rescue breathing during a medical emergency?

(A) Victim is having difficulty breathing
(B) Victim is bleeding heavily and losing consciousness
(C) Victim stops breathing and is unconscious
(D) Victim is conscious and has an obstruction in the airway

**86.** Select the most accurate statement concerning the administration of CPR on victims by first responders.

(A) Victim is in cardiac arrest
(B) Victim is in cardiac arrest and lacks a pulse
(C) Victim is in cardiac arrest, has a pulse, and is breathing
(D) Victim is in cardiac arrest, lacks a pulse, and is not breathing

**87.** What is the compression/ventilation ratio for two first responders performing CPR on a 9-month old baby?

(A) 30:2
(B) 15:2
(C) 100:1
(D) 100:2

**GO ON TO THE NEXT PAGE**

**88.** The rule of nines assesses the percentage of burn on an adult victim at a fire scene. It is used to help guide treatment decisions and determine transfer to a burn unit. If both legs (anterior and posterior), the groin, chest, and abdomen were burned, this would involve what percentage of the body?

(A) 48%
(B) 55%
(C) 63%
(D) 72%

**89.** Which safety practice information is not correct pertaining to automated external defibrillator usage by first responders?

(A) Before administration of an electrical shock, rescuers place their hands on the shoulders of the victim
(B) Remove metal necklaces and underwire bras from victims
(C) Before using AED, check for accumulation of water in the work area
(D) If the victim's chest is wet, dry it

**90.** A fire officer arriving at a fire in a vacant warehouse notices a blue, late-model truck leaving the scene in a hurry. The officer wants to mention the truck in the fire report. What is the most effective way for the officer to account for the vehicle?

(A) As our apparatus arrived at the fire scene, I viewed a blue, late-model truck rushing away
(B) A truck should not have been on the street where the warehouse was located
(C) Inside the truck may have been the persons who started the fire
(D) I saw a vehicle leaving the fire scene upon our arrival

**91.** A fire marshal is drafting information to formulate a report, which will include the following five sentences:

1. Upon arrival I noticed the side door to the fire building was open.
2. My partner and I received information at headquarters from an anonymous source that a fire was burning inside a nearby building on the third floor.
3. We interviewed occupants of the building.
4. My partner discovered an apparent fire fatality on the third floor.
5. I summoned the police department to the scene.

**What is the most logical order for these sentences to appear in the report?**

(A) 5, 3, 4, 1, 2
(B) 2, 1, 4, 5, 3
(C) 4, 5, 2, 3, 1
(D) 2, 4, 5, 1, 3

**GO ON TO THE NEXT PAGE**

**92.** An EMT is distributing instructional brochures for performing rescue breathing on a collapsed victim. The information provided can be used by civilians before the arrival of first responders. The following five steps are included in the brochure, but they are not in sequential order:

1. Open the victim's airway
2. Call 911
3. Ascertain responsiveness
4. View and feel for respiratory movement
5. Pinch the nose of the victim shut while performing mouth-to-mouth rescue breathing

**Which step should be performed prior to ascertaining responsiveness?**

(A) Open the victim's airway
(B) Pinch the nose of the victim
(C) View and feel for respiratory movement
(D) Call 911

Answer **Questions 93–100** based on the fire scene drawn here. The ground floor for all buildings is considered the first floor.

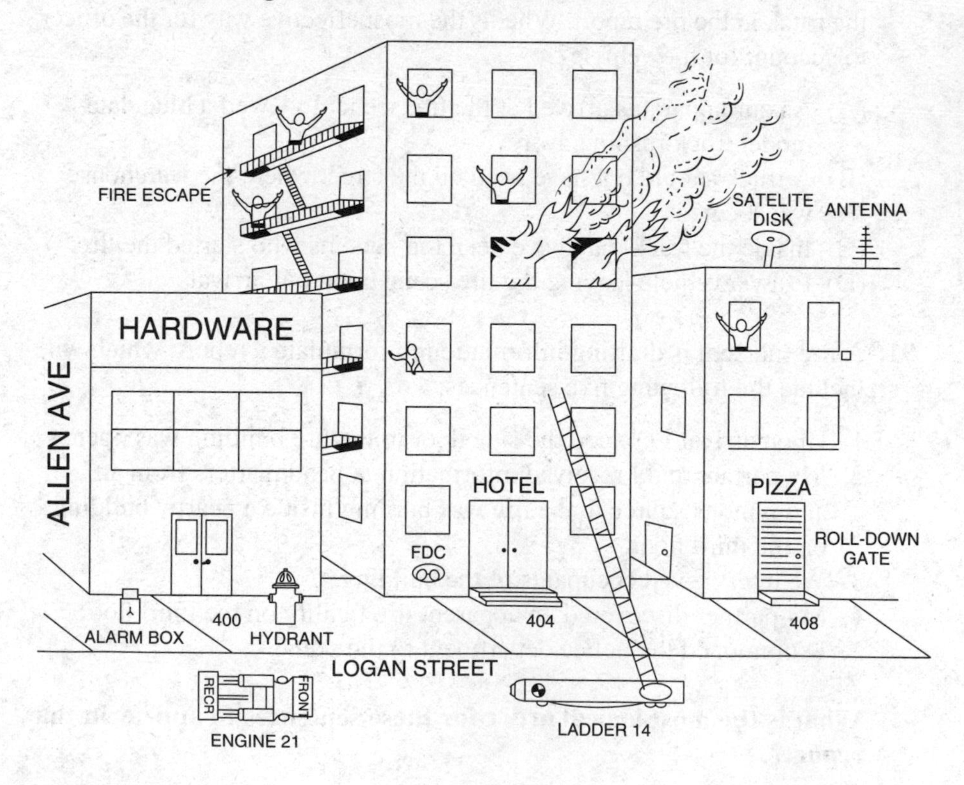


**GO ON TO THE NEXT PAGE**

**93.** How is the wind affecting the rescue effort?

(A) Blowing the flames back into the building
(B) Pushing fire/smoke away from the hotel occupant at the fifth-floor window
(C) Endangering occupants on the roof of 408 Logan Street
(D) Threatening to spread the flames into the hardware store

**94.** Where is the main body of fire located?

(A) Fourth-floor rear of the six-story hotel
(B) Two floors below the roof of the hotel
(C) Fourth-floor front of the six-story hotel
(D) One floor below the top floor of the six-story hotel

**95.** What occupant of the hotel is in the most danger from fire?

(A) Third-floor occupant at the left-front window
(B) Occupant on the fifth-floor fire escape
(C) Occupant at the fifth-floor, middle-front window
(D) Occupant on the sixth-floor fire escape

**96.** Where is the fire alarm box located?

(A) Alley between the hardware store and the hotel
(B) Near intersection of Logan Street and Allen Avenue
(C) Entrance to the hotel
(D) Entrance to 408 Logan Street

**97.** Which building most likely has a sprinkler/standpipe fire protection system?

(A) Hotel
(B) Pizza store
(C) Hardware store
(D) All of the above

**98.** Engine 21 should perform what initial action upon arrival at the scene of the fire?

(A) Stretch an attack hose line into the hardware store
(B) Connect to the fire hydrant and stretch a supply hose line into the FDC
(C) Stretch an attack hose line up Ladder 14's aerial
(D) Stretch an attack hose line up the fire escape

**GO ON TO THE NEXT PAGE**

**99.** What is the most difficult problem Ladder 14 faces regarding rescue operations?

(A) Positioning the aerial ladder to the fifth floor of the fire building
(B) Removing the occupant at the left-front, third-floor window of the hotel
(C) Laddering the roof of 408 Logan Street
(D) Laddering the fire escape of the hotel

**100.** A firefighter is ordered to go to the roof of the hotel. What is the best way for the firefighter to reach this destination?

(A) Access the fire escape in the alley between the hardware store and the hotel
(B) Access the roof of the hardware store and then jump over to the fire escape
(C) Use the roof of 408 Logan Street
(D) Use the aerial ladder after rescue operations are completed

**STOP. THIS IS THE END OF PRACTICE EXAM 6**

# Answer Key

1. D
2. A
3. B
4. C
5. C
6. A
7. B
8. D
9. C
10. B
11. D
12. A
13. C
14. D
15. B
16. C
17. A
18. B
19. D
20. A
21. C
22. A
23. D
24. A
25. B
26. B
27. C
28. C
29. D
30. A
31. B
32. C
33. B
34. C
35. D
36. D
37. B
38. B
39. C
40. C
41. A
42. C
43. B
44. A
45. D
46. D
47. A
48. B
49. A
50. C
51. D
52. A
53. C
54. D
55. B
56. B
57. A
58. C
59. D
60. A
61. B
62. B
63. C
64. B
65. D
66. C
67. B
68. A
69. A
70. D
71. C
72. C
73. A
74. B
75. B
76. D
77. C
78. B
79. A
80. D
81. A
82. D
83. B
84. A
85. C
86. D
87. B
88. B
89. A
90. A
91. B
92. D
93. B
94. C
95. C
96. B
97. A
98. B
99. A
100. D

# ANSWER EXPLANATIONS

Following you will find explanations to specific questions that you might find helpful.

**26.** B

$$\begin{array}{r} 6.2 \\ +\,3.8 \\ \hline 10.0 \end{array}$$

**27.** C

$$\begin{array}{r} 7.3 \\ \times\,4.5 \\ \hline 365 \\ 2920 \\ \hline 32.85 \end{array}$$

**28.** C Convert 0.668 to percent; move the decimal point two places to the right: 66.8%

**29.** D Convert 22.6% to a decimal; move the decimal point two places to the left: 0.226

**30.** A Part = % · Whole ÷ 100
Part = (20)(30) ÷ 100
Part = 600 ÷ 100 = 6

**31.** B % = Part ÷ Whole · 100
% = 5 ÷ 40 · 100
% = 0.125 · 100 = 12.5

**32.** C Whole = Part ÷ % · 100
Whole = 3 ÷ 30 · 100
Whole = 0.1 · 100 = 10

**34.** C $C = 5/9(F - 32)$
$C = 5/9(50 - 32)$
$C = 5/9(18) = 10$

**35.** D $F = 9/5(C) + 32$
$F = 9/5(38) + 32$
$F = 1.8(38) + 32$
$F = 68.4 + 32 = 100.4$

**36.** D $45 + 60 + 80 + 90 + 135 = 410$
$410/5 = 82$

**37.** B $d = r \cdot t$
When objects are moving in opposite directions, regardless of whether the objects are moving toward each other or away from each other, the rate at which the distance between them is changing is the sum of their individual rates.

Bus A (50 mph) + Bus B (70 mph) = 120 mph
t = number of hours the buses will travel after passing each other
240 miles = 120 mph · $t$

$$\frac{240 \text{ miles}}{120 \text{ mph}} = t$$

$2 = t$

**38.** B 60 students
10:2 ratio (pass/fail)
10 + 2 = 12
10/12 · 60 = 50 (number of students who passed)
2/12 · 60 = 10 (number of students who failed)

**40.** C $A = s^2$
$A = 5 \text{ m}^2 = 25 \text{ m}^2$

**41.** A P = 2L + 2W
P = 2(11 in) + 2(7 in) = 22 in + 14 in = 36 in

**42.** C $V = l \cdot w \cdot h$
$V = 3 \text{ in} \cdot 5 \text{ in} \cdot 8 \text{ in} = 120 \text{ in}^3$

**43.** B $V = \pi \cdot r^2 \cdot h$
$V = 3.14 \cdot 3 \text{ ft}^2 \cdot 5 \text{ ft}$
$V = 3.14 \cdot 9 \text{ ft}^2 \cdot 5 \text{ ft} = 141.3 \text{ ft}^3$

**44.** A $SA = 6 \cdot s^2$
$SA = 6 \cdot 4 \text{ mm}^2 = 6 \cdot 16 \text{ mm}^2 = 96 \text{ mm}^2$

**45.** D Fire District A = 35 injuries

**46.** D Fire District A = 35/70 = 0.50 · 100 = 50%
Fire District B = 20/60 = 0.33 · 100 = 33%
Fire District C = 25/60 = 0.41 · 100 = 41%
Fire District D = 15/75 = 0.20 · 100 = 20%

**47.** A Fire District A = 17/35 = 0.48 · 100 = 48%
Fire District B = 7/20 = 0.35 · 100 = 35%
Fire District C = 6/25 = 0.24 · 100 = 24%
Fire District D = 3/15 = 0.20 · 100 = 20%

**48.** B 35 + 20 + 25 + 15 = 95

**49.** A January/February = 125 + 60 = 185
March/April = 75 + 80 = 155
May/June = 35 + 55 = 90
July/August 100 + 15 = 115

**50.** C July = 100
August =15
September = 260
October = 55

**51.** D July/August = 115
January/February = 185
May/June = 90
September/October = 315
November = 65 + December = 50 = 115

**52.** A January/February = 185
March/April = 155
May/June = 90
July = 100
185 + 155 + 90 + 100 = 530

**53.** C February = 60 /March = 75 15 fire difference
April = 80 /May = 35 45 fire difference
August = 15/September = 260 245 fire difference
September = 260/October = 55 205 fire difference

**55.** B Rayon is a synthetic (manmade) material.

**57.** A Wood – 7,000 BTU/lb
Coal – 12,000 BTU/lb
Flammable liquids – 16,000 BTU/lb
Flammable gas – 23,000 BTU/lb

**60.** A Fulcrum is between the effort (force) and the resistance force (load).

**61.** B MA = load/effort = 90 lbs/30 lbs = 3

**62.** B MA = wheel radius/axle radius = 8 in/2 in = 4

**63.** C Calculated visually by counting the number of supporting ropes.

**64.** B Unlike gears, belt drives rotate in the same direction.

**65.** D When computing gear ratio, always compare the larger gear rotating once to the smaller gear, regardless of whether it is a driver or driven gear.
Gear 2 = 40 teeth
Gear 1 = 20 teeth
1:2
Gear 1 rotates two times faster than Gear 2
Given: Gear 1 rotates at 60 rpm
Gear 2 = 60 rpm/2 = 30 rpm

**68.** A A gate valve is an appliance (not a fitting) used to open and close the flow of water from a hydrant.

**69.** A Ranges in length from 8 feet to 14 feet

**71.** C

| | | |
|---|---|---|
| 1 | 2 | 3 |
| 2 | 3 | 1 |
| 3 | 1 | 2 |

**73.** A

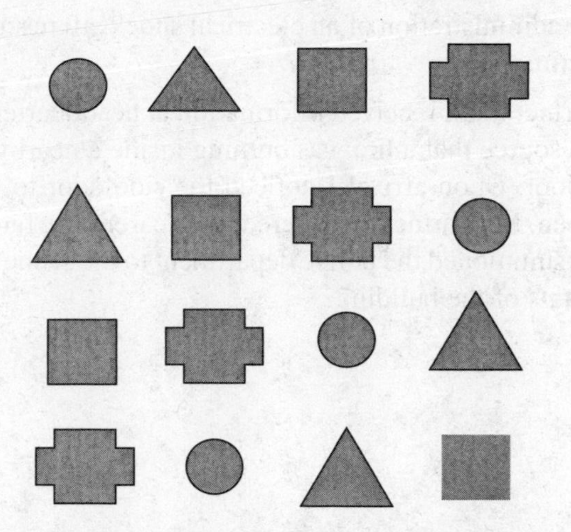

**74.** B The driver of the speeding vehicle steps out of the car with license and registration before the officer can exit his squad car. Driver does not initially comply with the police officer's order to return to his vehicle. Driver asks the officer if they can settle the issue immediately without his having to present more information.

**75.** B The apparatus was pulling out from a parked position at the curb when struck. The Ford was traveling approximately 20 miles per hour over the speed limit just prior to the accident. The late-model Ford struck the apparatus in the rear, backed up, and then continued on its way down the road. The license plate number of the vehicle was recorded. Two firefighters sitting inside the apparatus were treated by EMTs for bruises.

**76.** D Steps taken: Medical treatment is administered and an ambulance is requested to respond.
Next step should be: Notify Borough Dispatch if the victim is wearing a Medic-Alert emblem.

**77.** C 400 ft $\cdot$ 5 psi = 200 psi

**79.** A Given: 300 ft of hose
200 gpm corresponds to 5 at the tip of the finger
At the base of each finger is 12
$5 \cdot 12 = 60$ psi for each 100 ft of hose
$60 \times 3 = 180$ psi

**80.** D (A) Laundry – Ordinary hazard
Hospital – Light hazard
(C) Automobile showroom – Ordinary hazard
(D) Aircraft hangar – Extra hazard

**88.** B Both legs (anterior and posterior) – 18% + 18% = 36%
Groin = 1%
Chest = 9%
Abdomen = 9%
Total = 55%

**89.** A Before administration of an electrical shock, all rescuers stay clear of the victim.

**91.** B My partner and I received information at headquarters from an anonymous source that a fire was burning inside a nearby building on the third floor. Upon arrival I noticed the side door to the fire building was open. My partner discovered an apparent fire fatality on the third floor. I summoned the police department to the scene. We interviewed occupants of the building.

# Informational Resources for Careers in Firefighting

The following resources can be found on the Internet via the World Wide Web to help the interested learn more about being a firefighter. At these websites, you will find information concerning the nature of the work, firefighting and related career employment opportunities, job outlook information, and colleges, universities, and academies offering fire science educational courses. Also included in this summary of resources are Internet search engines that can be used to research the information mentioned in this book concerning firefighting and firematics. If you input key words or phrases, the search engines will lead you to websites related to many topics about firefighting. The websites listed are just a few of many that can provide information, but their listing is not an endorsement. Feel free to explore and locate others that will provide you with a wealth of valuable information about the world of the firefighter.

## AUTHOR'S WEBSITE

**www.firefighterexams.org** This website includes updates to study material covered in this book keeping it contemporary and valid. It also provides the reader with announcements of firefighter examinations in major U.S. cities. Up-to-date first responder, firematic, and firefighting technology information can be garnered from this website keeping firefighters and prospective firefighters abreast of the latest happenings in the fire service. Pertinent articles written by Chief Spadafora are presented to enhance reader insight of key concepts regarding fire department organization, managerial leadership, training, fire science curriculum, and operational firefighting strategy and tactics.

## WORLD WIDE WEBSITES—EMPLOYMENT OPPORTUNITY SERVICES

The following are subscription sites that provide employment information for fire service positions throughout the country.

www.firerecruit.com

www.firecareers.com

The following site is the Department of Interior's Fire Integrated Recruitment Employment Systems (FIRES) site.

www.firejobs.doi.gov/

The following site provides wildland firefighting employment information, both help wanted and jobs wanted, as well as federal agency wildland firefighting links and information.

www.wildlandfire.com/jobs.htm

## BOOKS AND MANUALS

*A Handbook on Women in Firefighting: The Changing Face of the Fire Service*. Federal Emergency Management Administration, January 1993.

*Careers in Firefighting*. Mary Price Lee and Richard S. Lee. New York: Rosen Publishing Group, 1992.

*Choosing a Career as a Firefighter*. Walter Oleksy. New York: Rosen Publishing Group, 2000.

*Essentials of Firefighting, and Fire Department Operations*, 6th ed. International Fire Service Training Association, 2013.

*Fire and Emergency Services Orientation and Terminology*, 5th ed. International Fire Service Training Association, 2011.

*Firefighter's Handbook on Wildland Firefighting: Strategy, Tactics and Safety*. William C. Teie, 2nd Edition, Deer Valley Press, 2005.

*Great Careers with a High School Diploma: Public Safety, Law, and Security*. Jon Sterngass. New York: Infobase Publishing, 2008.

*Leadership for the Wildland Fire Officer: Leading in a Dangerous Profession*. William C. Teie, Brian F. Weatherford, Timothy M. Murphy, and Dave A. Hubert. Rescue, CA: Deer Valley Press, 2010.

*Occupational Outlook Handbook*, 2015 Edition. Bureau of Labor Statistics, U.S. Department of Labor. www.bls.gov/oco/.

*On Fire: A Career in Wildland Firefighting and Incident Management Team Response*. Thomas C. Cable, Xlibris Corporation, 2012.

*Real Resumes for Firefighting Jobs*. Edited by Anne McKinney. Fayettville, NC: Prep Publishing, 2004.

*The Fire Service Joint Labor Management Wellness-Fitness Initiative*, 3rd ed. International Association of Fire Fighters/International Association of Fire Chiefs, 2008. www.iafc.org/files/healthWell_WFI3rdEdition.pdf.pdf.

*Training Catalog*. NYS Division of Homeland Security and Emergency Services, Office of Fire Prevention and Control, 2014. www.dhses.ny.gov/ofpc/training/documents/trainingcatalog.pdf.

## ACCREDITED COLLEGES OFFERING ONLINE FIRE SCIENCE DEGREE PROGRAMS

**American Military University**
Offers A.S. in Fire Science and B.S. in Fire Science Management
www.amu.apus.edu/index.html

**American Public University**
Offers A.S. in Fire Science and B.S. in Fire Science Management
www.apus.edu/

**Columbia Southern University**
Offers B.S. in Fire Science
www.colsouth.edu

**Empire State College**
Offers B.S. in Fire Service Administration & Emergency Management, or B.S. in Emergency Management & Fire Service Administration
www.esc.edu

**Kaplan University**
Offers B.S. in Fire Science
kaplanuniversity.edu

**University of Florida**
M. E. Rinker School of Building Construction
Offers B.A./M.A. in Fire & Emergency Management
www.bcn.ufl.edu

## ADDITIONAL FIRE SCIENCE DEGREES ONLINE

**WorldWideLearn.com**
Provides information concerning online degree programs. Featured education partners include the University of Maryland.
www.worldwidelearn.com/

**Universities.com**
Provides information concerning online degree programs. Featured education partners include the University of Maryland and the University of Phoenix.
www.universities.com/

## FIRE AND EMERGENCY SERVICES HIGHER EDUCATION (FESHE)

A FESHE recognition certificate is an acknowledgment that particular collegiate division of emergency services degree programs meet the minimum standards of excellence established by FESHE professional development committees and the National Fire Academy. The following website provides a listing of recognized U.S. programs.

www.usfa.fema.gov/training/prodev/about_feshe.html

## SEARCH ENGINES

AOL Search at www.search.aol.com
Ask.com at www.ask.com
Google at www.google.com
HotBot at www.hotbot.com
Lycos at www.lycos.com
MSN Search at www.msn.com
Yahoo at www.yahoo.com

## ORGANIZATIONS

**International Association of Fire Fighters**
1750 New York Avenue NW
Washington, DC 20006
www.iaff.org

**International Fire Service Training Association/Fire Protection Publications**
Oklahoma State University Campus
Stillwater, OK 74078-4082
www.ifsta.org/

**National Fire Academy**
(U.S. Fire Administration) Emmitsburg, MD 21727
www.usfa.fema.gov/training/nfa/index.html

**National Fire Protection Association**
1 Batterymarch Park
Quincy, MA 02169-7471
www.nfpa.org/

**U.S. Fire Administration**
16825 South Seton Avenue
Emmitsburg, MD 21727
www.usfa.fema.gov/

# TRI-STATE (NEW YORK, NEW JERSEY, CONNECTICUT) CENTERS OF LEARNING

## New York

**Borough of Manhattan Community College (CUNY)**
199 Chambers Street
New York, NY 10007
212-220-8000
www.bmcc.cuny.edu/

**Hudson Valley Community College (SUNY)**
80 Vandenburgh Avenue
Troy, NY 12180-6025
518-629-4822
www.hvcc.edu/

**John Jay College of Criminal Justice (CUNY)**
899 Tenth Avenue
New York, NY 10019
212-237-8000
www.jjay.cuny.edu

**Metropolitan College of New York**
60 West Street
New York, NY, 10006
212-343-1234
www.mcny.edu/

**Monroe Community College (SUNY), Public Safety Training Center**
1190 Scottsville Road
Rochester, NY 14624
585-292-2000
www.monroecc.edu/depts/pstc/

**Frederick L. Warder Academy of Fire Science**
600 College Avenue
Montour Falls, NY 14865-9634
607-535-7136
www.dhses.ny.gov/ofpc/training

**New York Wildfire and Incident Management Academy**
New York State Department of Environmental Conservation
Westhampton Beach, NY 11978
631-769-1556
www.nywima.com

**Rockland Community College (SUNY)**
145 College Road
Suffern, NY 10901-3611
845-574-4000
www.sunyrockland.edu/

**Suffolk County Community College–Ammerman Campus (SUNY)**
533 College Road
Selden, NY 11784
631-451-4110
www.sunysuffolk.edu

## NEW JERSEY

**Bergen County EMS Training Center**
281 Pascack Road
Paramus, NJ 07652
201-343-3407
www.emsregistration.bergen.org/

**Burlington County Emergency Services Training Center**
53 Academy Drive
Westhampton, NJ 08060
609-702-7157

**Passaic County Community College**
One College Boulevard
Patterson, NJ 07505
973-684-6868
www.pccc.cc.nj.us

**Sussex County Community College**
One College Hill
Newton, NJ 07860
973-300-2100
www.sussex.edu/

## Connecticut

**Connecticut Department of Emergency Services and Public Protection**
1111 Country Club Road
Middletown, CT 06457
www.ct.gov/despp/cwp/view.asp?a=4156&q=487894

**University of New Haven**
300 Boston Post Road
West Haven, CT 06516-1916
203-932-7000
www.newhaven.edu/

# Glossary of Firefighting Terms

**Accelerant**—a substance (usually a liquid) used by arsonists to enhance burning.

**Aerial ladder**—a fire truck having an extension ladder that is raised and lowered using the power of the truck.

**Air tanker**—a fixed-wing aircraft capable of extinguishing wildland fires utilizing fire suppressants or fire retardants.

**Airborne pathogens**—disease-causing microorganisms.

**Alarm system**—equipment designed to transmit a warning signal and/or detect smoke, heat, fire, or carbon monoxide in order to alert occupants and dispatch resources.

**Apparatus**—commonly refers to fire service vehicles (engines, ladders, rescue truck, water tenders, etc.).

**Appliance**—a device (ball valve, gate valve, wye, water thief, siamese) connected to a hose line or hydrant used to control, augment, divide, or discharge a water stream or fire extinguishing agent.

**Arson**—a crime involving starting a fire with the intent to kill, injure, defraud, or destroy property.

**Attack hose**—a small-diameter hose stretched off engine apparatus by firefighters to extinguish fires inside structures.

**Automated external defibrillator (AED)**—a device designed to analyze the cardiac rhythm of a patient and determine if defibrillation is needed and to apply a measured dose of electrical current to restore normal rhythm of the heart

**Backdraft**—a fire phenomenon caused by the sudden influx of air into a compartment or room that mixes with flammable gases (carbon monoxide, hydrogen, methane) already above their ignition temperature to create an explosion.

**Backfire**—a fire intentionally set by wildland firefighters along the inner edge of the fireline to consume fuel in the path of a fire.

**Bambi bucket**—a water-carrying device slung below a helicopter.

**Boiling liquid expanding vapor explosion (BLEVE)**—generally relating to the failure of a pressure vessel containing liquefied gas as a result of fire impinging on the container or structural damage from impact.

**Boiling point**—temperature of a liquid at which it will liberate the most vapors.

**Booster hose**—a small-diameter rubberized hose carried on a reel of an engine apparatus.

**Booster tank**—an internal water container found within engine apparatus.

**British thermal unit (BTU)**—the amount of heat energy required to raise the temperature of one pound of water (measured at 60 degrees Fahrenheit at sea level) one degree Fahrenheit. One BTU equals 1.055 kilojoules (kJ).

**Brush**—refers to vegetation consisting of shrubs, plants, bushes, and small trees.

**Bunker gear**—firefighter protective clothing consisting of coat (jacket) and pants. The term can also refer to the entire firefighter ensemble (helmet, gloves, hood, and boots).

**Calorie**—a heat energy unit. The amount of heat energy required to raise the temperature of one gram of water (measured at 15 degrees Celsius at sea level) one degree Celsius.

**Carbon dioxide**—extinguishing agent (gas) used to smother and cool a fire.

**Carbon monoxide**—a flammable gas that is a deadly by-product of combustion.

**Cardiopulmonary resuscitation (CPR)**—application of ventilations and external cardiac compressions used on patients in cardiac arrest.

**Celsius degree**—temperature unit pertaining to the Celsius (centigrade) temperature scale.

**Chief officer**—superior officer, generally taking on the role of incident commander at a firefighting incident or emergency.

**Clean agent**—an electrically nonconductive, volatile, or gaseous fire-extinguishing agent that does not leave a residue upon evaporation.

**Combustible**—material that will ignite and burn.

**Community Emergency Response Team (CERT)**—citizen corps program designed to train civilians to be better prepared to respond to emergency situations and assist first responders.

**Company officer**—company leader, generally having the rank of lieutenant or captain.

**Conduction**—the transfer of heat through a medium (solid, liquid, or gas).

**Confined space**—an area (tunnel, trench, storage tank, sewer pipe) not designed for human habitation due to physical dimensions and lack of natural ventilation.

**Confinement**—operations designed to control a fire to a manageable area.

**Conflagration**—a fire of great magnitude that covers a wide area and crosses natural boundaries.

**Convection**—the transfer of heat through a circulating medium (air).

**Critical incident stress debriefing**—discussions among first responders conducted by professionals (counseling personnel) after particularly stressful operations. It is intended to help firefighters cope with their emotions.

**Decontamination**—removal of harmful substances from victims. It usually consists of removing outer clothing and washing down the victim.

**Defibrillation**—delivery of a measured dose of electrical current in order to regain normal rhythm of the heart.

**Deflagration**—an explosion that propagates at a speed less than the speed of sound.

**Demobilization**—removal of personnel and equipment from working at a fire or emergency operation.

**Detonation**—an explosion that propagates at a speed greater than the speed of sound.

**Dilution**—extinguishing method using water on a water-soluble material to lower its concentration and raise its flash point.

**Dispatcher**—a person employed to coordinate the receipt, confirmation, and transmittal of fire and emergency calls and alarm signals.

**Drafting**—use of suction hose and engine apparatus pump to lift water from below the level of the pump.

**Drip torch**—handheld device with a fuel font, burner, and igniter containing gasoline and diesel fuel used by wildland firefighters to ignite a backfire.

**Dry chemical**—a fire-extinguishing agent that interferes with the chemical chain reaction of combustion.

**Dry powder**—a fire-extinguishing agent used on combustible metals.

**Dry standpipe**—a standpipe fire protection system that is not filled with water until needed in firefighting.

**EMS**—emergency medical service.

**EMT**—emergency medical technician.

**Emulsification**—extinguishment method using water to cause agitation of insoluble liquids to produce a vapor-inhibiting froth.

**Endothermic reaction**—a chemical reaction that absorbs heat.

**Engine (pumper) apparatus**—a fire service vehicle consisting of a water pump, portable water tank, various lengths and sizes of hose and applicable appliances, nozzles, tips, and fittings.

**Engine company**—firefighters assigned to work on an engine (pumper) apparatus.

**Exothermic reaction**—a chemical reaction that gives off heat.

**Explosion**—rapid expansion of gases that have premixed prior to ignition.

**Extension ladder**—a portable ladder with one or more movable sections that can be extended to a desired height.

**Extinguisher**—a firefighting device consisting of a metal container containing extinguishing agent under pressure.

**Fahrenheit degree**—temperature unit pertaining to the Fahrenheit temperature scale.

**Fire (combustion)**—rapid, self-sustaining oxidation reaction with the emission of heat and light.

**Fire department connection** (FDC)—device that a pumper connects to in order to supply and augment the water flow in a standpipe and/or sprinkler fire protection systems. Combines two hose lines into one. Also referred to as a siamese.

**Firefighter**—first responder who operates at fires and emergencies performing a multitude of tasks that include fire suppression, utility emergency mitigation, and medical evaluation and treatment.

**Fireground**—area in and around the operational jurisdiction of firefighters.

**Firehouse**—station where firefighters work and fire apparatus are stored.

**Fire hydrant**—a source of water supply to firefighters consisting of one or more valves and outlets usually connected to a municipal water supply.

**Fire inspector**—a person employed to enforce the Fire Code.

**Fire line**—the removal of brush and foliage, by hand or machine, to form a barrier against fire spread.

**Fire marshal**—a person designated to prevent and investigate fires.

**Fire shelter**—a tentlike personal life safety device designed to be deployed quickly and entered by wildland firefighters when endangered.

**Fire tetrahedron**—model used to represent the growth of ignition to fire. It expands on the fire triangle by adding a fourth factor (chemical chain reaction).

**Fire triangle**—model used to represent the three factors—oxygen, fuel, and heat—necessary for ignition.

**Fittings**—devices (increasers, reducers, double male/female connections, adapters) used in conjunction with hose line couplings to solve hose connection problems.

**Flammable range**—the percentage mixture of vapors in air that will sustain combustion.

**Flashover**—fire phenomenon that occurs when all the contents of a room or compartment reach their ignition temperature and simultaneously burst into flames.

**Flash point**—lowest temperature at which a substance/material will emit a vapor that is ignitable in air.

**Foam**—an extinguishing agent created by introducing air into a mixture of water and foam concentrate.

**Fog stream**—a hose line stream characterized by a wide-pattern of small droplets of water.

**Forcible entry**—gaining access to areas through locked doors and windows with the use of specialized tools and equipment.

**Free-burning (fully developed) phase**—the second phase of fire development.

**Friction loss**—reduction in the amount of water flowing through hose or piping as a result of inner lining resistance.

**Fuel**—material that will burn.

**Fulcrum—**a pivot point or support on which a lever turns.

**Gas**—considered the third state of matter.

**GPM**—gallons per minute.

**Ground ladder**—portable ladder designed to be utilized manually on the fireground.

**Halligan**—forcible entry tool with a pointed pick and adze at right angles at one end of the shaft and a fork at the other end.

**Halyard**—a rope attached to extension ladders for use in raising and lowering the fly ladder.

**Hard suction hose**—a noncollapsible hose used for drafting water.

**Hazardous material (hazmat)**—substance (solid, liquid, or gas) that because of its physical and chemical characteristics is dangerous to life and the environment.

**Heat**—a measure of the quantity of energy inside a substance.

**Helitack**—specialized wildland fire crews that use helicopters to fight fires and provide for movement of personnel and equipment.

**Hellfighters**—personnel hired by oil well control companies to extinguish oil well fires.

**Hose**—cotton, rubber, or synthetic conduits used by firefighters for moving water under pressure.

**Hose bed**—part of engine apparatus designed to hold various types of hose for ready use at firefighting operations.

**Hose bridge**—a ramp used to allow vehicles to pass over hose without damaging it.

**Hose couplings**—ends of fire hose used to connect to other lengths of hose or to hydrants, engine apparatus pumps, and appliances.

**Hose roller**—device designed to be attached to the roof or windowsill to facilitate hoisting and lowering of hose line.

**Hose spanner**—tool used to loosen and tighten hose line couplings.

**Hose strap**—a tool (rope with eye-loop at one end and metal hook at the other end) used to support the weight of hose couplings when hose is stretched vertically up stairwells and fire escapes.

**Hotshots**—specialized wildland fire crews that are mainly used to build firelines by hand.

**Hydrant wrench**—a tool used to operate and open and shut down a hydrant.

**Hydraulics**—the study of water pressure, flow, friction loss, and water supply systems.

**Hydraulic spreader (Jaws of Life)**—mechanical levering tool powered by a hydraulic pump engine. It is used by firefighters to extricate trapped victims inside motor vehicles.

**Ignition temperature**—minimum temperature a material must be heated to for it to ignite and be self-sustaining without an external input of heat.